E. D. Pittman · M. D. Lewan (Eds.)

Organic Acids in Geological Processes

With 151 Figures

Springer-Verlag
Berlin Heidelberg New York
London Paris Tokyo
Hong Kong Barcelona
Budapest

Dr. Edward D. Pittman
Consulting Petrologist-Sedimentologist
222 Bowstring Drive
Sedona, Arizona 86336, USA

Dr. Michael D. Lewan
U.S. Geological Survey
P.O. Box 25046/M.S. 977
Denver, Colorado 80225, USA

ISBN 3-540-56953-7 Springer-Verlag Berlin Heidelberg New York
ISBN 0-387-56953-7 Springer-Verlag New York Berlin Heidelberg

Library of Congress Cataloging-in-Publication Data. Organic acids in geological processes / E.D. Pittman, M.D. Lewan (eds.). p. cm. Includes bibliographical references and index. ISBN 3-540-56953-7 (Berlin: acid-free). − ISBN 0-387-56953-7 (New York: acid-free) 1. Organic acids. 2. Organic geochemistry. I. Pittman, Edward D. II. Lewan, M.D. (Michael D.), 1948− . QE516.5.O65 1994 551.9 − dc20 94-3819 CIP

© Springer-Verlag Berlin Heidelberg 1994
Printed in Germany

The use of general descriptive names, registered names, trademarks, etc. in this publication does not imply, even in the absence of a specific statement, that such names are exempt from the relevant protective laws and regulations and therefore free for general use.

Typesetting: Best-set Typesetter Ltd., Hong Kong
SPIN: 10122121   32/3130-5 4 3 2 1 0 − Printed on acid-free paper

Preface

In May of 1991, Victor Van Buren, who was then with Springer-Verlag in New York City, asked us for timely topics in the earth sciences that would be appropriate for publication as a book. We all quickly agreed that recent interest and research activity on the role of organic acids in geological processes would make a timely book on this diverse and controversial topic. As coeditors, we outlined chapter topics for such a book that maintained a good balance between geological and geochemical interests. Specific authors were then sought for each of the chapter topics. We had exceptional success in getting leading researchers as authors, and their response was universally enthusiastic. This approach has been most gratifying in that it provides a cohesion and conciseness that is not always present in books representing compilations of papers from symposia. This book does not resolve the controversies that exist regarding the significance of organic acids in geological processes. However, it does present both sides of the controversies in terms of available data and current interpretations. Readers may judge for themselves and envisage research necessary to resolve these controversies in the future. We thank the authors of this book for their participation, dedication, and cooperation. We are also grateful for support from Dr. Wolfgang Engel and his staff at Springer-Verlag (Heidelberg) in expediting the editing and publication of this book in a timely manner. The use of trade or company names in any of the chapters of this book is only for identification purposes and is not an endorsement by authors or editors or their employers or sponsors.

March 15, 1993

Edward D. Pittman
222 Bowstring Drive
Sedona, Arizona 86336, USA

Michael D. Lewan
US Geological Survey
Box 25046 MS977
Denver, Colorado 80225, USA

Table of Contents

Chapter 1   Introduction to the Role of Organic Acids in Geological Processes

Michael D. Lewan[1] and Edward D. Pittman[2]

Summary

This chapter is intended to provide sufficient information to allow one to read the following chapters in any order they prefer. The objective and need for this book are presented in the introductory section. Nomenclature for organic acids is presented on an elementary level for nongeochemists. A historical account is then given on the role of organic acids in geological processes, which is followed by a brief synopsis of each chapter. The final section explores the needs for future research in terms of natural system studies and laboratory experimental studies.

1   Introduction

Organic acids have had a rich and diverse presence in the earth sciences and particularly in geochemistry. Fatty and amino acids extracted from meteorites have provoked controversies regarding the origin of life and concerns on terrestrial contamination (Hayes 1967; Cronin et al. 1988). Fatty acids have been used to interpret the precursory organisms responsible for organic matter deposited in modern sediments and the level of thermal diagenesis experienced by organic matter buried in sedimentary rocks (Parker 1969). These organic acids also have been considered a source of natural gas (Carothers and Kharaka 1978) and petroleum (Shimoyama and Johns 1971). In addition, certain types of carboxylic acids in petroleum have been advocated as indicators of migration distance (Jaffé et al. 1988). Amino acids have been used to age date fossil remains (Lee et al. 1976) and marine sediments (Kvenvolden et al. 1973). Considerable effort has been placed on understanding the structural complexity of high-molecular-weight

[1] US Geological Survey, Box 25046, MS977, Denver, Colorado 80225, USA
[2] 222 Bowstring Dr., Sedona, Arizona 86336, USA

polymeric acids peculiar to some formation waters (Barden et al. 1984) and humic acids in soils (Stevenson 1982).

Several comprehensive books have been published on the geochemical aspects of organic acids. Lochte and Littmann (1955) wrote *The Petroleum Acids and Bases*, which presents analytical methods and the identities of organic acids and bases found in petroleum. Hare et al. (1980) edited the book entitled *Biogeochemistry of Amino Acids*. This compilation of papers emphasized natural occurrence, analytical methods for study, experimental results on racemization, and age-dating applications of amino acids. The book entitled *Organic Geochemistry of Natural Waters* by Thurman (1985) gives a thorough account of the types and abundances of organic acids in natural surface and near-surface waters. Aiken et al. (1985) have edited a book entitled *Humic Substances in Soil, Sediment, and Water*, in which natural occurrences and chemistry of humic and fulvic acids are discussed in detail.

Whereas all of these mentioned books provide pertinent information on natural occurrences, analyses and isolations, or structural characterizations of organic acids, there remains a need for a book concerning the influence of organic acids on geological *processes*. Organic acids have been invoked as influential participants in soil formation, surface weathering, subsurface porosity generation, and ore formation. Although controversy surrounds the alleged role of organic acids in some of these geological processes, a collection of data and interpretations may provide insights and research directions for future resolution of these controversies. This book is an attempt to bring together current knowledge on the role of organic acids in geological processes. No overall consensus is sought in this book, and the following chapters are authored by dedicated researchers representing a diversity of interests, approaches, and hypotheses concerning organic acids.

2 Organic Acid Nomenclature

Organic acids are molecules consisting of a carbon-to-carbon bonded framework containing at least one functional group capable of relinquishing a proton. The functional group most commonly responsible for acidity in naturally occurring organic acids is the carboxylic group ($-COOH$). This functional group typically relinquishes a proton in water between pH values of 3 and 10 depending on the overall chemical character of the organic acid. Natural surface and subsurface waters have pH values within this range. As a result, these dissolved organic molecules may occur in significant concentrations as both a protonated acid and a deprotonated anion. An acid anion is named by omitting the word acid and replacing the *-ic* suffix of the acid name with an *-ate* suffix (e.g., acetic acid/acetate). This nomenclature is essential when describing unique properties or reactivities of an anion relative to its acid. However, it may be inferred in the absence of discussion on water chemistry and speciation that the use of an acid name alone collectively refers to both the acid and its anion irrespective of their relative concentrations. More specific nomenclature for organic acids is determined by the composition and structure of the molecule hosting the carboxylic group.

Aliphatic framework molecules most common in organic acids include alkanes (saturated hydrocarbons) and alkenes (unsaturated hydrocarbons). These saturated and unsaturated aliphatic carboxylic acids may be acyclic (straight or branched chains) or alicyclic (aliphatic rings). Acyclic aliphatic monocarboxylic acids are also referred to as fatty acids (Table 1). The first five saturated acids (formic to valeric) of this type are sometimes referred to as short-chain, low-molecular-weight, or volatile fatty acids. Although a nomenclature for these acids has been established by the International Union of Pure and Applied Chemistry (IUPAC), the convention of using the trivial names for the first five saturated acids has remained. Similarly, trivial names are used for the aliphatic dicarboxylic acids (Table 2) that are saturated with two to four carbon atoms (C_2-C_4) and unsaturated with four carbon atoms (C_4). Alicyclic carboxylic acids contain one or more saturated or partially unsaturated rings. These acids most commonly occur

Table 1. Examples of trivial names and IUPAC names for acyclic (i.e., straight or branched chains) aliphatic monocarboxylic acids (i.e., fatty acids). Preferred names are italicized

Degree of saturation	Trivial name	IUPAC name[a]	Chemical formula[b]
Saturated	*Formic*	Methanoic	$HCOOH$
	Acetic	Ethanoic	CH_3COOH
	Propionic	Propanoic	CH_3CH_2COOH
	Butyric	Butanoic	$CH_3(CH_2)_2COOH$
	Valeric	Pentanoic	$CH_3(CH_2)_3COOH$
	Isovaleric	*3-Methylbutanoic*	$CH_3[CH_3]CHCH_2COOH$
	Caproic	*Hexanoic*	$CH_3(CH_2)_4COOH$
	Isocaproic	*4-Methylpentanoic*	$CH_3[CH_3]CH(CH_2)_2COOH$
	Enanthic	*Heptanoic*	$CH_3(CH_2)_5COOH$
	Caprylic	*Octanoic*	$CH_3(CH_2)_6COOH$
	Capric	*Decanoic*	$CH_3(CH_2)_8COOH$
	Lauric	*Dodeaanoic*	$CH_3(CH_2)_{10}COOH$
Unsaturated	Acrylic	*Propenoic*	$CH_2:CHCOOH$
	Cratonic	*2-Butenoic*	$CH_3CH:CHCOOH$
	Allylacetic	*4-Pentenoic*	$CH_2:CH(CH_2)_2COOH$
	Hydrosorbic	*3-Hexenoic*	$CH_3CH_2CH:CHCH_2COOH$
	Dehydracetic	*4-Hexenoic*	$CH_3CH:CH(CH_2)_2COOH$
	Oleic	*cis-9-Octadecenoic*	$CH_3(CH_2)_7CH:CH(CH_2)_7COOH$
	Vaccenic	*trans-11-Octadecenoic*	$CH_3(CH_2)_5CH:CH(CH_2)_9COOH$
Polyunsaturated	Sorbic	*2,4-Hexadienoic*	$CH_3CH:CHCH:CHCOOH$
	Linoleic	*cis,cis-19,12-Octadecadienoic*	$CH_3(CH_2)_3(CH:CH)_3(CH_2)_7COOH$
	Linolenic	*cis,cis,cis-9,12,15-Octadecadiernoic*	$CH_3(CH_2CH:CH)_3(CH_2)_7COOH$

[a] International Union of Pure and Applied Chemistry.
[b] [ ], branched group and :, double bond.

Table 2. Examples of trivial names and IUPAC names for acyclic (i.e., straight or branched chains) aliphatic dicarboxylic acids. Preferred names are italicized

Degree of saturation	Trivial name	IUPAC name[a]	Chemical formula
Saturated	*Oxalic*	Ethanedioic	$HOOCCOOH$
	Malonic	Propanedioic	$HOOCCH_2COOH$
	Succinic	Butanedioic	$HOOC(CH_2)_2COOH$
	Glutaric	*Pentanedioic*	$HOOC(CH_2)_3COOH$
	Adipic	*Hexanedioic*	$HOOC(CH_2)_4COOH$
	Pimelic	*Heptanedioic*	$HOOC(CH_2)_5COOH$
	Suberic	*Octanedioic*	$HOOC(CH_2)_6COOH$
Unsaturated	*Maleic*	*cis*-Butenedioic	$HOOCCH:CHCOOH$
	Fumaric	*trans*-Butenedioic	$HOOCCH:CHCOOH$

[a] See Table 1 footnotes.

in petroleum rather than formation waters, and they are sometimes referred to as naphthenic acids (Lochte and Littmann 1955). Alicyclic carboxylic acids containing only one aliphatic ring typically subscribe to the IUPAC nomenclature (e.g., cyclohexane carboxylic acid), and those containing two or more aliphatic rings subscribe to the trivial nomenclature (e.g., oleanolic acid).

Aromatic carboxylic acids contain one or more conjugated six-carbon rings (i.e., benzene) with one or more carboxylic groups. The acids containing one benzene ring typically are referred to by their trivial names, but acids with two or more fused benzene rings are referred to by their IUPAC or trivial names. Examples of aromatic carboxylic acids and their nomenclatures are shown in Fig. 1a. Hydroxyl ($-OH$) and methoxyl ($-OCH_3$) functional groups may be directly bound to the benzene rings of these carboxylic acids (Fig. 1b). It becomes apparent from the examples of these highly functionalized acids in Fig. 1b that the nomenclature becomes more ungovernable as the number and types of functional groups increase. As a result, most high-molecular-weight and highly functionalized organic acids that are relevant to geological processes are referred to by their trivial names. Extreme examples of this convention are the humic and fulvic acids, which are operational names based on acid insolubility and solubility of nonvolatile, high-molecular-weight (500 to 5000 amu), highly functionalized, organic acids with a structural framework of aliphatic and aromatic components.

3   History of Organic Acids in Geological Processes

According to Zinger and Kravchik (1972), the first published data on organic acids in oil-field waters were by Potylitsin in 1882. The first English language author on the subject appears to have been Rogers (1917). A hiatus in the

(a) AROMATIC CARBOXYLIC ACIDS

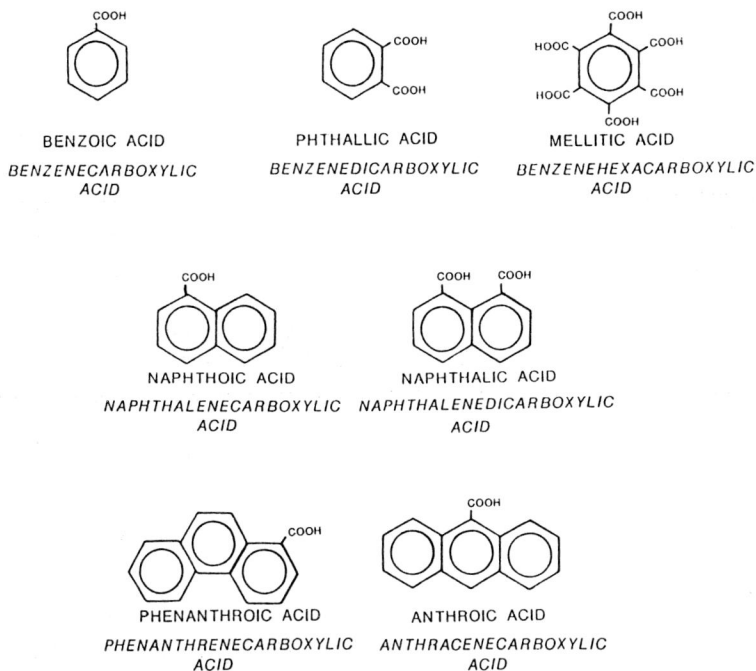

BENZOIC ACID
BENZENECARBOXYLIC ACID

PHTHALLIC ACID
BENZENEDICARBOXYLIC ACID

MELLITIC ACID
BENZENEHEXACARBOXYLIC ACID

NAPHTHOIC ACID
NAPHTHALENECARBOXYLIC ACID

NAPHTHALIC ACID
NAPHTHALENEDICARBOXYLIC ACID

PHENANTHROIC ACID
PHENANTHRENECARBOXYLIC ACID

ANTHROIC ACID
ANTHRACENECARBOXYLIC ACID

(b) HYDROXY/METHOXY AROMATIC CARBOXYLIC ACIDS

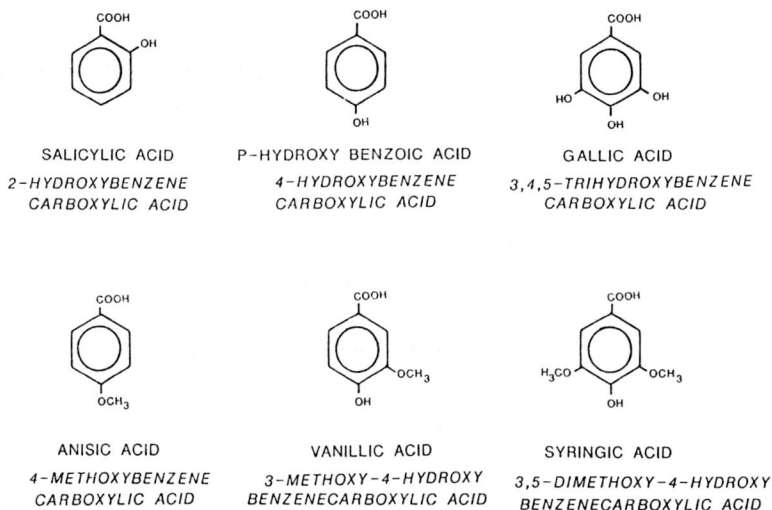

SALICYLIC ACID
2-HYDROXYBENZENE CARBOXYLIC ACID

P-HYDROXY BENZOIC ACID
4-HYDROXYBENZENE CARBOXYLIC ACID

GALLIC ACID
3,4,5-TRIHYDROXYBENZENE CARBOXYLIC ACID

ANISIC ACID
4-METHOXYBENZENE CARBOXYLIC ACID

VANILLIC ACID
3-METHOXY-4-HYDROXY BENZENECARBOXYLIC ACID

SYRINGIC ACID
3,5-DIMETHOXY-4-HYDROXY BENZENECARBOXYLIC ACID

Fig. 1. Examples of **a** aromatic carboxylic acids and **b** hydroxyl/methoxy aromatic carboxylic acids. Trivial names are given in *regular type* and IUPAC names are *italicized*

study of naturally occurring organic acids followed until it was recognized in the 1940s that organic acids in formation waters caused significant corrosion to oil-field equipment and production pipes (Menaul 1944; Greco and Griffin 1946). This corrosion was particularly apparent in gas-condensate wells producing from depths in excess of 1.5 km at temperatures greater than 70 °C (Lowe 1953). The negative economic impact of this corrosion on production costs prompted the Natural Gasoline Association of America to establish the Corrosion Committee to collect and administer industrial contributions for funding research concerning this corrosion problem. Results of studies funded by this initiative were collectively published under the editorship of Prange et al. (1953). Papers in the book that provided particularly relevant data and observations on the corrosive role of organic acids are by Shock (1953), Chesney (1953), and Prange (1953). As an example, Fig. 2 shows the combined influence of CO_2 partial pressure and concentrations of

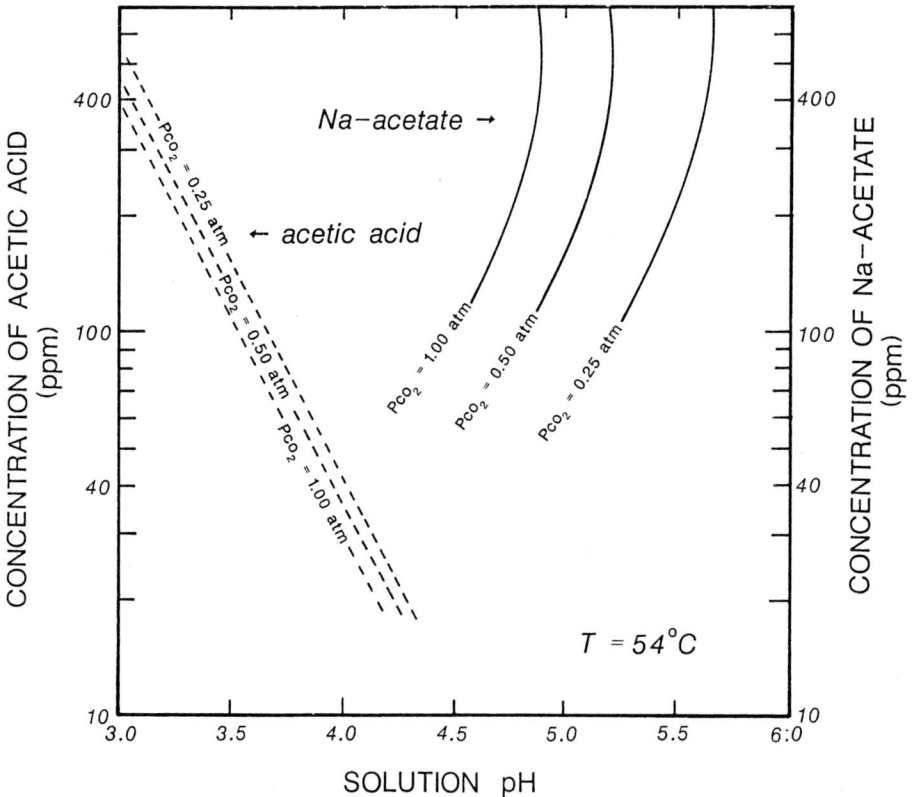

Fig. 2. Diagram of experimental data showing the influence of CO_2 partial pressure and concentrations of acetic acid (*dashed lines*) and sodium acetate (*solid lines*) on the pH of a water at 54 °C. (After Shock 1953)

acetic acid and sodium acetate on the pH of water at 54 °C in laboratory experiments summarized by Shock (1953). Later studies by Russian investigators also attributed corrosion of production field pipes to organic acids (Obuchova and Kutovaya 1968).

Within the geochemistry community, early efforts were directed toward documentation of organic acid occurrences or their proximity indicators to petroleum accumulations (e.g., Collins et al. 1961; Gullikson et al. 1961; Bykova and Nikitina 1964; Shvets and Seletskiy 1968; Shvets 1970; Bykova et al. 1971; Zinger and Kravchik 1972; Kudryakov 1974; Kartsev 1976). Willey et al. (1975) were probably the first to use modern analytical techniques to quantify organic acids. They showed that the measured alkalinity values of waters from Kettleman North Dome oil field, California, were dominated (75 to 100%) by organic acids. Failure to consider this contribution by organic acids to alkalinity results in overestimating carbonate concentrations. Equilibrium calculations based on this overestimate result in exaggerated saturation states for carbonate minerals. Matusevich and Shvets (1973) and Carothers and Kharaka (1978) demonstrated a correlation between organic acids in formation waters from oil fields and temperature. At temperatures below about 80 °C, the acids are assumed to be consumed by bacteria, whereas at temperatures exceeding 140 °C, the acids undergo decarboxylation. Organic acids are most abundant around 90 °C (Fisher 1987). Carothers and Kharaka (1978) believe organic acids are assumed to undergo decarboxylation while in reservoir rocks to form methane. Other papers of significance in the study of organic acids in oil-field waters include Hatton and Hanor (1984), Hanor and Workman (1986), Kharaka et al. (1986), Means and Hubbard (1987), Barth (1987), Fisher (1987), MacGowan and Surdam (1988), and Fisher and Boles (1990).

Surdam and his students (1984) were the first to recognize that organic acids have the potential to dissolve minerals, including aluminous silicates, to create secondary porosity in the subsurface. Until this time, secondary porosity was believed to originate primarily through the action of carbon dioxide (Schmidt and McDonald 1979). Lundegard et al. (1984) and Bjørlykke (1984) showed that the volume of available carbon dioxide from decarboxylation of organic matter was inadequate to explain the volume of secondary porosity believed to exist in sandstones of the Frio Formation in the Gulf Coast, USA. Surdam knew, from Huang and Keller's 1970 work, that organic acids dissolve minerals in soils and suggested that organic acids derived from kerogen in source rocks had the capability to dissolve carbonate and silicate minerals in subsurface sandstones. Organic acids are known to complex metals, which is of significance to diagenesis and the origin of hydrothermal ore deposits. Surdam, in later work (MacGowan and Surdam 1988; Surdam et al. 1989), stressed the importance of dicarboxylic acids in creating secondary porosity in sandstones. Surdam et al. (1984, 1989), MacGowan and Surdam (1988, 1990) and MacGowan et al. (1990) have reported on the relationship of organic and inorganic reactions during di-

agenesis of sandstones. They advocate the importance of organic acids in
creating secondary porosity in sandstones. This view has been challenged by
Giles and his coworkers (Giles and Marshall 1986; Giles and de Boer 1989,
1990).

Because of the interest in organic acids and their significance to secondary
porosity, a variety of laboratory studies have been conducted. Organic
matter has been heated to generate organic acids (e.g., Kawamura and
Ishiwatari 1985; Kawamura et al. 1986; Eglinton et al. 1987; Kawamura and
Kaplan 1987; Lundegard and Senftle 1987; Barth et al. 1988). Crossey
(1991) experimentally investigated the thermal degradation of organic acids.
Laboratory studies of the effect of organic acids on the dissolution of
aluminosilicates were made by Surdam et al. (1984), Mast and Drever
(1987), Bevan and Savage (1989), Hajash et al. (1989), and Stoessell and
Pittman (1990). The experiments of Stoessell and Pittman (1990) did not
produce as much dissolution of feldspar as the other studies. Fein (1991a)
experimentally evaluated acetate complexing and incorporated the results in
thermodynamic models, which indicated that the presence of acetate in
formation waters cannot significantly increase porosity in sandstones through
mineral dissolution. A companion study by Fein (1991b) indicated that Al-
oxalate complexing is more important than Al-acetate complexing.

Harrison and Thyne (1992) used geochemical modeling (EQ3/6) to evaluate the effects of organic acids. They concluded that organic acid anions are
ineffective at neutral to alkaline pH in modifying aluminosilicate minerals,
but may be effective over a wide range of pH in dissolving carbonate
minerals. Lundegard and Land (1989) also used geochemical modeling to
investigate the effectiveness of acetate anions in buffering the pH of natural
waters. They concluded that for the Gulf Coast Basin, an increase in partial
pressure of carbon dioxide would have promoted calcite dissolution despite
the buffering effects of the acetate. Lundegard and Kharaka (1990) reached
the same conclusion.

The kinetics of the decarboxylation reaction for acetic acid anions have
been studied by Kharaka et al. (1983) and Palmer and Drummond (1986).
Apparently, the decarboxylation reactions are slow under diagenetic conditions. Shock (1988) has investigated the stability of organic acids in
sedimentary basins on the basis of thermodynamic metastability.

There has been considerable interest in the role of organic acid anions in
weathering and groundwater. Antweiler and Drever (1983) presented evidence that organic acid complexes increase the mobility of aluminum and
iron in the weathering of volcanic ash. The dissolution of quartz by organic
acids in groundwaters (Bennett and Siegel 1987; Bennett 1991) and of
quartz and aluminosilicates in peat bogs (Bennett et al. 1991) has been
documented. Laboratory dissolution of quartz in dilute aqueous solutions of
organic acids at 25 °C suggests that organic acids in organic-rich soils and
weathering zones are capable of complexing silica at neutral pH (Bennett et
al. 1988). The analytical work of Marley et al. (1989) using laser Raman and

Fourier transform infrared spectroscopies appears to support the observation that organic dicarboxylic acids can lead to enhanced solubility of quartz.

During the 1980s, there was considerable research activity on the hydrothermal transport of metals by organic acid complexes. Papers of interest include Giordano and Barnes (1981), Giordano (1985, 1989), Drummond and Palmer (1986), Hennet et al. (1988), and Yang et al. (1989).

4 Brief Synopsis of Chapters

Chapter 2, by MacGowan and Surdam, covers the procedures and problems related to the collection of formation waters and analysis of organic acid anions. Early studies documented only monocarboxylic acid anions in oil-field waters. MacGowan and Surdam (1988) showed later that dicarboxylic acid anions can also be abundant (up to 2540 ppm malonate) in oil-field waters. Other laboratories have not confirmed these high values for carboxylic acid anions. This may be due to sampling or analytical procedural problems. Ion chromatography exclusion currently is the most popular analytical technique for analysis of carboxylic acid anions in sedimentary formation waters. MacGowan and Surdam show that this technique is difficult to apply and requires patience and time. It is almost an art form and requires multiple runs. In addition, the resampling of wells through time may yield inconsistent values of organic acids. For example, MacGowan and Surdam (Chap. 2, Table 1) sampled one well four times and obtained acetate concentrations ranging from 56 to 2435 ppm. This variation is to be expected according to MacGowan and Surdam because when a sample is taken from the separator the mixture of oil and water varies, which can lead to varying results.

Chapter 3, by Lundegard and Kharaka, provides an overview of the distribution and nature of organic acids in the subsurface. At temperatures less than 40 and greater than 180 °C, concentrations of organic acid anions are less than a few hundred mg/l. At intermediate temperatures, there is a crude relationship between temperature and the maximum observed concentrations. Factors that contribute to the variation in quantity of organic acids are attributed to the richness and type of organic matter in the source rocks and the thermal history of the source rocks. Post-generation factors that affect organic acids in formation waters include variability of dilution, biodegradation, and effects of natural catalysis on decarboxylation. It is useful to consider the potential geochemical effects that the maximum organic acid concentrations have on the geochemical system. However, it is more important to consider the effect of "normal" concentrations because available data suggest that organic acid concentrations >3000 mg/l are rare. New data on the generation of organic acids from reservoired oil under thermal stress are also presented and discussed.

Chapter 4, by Lewan and Fisher, examines the role petroleum source rocks play in supplying organic acids to formation waters during their thermal maturation in the subsurface of subsiding sedimentary basins. The vastness and dynamics of sedimentary basins and the lack of data on subsurface waters associated with petroleum source rocks necessitate the use of laboratory pyrolysis experiments to gain insights on this potential source of organic acids. The authors discuss important experimental conditions that should be considered in laboratory simulations of organic matter maturation. Experimental data published in the literature are summarized and compared, along with new data from their laboratory not previously published. Bonding of organic acids in sedimentary organic matter appears to be noncovalent in character. Maximum conversion of sedimentary organic matter to short-chain carboxylic acids appears to be 2.5 wt%. Half of these organic acids may be lost during early diagenesis as a result of diffusion. The authors' calculations suggest that the sluggishness of this process prevents significant and timely quantities of organic acids from generating secondary porosity on a basinwide scale. Another implication presented in this chapter is that currently observed organic acids in formation waters associated with petroleum accumulations may have originally been carried as dissolved species in the petroleum and released into the associated formation waters as the migrating petroleum reequilibrates with cooler, lower salinity, or lower gas pressure conditions of a reservoir.

Chapter 5, by Pittman and Hathon, reviews published material-balance considerations for calculating whether sufficient acids can be generated from kerogen in shales to account for secondary porosity generally believed to be present in sandstones. This chapter also presents what is believed to be the first published material-balance calculations based on actual data on source rocks and sandstones for a given area (deep part of the Denver Basin), rather than using estimates or hypothetical considerations. Source rock geochemistry provided information on the volume and type of organic matter for each source rock. Point-count analysis for porous sandstones throughout the stratigraphic column yielded data on the amount and type of secondary porosity. Geophysical logs were used to determine the thickness of source rock and sandstone units for calculating volumes. The modeling is based on hydrous pyrolysis of thermally immature source rocks to establish the conversion efficiency of kerogen oxygen to mono- and dicarboxylic acids. Under optimum conditions for organic acid activity, the authors calculate that 1.72% secondary porosity could form in the sandstones of the basin from CO_2 and carboxylic acids; whereas, by point-count analysis, using a conservative approach to identifying secondary porosity, there is at least 2.49% secondary porosity in sandstones in the basin.

Chapter 6, by Drever and Vance, provides a useful overview of the distribution and significance of organic acids in soils. Low-molecular-mass acids (e.g., acetic, oxalic, formic) exist in dynamic balance within soils. They are rapidly produced and consumed by microorganisms. Concentrations of organic acids are generally highest in the organic layer at the top of the soil profile and decrease with depth. Organic acids affect the mineralogy of soils by complexing and transporting Fe and Al. There is considerable controversy regarding whether organic acids at natural concentrations significantly accelerate the rate of dissolution of primary silicate minerals. It is not likely that organic acids are important in the weathering of granitic rocks, but they may be important in the weathering of more mafic rocks. Organic acid concentrations may be significantly higher in microenvironments surrounding rootlets or fungal hyphae than in bulk soil solutions.

Chapter 7, by Bennett and Casey, covers the chemistry and mechanisms of the dissolution of silicate minerals by organic acids. Organic acids chelate silica and aluminum via ligand-exchange complexation mechanisms. Silica is complexed at near-neutral pH values, whereas aluminum is most effectively chelated at acidic pH values. The chemistry of solution complexation can be applied to silicate surfaces where ligand-promoted dissolution is commonly faster than dissolution in nonorganic systems. Steep chemical gradients and high concentrations of organic acids exist in the vicinity of microbial colonies, which suggest that microbial activity may be important in some weathering processes. Bennett and Casey present examples from peat bogs, underclays, and petroleum-contaminated groundwaters.

In *Chapter 8*, Hajash reviews experimental data on dissolution of feldspar. These data vary because of the initial solids, fluids, experimental systems, and procedures used by different investigators. Hajash concludes that most data indicate that carboxylic acids enhance the solubility of feldspar, especially at moderately acidic (pH 4 to 5) conditions. Dicarboxylic acids in some subsurface brines suggests the acids have been stabilized by complexation with metal ions. The kinetics for decarboxylation of mono- and dicarboxylic acids are more effective than monocarboxylic acids. Generally, the aluminum content of the reacted solutions varies inversely with pH and directly with the organic acid concentration. The effectiveness of organic acids in dissolution and buffering processes is a function of their type and concentration, pH, temperature, reaction kinetics, and flow rate. Mobility of aluminum during diagenesis may also be influenced by the presence of carbonate or mafic minerals and the activity of CO_2 or other dissolved components in formation waters.

Chapter 9, by Bell and Palmer, reviews experimental studies concerned with the decomposition of organic acids and their anions. In the absence of an effective catalytic surface, acetic acid may be expected to survive indefinitely. However, there are many potential catalysts. Laboratory determined activation energies for the decarboxylation of acetic acid varies from 34 to $170\,kJ\,mol^{-1}$ for stainless steel and titanium oxide, respectively. Reaction rates for the decarboxylation of dicarboxylic acids are extremely fast and uncomplexed acids are not likely to survive geologically significant lengths of time. However, the observed occurrence of dicarboxylic acids in some subsurface brines suggests the acids have been stabilized by complexation with metal ions. The kinetics for decarboxylation of mono- and dicarboxylic acids are first-order with respect to total carboxylate concentration, and their rate constants are greater in the acid form than in the anionic form. Rate constants at intermediate pH values follow a linear relation between the acid and anion forms for dicarboxylic acids and are greater than the acid form for monocarboxylic acids.

Chapter 10, by Shock, considers the application of thermodynamic calculations to geochemical processes involving organic acids. This chapter starts with a discussion of how to calculate standard state thermodynamic properties of aqueous organic acids. Thermodynamic calculations are then applied to decarboxylation reactions involving organic acids and to redox reactions between organic acids, CO_2, and hydrocarbons. The decrease in concentration of carboxylic acids with increasing temperature from 80 to 200 °C, as first published by Carothers and Kharaka (1978), does not exist when all currently available data are plotted. Shock believes there is no basis for drawing an upper temperature limit on this plot as Surdam et al. (1984) have done. High concentrations of acetic acid in oil-field brines are preserved in a metastable state with respect to the decarboxylation reaction and hydrogen fugacity is controlled by the presence of petroleum. The decarboxylation reaction of acetic acid in sedimentary basins is inhibited and cannot be in stable equilibrium with both CO_2 and CH_4. The potential for metastable equilibrium between acetic acid and CO_2 is large under geologically realistic conditions, but the potential for metastable equilibrium involving acetic acid and CH_4 is severely limited.

Chapter 11, by Giordano, is an overview of organic acid complexing in the genesis of ore deposits. Models for Mississippi Valley-type ore solutions and ore fluid for red-bed-related, base-metal deposits are evaluated in this chapter. These models are limited by availability of thermodynamic data, but their chief deficiency is lack of well-constrained geochemical parameters for the ore-forming systems. These models suggest that metal-organic complexes may be important in transporting lead, zinc, and other metals in ore fluids for these two major types of deposits. The significance of aliphatic-carboxylate complexes in transporting ore metals is inversely related to the activities of hydrogen sulfide, bisulfide, and chloride. Transport of metals and reduced sulfur in the same ore fluid may occur in the form of organosulfur complexes.

Chapter 12, by Harrison and Thyne, provides an overview of geochemical modeling for rock-water interactions in the presence of organic acids. Geochemical modeling is useful because it provides insight into the role of organic acids and their anions in subsurface formation waters by defining boundary conditions under which such interactions may be significant. These authors conclude that aluminum acetate complexes are of minor importance; whereas aluminum oxalate dominates the species distribution of Al under acidic conditions. They believe that organic acids contribute to the overall patterns of fluid-rock interaction, but appear unlikely to dominate such reactions except in restricted geochemical environments where their concentrations are in excess of typical values. Probable environments meeting this requirement include wetlands, gasoline-contaminated groundwaters, and sandstones adjacent to organic-rich shales.

Chapter 13, by Surdam and Yin, provides an overview of the relationship for organic-inorganic diagenesis and provides a detailed case history. They suggest that a sequential set of carbonate reactions characterizes many clastic source/reservoir rock systems during progressive burial. With increasing thermal exposure during burial, the sequence is: (1) formation of carbonate cements that preserve intergranular pore volume; (2) dissolution of early carbonate cements, which may enhance porosity and result in positive porosity anomalies; (3) formation of late carbonate cements, which may retard compaction; and (4) if tempertures are high enough and if quartz cementation is inhibited, dissolution of late carbonate cements, again enhancing porosity. Coupling the above sequence of carbonate reactions with parallel organic reactions, including generation and decarboxylation of organic acids, Surdam and Yin construct a predictive, process-oriented model. This model consists of three operations: (1) reaction pathway diagnostics; (2) kinetic modeling of organic reactions; and (3) simulation of the rock/water interactions in either time or temperature space. They believe the integration of the above three operations allows prediction of zones of carbonate dissolution in source/reservoir rock systems.

Chapter 14, by Giles, de Boer, and Marshall, evaluates the hypothesis that organic acids create significant secondary porosity in sandstones and enhance aluminum mobility in the subsurface. They believe the hypothesis fails because: (1) evidence does not support a sudden increase in secondary porosity pore volume over the temperature range where carboxylic acids are most abundant; (2) experimental data on the dissolution of feldspars by organic acids show little evidence to support enhanced dissolution or increased aluminum mobility except at unrealistically low pH values; (3) complexing of aluminum by organic acids is unlikely in natural formation waters because of the competition of other ions for the organic acid ligands; (4) acids generated in a shale source rock would quickly be neutralized either in the source rock (unless it was exceptionally rich in organic matter) or along migration pathways; and (5) mass-balance calculations suggest that source rocks would have to be unreasonably abundant to generate enough carboxylic acids to account for even a few percent of secondary porosity.

5 Future Research

Although the following chapters present current observations and views on the role of organic acids in geological processes, additional research is needed to elevate the current evolving hypotheses into working scientific concepts. The future research needed to reach this goal is discussed under two headings: (1) natural system studies and (2) laboratory experimental studies. Both types of studies are included in this book and their interdependence is unquestionable. Natural system studies provide the critical observations in natural settings that define the boundary conditions within which geological processes operate. Usually, the vastness and dynamics of the natural system result in limited and static data with wide knowledge gaps that allow for multiple interpretations. These knowledge gaps in natural system studies may be bridged in part by laboratory experimental studies. The success of bridging these knowledge gaps by laboratory experiments depends on designing experiments that best simulate the natural system.

5.1 Natural System Studies

1. There is general agreement that dicarboxylic acids have a more significant effect on dissolution of silicate minerals than monocarboxylic acids, but their presence and abundance in natural formation waters remain uncertain. A collaborative study involving different laboratories is needed to resolve this uncertainty. Each laboratory would analyze aliquots of formation waters collected from a series of wells representing different subsurface conditions and water chemistries. The analytical results would be reviewed by the participating laboratories with the objectives of resolving analytical problems and establishing accepted procedures and standards. This type of study could also be expanded to include analyses of monocarboxylic acids and more complex higher-molecular-weight organic acids.

2. Experimental data clearly demonstrate that organic acids can have a significant influence on mineral dissolution reactions when pH values are less than 5 (Fig. 2). The uncertainty here is whether natural waters, particularly subsurface formation waters, typically attain these low pH values. Drilling engineers in the late 1940s and early 1950s faced a similar uncertainty in explaining the high rate of corrosion to drill pipes in some condensate wells. Initially, they did not consider the corrosion process to be related to organic acids because of the high pH values (≥ 6) of the produced waters as measured in the laboratory. Subsequent studies showed that the pH values measured in laboratories under atmospheric conditions were not representative, and that pH values measured under subsurface conditions were low enough (<5) to invoke organic acids as important participants in the corrosion process (Shock 1953). Procedures and apparatus for measuring the pH of formation waters at wellhead pressures (Carlson 1949) and flow rates (Collins 1964) have been shown to provide pH values more representative of the subsurface waters. Downhole drill-stem testing equipment is currently available for sampling subsurface formation waters. Currently, these downhole tools may be equipped with optical sensors for in situ viewing of fluid phase behavior or resistivity probes for in situ estimates on water salinities

(Badry et al. 1993), but no tools are currently equipped with electrodes for in situ measurements of pH. pH electrodes that operate at high temperatures (e.g., 200–300 °C) are currently being developed (Bourcier et al. 1987), and research into their use during subsurface drill-stem tests of formation waters may provide the best results in determining this important variable.

3. The existing data base on the subsurface occurrence of organic acids consists almost exclusively of formation waters associated with petroleum accumulations. Between the years 1970 and 1990, more than 200000 exploration wells were drilled in the United States and only 13% of these wells encountered petroleum (Energy Information Administration 1991). Future research in natural system studies should make a concerted effort to expand the current data base with analyses of formation waters not associated with petroleum accumulation. Comparisons of organic acid types and abundances in formation waters associated and not associated with petroleum accumulations may provide new data relevant to the origin and emplacement of organic acids. In addition, a concerted effort to collect subsurface waters associated with humic coals and organic-rich shales at different levels of thermal maturity would provide critical data to our understanding of the origin of organic acids.

4. There appears to be general agreement that organic acids can accumulate in high enough concentrations to influence processes in soil and groundwater microenvironments or in subsurface rocks immediately adjacent to organic-rich rocks. Future research is needed to determine whether variations in the physical, chemical, and biological controls on these localized occurrences can influence neighboring processes that are more regionally operative. Natural system studies of this type will require more detailed sampling and consideration of microbial activity.

5. Although organic acids have been shown to be capable of enhancing metal transport in ore fluids at 75 to 150 °C, their occurrence in ore fluids remains to be established. Future research directed at identifying and quantifying organic acids in fluid inclusions of ore minerals or gangue is needed. This type of data would provide the geochemical constraints necessary to test and develop models for ore-forming systems. A method for analyzing short-chain organic acids in fluid inclusions has been developed (Hofstra and Emsbo 1992). This technique also could be profitably applied to fluid inclusions in mineral cements of petroleum-bearing strata.

5.2 Laboratory Experimental Studies

1. The hypothesis that organic acids in formation waters are derived from coexisting petroleum as a result of reequilibration of the two immiscible phases under reservoir conditions needs further study. Although partition coefficients for organic acids in water and pure hydrocarbons are available at 25 °C and 1 atm (Leo et al. 1971), partition coefficients for organic acids in

water and petroleum are not available under reservoir conditions. The laboratory experimental study by Knaepen et al. (1990) has shown that the partition coefficients for ethyl acetate in water and petroleum can be significantly different with increasing water salinity, dissolved gas content, and temperature. Future research using a similar experimental approach to determine partition coefficients for organic acids under reservoir conditions is needed. Other variables worth considering include the composition of petroleum and the factors that influence the time required for partition coefficients to reach equilibrium.

2. Laboratory experiments have demonstrated that mono- and dicarboxylic acids, in low salinity solutions with pH values lower than 5, enhance solubility and increase dissolution rates of feldspars and clay minerals at temperatures between 70 and 100 °C. Future laboratory experiments should be conducted with brines of various salinities and compositions, dissolved CO_2 at different partial pressures, and different mineral assemblages representative of subsurface conditions. These experiments will more closely simulate the natural system and provide data on the effects of cation competition for complexing sites, salinity on silica solubility, enhanced acidity due to dissolved CO_2, and mineral interactions on overall dissolution rates.

3. Experimental studies concerned with the stability of mono- and dicarboxylic organic acids have focused on decarboxylation as the primary reaction responsible for their destruction. These studies have typically evaluated kinetics of the assumed decarboxylation reaction by monitoring changes in the organic acid concentrations. It is essential that future experimentation identify and monitor the resulting products, as well as the loss of reactant. This approach will better define the destructive reactions, and assist in evaluating the significance of organic acid destruction through oxidation and condensation reactions. The importance of catalytic surfaces in the destruction of monocarboxylic organic acids suggests that future experiments be conducted in the presence of mineral surfaces common in the natural system (e.g., quartz, clay minerals, and feldspars). Other potentially significant variables worth considering in future experiments are differences in stability between different metal-organic complexes and between organic acids dissolved in water and those dissolved in petroleum.

4. From a petrographic perspective, the recognition of secondary porosity in sandstones is based on textural evidence visible in thin sections using criteria presented by Schmidt and McDonald (1979). There is no problem recognizing secondary porosity in thin sections when the porosity is intragranular or moldic. However, a problem arises when one attempts to determine how much, if any, intergranular porosity is secondary in origin due to dissolution of cement (e.g., carbonates). Textural evidence gained from the scanning electron microscope is helpful in some situations (Burley and Kantorowicz 1986; Larese and Pittman 1987), but there is currently no known way to quantify the amount of secondary porosity. As a result, petrologists subjectively interpret the textural evidence as indicative of sparse

or abundant intergranular secondary porosity depending on their experience, training, or bias. It is also possible that secondary porosity has been over-estimated due to the presence of artifact porosity, which mimics secondary porosity (Pittman 1992).

A replacement texture imprint commonly is visible under the scanning electron microscope on sand grain surfaces and/or on preexisting cements after exposure by leaching of carbonate cement from sandstone in the laboratory. Larese and Pittman (1988 unpubl. data) observed replacement textures in every rock that they studied using this technique, which commonly requires the use of high magnification; however, it is unknown if a replace-ment imprint is visible in every pore. If so, then this technique could be used to quantify the amount of secondary intergranular porosity in naturally porous sandstone. Certainly, some method is needed for quantifying the amount of secondary intergranular porosity.

References

Aiken GR, McKnight DM, Wershaw RL, McCarthy P (1985) Humic substances in soil, sediment and water. Wiley, New York, 692 pp

Antweiler RC, Drever JI (1983) The weathering of a late Tertiary ash: importance of organic solutes. Geochim Cosmochim Acta 47: 623–629

Badry R, Head E, Morris C, Travoulay I (1993) New wireline formation tester techniques and applications. SPWLA Annu Symp, Calgary, Alberta, June 13–16, 1993, pp H1–H15

Barden RE, Logan ER, Branthaver JF, Neet KE (1984) The average molecular weight and shape of the "polymeric acids" found in black trona water from the Green River Basin. Org Geochem 5: 217–225

Barth T (1987) Multivariate analysis of aqueous organic acid concentrations and geological properties of North Sea reservoirs. Chemometrics Intelligent Lab Syst 2: 155–160

Barth T, Borgund AE, Hopland AL, Graue A (1988) Volatile organic acids produced during kerogen maturation – amounts, composition and role in migration of oil. Adv Org Geochem 13: 461–465

Bennett PC (1991) Quartz dissolution in organic-rich aqueous systems. Geochim Cosmochim Acta 55: 1781–1797

Bennett PC, Siegel DI (1987) Increased solubility of quartz in water due to complexing by organic compounds. Nature 326: 684–686

Bennett PC, Melcer ME, Seigel DI, Hassett JP (1988) The dissolution of quartz in dilute aqueous solutions of organic acids at 25 °C. Geochim Cosmochim Acta 52: 1521–1530

Bennett PC, Siegel DI, Hill BM, Glaser PH (1991) Fate of silicate minerals in a peat bog. Geology 19:328–331

Bevan J, Savage D (1989) The effect of organic acids on the dissolution of K-feldspar under conditions relevant to burial diagenesis. Mineral Mag 53: 415–425

Bjørlykke K (1984) Formation of secondary porosity: how important is it? In: McDonald DA, Surdam RC (eds) Clastic diagenesis. Am Assoc Pet Geol Mem 37: 277–286

Bourcier WL, Ulmer GC, Barnes HL (1987) Hydrothermal pH sensors of ZrO_2, Pd hydrides, and Ir oxides. In: Ulmer CG, Barnes HL (eds) Hydrothermal experimental techniques. Wiley, New York, pp 157–188

Burley SD, Kantorowicz JD (1986) Thin section and S.E.M. textural criteria for the recognition of cement-dissolution porosity in sandstones. Sedimentology 33: 587–604

Bykova EL, Nikitina JB (1964) Water-soluble organic matter of groundwater and surface water of south Yakutia. Geokhimiya 12: 1298–1304 (in Russian)

Bykova EL, Melkanovitskaya SG, Shvets VM (1971) Distribution of organic acids in underground waters. Sov Geol 14: 135–142 (in Russian)

Carlson HA (1949) Corrosion in natural gas-condensate wells – pH and carbon dioxide content of well waters at wellhead pressure. Ind Eng Chem 41: 644–645

Carothers WW, Kharaka YK (1978) Aliphatic acid anions in oil-field waters – implications for origin of natural gas. Am Assoc Petrol Geol Bull 62: 2441–2453

Chesney RB (1953) Laboratory work on corrosion in carbon dioxide and organic acids. In: Prange FA, Edwards WHJ, Greco EC, Griffith TE, Grimshaw JA, Nathan CC, Shock DA (eds) Condensate well corrosion. Nat Gasoline Assoc Am, Tulsa, pp 157–163

Collins AG (1964) Eh and pH of oilfield waters. Prod Monthly 28: 11–12

Collins AG, Pearson DH, Attaway DH, Watkins JW (1961) Methods of analyzing oil field waters. US Bur Mines Rep Invest 5819: 11–17

Cronin JR, Pizzarello S, Cruikshank DP (1988) Organic matter in carbonaceous chondrites, planetary satellites, asteroids and comets. In: Kerridge JF, Matthews MS (eds) Meteorites and the early solar system. University of Arizona Press, Tucson, pp 819–857

Crossey LJ (1991) Thermal degradation of aqueous oxalate species. Geochim Cosmochim Acta 56: 1515–1527

Drummond SE, Palmer DA (1986) Thermal decarboxylation of acetate, part II. Boundary conditions for the role of acetate in the primary migration of natural gas and the transportation of metals in hydrothermal systems. Geochim Cosmochim Acta 50: 825–833

Eglinton TI, Curtis CD, Rowland SJ (1987) Generation of water-soluble organic acids from kerogen during hydrous pyrolysis: implications for porosity development. Mineral Mag 51: 495–503

Energy Information Administration (1991) US crude oil, natural gas, and natural gas liquids reserves, 1990 annual report. DOE/EIA-0216(90), Distribution Category UC-98, Washington DC, 109 pp

Fein JB (1991a) Experimental study of aluminum-, calcium-, and magnesium-acetate complexing at 80 °C. Geochim Cosmochim Acta 55: 955–964

Fein JB (1991b) Experimental study of aluminum-oxalate complexing at 80 °C: implication for aluminum mobility in sedimentary basin fluids. Geology 19: 1037–1040

Fisher JB (1987) Distribution and occurrence of aliphatic acid anions in deep subsurface waters. Geochim Cosmochim Acta 51: 2459–2468

Fisher JB, Boles JR (1990) Water-rock interaction in Tertiary sandstones, San Joaquin Basin, California, USA: diagenetic controls on water composition. Chem Geol 82: 83–101

Giles MR, de Boer RB (1989) Secondary porosity: creation of enhanced porosities in the subsurface from the dissolution of carbonate cements as a result of cooling formation waters. Mar Pet Geol 6: 261–269

Giles MR, de Boer RB (1990) Origin and significance of redistributional secondary porosity. Mar Pet Geol 7: 378–397

Giles MR, Marshall JD (1986) Constraints on the development of secondary porosity in the subsurface: re-evaluation of processes. Mar Pet Geol 3: 243–255

Giordano TH (1985) A preliminary evaluation of organic ligands and metal organic complexing in Mississippi Valley-type ore solutions. Geochim Cosmochim Acta 80: 96–106

Giordano TH (1989) Anglesite ($PbSO_4$) solubility in acetate solutions: the determination of stability constants for lead acetate complexes to 85 °C. Geochim Cosmochim Acta 53: 359–366

Giordano TH, Barnes HL (1981) Lead transport in Mississippi Valley-type ore solutions. Econ Geol 76: 2200–2211

Greco EC, Griffin HT (1946) Laboratory studies for determination of organic acids as related to internal corrosion of high pressure condensate wells. Corrosion 2: 138–152

Gullikson DM, Carraway WH, Gates BL (1961) Chemical analysis and electrical resistivity of selected California oil-field waters. US Bur Mines Rep Invest 5736: 1–21

Hajash A, Mahoney AJ, Elias BP (1989) Role of carboxylic acids in the dissolution of silicate sands: an experimental study at 100 °C, 345 bars. Geol Soc Am Abstr 21: A49

Hanor JS, Workman AL (1986) Distribution of dissolved volatile fatty acids in some Louisiana oil field brines. Appl Geochem 1: 37–46

Hare PE, Hoering TC, King K Jr (1980) Biogeochemistry of amino acids. Wiley, New York, 558 pp

Harrison WJ, Thyne GD (1992) Predictions of diagenetic reactions in the presence of organic acids. Geochim Cosmochim Acta 56: 565–586

Hatton RS, Hanor JS (1984) Dissolved volatile fatty acids in subsurface hydropressured and geopressured brines: a review of published literature on occurrence, genesis, and thermochemical properties. US Dep Energy Rep DOE/NV/10174-3, June 25, 1984, Washington DC, pp 348–454

Hayes JM (1967) Organic constituents of meteorites – a review. Geochim Cosmochim Acta 31: 1395–1440

Hennett R JC, Crerar DA, Schwartz J (1988) organic complexes in hydrothermal systems. Econ Geol 83: 742–764

Hofstra AH, Emsbo P (1992) A new method to analyze anions and cations in fluid inclusions using ion chromatography – applications to ore genesis. Geol Soc Am Abstr Prog 24: A144

Huang WH, Keller WD (1970) Dissolution of rock-forming silicate minerals in organic acids: simulated first-stage weathering of fresh mineral surfaces. Am Mineral 55: 2076–2094

Jaffé R, Albrecht P, Oudin JL (1988) Carboxylic acids as indicators of oil migration. II. Case of the Mahakam Delta Indonesia. Geochim Cosmochim Acta 52: 2599–2607

Kartsev AA (1976) Hydrogeology of oil and gas deposits. Natl Tech Inf Serv Rep TT73-58022, 323 pp

Kawamura K, Ishiwatari R (1985) Conversion of sedimentary fatty acids from extractable (unbound + bond) to tightly bound form during mild heating. Org Geochem 8: 197–201

Kawamura K, Kaplan IR (1987) Dicarboxylic acids generated by thermal alteration of kerogen and humic acids. Geochim Cosmochim Acta 51: 3201–3207

Kawamura K, Tannenbaum E, Huizinga BJ, Kaplan IR (1986) Volatile organic acids generated from kerogen during laboratory heating. Geochem J 20: 51–59

Kharaka YK, Law LM, Carothers WW, Goerlitz DF (1976) Role of organic species dissolved in formation waters from sedimentary basins in mineral diagenesis. In: Gautier DL (ed) Roles of organic matter in sediment diagenesis. Soc Econ Paleontol Mineral Spec Publ 38, pp 111–122

Kharaka YK, Carothers WW, Rosenbauer RJ (1983) Thermal decarboxylation of acetic acid: implications for origin of natural gas. Geochim Cosmochim Acta 47: 397–402

Knaepen WAI, Tijssen R, van den Berger EA (1990) Experimental aspects of partitioning tracer tests for residual oil saturation determination with FIA-based laboratory equipment. SPE Reservoir Eng 5: 239–244

Kudryakov VA (1974) Genetic significance of organic matter in subsurface waters in oil geology. Geol Nefti i Gaza 7: 66–68 (in Russian)

Kvenvolden KA, Peterson E, Wehmiller J, Hare PE (1973) Racemization of amino acids in marine sediments determined by gas chromatography. Geochim Cosmochim Acta 37: 2215–2225

Larese RE, Pittman ED (1987) Indirect evidence of secondary porosity in sandstones (Abstr). Am Assoc Petr Geol Bull 71: 581

Lee C, Bada JL, Peterson E (1976) Amino acids in modern and fossil woods. Nature 259: 183–186

Leo A, Hansen C, Elkins D (1971) Partition coefficients and their uses. Chem Rev 71: 525–616

Lochte HL, Littmann ER (1955) The petroleum acids and bases. Chemical Publishing Co, New York, 368 pp

Lowe WF (1953) History of the condensate well corrosion committee. In: Prange FA, Edwards WH, Greco EC, Griffith EE, Grimshaw JA, Nathan CC, Shock DA (eds) Condensate well corrosion. Nat Gasoline Assoc Am, Tulsa, pp 1–9

Lundegard PD, Kharaka YK (1990) Geochemistry of organic acids in subsurface waters. In: Melchoir D, Bassett R (eds) Chemical modeling in aqueous systems, II. Am Chem Soc, Washington DC, pp 169–189

Lundegard PD, Land LS (1989) Carbonate equilibria and pH buffering by organic acids – response to changes in P_{CO_2}. Chem Geol 74: 277–287

Lundegard PD, Senftle JT (1987) Hydrous pyrolysis: a tool for the study of organic acid synthesis. Appl Geochem 2: 605–612

Lundegard PD, Land LS, Galloway WE (1984) Problem of secondary porosity: Frio Formation (Oligocene), Texas Gulf Coast. Geology 12: 399–402

MacGowan DB, Surdam RC (1988) Difunctional carboxylic acid anions in oilfield waters. Org Geochem 12: 245–259

MacGowan DB, Surdam RC (1990) Importance of organic-inorganic reactions to modeling water-rock interactions during progressive clastic diagenesis. In: Melchoir D, Bassett R (eds) Chemical modeling in aqueous systems, II. Am Chem Soc, Washington DC, pp 494–507

MacGowan DB, Surdam RC, Ewing RE (1990) The effect of carboxylic acid anions on the stability of framework mineral grains in petroleum reservoirs. Soc Petrol Eng, Formation Evaluation, June, pp 161–166

Marley NA, Bennett P, Janecky DR, Gaffney JS (1989) Spectroscopic evidence for organic diacid complexation with dissolved silica in aqueous systems. 1. Oxalic acid. Org Geochem 14: 525–528

Mast MA, Drever JI (1987) The effect of oxalates on the dissolution rates of oligoclase and tremolite. Geochim Cosmochim Acta 51: 2559–2568

Matusevich VM, Shvets VM (1973) Significance in petroleum prospecting of organic acids dissolved in ground waters of the western Siberian lowland. Geol Nefti i Gaza 10: 63–69 (in Russian)

Means JL, Hubbard N (1987) Short-chain aliphatic acid anions in deep subsurface brines: a review of their origin, occurrence, properties, and importance and new data on their distribution and geochemical implications in the Palo Duro Basin, Texas. Org Geochem 11: 177–191

Menaul PL (1944) Causative agents of corrosion in distillate field. Oil Gas J, Nov 11: 80–81

Obuchova ZP, Kutovaya AA (1968) Study on distribution of organic acids in condensation waters in the area of gas-condensate deposits. Korroziyi i Zashchita v Neftedoby-vayushchei Promyshlennost Nauch-Fekh 4: 16–19 (in Russian)

Palmer DA, Drummond SE (1986) Thermal decarboxylation of acetate. Part 1. The kinetics and mechanism of reaction in aqueous solution. Geochim Cosmochim Acta 50: 813–823

Parker PL (1969) Fatty acids and alcohols. In: Eglinton G, Murphy MTJ (eds) Organic geochemistry. Springer, Berlin Heidelberg New York, pp 357–373

Pittman ED (1992) Artifact porosity in thin sections of sandstone. J Sediment Petrol 62: 734–737

Prange FA (1953) Metallurgical factors. In: Prange FA, Edwards WH, Greco EC, Griffith TE, Grimshaw JA, Nathan CC, Shock DA (eds) Condensate well corrosion. Nat. Gasoline Assoc Am, Tulsa, pp 99–115

Prange FA, Edwards W, Greco EC, Griffith TE, Grimshaw JA, Nathan CC, Shock DA (1953) Condensate well corrosion. Nat Gasoline Assoc Am, Tulsa, 203 pp

Rogers GS (1917) Chemical relations of the oil-field waters in San Joaquin Valley, California. US Geol Surv Bull 653: 119 pp

Schmidt V, McDonald DA (1979) The role of secondary porosity in the course of sandstone diagenesis. In: Scholle PA, Schluger PR (eds) Aspects of diagenesis. Soc Econ Paleontol Mineral Spec Publ 26, pp 175–207

Shimoyama A, Johns WD (1971) Catalytic conversion of fatty acids to petroleum-like paraffins and their maturation. Nat Phys Sci 232: 140–144

Shock DA (1953) Acidity of condensate well waters. In: Prange FA, Edwards WH, Greco EC, Griffith TE, Grimshaw JA, Nathan CC, Shock DA (eds) Condensate well corrosion. Nat Gasoline Assoc Am, Tulsa, pp 143–157

Shock EL (1988) Organic acid metastability in sedimentary basins. Geology 16: 886–890

Shvets VM (1970) Concentration and distribution of organic substances in underground waters. Dokl Akad Nauk SSSR 201: 453–456 (in Russian)

Shvets VM, Seletskiy UB (1968) Organic substances in the thermal waters of southern Kamchtka. Dokl Akad Nauk SSSR 182: 441–444 (in Russian)

Stevenson FJ (1982) Humus chemistry: genesis, composition, reactions. Wiley-Interscience, New York, 443 pp

Stoessell RC, Pittman ED (1990) Secondary porosity geochemistry revisited: feldspar dissolution by carboxylic acids and their anions. Am Assoc Petrol Geol Bull 74: 1795–1805

Surdam RC, Boese SW, Crossey LJ (1984) The chemistry of secondary porosity. In: McDonald DA, Surdam RC (eds) Clastic diagenesis. Am Assoc Petrol Geol Mem 37, pp 127–149

Surdam RC, Crossey LJ, Hagen ES, Heasler HP (1989) Organic-inorganic interactions and sandstone diagenesis. Am Assoc Petrol Geol Bull 73: 1–23

Thurman EM (1985) Organic geochemistry of natural waters. Nijhoff/Junk, Boston, 497 pp

Willey LM, Kharaka YK, Presser TS, Rapp JB, Barnes I (1975) Short chain aliphatic anions in oil field waters and their contribution to the measured alkalinity. Geochim Cosmochim Acta 39: 1707–1711

Yang MM, Crerar DA, Irish DE (1989) Raman spectroscopic study of lead and zinc acetate complexes in hydrothermal solutions. Geochim Cosmochim Acta 53: 319–326

Zinger AS, Kravchik TE (1972) Simplest organic acids in ground-water of the lower Volga region (genesis and possible use in prospecting for oil). Dokl Akad Nauk SSSR 203: 693–696 (in Russian)

Chapter 2 Techniques and Problems in Sampling and Analyzing Formation Waters for Carboxylic Acids and Anions

Donald B. MacGowan[1] and Ronald C. Surdam[1]

Summary

Carboxylic acids and anions (CAA) have been known to exist in sedimentary formation waters since before the turn of the century. They have been proposed as essential agents in various geochemical processes such as metal ore migration and deposition, as precursors to natural gas, and as the mediators of organic-inorganic diagenesis. If the importance of CAA in geochemical processes is to be critically evaluated, viable samples of formation waters must be taken, and accurate and precise measurements of the CAA in the samples made. Key among sampling procedures designed to ensure sample viability are: flushing all production lines, sample lines, and sample containers with the formation fluid before the sample is taken; filtration of the sample through 0.1 μm at the time of sampling; addition of preservatives; and protection of the sample from light and elevated temperature. Accurate and precise measurements of the concentration of CAA in formation waters may be made by various analytical techniques, the most popular of which is ion chromatography. However, in complex mixtures this analysis is nontrivial; adjustments to the analytical technique may have to be made, and multiple runs performed on a single sample. Generally, a single sample run is an insufficient basis for meaningful interpretation.

1 Introduction

Carboxylic acids and anions (CAA) have been known to exist in sedimentary basin formation waters since before the turn of the century, and have variously been proposed as essential agents in a number of geological processes, most recently as mediators of organic-inorganic clastic diagenesis (see review in MacGowan and Surdam 1990).

---

[1] Department of Geology and Geophysics, University of Wyoming, Laramie, Wyoming, USA

Because of their reactivity, there is no doubt of the importance of CAA in the geochemistry of formation waters. To meaningfully evaluate their activity in geochemical processes, viable and representative samples of formation waters must be collected, and they must be accurately and precisely analyzed. Methods for procuring viable formation water samples from oil wells are discussed below, and common techniques applied to analyzing CAA in formation water samples are reviewed and evaluated, inasmuch as the separation of weakly charged CAA from high-ionic-strength formation waters is a nontrivial, and sometimes awkward, analysis. Techniques and problems associated with the most popular method of analyzing CAA in formation waters are then discussed, highlighting ion chromatography exclusion (ICE). Details of analyzing the results of ICE determinations are discussed last.

2 Sampling Formation Waters from Petroleum Wells

Great care must be taken to select for sampling only wells that show the desired formation water characteristics. It is necessary to select wells that are perforated over known intervals, produce from a single, unique unit, and have not been affected by well workover, water-flooding, or any other stimulation or enhanced oil recovery technique (acidification, etc.) if the goal of the study is to understand the interaction between interstitial fluids and the formation. It has been shown that workovers and enhanced recovery treatments are chemically detectable in formation water samples for some time after production recommences (Smith et al. 1991). It may be difficult to locate an unaffected well in a mature producing area. Drill stem test (DST) samples are preferred to production samples, if it can be shown that the water recovered is a formation water unaltered by drilling fluids. On occasion, flow is insufficient during a DST to produce water from a formation not contacted by drilling fluids. If the sample is to be used in studies of scale formation or other production studies, it is not necessary or desirable to apply these criteria to well selection.

Samples should be taken as close to the well head as possible, preferably ahead of the separator. If the sample must be taken from the separator, the separator should be thoroughly flushed with production fluid to remove any fluids or solids that it contains before the sample is taken. The produced fluid should be allowed to flow freely into and through a large polycarbonate carboy. When the carboy has been sufficiently flushed, the bottom valve should be closed, and the carboy filled to overflowing and immediately capped. Care should be taken to protect the sample from the atmosphere. Most oil wells that produce water also produce a sufficient volume of oil to insulate the water sample from the atmosphere, once in the carboy; otherwise, the carboy head-space should be purged with argon or nitrogen.

Suspension of a thermometer in the production stream allows measurement of the temperature of the production fluid at the time of the sampling; this is important if the results of the analyses will be used later for thermodynamic modeling.

At the time of sampling, samples should be filtered through $0.1\,\mu$m; if this is impractical, samples can be filtered through glass wool packed in tripled coffee filters, and then refiltered through $0.1\,\mu$m as soon as possible. However, waiting to filter out the clay-sized or colloidal particles may allow colloidal and chemical reactions to alter the chemistry of the water as it attains equilibrium with surface conditions and added preservatives. Further sampling and filtration schemes are discussed by Lico et al. (1982). It is also imperative to determine pH and dissolved oxygen at the time of sample collection, as these quantities cannot be accurately determined after shipping and storage (an accurate measurement of sample temperature is also required). Alkalinity must also be measured at the site, or as soon after returning the sample to the laboratory as possible. It is not clear what effect shipping and storage have on carbonate alkalinity, but postsampling precipitation of sulfides and aluminum oxides may appreciably alter both the measured pH and measured alkalinity.

Should an oil-water emulsion be collected, it should be filtered as above, and then allowed to stand in a separatory funnel or carboy for a few hours. If the emulsion persists, it is considered best to preserve the sample as below, and either allow the emulsion to break on its own with time, or use a centrifuge or surfactant/emulsion breaker in the laboratory. It is unclear how emulsion breakers may alter the chemistry of the water, for many organic-metal complexes have been demonstrated to partition into the oil phase, once formed (Barth et al. 1990; MacGowan and Surdam 1990). It is often difficult to filter an emulsion through $0.1\,\mu$m; if this is the case, it should be filtered as finely as possible, and then the water refiltered through $0.1\,\mu$m after the emulsion has been broken. Acid preservation of an emulsion that has not been filtered through $0.1\,\mu$m should *not* be undertaken, as dissolution of clay-sized particles and colloids will render metal analyses highly inaccurate.

The biodegradation of short-chain CAA in aqueous environments at surface conditions is well documented. Additionally, data presented below indicate that even laboratory standards preserved with $HgCl_2$ show a measurable decrease (i.e., greater than absolute or relative error) in CAA concentration in as little as 260 h, attributable to the action of bacteria not poisoned by $HgCl_2$ and to photoactive degradation. The degradation of CAA in unpreserved formation waters has previously been documented by several authors (see MacGowan and Surdam 1988, 1990). Therefore, it is essential that all samples collected for CAA analysis be preserved, either with $HgCl_2$ or hexacetylpyridinium chloride. Hexacetylpyridinium chloride is preferred to $HgCl_2$ inasmuch as it poisons more bacteria and does not cause oxidation of reduced carbon species or metals, both problems associated with the use of $HgCl_2$ (P. Bennett, Department of Geology, University of Texas, Austin, 1990, pers. comm.; MacGowan and Surdam 1990). However, hexacetyl-pyridinium chloride may interfere with analyses involving refractive index or UV-VIS detection. Samples should be stored in the dark and as near 4 °C as possible until analysis. Inasmuch as sample degradation does occur even in preserved samples, immediate analysis is imperative to ensure sample viability.

Preservative action must be taken at the sampling site, as soon after filtration as possible. One aliquot of 25 to 50 ml of the water should be collected for each preservative step required; in addition, 50 to 100 ml of the water should be collected untreated. Polycarbonate or borosilicate bottles sealed with either Parafilm or Teflon tape under the bottle top are preferred for sample storage. Before leaving on the sampling trip, the bottles should be prewashed with 1:10 nitric acid:distilled water and then rinsed with deionized/distilled water. It has been found useful to place the required preservative in the bottles before leaving on the sampling trip, and to prelabel the sample bottles. In practice it has been found useful to tape bottle tops on the bottles with vinyl tape to prevent them from loosening during shipping. Additionally, steps must be taken to ensure that the sample label is written in indelible ink and securely attached to the sample bottle; if a single sample in the shipment leaks, it may cause the rest of the samples to lose their identification.

In addition to analyses for CAA, other analyses that require different preservation steps may be desired. Listed below are the most common analyses and the preservation steps that they require; also, see lists in Lico et al. (1982). Samples for chemical oxygen demand and total organic carbon should be treated with 2 ml of concentrated H_2SO_4 per liter of sample. Samples for phenols should have 1 g of copper sulfate added per liter of sample, and then should be acidified to pH 4 with phosphoric acid. Samples for dissolved metals should have 3 ml of 1:1 nitric acid added per liter of sample. Samples for sulfide analysis should have 2 ml of zinc acetate added per liter of sample. Samples for nonmetal anions and sulfate analysis need no treatment.

It has been the experience of the authors that serial samples of single wells can yield vastly different concentrations of CAA in the formation waters (see below). This is not surprising, given the results of CAA-oil-water partitioning experiments (see Barth et al. 1990; MacGowan and Surdam 1990): samples from the same well may vary greatly in composition through time. As an example of this, Table 1 lists acetate concentrations from various analyses of samples from the Paloma field (San Joaquin Basin, California) 24×11 well taken at four different times from 1982 through 1987. These and other data (see Tables 1 and 2 in MacGowan and Surdam 1990) show that serial samples from the same well may give results that vary radically from one another. Given that wells may make oil one day and gas or water the next, there is a chaotic mix produced ahead of the separator.

Table 1. Acetate concentrations in Paloma field well 24×11

Sample	Acetate (ppm)
1982 Arco sample unpreserved	2435
1984 Arco sample preserved	1640
1986 Split of J.B. Fisher and J.R. Boles sample[a]	56.0
1987 R.C. Surdam and J.R. Boles sample	1040

[a] Previously published by Fisher and Boles (1990), but unpublished by this group.

The results of aqueous CAA extraction from crude oil experiments (Barth et al. 1990; MacGowan and Surdam 1990) indicate that there are significant differences in CAA concentration between aqueous extracts and formation waters. Thus, it is not surprising that as fluids are produced from an oil well, varying concentrations through time of all dissolved organics (gas, liquid hydrocarbons, and CAA) occur.

3   Methods for the Analysis of Carboxylic Acids and Ions in Formation Waters

Several methods have been employed by various researchers to quantify the short-chain carboxylic acids and anions in solutions of geochemical interest. Before the advent of gas chromatography, many researchers used titrimetric methods to quantify CAA in formation waters. However, in practice, formation waters that contain high concentrations or complex mixtures of CAAs yield titrimetric results that are uninterpretable (e.g., a Gran Mals plot that yields a line rather than a set of curves with clearly defined titration inflection points).

Esterification of carboxylic acids followed by hydrocarbon extraction and gas chromatography has been widely used (e.g., Willey et al. 1975; Carothers and Kharaka 1978; Kharaka et al. 1985). The main advantage of this method is the ability to detect and quantify some of the longer-chain acids that are not detectable by other methods. The disadvantages are that (1) many of the esters of these carboxylic acids are more hydrophilic than hydrophobic, and thus are never extracted into the analate; and (2) the short-chain difunctional acid esters are apt to either precipitate as insoluble Ca or Mg salts during evaporative concentration of the extract, or totally decarboxylate at the injection port, and thus are either not analyzed or are analyzed as CO_2, respectively (Kawamura et al. 1986; Alltech Associates 1985). This method is also expensive and time-consuming.

Gas chromatography-mass spectroscopy (GC-MS) has also been successfully applied to the anlaysis of aqueous CAA in formation water (Kharaka et al. 1985). This method has been shown to identify many more dissolved organic species than other methods discussed in this chapter, but shares the drawbacks of standard gas chromatography discussed above.

Other researchers have used capillary isotachophoresis for this analysis (Barth 1987a,b). This method has two great advantages over esterification/ extraction/gas chromatography, GC-MS, or titration: it is rapid and it avoids the loss of any sample components. Its major weaknesses are that identification must be done by standard addition, and that the method experiences a decline in detectability of sample components with increasing carbon number; ideally, this limits the method to identification of species with six or fewer carbon atoms per molecule (Barth 1987a,b).

Ion chromatography (IC), introduced by Small et al. (1975), has been used to help overcome many problems in the analysis of aqueous ionic species, including the analysis of carboxylic acid anions in formation waters of relatively high ionic strength (Surdam et al. 1984; Crossey 1985; MacGowan et al. 1986; Fisher 1987; MacGowan and Surdam 1988, 1990; Fisher and Boles 1990) and the detection of carboxylic acid anions in the aqueous phase of hydrous pyrolsates (Lundegard and Senftle 1987; MacGowan and Surdam 1987).

IC has many advantages over other methods of analysis. These include small sample size, rapidity of analysis, high sensitivity, specificity in analysis of similar ions, and multiple determinations in a single chromatographic run (Dionex 1979). In addition, the necessary reagents, chromatographic columns, instruments, etc. are universally available and relatively inexpensive. There are several limitations to this experimental technique: ions with pK_a (negative log of the acid dissociation constant) higher than 7 (such as cyanate, silicate, and phenates) have poor detection limits even though they may be separated by IC; irreversible reactions between aqueous species and the resin may occur (such as precipitation of heavy and transition metals and metal/organic complexes as hydroxides); and channeling and destruction of the resin bed may be caused by organic solvent concentrations higher than about 10% (Dionex 1979). These limitations are generally of no concern in the analysis of formation waters or the aqueous phase of hydrous pyrolysates.

The major drawbacks of this technique are similar to those of isotachophoresis. First, the method is not preparative, and therefore relies upon standard addition for peak identification. This means that a rather large stock supply of standards must be kept on hand, and that even preserved standards need to be prepared frequently (see below). Second, the method also experiences a decline in detectability of weak organic acid species as pK_a, molecular sphericity, and carbon number all increase. This can be overcome to a certain degree by the addition of organic solvents to the sample; however, the practical limit of this method for analysis of carboxylic acids is molecules of carbon number eight or less. Due to the many organic and inorganic interferences, the method can be more time-consuming than reported analysis times for isotachophoresis (Barth 1987a,b), but is much faster and involves less laboratory preparation than esterification/extraction/

gas chromatography or GC/MS. Additionally, no sample components are lost during analysis, as they are in the GC or GC/MS methods.

Ion chromatography exclusion (ICE) is a type of ion chromatography employed by many research groups. ICE is the most popular analytical technique for the analysis of CAA in sedimentary formation waters, and therefore this chapter will focus on the theory and methodology of ICE analysis. ICE eliminates most ionic matrix effects and interferences, and allows a more accurate identification and quantification of weak acids and bases (such as the carboxylic acids and anions) in high-ionic-strength solutions. This technique combines separation of ionic species in a column packed with ion exchange resin, suppression of the highly conductive eluent cationic species across a semipermeable membrane in a suppressor column, and detection of eluted species by an electrical conductivity or UV detector. In addition to the disadvantages of regular IC analyses, the disadvantages of ICE are that some inorganic species overlap the weak acid elution times and that it is sometimes difficult to differentiate between organic species with similar elution times at very low concentration. These difficulties can be generally overcome by minor modification of the analytical technique, as discussed below.

More complex ICE analytical schemes than the one described below for the detection of CAA have been developed; these involve the use of gradient elution pumps, multiple separator columns and precolumns, and relatively exotic reagents (P. Bennett, Dept. Geology, University of Texas, Austin, 1990, pers. comm.; S. Boese, Dept. Geology and Geophysics, University of Wyoming, 1987, pers. comm.; D.B. MacGowan, unpubl. data). These techniques lend themselves to better peak resolution for the complex analysis of aqueous organic species not generally detectable by the standard ICE method outlined below. Also, the technique of freeze drying the water sample, followed by critical-point CO_2 extraction and CO_2 critical-point chromatography, has shown promise (D.B. MacGowan, unpubl. data). However, these techniques are much more time-consuming, complex, and expensive and are not mature with respect to analyzing for CAA in low concentration in high-ionic-strength solutions. They also tend to be quite specific to certain analytical problems (e.g., high-sulfate groundwater vs. high-bicarbonate groundwater). For these reasons, they are not yet widely used, and will be discussed elsewhere.

4   Theory and Principles of Ion Chromatography Analysis

4.1   Ion-Exchange Principles

The ion exchange resins used in the IC instrument generally consist of a spherical polystyrene matrix, usually cross-linked with divinylbenzene or a similar vinyl aromatic to provide structural rigidity, to which functional groups (such as sulfonic acid or quaternary ammonium) are chemically bonded. These functional groups provide exchange sites where the sample flows through the resin pores.

The formal ion exchange capacity of the resin bed is defined as the number of active exchange sites per mass or volume of resin; it is normally expressed as milliequivalents per gram of resin. The type and distribution of these exchange sites, as well as the resin bead diameter, all strongly affect the exchange capacity of the resin. This capacity is the principal factor affecting separation of aqueous species during analysis: as the capacity increases, increasingly stronger eluents are required to obtain peaks with sufficient height for quantification.

Ion exchange reactions proceed both reversibly and stoichiometrically; for every ion removed from solution at an exchange site, one charge-equivalent counter ion from the resin enters the solution. Ions of like charge and character will compete with each other for these exchange sites. The apparent exchange constant (K_q) is a measure of the affinity of a sample ion for the exchange site, and is also a measure of the selectivity of a particular resin for that sample ion. The greater the selectivity, the longer the sample ion remains attached to the exchange site, and the later it will elute from the column. Determination of the selectivity of a resin for a given ionic species is difficult, but the following generalizations can be useful:

1. The greater the valence of the sample ion, the greater its attraction for an ion exchange site.
2. For different ions of the same valence, the larger its ionic radius, the more strongly the sample ion is attracted to an ion exchange site.

The different selectivities of the resin for various ionic species cause them to be eluted from the separator column at different times. This is the charac-

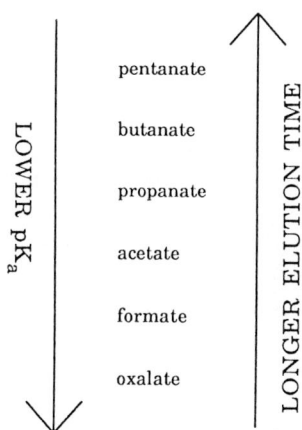

Fig. 1. Relationship of pK_a to elution time for short-chain CAA

teristic of ion chromatography that allows identification and quantification of the various components in an aqueous solution. The selectivity sequence (or elution order) is the order in which sample ions in a mixture will elute from the separator column. Figure 1 shows some relative affinities and selectivity sequences for selected organic aqueous ions (Dionex 1979).

4.2   Ion Chromatography Exclusion (ICE) Principles

Ion chromatography exclusion uses a resin similar to those used for IC. The resin counter ion is generally the same as the eluent ion, to minimize competing exchange reactions. Because the charged functional groups attached to the resin matrix give the liquid trapped in the resin pores a more strongly ionic character than the liquid surrounding the beads, the bead surface acts as a semipermeable membrane separating the two liquids. Donnan exclusion prevents highly charged species (such as most inorganic species) from entering the resin pores, and they pass through the column and elute in the void volume (or salt peak). Nonionic or weakly ionic species, such as the organic acids and anions, are unaffected by the semipermeable membrane and are free to diffuse in and out of the resin pores. The primary separation of various aqueous species of noncharged or weakly charged ions is diffusion in and out of the resin pores. The rate of diffusion is controlled by eluate pK_a, eluent pH, resin hydrophilicity and other characteristics, and temperature. Generally, for the short-chain organic acids and anions, pK_a can be used to predict elution order (see Fig. 2); however, as carbon number and sphericity increase, van der Waals forces between the sample molecules and the resin may cause the elution order to differ from that predicted on the basis of pK_a. Additionally, elution time may be altered by changing the eluent strength; elution order can be changed by the addition of an organic solvent to the sample (see below) (Dionex 1982a,b, 1984).

4.3   Eluent Suppression

Until a method of suppression of the highly conductive eluents used in IC and ICE separations was developed, conductimetric detection was not possible. Eluent conductivity suppression is achieved either by a high-capacity ion exchanger using a metal (such as silver) to precipitate high-conductivity anions, or by a semipermeable microfiber suppressor. Usually, a silver cation exchange column is used, which exchanges H^+ in the eluent for Ag^+ on the resin; this allows the Ag^+ from the exchange reaction and the Cl^- from the eluent to precipitate out of solution, leaving a very low-conducting effluent solution. The problem with using this technique when analyzing formation water is that Cl^- from the sample is also removed; the continuous

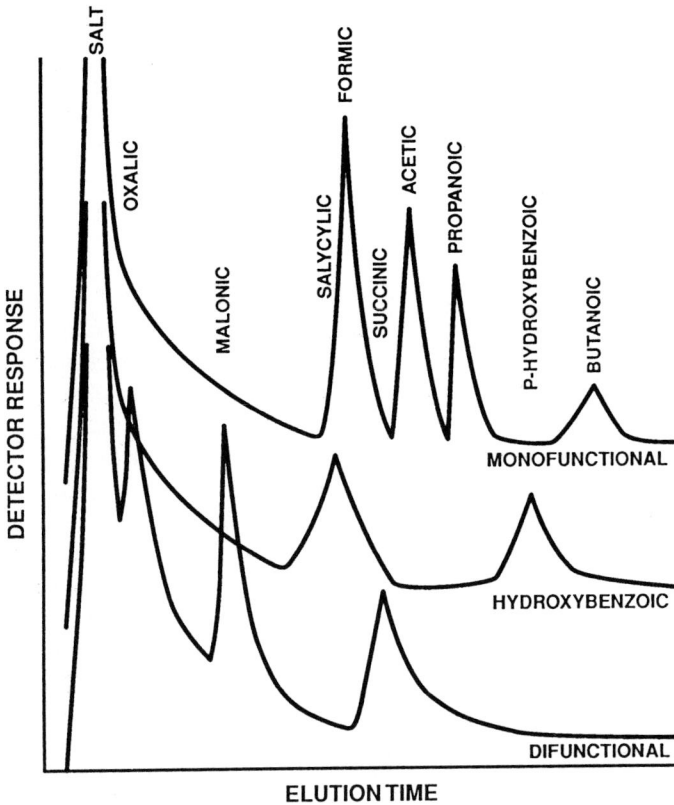

Fig. 2. Schematic chromatogram showing relative peak heights and elution times for monofunctional, hydroxybenzoic, and difunctional acids

running of water samples with total dissolved solids (TDS) values ranging from 5000 ppm to 250 000 ppm causes a drastic shortening of the suppressor exchange column life.

Because the silver-form exchange column is not easily regenerated and is fairly expensive, the microfiber suppressor is preferred. The suppression mechanism in this column is as follows. The suppressor microfiber (through which the eluent stream flows) is semipermeable and is immersed in a regenerant stream of tetrabutylammonium hydroxide (TBA$^+$OH$^-$), or similar, solution; H$^+$ in the eluent is exchanged across the membrane for a TBA$^+$ in the regenerant solution. Because the TBA$^+$/Cl$^-$ pair has a much lower conductivity than the H$^+$/Cl$^-$ eluent pair, a low-conducting eluate stream is obtained. The semipermeable fiber can be viewed as having three distinct regions: (1) totally expended fiber where the eluent first enters; (2) a region of "dynamic equilibrium" where the suppression reaction occurs; and (3) a totally regenerated region where the eluent stream exists. The length of these three regions is dependent upon the eluent strength, the composition of the analate, the regenerant concentration, the fiber length, and the flow rate. Once established, the "dynamic equilibrium" is maintained as long as the pertinent operating parameters are not altered (Dionex 1982a,b, 1984).

4.4   Detection of Eluted Ions

There are three common methods for detecting eluted species in ion chromatography: refractive index measurement, photometric (usually UV-VIS) absorbance measurement, and conductivity measurement. Refractive index measurement is limited by its low sensitivity and inability to differentiate between some eluent and sample ions. Photometric absorbance, particularly UV, is of great utility in the analysis of organic ions. It is limited to ions that absorb electromagnetic radiation in an easily detectable range, either as molecules or as complexes, and that have distinctive absorbance bands. Photometric absorbance is also limited in sensitivity by impurities in the eluent. Also, many inorganic species do not absorb electromagnetic radiation, either as ions or complex ions. Further, photometric absorbance is not an effective method when organic regenerants are used in eluent suppression (such as in the analytical technique described above), because of the resultant high background.

Most commonly, electrical conductivity measurement is used to detect eluted species in ion chromatographic analyses. The method is both sensitive and universally responsive. At low concentrations of sample ions, the response is an almost linear function of concentration. Detector response to the carboxylic acids and anions is based generally on the charge-to-mass ratio: the higher the ratio, the higher the response. Thus, formic acid (monovalent and small) has a higher detector response than does the next higher carboxylic acid homologue, acetic acid; and the detector response of acetic acid is much higher than that of propanoic acid (Fig. 2; Dionex 1982a,b, 1984).

5   Analysis of Results

The analysis of the results of ICE chromatograms takes patience, intuition, and time. There are several factors to be taken into account when unraveling the chromatograms: there are interferences from inorganic species, overlapping peaks, and shifting baselines; and elution times can be shifted downfield in high-ionic-strength samples by the salt peak. Frequently, one must analyze aliquots of a sample several times in order to identify most of the peaks in a chromatogram; however, there will be unidentifiable peaks.

Because of the complexity of analyzing the chromatograms, simple batch-type, single-sample runs can lead to misidentification and the misquantification of results. The analyst is cautioned from basing a strong interpretation on a single analysis. Techniques are described below for improving peak resolution and identification in ICE analysis.

5.1 Standards and Peak Identification

Standards are run in triplicate at the beginning and at the end of each ICE analytical session. From these standards, calibration curves are derived and a measure of relative error can be made. Standards are run in 25, 50, and 125 ppm concentrations, as detector response is considered linear over this range for all carboxylic acids analyzable by this methodology. If peaks resulting from species other than those in the prepared standards are encountered, they must be identified (generally by serial addition of knowns to the sample), and special standard solutions for these species must be prepared and analyzed, during the same analytical session.

The standards, and the stock solutions from which they are prepared, must be preserved with 200 ppm mercuric chloride or hexacytylpyridinium chloride as a bactericide (MacGowan and Surdam 1988, 1990). Table 2 illustrates the degradation of standards preserved with $HgCl_2$ through time for a preserved 100-ppm acetate standard, expressed as percent difference from the initial analysis, over a period of about 1200 h. As can be seen, by about 260 h measurable and significant degradation of the standard has occurred. This degradation is likely due to photochemical reactions and bacteria not poisoned by $HgCl_2$. Many researchers now use hexacetyl-pyridinium chloride as a preservative for this reason (J.B. Fisher, Amoco

Table 2. Degradation of acetate in a 100-ppm standard through time (absolute error on initial analysis ± 4.6%)

Time (h)	Difference from initial analysis (%)
0	–
24	+1.6
48	−0.5
100	−1.2
150	−2.0
180	−4.2
260	−6.9
324	−8.7
648	−10.3
936	−8.7
1220	−14.6

Production Co., Tulsa, Oklahoma, 1987, pers. comm.; P. Bennett, University of Texas, Austin, 1990, pers. comm.; MacGowan and Surdam 1990). Thus, it is imperative that samples, and standards, be preserved. The standards must be stored in the dark, as they are susceptible to decay by exposure to ultraviolet light.

The elution order for these short-chain CAA acids under the given standard chromatographic parameters is as follows: oxalic, malonic, o-hydroxybenzoic, formic, succinic, acetic, propanoic, p-hydroxybenzoic, butanoic, pentanoic, hexanoic, heptanoic, octanoic (Fig. 2). Peaks in sample chromatograms are identified on the basis of elution time and also by internally spiking the sample with knowns in serial runs. Each reported acid anion present in a sample should be identified by internally spiking the sample. This method has been used for ICE and other carboxylic acid anion analyses in solutions of geochemical interest by Surdam et al. (1984), Crossey (1985), Fisher (1987), MacGowan and Surdam (1988, 1990), Lundegard and Senftle (1987), and Fisher and Boles (1990), among others.

5.2   Interference with Inorganic Species

Several late-eluting inorganic species that have some affinity for the exchange resin interfere with the analysis of the early-eluting organic species (Fig. 3). In very dilute samples, such as the aqueous phase of hydrous pyrolysates, the "water dip" can interfere with the detection of oxalate. The water dip is really an inverse peak, caused when the sample has a lower conductivity than the eluent. There are three methods for avoiding this. First, one can run the sample again using distilled water as an eluent. However, this may cause other organic species to have poorer elution characteristics and peak shape. Second, several milligrams of NaCl can be added to the sample; this will raise the sample conductivity well above that of the eluent. Third, methanol or isopropanol can be added to the eluent (1 to 5% by volume). This will cause the resin to become more hydrophilic and will greatly promote exchange on the resin with noncharged or weakly charged species, thus increasing their elution time. This has the effect of "pulling" the oxalate peak downfield of the water dip. The disadvantage of this method is that it also changes the elution order of the organic acid anions; it is then possible to misidentify other peaks. One must also take care not to exceed 10% organic solvent by volume in the eluent, or channeling and destruction of the resin may result.

Conversely, in samples with high TDS (in excess of 15 000 ppm), the salt peak may swamp out the early-eluting organic acid peaks. One way to avoid this is by serial dilution of the sample. However, if the difunctional carboxylic acids are in low concentration, they may not be detected at all. Another method is to use a "stripper" column that removes the strongly charged species from the eluent stream before it passes through the detector. How-

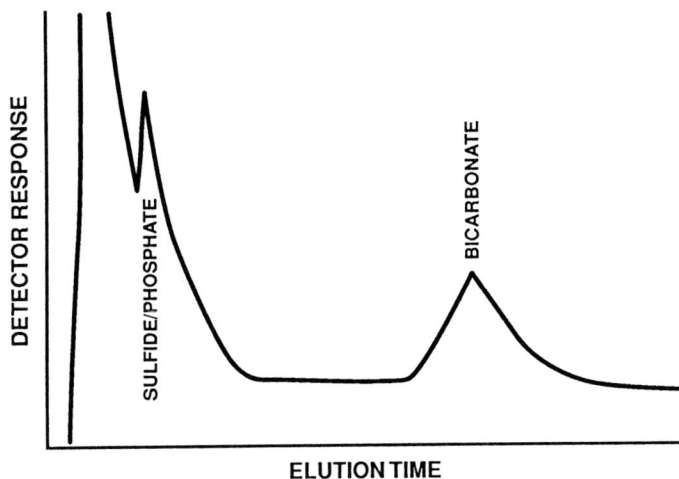

Fig. 3. Schematic chromatogram showing relative elution times and peak shapes for inorganic anions interfering with ICE analysis of CAA

ever, running high-ionic-strength formation water continually through the system greatly reduces the life of the stripper column.

Sulfide, sulfate, phosphate, and ammonium all elute between oxalate and malonate under the given chromatographic conditions (Fig. 3). They can completely swamp out the oxalate and malonate peaks if they are in sufficient concentration. Changing the eluent pH (by making the eluent stronger) will increase the retention time of the weakly charged organic acids over that of the inorganic anions (Fig. 4). However, this has the effect of increasing eluent background, and if oxalate and malonate are in low concentrations, they will be lost in the background. Adding organic solvents (1 to 5% methanol or isopropanol) to the solvent will increase the resin hydrophilicity, thus increasing the retention times of the organic acids over those of the inorganic anions. This has the added advantage of enhancing peak shape for the longer-chain acids as well, but may cause the elution order of the organic acids to be altered. Finally, one can heat the column to about 45 °C. This will increase the activity of the resin and will increase the retention time of the weakly charged species. In some complex analyses, it is necessary to combine some or all of these methods to effectively separate the late-eluting inorganic species from the early-eluting organic species.

In concentrations of more than about 100 ppm, bicarbonate species can cause interference with the detection of acetate. The bicarbonate peak is generally broad, whereas the acetate peak is generally quite sharp, so it is fairly easy, even at a low acetate concentration, to visually determine if this interference is occurring. As with the other inorganic anion interferences, lowering the pH of the eluent will cause the weakly charged species (in this case acetate) to be retained on the column longer relative to the more

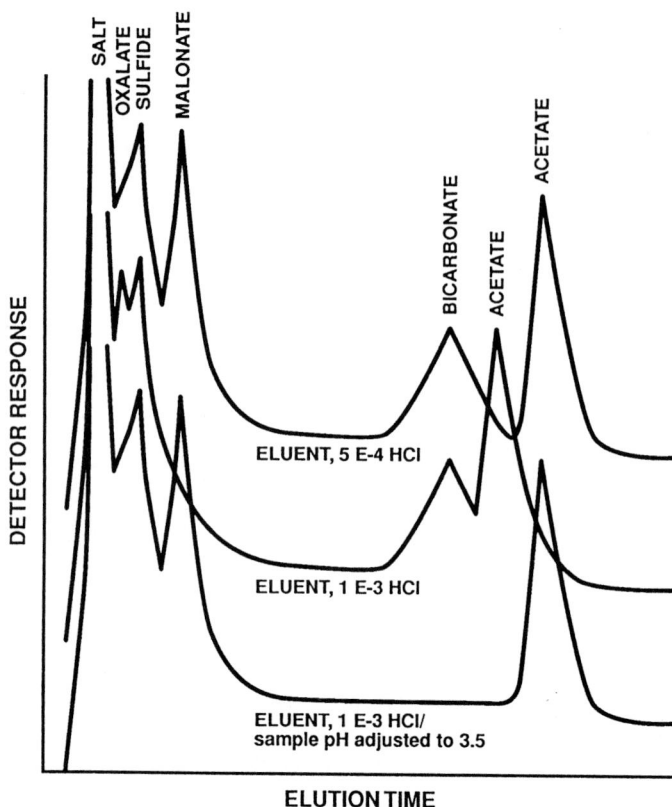

Fig. 4. Schematic chromatogram illustrating how the HCO_3^- and acetate peaks may be separated by manipulation of the ICE analytical technique

strongly charged species (in this case bicarbonate) (Figs. 3 and 4). It is commonly difficult, however, to completely resolve these two peaks without raising the background conductivity to unacceptable levels. Raising the column temperature or adding organic solvents also seems occasionally to be inadequate to completely resolve these two peaks. In such a case, the sample must be acidified to a pH of around 3 (which should drive off the bicarbonate as CO_2) before analysis.

5.3 Inferences Between Organic Anions

Several organic acids (particularly the long-chain monofunctional and di-functional aliphatic acids) overlap in elution time under these analytical conditions. The methods for resolving these peaks are similar to those given above.

o-Hydroxybenzoic acid elutes between malonate and formate but overlaps the formate peak (Fig. 2). By raising the column efficiency (heating or adding organic solvents or increasing the eluent strength), one can generally completely resolve malonate and o-hydroxybenzoate. However, to resolve o-hydroxybenzoate and formate, it is necessary to decrease the column efficiency by lowering the eluent strength. Succinic acid elutes between formate and acetate and must be resolved from them in the same manner as o-hydroxybenzoic acid is resolved from malonate and formate.

5.4 Quantification of Results

Chromatograms are quantified by integrating detector response peak area, either digitally or on a chart recorder. It is essential to accurately track the baseline through the chromatogram to ensure accurate measurement of peak area. Analyses of standards are used to generate a calibration curve. Peak areas for samples are then compared with these standard curves to determine the concentration of carboxylic acids in the sample. Multiple standard runs during the chromatographic session allow for the calculation of error bars on the standard curves, from which the absolute and relative errors for each sample can be calculated. Error for the method described above rarely exceeds 5.5% absolute.

The hydroxybenzoic acids and the longer-chain aliphatic carboxylic acids can give poor detector response and peak shape for quantification. If they are important constituents of a sample, it is a good idea to rerun the sample with 1 to 5% methanol or isopropanol in the eluent to improve detector response and peak shape. It is imperative that the new relative elution times for the carboxylic acid anions be determined at this time so that the misidentification of peaks cannot occur.

6 Conclusions

It is important that viable, representative samples of formation waters be collected and that accurate and precise determinations of the concentrations of carboxylic acids and anions be made, so that the contribution to geochemical processes of CAA can be critically evaluated. Important sampling procedures include screening wells to ensure that the water collected is representative of the conditions under investigation; flushing all production and sampling lines, carboys, and sample containers to ensure that all foreign materials have been swept away; and protecting the samples from the atmosphere, elevated temperature, and light. A sample should immediately be filtered through $0.1\,\mu$m and appropriate preservatives added. Samples for CAA analysis should be preserved with either $HgCl_2$ or hexacetylpyridinium chloride to reduce bacterial degradation. Samples must be adequately sealed and permanently labeled at the time of sampling. They should be stored in the dark at 4 °C. Other analyses that must be done at the time of sampling include sample temperature, pH, alkalinity, and dissolved oxygen content. Samples should be analyzed for CAA as soon as possible, because even preserved samples are subject to degradation with time.

Accurate and precise measurements of the concentration of CAA in formation waters can be made using various analytical techniques. Ion chromatography exclusion (ICE) is the technique of choice today due to its advantages over other analytical techniques: small sample size, rapid analysis, high sensitivity, specificity in analysis of similar ions, and multiple determinations in a single-sample run. Additionally, the necessary reagents, chromatographic columns, instruments, etc. are universally available and relatively inexpensive. However, interpretation of the results of ICE analysis is nontrivial. Samples may have to be run several times with various adjustments to the analytical parameters to ensure the correct interpretation of results. Generally, a single-sample run is an insufficient basis for meaningful interpretation.

Acknowledgments. The authors wish to acknowledge the helpful discussions and advice of T.L. Dunn, J.I. Drever, and S. Boese (Department of Geology and Geophysics, University of Wyoming), J.B. Fisher (Amoco Production Co., Tulsa, Oklahoma), as well as the technical support staffs at Dionex, Alltech, and Supelco. L.J. Crossey (Department of Geology, University of New Mexico) made a significant contribution in the initial investigations of analytical techniques. David Copeland (Department of Geology and Geophysics, University of Wyoming) edited the manuscript and made many helpful suggestions. Support for various portions of this research came from grants from Texaco USA, Phillips Petroleum, Arco Oil and Gas, Mobil Oil, and Amoco Production Co., as well as the Gas Research Institute through contracts no. 5098-260-1894 and no. 5091-221-2146.

References

Alltech Associates Inc (1985) Gas-Chrom Newsletter, Bull 72. Deerfield, IL, 12 pp
Barth T (1987a) Quantitative determination of volatile carboxylic acids in formation waters by isotachophoresis. Anal Chem 59: 2232–2237
Barth T (1987b) Multivariate analysis of aqueous organic acid concentrations and geologic properties of North Sea reservoirs. Chemometrics Intelligent Lab Syst 2: 155–160
Barth T, Borgund AE, Riis M (1990) Organic acids in reservoir waters – relationship with inorganic ion composition and interactions with oil and rock. Org Geochem 16: 489–496
Carothers WW, Kharaka YK (1978) Aliphatic acid anions in oil field waters – implications for origin of natural gas. Am Assoc Pet Geol Bull 62: 2441–2453

Crossey LJ (1985) The origin and role of water-soluble organic compounds in clastic diagenetic systems. PhD Dissertation, University of Wyoming, Laramie, WY, 134 pp

Dionex Corp (1979) Ion chromatograph 10, ion chromatograph 14. Operation and maintenance manual. Sunnyvale, CA, 150 pp

Dionex Corp (1982a) Anion fiber suppressor. Technical Note 14. Sunnyvale, CA, 8 pp

Dionex Corp (1982b) Anion fiber suppressor. Technical Note 14R. Sunnyvale, CA, 8 pp

Dionex Crop (1984) Anion fiber suppressor-2. Technical Note 17. Sunnyvale, CA, 5 pp

Fisher JB (1987) Distribution and occurrence of aliphatic acid anions in deep subsurface waters. Geochim Cosmochim Acta 51: 2459–2468

Fisher JB, Boles JR (1990) Water-rock interaction in Tertiary sandstone, San Joaquin Basin: diagenetic controls on water composition. Chem Geol 82: 83–101

Kawamura K, Tannebaum E, Huizinga BJ, Kaplan IR (1986) Volatile organic acids generated from kerogen during laboratory heating. Geochem J 20: 51–59

Kharaka YK, Hull RW, Carothers WW (1985) Water-rock interactions. In: Relationship of organic matter and mineral diagenesis. Soc Econ Paleontol Mineral Short Course Notes 17, pp 79–176

Lico MS, Kharaka YK, Carothers WW, Wright UA (1982) Methods for the collection and analysis of geopressured, geothermal, and oil field waters. U S Geol Surv Water Supply Pap 2194, 21 pp

Lundegard PD, Senftle JT (1987) Hydrous pyrolysis – a tool for the study of organic-acid synthesis. Appl Geochim 2: 605–612

MacGowan DB, Surdam RC (1987) The role of carboxylic acid anions in formation waters, sandstone diagenesis and petroleum reservoir modeling. Geol Soc Am Programs Abstr 19: 753

MacGowan DB, Surdam RC (1988) Difunctional carboxylic acid anions in oil field waters. Org Geochem 12: 245–259

MacGowan DB, Surdam RC (1990) Carboxylic acid anions in formation waters, San Joaquin Basin and Louisiana Gulf Coast, USA: implications for clastic diagenesis. Appl Geochim 5: 687–701

MacGowan DB, Fischer KJ, Surdam RC (1986) Alumino-silicate dissolution in the subsurface: experimental simulation of a specific geologic environment. Am Assoc Pet Geol Bull 70: 1048 (Abstr)

Small H, Stone TS, Bauman WC (1975) Novel ion exchange chromatographic method using conductimetric detection. Anal Chem 47: 1801 (Abstr)

Smith LK, MacGowan DB, Surdam RC (1991) Scale prediction during CO_2 huff "n" puff enhanced recovery, Crooks Gap field, Wyoming. Pap SPE-21838, Proc Soc Petroleum Engineers, Rocky Mountain Regional Meet Low-Permeability Reservoir Symp, Denver. Soc Petroleum Engineers, Richardson, Texas, pp 337–344

Surdam RC, Boese SW, Crossey LJ (1984) The chemistry of secondary porosity. In: McDonald D A, Surdam R C (eds) Clastic diagenesis. Am Assoc Pet Geol Mem 37, pp 127–149

Willey LM, Kharaka YK, Presser TS, Rapp JB, Barnes I (1975) Short chain aliphatic acid anions in oilfield waters and their contribution to the measured alkalinity. Geochim Cosmochim Acta 39: 1707–1711

Chapter 3 Distribution and Occurrence of Organic Acids in Subsurface Waters

Paul D. Lundegard[1] and Yousif K. Kharaka[2]

Summary

This chapter provides a summary of data on the occurrence of dissolved organic acid anions in subsurface waters, especially in formation waters associated with petroleum. In addition, it discusses general aspects of the origin and survivability of organic acid anions in the subsurface.

Published data on the concentration of dissolved organic acid anions can collectively be characterized as showing considerable variation. Maximum concentrations and the highest average concentrations of organic acid anions in a basin commonly occur at reservoir temperatures between 80 and 140°C, but temperature alone is a poor predictor of the concentration in individual samples. Furthermore, it is clear that organic acid anion concentrations greater than about 3000 mg/l acetate-equivalent are rare.

Acetate is by far the most abundant species contributing to the organic alkalinity of formation waters in sedimentary basins, although in some formation waters from subsurface temperatures less than about 80°C other monocarboxylic organic acids may occur in higher concentrations. Much fewer data are available on the concentrations of dicarboxylic acid anions and some reported results are contradictory. In general, dicarboxylic acid anions are much less abundant than monocarboxylic acid anions and typically occur at concentrations less than 100 mg/l. Succinate and methyl-succinate appear to be the most abundant dicarboxylic species. The concentrations of some dicarboxylic acid anions are probably limited by the low solubility of their calcium salts and their greater susceptibility to thermal decomposition. The dominant species contributing to the organic alkalinity of formation waters have been identified, but waters exist in which organic acid anions do not account for all of the dissolved organic carbon. Little work has been done to characterize and quantify these other species contributing to the dissolved organic carbon.

Organic alkalinity, contributed mainly by short-chain monocarboxylic acid anions, is responsible for a variable fraction of total formation water

[1] Unocal Science and Technology, 376 S. Valencia Ave., Brea, California 92621, USA
[2] US Geological Survey, 345 Middlefield Rd., Menlo Park, California 94025, USA

alkalinity. The fraction of total alkalinity contributed by organic acid anions varies with respect to temperature and from basin to basin. In Cenozoic formations it is common for organic alkalinity to exceed bicarbonate alkalinity in the temperature range from 80 to 140 °C. At temperatures less than 80 °C and greater than 140 °C bicarbonate generally dominates the alkalinity of formation water. Formation waters in Miocene reservoirs in the San Joaquin Basin, California, are exceptional in their very high concentrations of organic acid anions and in the dominance of organic alkalinity over bicarbonate alkalinity.

Present-day concentrations of organic acid anions are influenced by factors that control the competing processes of organic acid production and destruction. Together, field and experimental data allow identification of the important factors and provide insight into the processes they control. Field data discussed here as well as experimental data discussed in other chapters demonstrate that thermal alteration of kerogen in shales is the dominant process by which aqueous organic acid anions are produced in basins. Consideration of the effects of burial on shale porosity and thermal history shows that high geothermal gradients will increase the rate of organic acid production from kerogen, whereas low geothermal gradients will tend to produce higher organic acid concentrations. In situ thermal alteration of reservoired oil may be locally significant in producing moderate concentrations of aqueous organic acids but is unlikely to be a process of diagenetic significance. The association of biodegraded oils with lower concentrations of organic acid anions and the generally low concentrations of organic acid anions in shallow formation waters indicate that biological processes are important in the destruction of these species. Field studies indicate that organic acid anions undergo destruction with progressive thermal exposure, and at temperatures of roughly 100 °C, acetate has a half-life of tens of millions of years.

1 Introduction

Since the widespread occurrence in formation waters of sedimentary basins began to be recognized (Carothers and Kharaka 1978), the origin, distribution, and geochemical significance of low molecular weight organic acids have become intensively studied fields (Gautier 1986; Lundegard and Kharaka 1990). The most abundant organic acid anion in formation waters is acetate, although a variety of monocarboxylic and dicarboxylic species have been reported. The highest concentrations are present in relatively young basins at temperatures between approximately 80 and 140 °C (Germanov and Mel'kanovitskaya 1975; Willey et al. 1975; Carothers and Kharaka 1978; Workman and Hanor 1985; Hanor and Workman 1986; Kharaka et al. 1986; Fisher 1987; Means and Hubbard 1987; Fisher and Boles 1990; Lundegard and Kharaka 1990; Lundegard and Trevena 1990; MacGowan and Surdam 1990).

Geochemical interest in these organic species stems from their potential importance in water-rock interaction in sedimentary basins (Willey et al. 1975; MacGowan and Surdam 1990), groundwaters (Thurman 1984), and hydrothermal systems (Simoneit 1984). In particular, these species can act as sources of protons and CO_2, and as pH and Eh buffering agents (Carothers and Kharaka 1978; Crossey et al. 1986; Kharaka et al. 1986; Lundegard and Land 1989). They may also form geochemically important complexes with Ca, Al, Fe, Pb, and Zn (Giordiano and Barnes 1981; Lundegard and Kharaka 1990; Harrison and Thyne 1992). Organic acid anions have been studied for their use as proximity indicators for petroleum (Kartsev 1974; Carothers and Kharaka 1978) and as possible precursors for natural gas (Kharaka et al. 1983; Drummond and Palmer 1986).

2 Occurrence of Dissolved Organic Acids

2.1 Sample Collection

Routine field procedures often can serve as a guide to the detection of organic alkalinity in formation waters. The presence of organic acid anions in samples of sedimentary formation water is suggested by: (1) total titration alkalinities greater than about 8 mEq/l; (2) the lack of effervesence upon acidification to pH 2; (3) titration curve inflection points that fall near pH 3.5, which is characteristic of acetate and other short-chain aliphatic acids, rather than pH 4.5, which is characteristic of bicarbonate (Fig. 1); and (4) the characteric rancid odor of butyric acid produced when samples are acidified.

Many of the organic acid species that occur in subsurface waters are considered volatile organic acids because they can be distilled from water (American Association of Public Health 1985). This volatility is a potential concern when sampling waters for dissolved organic acids, especially for low pH (less than approximately 5) waters. Whereas aeration or agitation of water during sampling should be avoided, volatilization does not appear to cause significant reduction in organic acid concentration in most cases because pH values of the well head are generally greater than 5 and organic acids are present as anions. Of greater concern is possible bacterial degradation of dissolved organics, especially for acetate and dicarboxylic acid anions. It is known that bacteria will gradually consume dissolved organic acids during sample storage unless preventive measures are taken (Kharaka et al. 1985; MacGowan and Surdam 1988). Field samples for organic acid analysis should be filtered through 0.45-μm or preferably 0.1-μm filters and stored in sterilized amber glass bottles. Refrigeration at temperatures of about 4 °C

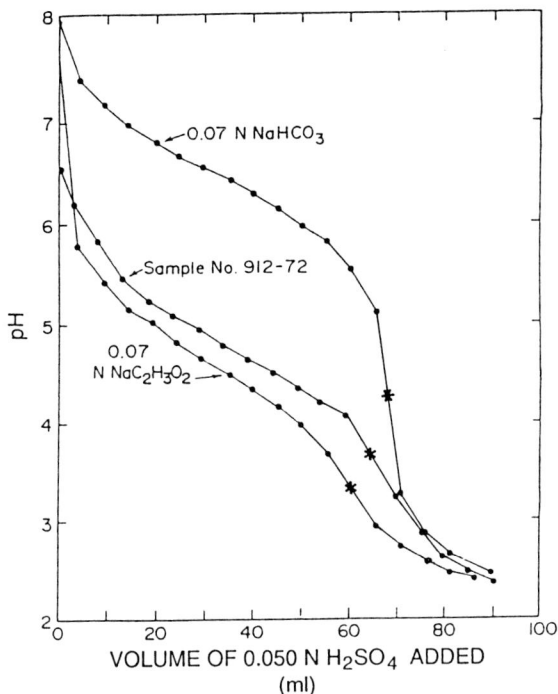

Fig. 1. Titration curves for sodium-acetate and sodium-bicarbonate (after Carothers and Kharaka 1978). Note that the inflection point for sodium acetate occurs at a lower pH than for sodium bicarbonate

greatly retards bacterial degradation and is strongly recommended for sample storage longer than 1 day. Whether refrigeration is used or not, a bactericide such as mercuric chloride (40 mg/l Hg), sodium azide, or zephrin chloride should always be used (Kharaka et al. 1985).

2.2 Field Data

The median concentration of dissolved organic species in shallow ground-waters unaffected by mixing with upward-migrating formation waters is about 1 mg C/l (Thurman 1984). An important portion (20 to 40%, and up to 90%) of this dissolved carbon consists of high molecular weight fulvic and humic acids. The proportion of these humic substances in organic species dissolved in the waters of the unsaturated zone is even higher (50 to 60%). The median concentrations of dissolved organic carbon are also higher, ranging from 20 mg C/l in the A soil horizon to 2 mg C/l in the C soil horizon. The concentrations of dissolved organic species in waters produced from oil and gas wells (ca. 40 to 200 °C), the main source of data on waters in sedimentary basins, are much higher than in shallow groundwaters. Concentrations may reach hundreds to thousands of milligrams C/l. Prior to the identification of dissolved organic acid anions in the formation water of sedimentary basins, these organic species were unknowingly recorded as part of the bicarbonate alkalinity determined by acid titration.

Data on the concentration of dissolved organic acid anions in natural formation water come predominantly from analysis of samples from Cenozoic clastic reservoirs (Fig. 2), although sampled strata range in age from Pennsylvanian to Pleistocene (Lundegard and Kharaka 1990). These data show that: (1) concentrations in subsurface waters are highly variable, both for a given temperature range and age; (2) concentrations greater than 3000 mg/l (0.05 m) acetate-equivalent are rare, an important observation for geochemical modeling; (3) maximum concentrations vary with temperature, and generally show a peak somewhere in the range 80–140 °C; and (4) concentrations in Cenozoic formations are generally higher than in Mesozoic or Paleozoic formations.

In Fig. 2, organic alkalinity analyses of 513 samples are plotted against reservoir temperature. Treated as a whole, this data base shows that average organic acid anion concentrations of 1000 mg/l can occur in the temperature range from 80 to 160 °C, depending on a variety of factors. The large amount of variation demonstrates that, in general, temperature is not of overriding importance and other factors must influence organic acid anion concentrations. It is only when the variation in some of these factors is reduced, as when data from a single formation in a single basin is considered, that the influence of temperature on the average organic acid anion concentration becomes apparent (Fig. 3). Studies of individual basins show that the average organic acid concentration in this temperature range is related to the age of the reservoir sampled (Fig. 4). Whereas variations in source material influence concentrations, this relationship between geologic time and organic acid anion concentration is apparently caused by the kinetics of organic acid decarboxylation and will be discussed later.

Maximum organic acid anion concentration increases from less than a few hundred mg/l at temperatures less than 40 °C, to a peak of a few thousand mg/l at approximately 100 °C, and then declines rapidly again with further increases in temperature. Extrapolation of available data suggests that maximum concentrations probably do not exceed 100 mg/l at temperatures

---

Fig. 2A,B. Organic alkalinity versus temperature for data compiled from the literature (N = 513). *Lines* indicate average for each 20° interval. **A** Semi-log plot. **B** Arithmetic plot. Data from Carothers and Kharaka (1978), Kharaka et al. (1985), Lundegard (1985), Workman and Hanor (1985), Crossey et al. (1986), Hanor and Workman (1986), Fisher (1987), Means and Hubbard (1987), MacGowan and Surdam (1988), Lundegard and Trevena (1990), MacGowan and Surdam (1990), Fisher and Boles (1990), Barth and Riis (1992)

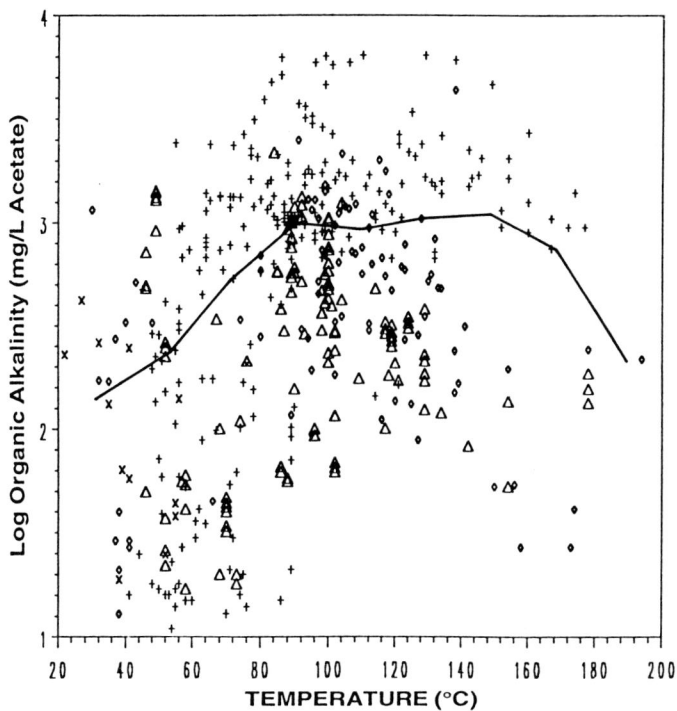

A

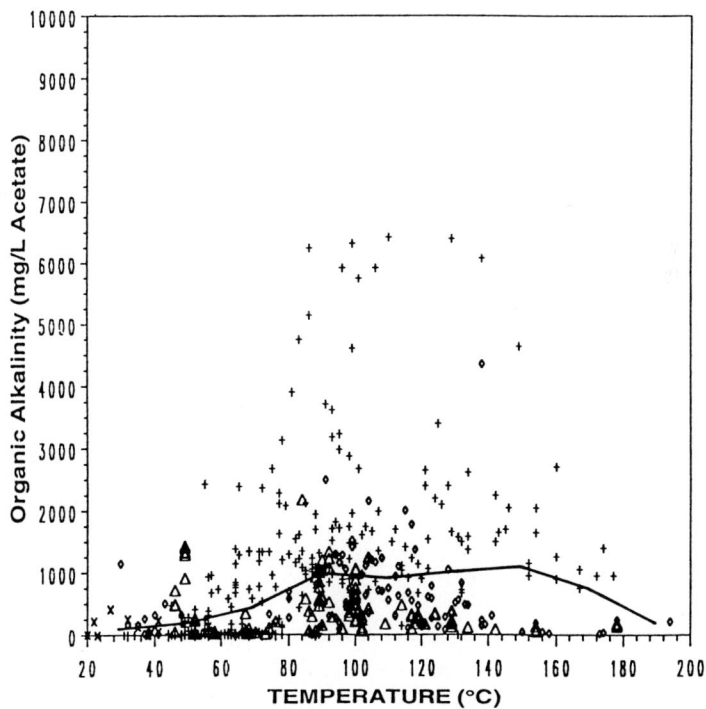

B

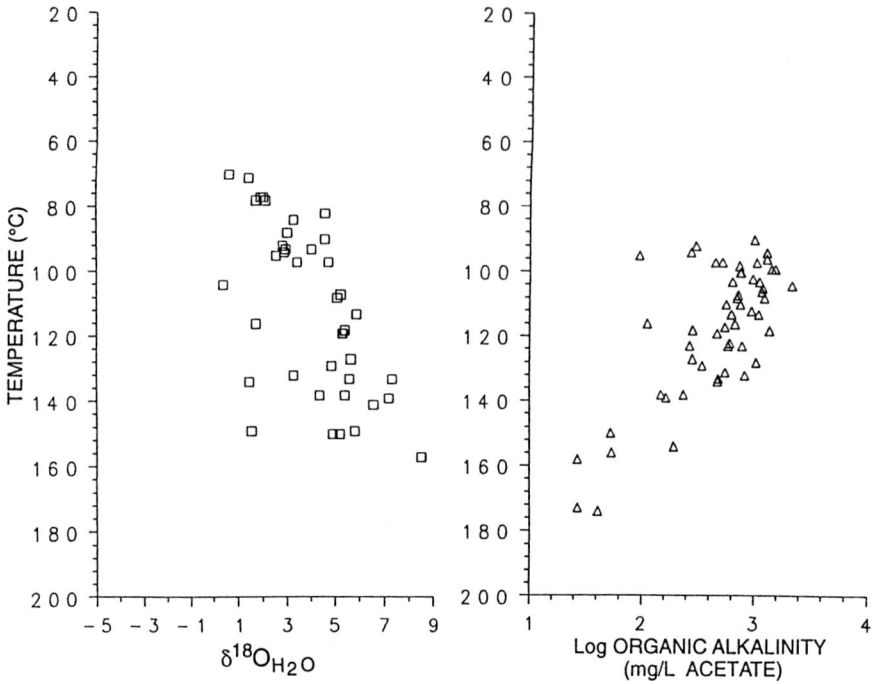

Fig. 3. Texas Gulf Coast Oligocene. *Right* Organic alkalinity versus temperature. *Left* Oxygen isotopic composition of formation water versus temperature. Data from Carothers and Kharaka (1978), Lundegard (1985), and Land (unpubl. data)

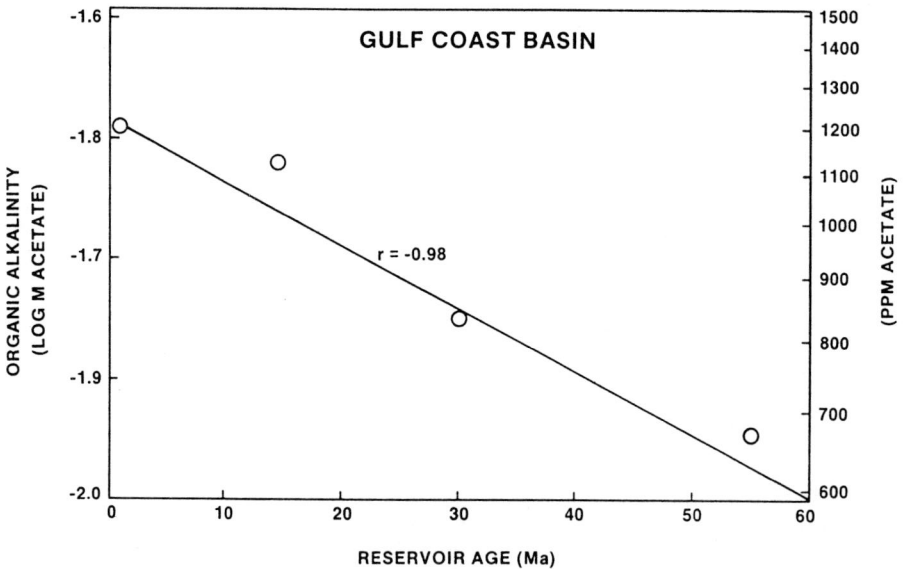

Fig. 4. Average organic alkalinity between 80 and 120 °C versus reservoir age for formation waters from the Gulf of Mexico Basin (after Lundegard and Land 1986). *Regression line* defines a psuedo-half-life of approximately 60 Ma

greater than 200 °C. Water samples from two geothermal wells (temperatures greater than 250 °C) at Salton Sea, California, showed no detectable organic acids (Y.K. Kharaka, unpubl. data). Early in the study of organic acids in subsurface waters it appeared that the distinction between these temperature zones was fairly distinct (Carothers and Kharaka 1978). Three zones were identified. Zone 1 was characterized by temperatures less than 80 °C and organic acid concentrations less than 60 mg/l. Zone 2 was characterized by temperatures between 80 and 200 °C, much higher concentrations of organic acids, and decreasing concentrations with increasing temperature. In zone 3, at temperatures above 200 °C, organic acid anion concentrations were very low or zero. The collection of additional data since the work of Carothers and Kharaka (1978) has confirmed that organic acid concentrations are very low to zero at temperatures above 200 °C, but has also shown that the boundary between zone 1 and zone 2 is not necessarily sharp (Fig. 2). Fisher's data (1987) from offshore Venezuelan oil reservoirs show a gradual increase in maximum organic acid anion concentrations as temperature increases towards 90 °C, and much higher concentrations than previously observed at these low temperatures. High concentrations (perhaps greater than 1000 mg/l) of organic acid anions at temperatures less than 80 °C may be indicative of upward migration from warmer zones of waters rich in organic acid anions. This interpretation was invoked to explain the occurrence of high organic acid anion concentrations in shallow, relatively cool formation waters from the High Island Field (offshore Texas; Kharaka et al. 1986) and from fields in the eastern Venezuelan Basin (offshore Trinidad; Fisher 1987).

At temperatures below 80 °C, the low and variable concentrations of organic acids are controlled by several factors. These include: (1) bacterial consumption of organic acids, especially acetate (Carothers and Kharaka 1978); (2) low generation rates of organic acids (Lundegard and Kharaka 1990); and (3) dilution of upward-moving waters rich in organic acid anions by mixing with surface water (Fisher 1987; Means and Hubbard 1987).

2.3 Organic vs. Bicarbonate Alkalinity

Organic acid anions can dominate the total alkalinity of subsurface waters, and the anions of short-chain aliphatic acids contribute the great majority of the organic alkalinity. Using data from Oligocene reservoirs in Texas and Miocene reservoirs in the San Joaquin Basin of California, Carothers and Kharaka (1978) showed that the relative abundance of bicarbonate and organic alkalinity varied with temperature. More recent data confirm this general trend, demonstrate significant variation from basin to basin, and show that the San Joaquin Basin is unusual in the very high organic acid concentrations and in the abundance of organic acid anions relative to bicarbonate (Fig. 5). We know of no other basin where the dominance of

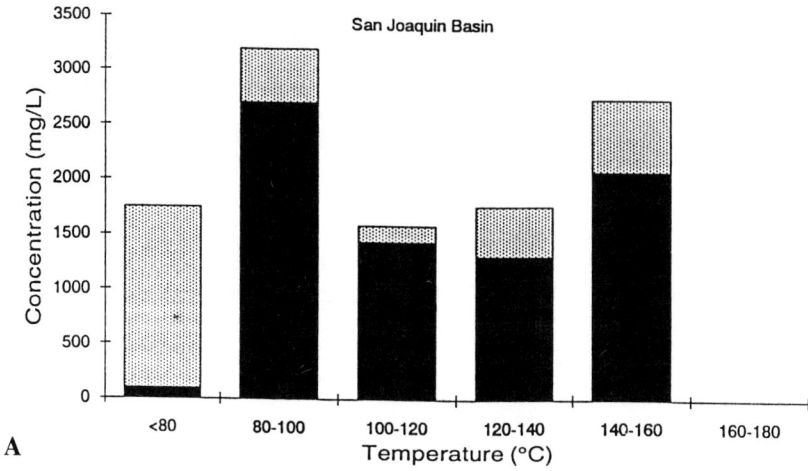

A

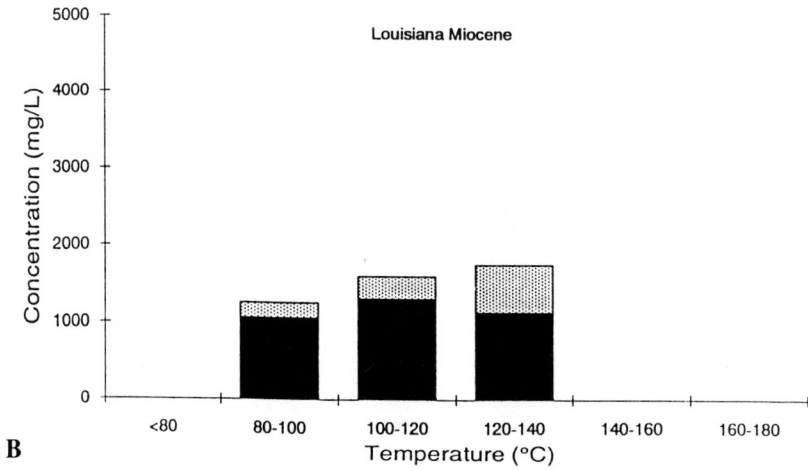

B

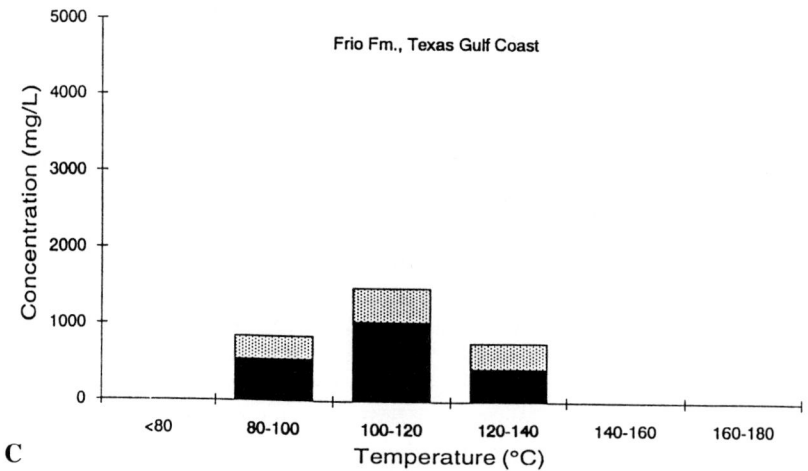

C

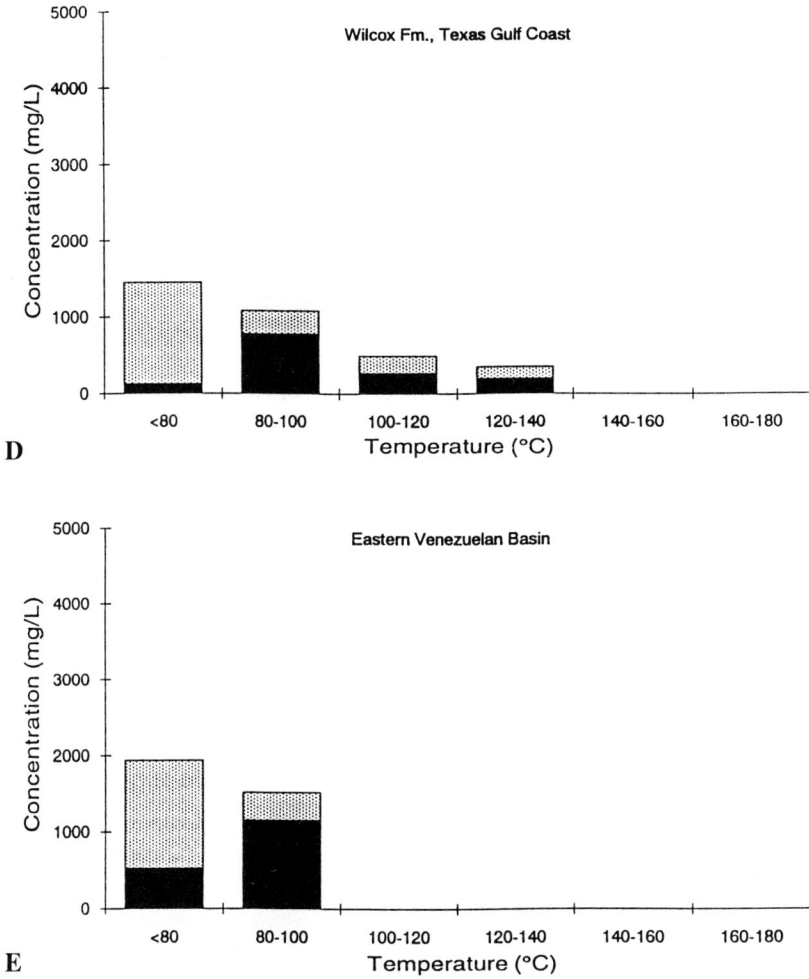

Fig. 5A–G. Relative abundance of organic and bicarbonate alkalinity versus temperature. *Black bars* represent equivalent acetate concentrations. *Stippled bars* represent equivalent bicarbonate concentrations. **A** San Joaquin Basin. Data from Carothers and Kharaka (1978) and Fisher and Boles (1990). **B** Louisiana Miocene. Data from Land et al. (1988). **C** Frio Formation (Oligocene), Texas Gulf Coast. Data from Carothers and Kharaka (1978) and Lundegard (1985). **D** Wilcox Formation (Eocene), Texas Gulf Coast. Data from Lundegard (1985). **E** Eastern Venezuelan Basin. Data from Fisher (1987). **F** Gulf of Thailand (Miocene). Data from Lundegard and Trevena (1990). **G** North Sea (mostly Jurassic). Data from Barth and Riis (1992)

F

G

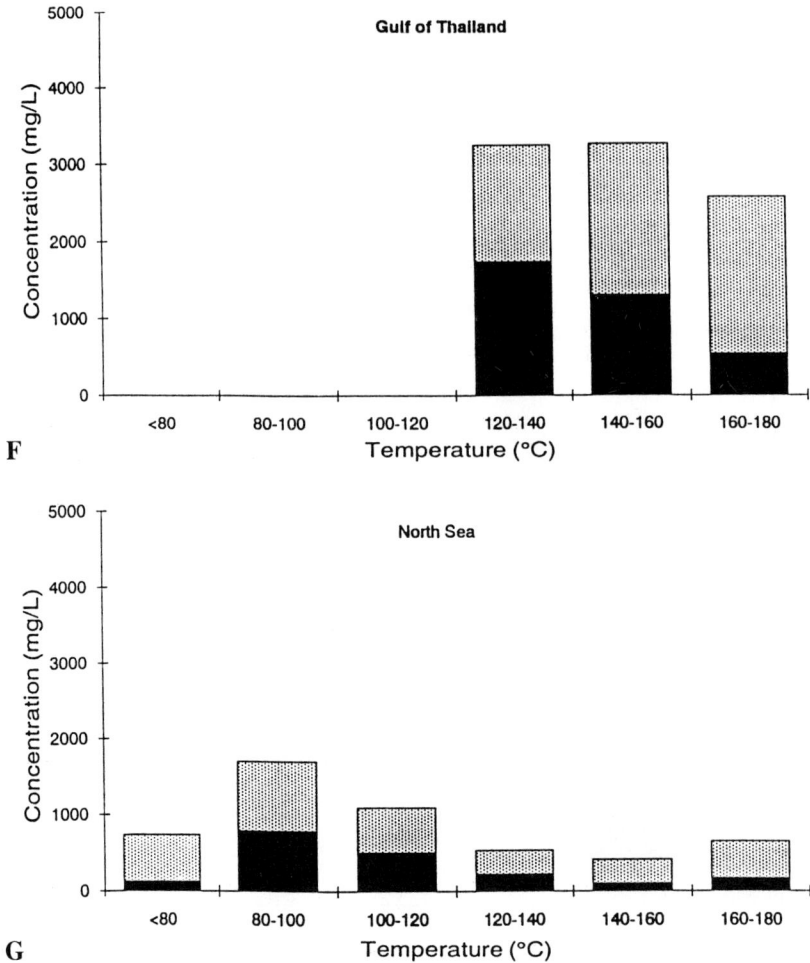

Fig. 5. *Continued*

organic alkalinity over bicarbonate alkalinity is more pronounced than in the San Joaquin Basin, California. There, organic alkalinity may contribute more than 90% of the total alkalinity of the formation water. Lundegard and Kharaka (1990) speculated that factors other than the relative generating capacity of kerogen control the high dissolved concentrations of organic acid anions in the San Joaquin Basin based on yields observed in hydrous pyrolysis experiments.

From the combined data that we have synthesized (Fig. 5), it can be said that in Cenozoic formations organic alkalinity commonly exceeds bicarbonate alkalinity in the temperature range 80 to 140°C. At temperatures below 80 and greater than 140°C, bicarbonate will probably dominate the total alkalinity of formation water in most basins. In pre-Cenozoic formations the importance of organic alkalinity relative to bicarbonate alkalinity is expected to be much less.

2.4   Relative Abundance of Monocarboxylic Acid Anions

We compiled from the literature 446 analyses of individual monocarboxylic acid concentrations in formation water. Acetate is by far the most abundant organic acid anion (Carothers and Kharaka 1978; Kharaka et al. 1985; Lundegard 1985; Workman and Hanor 1985; Crossey et al. 1986; Hanor and Workman 1986; Fisher 1987; Means and Hubbard 1987; MacGowan and Surdam 1988; Fisher and Boles 1990; MacGowan and Surdam 1990; Barth and Riis 1992). The abundance of other monocarboxylic aliphatic acid anions generally decreases with increasing carbon number (acetate > propionate > butyrate > valerate). However, at temperatures less than approximately 90 °C, the molar ratio of acetate to acetate plus propionate is highly variable (Fig. 6) and can range from 0.0 to 1.0. Above temperatures of approximately 90 °C, the ratio of acetate to acetate plus propionate is uniformly high, ranging from about 0.8 to 1.0. The dominance of propionate over acetate in low temperature waters has been attributed to the preferential bacterial consumption of acetate (Carothers and Kharaka 1978; Kharaka et al. 1985; Lundegard 1985; Fisher 1987). Upward-moving formation water rich in organic acids and dominated by acetate is envisaged to undergo dilution and bacterial alteration at depth where there is surface water influence. Recent data indicate that propionate dominance at low temperature is less common than previously thought (Workman and Hanor 1985; Hanor and Workman 1986; Kharaka et al. 1986; Fisher 1987; Means and Hubbard 1987). Hydrous pyrolysis experiments by Lundegard and Senftle (1987) showed that propionate dominance may even be the product of primary organic acid generation as well as bacterial consumption of acetate.

Acetate predominates over propionate for virtually all samples at temperatures exceeding 100 °C (Fig. 6). Above 100 °C the molar ratio of acetate to acetate plus propionate is greater than 0.8 ($M Ac^-/M Pr^- > 4$). At lower temperatures this ratio is far more variable, ranging from 0 to 1. Shock (1988, 1989) suggested that homogeneous equilibrium between acetic and propionic acids may be controlling their relative concentrations in sedimentary basins, according to the following reaction:

$$3CH_3COOH = 2CH_3CH_2COOH + O_2.$$

The equilibrium slope of the log of propionic acid activity versus the log of acetic acid activity at 100 °C and 300 bar was calculated by Shock (1988) to be 3/2. It is instructive to compare the analytical data on natural waters with this calculated slope. Since total solution composition and in situ pH are not known accurately for the published data set, it is not possible to

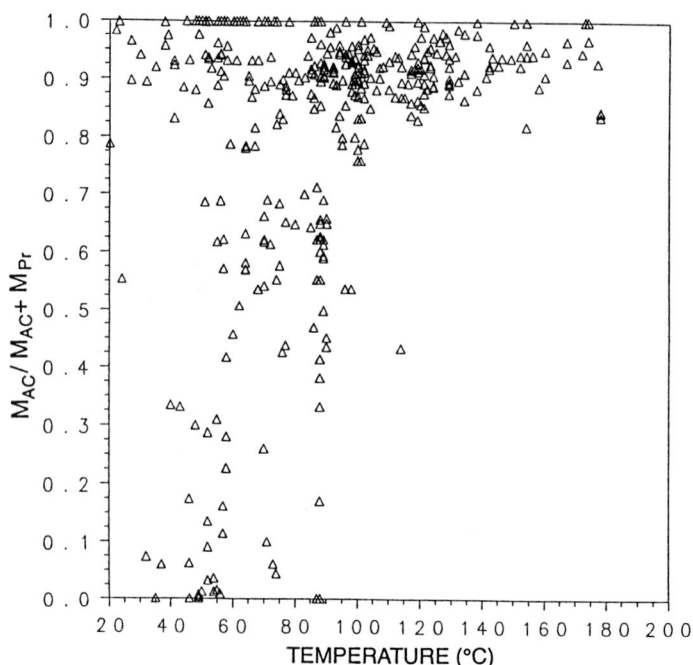

Fig. 6. Molar ratio of acetate/propionate + acetate versus temperature (N = 446). Data from Carothers and Kharaka (1978), Kharaka et al. (1985), Workman and Hanor (1985), Crossey et al. (1986), Hanor and Workman (1986), Fisher (1987), Means and Hubbard (1987), MacGowan and Surdam (1988), Lundegard and Trevena (1990), MacGowan and Surdam (1990), Fisher and Boles (1990), Barth and Riis (1992)

accurately calculate the activities of the free acids. Because propionic acid is slightly weaker than acetic acid, the activity ratio of propionic to acetic acid will be higher than the molality ratio of (propionic acid + propionate) to (acetic acid + acetate) for a given sample. However, failure to account for the effect of pH will not have a significant impact on the *slope* of the data. Using 82 analyses between 90 and 110 °C, we find that the data show some covariance but do not closely approximate a 3/2 slope (Fig. 7). Bearing in mind that Fig. 7 is a log-log plot, there is a large amount of scatter in the data. If the samples represented in Fig. 7 are actually tending toward equilibrium at a single oxygen fugacity, the concentrations of acetic and propionic acid in some samples must be too high or too low by a factor of at least 5. This does not mean that reactions between acetic and propionic acid are not taking place in basins, but it does imply that if such reactions are taking place they have not gone to equilibrium at a single fugacity of oxygen.

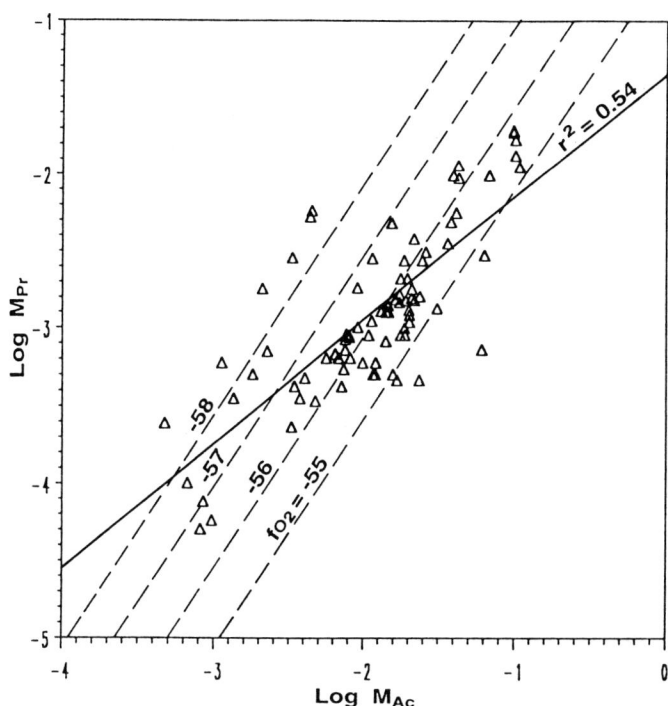

Fig. 7. Log molar concentration of total propionate (propionate + propionic acid) versus total acetate (acetate + acetic acid) (N = 82) for samples from 90 to 110 °C. *Solid line* is a least squares fit to the data. *Dashed lines* are contours of the log fugacity of oxygen from Shock (1989) and represent the calculated equilibrium between propionic and acetic acids at 100 °C and 300 bar. For data sources, see Fig. 6

2.5 Dicarboxylic Acid Anions

Data on the concentration of dicarboxylic acid anions in subsurface water are much more limited than those on monocarboxylic acid anions and some results are controversial. C_2 through C_{10} saturated acid anions have been reported in addition to maleic acid (cis-butenedioic acid) (Surdam et al. 1984; Kharaka et al. 1986; Barth 1987; MacGowan and Surdam 1988, 1990). Reported concentrations of dicarboxylic acid anions range widely from 0 to 2540 mg/l but mostly are less than a few 100 mg/l. Concentrations of these species in formation waters are probably limited by several factors, including the very low solubility of calcium oxalate and calcium malonate (Kharaka et al. 1986; Harrison and Thyne 1992), and the susceptibility of these dicarboxylic acid anions to thermal decomposition (MacGowan and Surdam 1988; Crossey 1991).

The concentration of dicarboxylic acid anions from the High Island field, offshore Texas, was reported by Kharaka et al. (1985). They employed GC-

Table 1. Concentrations of dicarboxylic acid anions in formation water in Pleistocene reservoirs from High Island Field, offshore Louisiana. (After Kharaka et al. 1985)

Dicarboxylic acid	Concentration (mg/l)		
	83-TX-3[a]	83-TX-10	83-TX12
Butanedioic	1.8	0.2	63
Pentanedioic	1.7	1.2	36
Hexanedioic	0.5	0.5	–
Heptanedioic	–	0.6	0.2
Octanedioic	–	0.7	5.0
Nonanedioic	–	1.4	6.0
Decanedioic	–	–	1.3

[a] – Below detection limit.

MS analysis of esterified organics to identify and quantify the dicarboxylic acids. The concentrations of dicarboxylic acid anions in the three samples with the highest concentrations of monocarboxylic acid anions are shown in Table 1. Succinate (butanedioic) and glutarate (penthanedioic) had the highest concentrations, 63 and 36 mg/l, respectively. Oxalate and malonate were below the detection limit of about 1 mg/l. Methybutanedioic and 2-methylpentaredioic acid anions were identified but not quantified.

Barth (1987), using isotachophoresis, reported concentrations of ethanedioic (oxalic) and propanedioic (malonic) acid anions of up to 38 and 10 mg/l, respectively, in water from North Sea oil wells. These species were found in one sample of produced water that was nearly devoid of monocarboxylic acid anions. She felt that the measured organic species were not representative of the natural formation water, attributing them instead to aerobic biodegradation of drilling mud components.

In a study of formation water chemistry in the Gulf of Thailand, Lundegard and Trevena (1990) analyzed 20 waters produced from reservoirs having temperatures between 120 and 177 °C. All samples were treated in the field with mercuric chloride to prevent bacterial growth. Monofunctional organic acid anions were abundant (up to 1500 mg/l acetate), but oxalate, the only dicarboxylic for which they analyzed, was never present above a concentration of 1 mg/l.

The highest concentrations of dicarboxylic acid anions are those reported by MacGowan and Surdam (1988) for the 12–31 well in the North Coles Levee field, San Joaquin Basin. They reported an oxalate concentration of 95 mg/l and a malonate concentration of 2540 mg/l in formation water from the 12–31 well. MacGowan and Surdam (1988) reported similarly high concentrations of oxalate and malonate for other wells in the San Joaquin and Santa Maria basins. Another analysis of organic acid concentrations in formation water from the 12–31 well was tabulated by MacGowan and

Surdam (1990). That analysis indicated much lower concentrations for oxalate and malonate, 16.5 and 42.1 mg/l, respectively. Fisher and Boles (1990) also analyzed formation water from the 12–31 well and from other wells in the North Coles Levee and neighboring fields. Their GC-MS analysis of the combined acid, base, and neutral methylene chloride extracts identified a wide variety of organic compounds, but they detected no oxalate and no malonate in their sample. Kharaka (unpubl. data) recently sampled the 12–31 well and other wells in the North Coles Levee field that had high total organic acid anion concentrations. Using columns appropriate for the analysis of dicarboxylic acid anions with a DIONEX ion chromatograph, the concentrations of oxalate, malonate, and maleate were all below the detection limit of about 0.5 mg/l. Succinate was detected at concentrations below 0.5 mg/l and methyl succinate was detected at concentrations of less than 5 mg/l. The concentrations of monocarboxylic acid anions measured were comparable to those reported by MacGowan and Surdam (1990) and Fisher and Boles (1990). Together, the data on the 12–31 well and neighboring wells reported by MacGowan and Surdam (1988, 1990), Fisher and Boles (1990), and Kharaka (unpubl. data) indicate that either the concentrations of dicarboxylic acid anions in formation water produced from the wells are fluctuating over a very wide range or the analytical accuracy of the various investigators is grossly different. We believe that the low concentrations of dicarboxylic acid anions reported by MacGowan and Surdam (1990) and Fisher and Boles (1990), and measured by Kharaka (unpubl. data), more closely represent the true concentrations in the formation water.

Available data indicate that dicarboxylic acid anion concentrations in formation waters are generally less than approximately 100 mg/l. Succinate and methyl succinate appear to be the most abundant dicarboxylic species.

2.6 Other Dissolved Organic Species

Few data exist on the types and concentrations of dissolved organic species other than the mono- and dicarboxylic acids. However, some data suggest that dissolved organic acids do not account for all of the dissolved organic carbon in formation waters (Kharaka et al. 1986; Fisher 1987). These data must be interpreted with some caution, however, because a generally unknown fraction of the measured organic carbon could be due to entrained oil or dissolved hydrocarbons.

In formation water from the High Island field, offshore Louisiana, Kharaka et al. (1986) identified, but did not quantify, a number of species. These included phenol, 2-, 3-, and 4-methylphenol, 2-ethylphenol, 3,4-, and 3-,5-dimethylphenol, cyclohexanone, and 1,2-, and 1,4-dimethylbenzene. Rapp (1976) identified several amino acids including serine, glycine, alanine, and aspartic acid in low concentrations (<0.3 mg/l) in oil field waters.

Unpublished data collected by Kharaka on formation water from oil and gas wells in the Sacramento Valley, California, indicate the following organic species: phenols (<20 mg/l), 4-methyl phenol (<2 mg/l) benzoic acid (<5 mg/l), 4-methyl benzoic acid (<4 mg/l), 2-hydroxybenzoic acid (<0.2 mg/l), 3-hydroxybenzoic acid (<1.2 mg/l), 4-hydroxybenzoic acid (<0.2 mg/l), and citric acid (<4 mg/l).

Fisher and Boles (1990) analyzed the dissolved organic matter in two formation water samples from the San Joaquin Basin by GC-MS analysis of combined acid, base, and neutral methylene chloride extracts. They identified various polar aliphatics (fatty acids to C_9 with various methyl and ethyl substituents), cyclics (phenols and benzoic acids), and heterocyclics (quinolines). They were able to quantify, at the ppm or sub-ppm level, phenol, methyl-substituted phenols, and benzoic acid.

Additional dissolved organic species will likely be discovered in formation waters as analytical procedures improve, but it appears that the dominant species contributing to total organic alkalinity have already been identified.

3 Origin of Major Species

3.1 Field Data

Variations in the organic acid concentrations of diverse waters with respect to temperature suggest a thermal influence on their origin (Fig. 2). In reservoirs from the Cenozoic maximum organic acid concentrations consistently occur in the temperature range 80 to 140 °C. If, as is conventionally believed, bacteria cannot survive in the subsurface at temperatures greater than approximately 80 °C (Davis 1967), the distribution of organic acids with respect to temperature suggests that bacterial processes are not important in their production. Bacterial processes dependent on surface water invasion for an energy source or for transport into the subsurface appear unimportant in organic acid production. Many waters with high organic acid concentrations are produced from deep, overpressured reservoirs, where oxygen isotope data indicate no mixing with meteoric water (Fig. 3; Lundegard 1985). Fisher (1987) reported that total aliphatic acid concentrations in formation waters of the Pliocene reservoirs of the Eastern Venezuelan Basin are lower in waters associated with biodegraded oils (Fig. 8). Barth and Riis (1992) also observed that in the North Sea biodegraded oil corresponds to lower organic acid anion concentrations. High organic acid concentrations found in association with non-biodegraded oils suggest that biodegradation is not a major acid-producing process. Carothers and Kharaka (1978) argued that at temperatures less than 80 °C, bacterial consumption of acetate is an active process because in that environment propionate sometimes shows a dominance over acetate. Fisher (1987), however, found no propionate

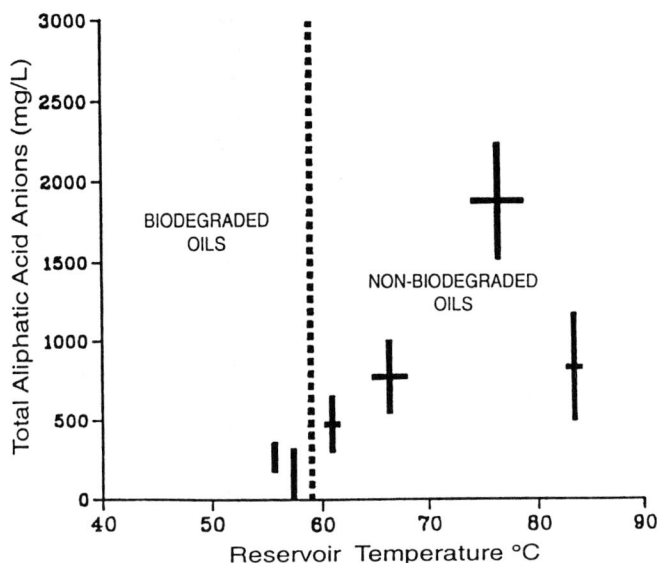

Fig. 8. Total aliphatic acid anions versus temperature for fields in the Eastern Venezuelan Basin (data provided by J.B. Fisher 1992). Temperature distribution of biodegraded oils is shown. *Vertical lines* Range of total aliphatic acid anion concentrations observed. *Horizontal lines* Range of temperature observed

dominance in waters associated with biodegraded oils in the Eastern Venezuelan Basin. Organic acid concentration in formation water also bears no relationship to whether samples are collected from oil or gas wells. Field data show, therefore, that thermal energy is required to produce the major organic acid species. Laboratory heating experiments with kerogens and shales confirm this conclusion (Surdam et al. 1984; Kawamura et al. 1986; Lundegard and Senftle 1987; Barth et al. 1988, 1989; Lundegard and Kharaka 1990; Lewan and Fisher this Vol.). Biological processes are unimportant in the production of organic acids but are important in their destruction at temperatures of less than approximately 80 °C.

4 Survivability of Organic Acids

The data in Fig. 2 show that maximum organic acid anion concentrations occur between 80 and 140 °C in sedimentary basins. This suggests that at higher or lower temperatures organic acid anions are either produced less abundantly or that they are somehow destroyed or diluted.

Table 2. Half-lives of acetic acid and acetate based on experimentally determined decomposition rates. (Data from
Kharaka et al. 1983; Palmer and Drummond 1986)

Solution	Reactor material[a]	Half-life at 100 °C (years)
Acetic acid	Treated Ti	3.7×10^{14}
	Ti	4.0×10^{8}
	SS	1.2×10^{1}
Sodium acetate	SS	3.9×10^{14}
	Au	1.6×10^{14}
	Ti	2.6×10^{12}

[a] Ti = titanium; SS = stainless steel; Au = gold.

4.1 Experimental Data

Acetic acid and other organic acids are not thermodynamically stable
under sedimentary conditions and will eventually decarboxylate to CO_2 and
alkanes (Shock 1988, this Vol.). Experimental studies of acetic acid decarboxylation show that the rate is extremely sensitive to temperature and the
types of catalytic surfaces available (Table 2; Kharaka et al. 1983; Palmer
and Drummond 1986). Extrapolated rate constants for acetic acid decarboxylation at 100 °C differ by more than 14 orders of magnitude between
experiments conducted in stainless steel and catalytically less active titanium
(Table 2; Palmer and Drummond 1986). Palmer and Drummond (1986)
showed that inherent (uncatalyzed) decomposition rates are similar for acetic
acid and acetate. However, in catalytic environments, their rates of decomposition differ markedly, and therefore a pronounced pH effect on the total
decarboxylation rate is observed. Further experiments are needed to investigate the influence of natural geologic materials on decarboxylation
rates.

Various experimental studies have shown that the dicarboxylic acid anions
are more susceptible to decarboxylation than are the monocarboxylic acid
anions (Lundegard and Senftle 1987; MacGowan and Surdam 1988; Crossey
1991; Lundegard et al. 1992).

4.2 Field Studies

Whereas experimental studies of organic acid decarboxylation have established some of the controls, the relevance of experimentally determined
rate constants for natural systems, where potential catalysts of many types
abound, is questionable. Field calibrations clearly are advantageous from

the standpoint of eliminating the effects of kinetic artifacts, but other limitations exist. In trying to relate the effects of time and temperature on decarboxylation rates in natural systems, the effects that variations in the type and abundance of organic matter can have on the production of organic acids, and therefore on their primary concentration, must be minimized.

In the Gulf Coast Basin, broadly similar sedimentary facies are represented throughout the Cenozoic section. As a result, organic matter type and abundance is reasonably similar. Using data from Eocene through Pleistocene reservoirs in the Gulf Coast Basin, Lundegard and Land (1989) showed that the average organic acid concentration in formation water over the temperature range 80–120 °C decreased with increasing age of the reservoir (Fig. 4). This trend represents the existence of fewer samples of low concentration in the younger reservoirs. These data indicated an approximate half-life of 60 million years for total acetate. In the Gulf Coast Basin maximum organic acid concentrations are not related to reservoir age in the same way as the average concentrations. Older reservoirs can have higher maximum concentrations than younger reservoirs. This fact probably reflects local variations in the primary generating capacities of the source rocks.

Combining data from three different basins, Kharaka et al. (1986) estimated half-lives of 27 to 51 million years at 100 °C. Field calibrations of these types are approximations and probably give maximum half-lives. The greatest unknown is the residence time of the dissolved organic acids at a particular temperature. This is a function of the thermal and hydrologic history of the basin, and is difficult to reconstruct. Nevertheless, the field data demonstrate a substantially shorter half-life than indicated by experimental determinations of the inherent stability of acetate (Table 2). However, at 100 °C, concentrations in Cenozoic waters have probably been reduced by at most a factor of two or three, and a much lower factor for late Cenozoic waters. As pointed out by Shock (1988), it is surprising how perserverant some organic acids are, considering how great their departure from equilibrium with respect to decarboxylation reactions is in most basins.

Low concentrations of organic acids at temperatures less than 80 °C occur for several reasons. One reason is the decreased organic acid generation from kerogen as a result of low thermal stress. The second reason is that dilution can occur where upward-moving acid-rich, formation waters mix with acid-poor waters of meteoric or other origins. This situation may exist in the Palo Duro Basin of west Texas (Means and Hubbard 1987) and the Pleistocene of offshore Louisiana (Kharaka et al. 1986). The third reason is that bacterial consumption of organic acids can occur at temperatures less than 80 °C (Carothers and Kharaka 1978).

5   Organic Acid Generation from Petroleum

It has been demonstrated experimentally that various low molecular weight organic acids can be generated from solid organic matter contained in shales (Surdam et al. 1984; Kawamura et al. 1986; Lundegard and Senftle 1987; Barth et al. 1989; Lewan and Fisher this Vol.). However, very little is known about the organic acid generating potential of liquid hydrocarbons, although for reasons discussed below it could be diagenetically important. Diagenetic interest in organic acids has focused on the impact of organic acids on dissolution and secondary porosity development in potential petroleum reservoirs. In this context, the proximity of the reservoir and the organic acid source is an important issue. If shales are the source of organic acids, one must consider whether it is possible for dissolved organic acids to be expelled from the shales and migrate to reservoir rocks without loss of their potential to perform diagenetic work. It is conceivable that a significant amount of the potential diagenetic work by these acids is done within the shale source rocks or along the migration pathway before ever reaching potential hydrocarbon reservoir rocks. On the other hand, organic acids that are locally generated from reservoired oil should be more likely to cause diagenetic alteration of the ambient reservoir rocks. Recently, Boles (1991) suggested that acids associated with hydrocarbons reservoired in the Stevens Sandstone caused local plagioclase dissolution in the North Coles Levee field in California. It would be useful to know, therefore, what yield of organic acids could be produced by in situ thermal alteration of a reservoired oil, and what effect this might have on the reservoir itself. It has also been suggested that certain complexes between metals and organic acid anions might partition into an oil phase (MacGowan and Surdam 1990), thereby enhancing aluminosilicate dissolution. While the content of oxygen in acid functional groups within oil is much less than it is within most immature kerogens (Constantinides and Arich 1967; Tissot and Welte 1978; Green et al. 1984), along migration conduits and in reservoir rocks the content of oil in pore space can be high.

Recently, the results of hydrous pyrolysis experiments with two oils were reported by Lundegard et al. (1992). Additional results are summarized by Kharaka et al. (1993). Two different oils, one from the Midway Sunset field, California, and one from the Kirkuk field, Iraq, were heated in the presence of water in gold-lined reactors at 300 °C for up to 2379 h and 200 °C for up to 5700 h. The Midway Sunset oil had an initial oxygen content of approximately 0.6 wt% and a titratable acidity of 0.95 mg KOH/g, whereas the Kirkuk oil had an initial oxygen content of approximately 0.2 wt% and a titratable acidity of 0.10 mg KOH/g. The concentrations of aqueous organic acid anions and headspace gases were monitored throughout the experiments. The dominant organic acid anions produced were acetate, propionate, and butyrate (Figs. 9 and 10). Small amounts of formate, succinate,

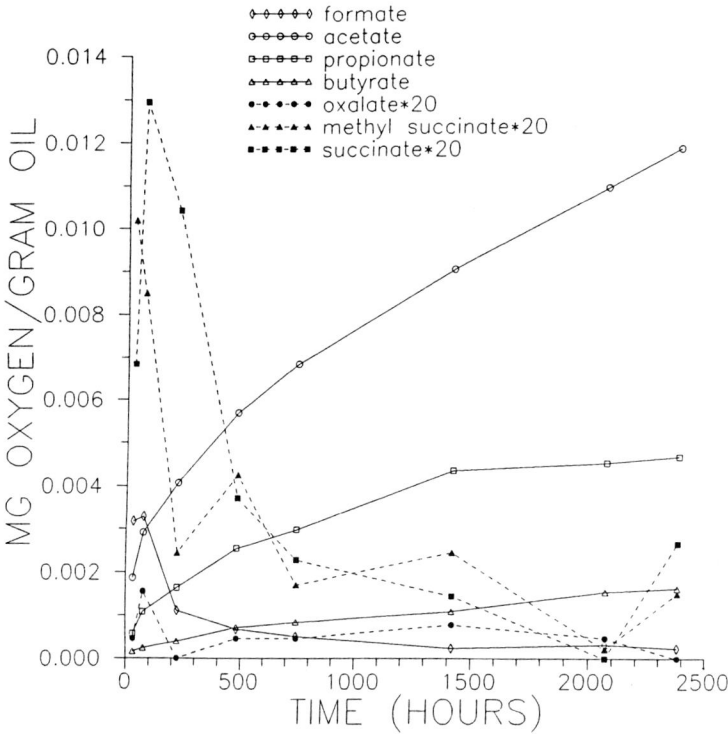

Fig. 9. Normalized yields of organic acid anions for hydrous pyrolysis of Midway Sunset oil versus heating time at 300 °C. (After Lundegard et al. 1992)

methyl succinate, and oxalate were also produced. Carbon dioxide was the dominant oxygen-containing product, as has been observed in similar studies of the hydrous pyrolysis of kerogen (Lundegard and Senftle 1987). The relative yield of organic acids for the two oils is much lower than yields observed during the hydrous pyrolysis of kerogen in shales. In fact, relative to the weight of unreacted oil or kerogen, the organic acid yields are approximately 100 times higher in kerogen-water experiments (Lundegard and Senftle 1987; Barth et al. 1989; Lundegard and Kharaka 1990).

These experiments demonstrate that the kinds of organic acids found in natural formation waters can be generated thermally from oils. The question arises as to what concentrations could actually be produced by in situ oil alteration and how much diagenetic work could be accomplished by these acids. If the maximum observed yields for the Midway Sunset oil are used, and a reservoir with 20% porosity and 80% oil saturation is assumed, then in situ oil alteration could produce 384 mg/l acetate, 209 mg/l propionate, and 95 mg/l butyrate. If oxygen in the oil is completely converted to organic acids and carbon dioxide in the same ratios as above, the resultant concen-

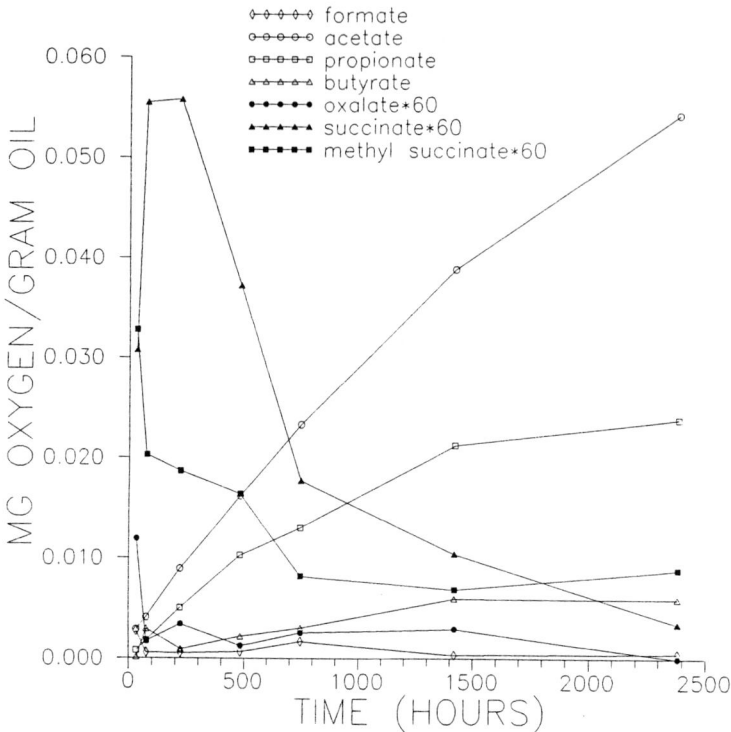

Fig. 10. Normalized yields of organic acid anions for hydrous pyrolysis of Kirkuk oil versus heating time at 300 °C. (After Lundegard et al. 1992)

trations would be 955 mg/l acetate, 520 mg/l propionate, and 236 mg/l butyrate. These concentrations are moderately high relative to those observed in formation waters. In order to assess the amount of diagenetic alteration that can be accomplished by organic acids generated in situ by thermal alteration of oil within the reservoir, several factors should be considered that probably limit the process to a one pore volume basis. First, accumulation of significant amounts of oil within a reservoir implies that an effective seal or barrier to transport through the reservoir exists. Second, oil is not an abundant fluid within basins. And finally, as oil thermal alteration proceeds it is likely that the formation of pyrobitumen will reduce the reservoir's permeability. If the maximum organic acid yields calculated above are invoked on a one pore volume basis, dissolution porosity produced by organic acid protons would only be a fraction of a percent. The results indicate that in situ thermal alteration of reservoired oils can produce modest concentrations of organic acids under favorable geologic conditions but this process is unlikely to be of great diagenetic significance.

The sequence of geologic events necessary to maximize in situ organic acid generation from reservoired petroleum is: (1) oil generation from a source rock; (2) migration of the oil to a reservoir; (3) exposure of the reservoired oil to a near-surface environment conducive to biodegradation so that the oxygen content of the oil can be increased; and (4) burial of the reservoir and its biodegraded oil to depths sufficient to cause thermal alteration and organic acid generation. Overall, this sequence of geologic events is more likely in tectonically active basins than in tectonically quiescent basins. It also might be expected that reservoirs beneath major unconformities and associated with updip tar mats (a possible product of oil biodegradation) will be more likely to experience significant in situ organic acid generation during burial.

6 Influence of Burial and Thermal History

6.1 Kinetics of Oxygen Loss from Kerogen

By analogy with the first-order kinetic behavior of oil generation from kerogen (Tissot and Espitalie 1975; Quigley and MacKenzie 1988), organic acid generation from kerogen should be affected by variations in heating rates, which are a function of geothermal gradients and burial rates. These factors will cause variations in the rates of organic acid generation with respect to time, depth, and temperature. Primary organic acid concentration will vary directly with these variations in generation rate and inversely with the ambient porosity at the time of generation. Primary concentration is considered here to be the concentration that exists before significant organic acid destruction takes place. High geothermal gradients should increase acid generation rates. But, because shale porosity tends to decrease exponentially with depth (Magara 1978), it is not obvious what net effect high geothermal gradients will have on the dissolved concentration of organic acids.

Studies of hydrocarbon generation from kerogen have demonstrated that models using only a single activation energy will not adequately represent the generation process over a range of heating rates (Larter 1988). This is presumably because the bonds in kerogen that are broken to yield hydrocarbons have variable bond energies, generally in the range of 50–60 kcal/mol. The kinetics of organic acid generation from kerogen are poorly known. However, one would expect that the bonds broken to produce, for example, acetic acid from kerogen will also vary in bond energy. Hence, activation energy distributions will probably be required to adequately describe the kinetics of organic acid generation. Attempts to estimate organic acid generation kinetics from hydrous pyrolysis experiments have so far yielded results that are not geologically meaningful (Barth et al. 1989). Organic acids and carbon dioxide are believed to be produced by the elimination of oxygen functional groups in kerogen. Oxygen functional groups in kerogen are lost at thermal ranks lower than those of principal hydrocarbon genera-

tion (Robin and Rouxhet 1978; Tissot and Welte 1978). Furthermore, carboxyl and carbonyl groups are the oxygen functions that are most readily
liberated from kerogen. Therefore, the activation energies for the elimination of these oxygen functional groups should generally be lower than those
for hydrocarbon generation mentioned above.

Any assumed kinetics for organic acid generation should be compatible
with the known relative rates of oxygen and hydrogen loss from kerogen,
and the field data on the temperature distribution of organic acids in basins.
Burnham and Sweeney (1989) recently published a kinetic description of the
reflectance of vitrinite, Type III kerogen, in which they accounted for
changes in the elemental composition of vitrinite with increasing maturity.
They derived kinetic parameters for these compositional changes, including
the loss of carbon dioxide from Type III kerogen. These parameters
represent the best current estimates of the kinetics of the loss of carboxyl
groups from kerogen and produce model results that are compatible with
the observed temperature distribution of organic acids in basins (Fig. 2). It
should be remembered, however, that while the Burnham and Sweeney
kinetics describe oxygen loss from kerogen they do not describe what fraction of the lost oxygen is in the form of organic acids. Hydrous pyrolysis
experiments show that most of the oxygen in kerogen is lost as carbon
dioxide, not organic acids (Lundegard and Senftle 1987; Lundegard and
Kharaka 1990). These kinetic parameters should therefore be used only as a
rough guideline and for looking at the relative effects of changing thermal
and burial conditions on organic acid generation. In order to emphasize this
point, modeled organic acid concentrations shown here are represented only
as relative, not absolute values.

6.2 Model Results

The model described here simulates the history of shaly source rock for an
assumed burial and thermal history. The results discussed are based on a
constant sedimentation rate of 333 m/m.y. The influence of thermal gradient
is investigated by comparing results for assumed gradients of either 20, 40,
or 60 °C/km. A Gulf Coast-type porosity gradient for shale is used, where
porosity $= 0.39e^{-0.000095 \times \text{depth (ft)}}$ (Magara 1978). For all model results, the
total cumulative yield is the same and is controlled by the assumed initial
oxygen content of the kerogen.

Higher thermal gradients cause oxygen to be liberated from kerogen at
shallower depths and over a narrower range of depths (Fig. 11). With a
higher thermal gradient, the peak generation rate occurs after less elapsed
time and at a slightly higher temperature. Accounting for the effect of
decreasing source rock porosity during burial, predicted organic acid concentrations were calculated. Higher thermal gradients produce higher acid
concentrations at shallow depths, but higher maximum concentrations are

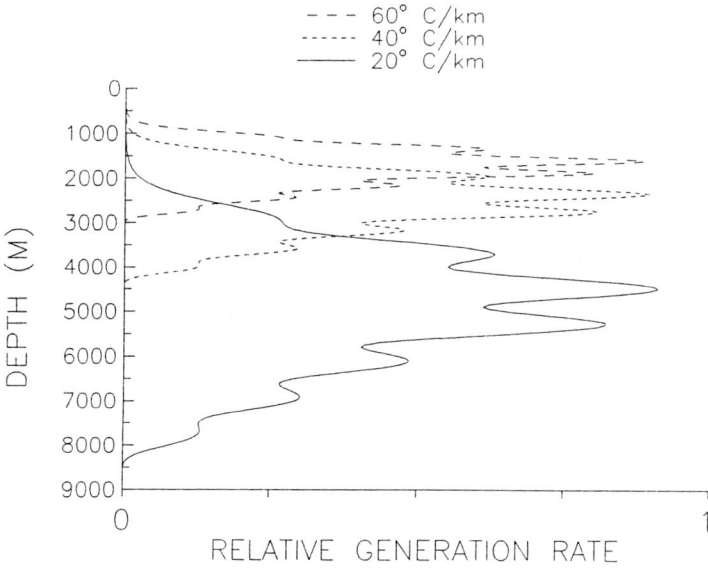

Fig. 11. Results of a mathematical model of organic acid generation in a compacting shale for three thermal gradients at a burial rate of 333 m/Ma. Plot shows relative generation rate versus burial depth. Higher thermal gradients result in peak generation rates at shallower depths and after less geologic time

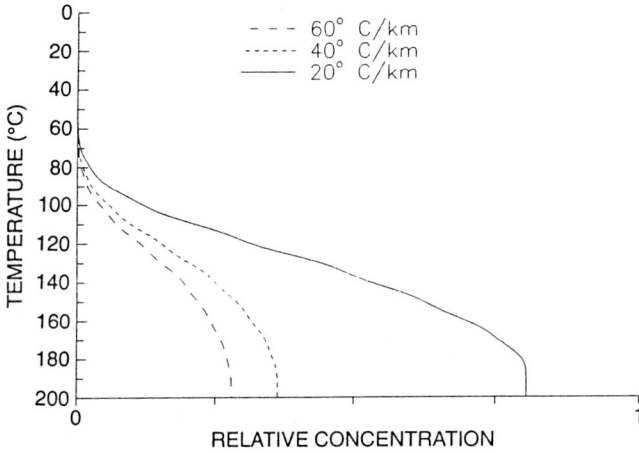

Fig. 12. Results of a mathematical model of organic acid generation in a compacting shale for three thermal gradients and a burial rate of 333 m/Ma. Plot shows relative concentration in pore water versus temperature. Lower thermal gradients result in principal organic acid generation at lower porosities, resulting in higher concentrations

eventually obtained when the thermal gradient is lower. At all temperatures, higher geothermal gradients actually result in lower predicted concentrations, as a result of the competing effects of generation rate and porosity (Fig. 12). When thermal gradients are higher, peak generation rates occur at shallower depths and higher porosities. Conversely, when the thermal gradients are lower, the lower porosity at the time of peak yield results in higher maximum acid concentrations. This analysis shows that high organic acid concentrations may, surprisingly, be favored by low thermal gradients. Maximum concentrations would presumably be achieved when geothermal gradients increase with time at a rate that results in rapid generation at depths where porosities are low. This situation might occur along convergent plate margins during subduction of spreading centers.

The model results discussed above express only the effects of time and temperature on organic acid generation from kerogen, and the effect of porosity on organic acid concentrations. They do not account for the effects of time and temperature on organic acid decarboxylation since decarboxylation kinetics under natural burial conditions are poorly known. Therefore, the results are best applied to Neogene source rocks where the field data discussed earlier indicate that the effect of thermal decarboxylation or decomposition on total acetate concentration is small, and where temperatures do not exceed approximately 150 °C.

7   Conclusions

More than anything else, the published literature on the concentrations of organic acids in natural formation water shows that their concentrations are highly variable. At temperatures of less than approximately 40 °C or greater than 180 °C, concentrations are less than a few hundred mg/l. At intermediate temperatures, there is a rough relationship between temperature and maximum observed concentrations. However, at any given temperature, organic acid concentration can vary by several orders of magnitude. Known or suspected factors that contribute to the variation in primary organic concentrations include the variability in the richness and type of organic matter in shales, source rock thermal history, and porosity history. Postgenerational factors include variability in dilution, biodegradation, and the effects of natural catalysis on decarboxylation and decomposition. The current understanding of these factors allows only rough predictions of organic acid concentrations as a function of space and time in basins.

It is particulary important that the variation in concentrations of organic acids in natural formation waters be kept in mind during geochemical modeling and diagenetic mass transfer calculations. It is useful to consider the potential geochemical effects of "maximum" organic acid concentrations in order to evaluate the limits of feasibility. However, it is more important to consider the possible effects of "normal" concentrations. Available data indicate that concentrations of total organic acid anions greater than 3000 mg/l acetate equivalent are rare. In this respect formation water in Miocene reservoirs of the San Joaquin Basin contains unusually high concentrations of organic acid anions, and has high ratios of organic alkalinity to bicarbonate alkalinity.

Acknowledgments. Tanya Barth kindly provided data on the concentration of organic acids in North Sea formation waters in advance of their publication. Bert Fisher provided us with Fig. 8 and Ed Pittman and Mike Lewan provided very helpful reviews of the manuscript. We thank Unocal Corporation for allowing this chapter to be published.

References

American Association of Public Health (1985) Standard methods for the examination of water and wastewater. American Public Health Association, Washington DC, 1268 pp

Barth T (1987) Quantitative determination of volatile carboxylic acids in formation waters by isotachophoresis. Anal Chem 59: 2322–2327

Barth T, Riis M (1992) Interactions between organic acid anions in formation waters and reservoir mineral phases. Org Geochem 19: 455–482

Barth T, Borgund AE, Hopland AL, Graue A (1988) Volatile organic acids produced during kerogen maturation – amounts, composition and role in migration of oil. Org Geochem 13: 461–465

Barth T, Borgund AE, Hopland AL (1989) Generation of organic compounds by hydrous pyrolysis of Kimmeridge oil shale – bulk results and activation energy calculations. Org Geochem 14: 69–76

Boles JR (1991) Plagioclase dissolution related to oil residence time, North Coles Levee field, California. Am Assoc Pet Geol Bull 75: 544 (Abstr)

Burnham AK, Sweeney JJ (1989) A chemical kinetic model of vitrinite maturation and reflectance. Geochem Cosmochim Acta 53: 2649–2657

Carothers WW, Kharaka YK (1978) Aliphatic acid anions in oil-field waters – implications for the origin of natural gas. Am Assoc Pet Geol Bull 62: 2441–2453

Constantinides G, Arich G (1967) Non-hydrocarbon compounds in petroleum. In: Nagy B, Colombo U (eds) Fundamental aspects of petroleum geochemistry. Elsevier, New York, pp 77–108

Crossey LJ (1991) The thermal stability of organic acids in sedimentary basins. Am Assoc Pet Geol Bull 75: 557 (Abstr)

Crossey LJ, Surdam RC, Lahann R (1986) Application of organic/inorganic diagenesis to porosity prediction. In: Gautier DL (ed) Roles of organic matter in sediment diagenesis. Soc Econ Paleontol Mineral Spec Publ 38, pp 147–155

Davis JB (1967) Petroleum microbiology. Elsevier, New York, 604 pp

Drummond SE, Palmer DA (1986) Thermal decarboxylation of acetate. Part 1. The kinetics and mechanism of reaction in aqueous solution. Geochim Cosmochim Acta 50: 813–823

Fisher JB (1987) Distribution and occurrence of aliphatic acid anions in deep subsurface waters. Geochim Cosmochim Acta 51: 2459–2468

Fisher JB, Boles JR (1990) Water-rock interaction in Tertiary sandstones, San Joaquin Basin, California, USA: diagenetic controls on water composition. Chem Geol 82: 83–101

Gautier DL (ed) (1986) Roles of organic matter in sediment diagenesis. Soc Econ Paleontol Mineral Spec Publ 38, 203 pp

Germanov AI, Mel'kanovitskaya SG (1975) Organic acids in hydrothermal ores of polymetallic deposits and in ground waters. Akad Nauk SSSR Dokl 225: 2200–2211

Giordano TH, Barnes HL (1981) Lead transport in Mississippi Valley-type ore solutions. Econ Geol 76: 2200–2211

Green JB, Hoff RJ, Woodward PW, Stevens LL (1984) Separation of liquid fossil fuels into acid, base and neutral concentrates. 1. An improved nonaqueous ion exchange method. Fuel 63: 1290–1301

Hanor JS, Workman AL (1986) Distribution of dissolved volatile fatty acids in some Louisiana oilfield brines. Appl Geochem 1: 37–46

Harrison WJ, Thyne GD (1992) Predictions of diagenetic reactions in the presence of organic acids. Geochim Cosmochim Acta 56: 565–586

Kartsev VV (1974) Hydrogeology of oil and gas deposits. Nat Tech Inf Serv Rep TT73-58022, 323 pp

Kawamura K, Tannenbaum E, Huizinga BJ, Kaplan IR (1986) Volatile organic acids generated from kerogen during laboratory heating. Geochem J 20: 51–59

Kharaka YK, Carothers WW, Rosenbauer RJ (1983) Thermal decarboxylation of acetic acid: implications for the origin of natural gas. Geochim Cosmochim Acta 47: 397–402

Kharaka YK, Hull RW, Carothers WW (1985) Water-rock interactions in sedimentary basins. Soc Econ Paleontol Mineral Short Course 17: 79–176

Kharaka YK, Law LM, Carothers WW, Goerlitz DF (1986) Role of organic species dissolved in formation waters from sedimentary basins in mineral diagenesis. In: Gautier DL (ed) Roles of organic matter in sediment diagenesis. Soc Econ Paleontol Mineral Spec Publ 38: 111–122

Kharaka YK, Lundegard PD, Ambats G, Evans WC, Bischoff JL (1993) Generation of aliphatic acid anions and carbon dioxide by hydrous pyrolysis of crude oils. Appl Geochem (in press)

Land LS, MacPherson GL, Mack LE (1988) The geochemistry of saline formation waters, Miocene offshore Louisiana. Gulf Coast Assoc Geol Soc Trans 38: 503–511

Larter S (1988) Some pragmatic perspectives in source rock geochemistry. Mar Pet Geol 5: 194–204

Lundegard PD (1985) Carbon dioxide and organic acids: origin and role in burial diagenesis (Texas Gulf Coast Tertiary). PhD Dissertation, The University of Texas at Austin, Austin, TX, 145 pp

Lundegard PD, Kharaka YK (1990) Geochemistry of organic acids in subsurface waters – field data, experimental data, and models. In: Melchior DC, Bassett RL (eds) Chemical modeling of aqueous systems II. American Chemical Society, Washington DC, pp 169–189

Lundegard PD, Land LS (1986) Carbon dioxide and organic acids: their role in porosity enhancement and cementation, Paleogene of the Texas Gulf Coast. In: Gautier DL (ed) Roles of organic matter in sediment diagenesis. Soc Econ Paleontol Mineral Spec Publ 38: 129–146

Lundegard PD, Land LS (1989) Carbonate equilibria and pH buffering by organic acids – response to changes in pCO_2. Chem Geol 74: 277–287

Lundegard PD, Senftle JT (1987) Hydrous pyrolysis: a tool for the study of organic acid synthesis. Appl Geochem 2: 605–612

Lundegard PD, Trevena AS (1990) Sandstone diagenesis in the Pattani Basin (Gulf of Thailand): history of water-rock interaction and comparison with the Gulf of Mexico. Appl Geochem 5: 669–685

Lundegard PD, Kharaka YK, Rosenbauer RJ (1992) Petroleum as a potential diagenetic agent: experimental evidence. In: Kharaka YK, Maest AS (eds) Water-rock interaction. Balkeema, Rotterdam, pp 329–335

MacGowan DB, Surdam RC (1988) Difunctional carboxylic acid anions in oilfield waters. Org Geochem 12: 245–259

MacGowan DB, Surdam RC (1990) Carboxylic acid anions in formation waters, San Joaquin Basin and Louisiana Gulf Coast, USA – implications for clastic diagenesis. Appl Geochem 5: 687–701

Magara K (1978) Compaction and fluid migration – practical petroleum geology. Elsevier, New York, 319 pp

Means JL, Hubbard NJ (1987) Short-chain aliphiatic acid anions in deep subsurface brines: a review of their origin, occurrence, properties, and importance and new data on their distribution and geochemical implications in the Palo Duro Basin, Texas. Org Geochem 11: 177–191

Palmer DA, Drummond SE (1986) Thermal decarboxylation of acetate. Part I. The kinetics and mechanism of reaction in aqueous solution. Geochem Cosmochim Acta 50: 813–823

Quigley TM, MacKenzie AS (1988) The temperature of oil and gas formation in the subsurface. Nature 333: 549–552

Rapp JB (1976) Amino acids and gases in some springs and an oil field in California. J Res US Geol Surv 4: 227–232

Robin PL, Rouxhet PG (1978) Characterization of kerogens and study of their evolution by infrared spectroscopy: carbonyl and carboxyl groups. Geochim Cosmochim Acta 42: 1341–1349

Shock EL (1988) Organic acid metastability in sedimentary basins. Geol 16: 886–890

Shock EL (1989) Corrections to "Organic acid metastability in sedimentary basins". Geology 17: 572–573

Simoneit BRT (1984) Hydrothermal effects on organic matter – high vs. low temperature components. Org Geochem 6: 857–864

Surdam RC, Boese SW, Crossey LJ (1984) The chemistry of secondary porosity. In: McDonald DA, Surdam RC (eds) Clastic diagenesis. Am Assoc Pet Geol Mem 37, pp 127–150

Thurman EM (1984) Humic substances I. Geochemistry, characterization and isolation. Wiley, New York, 380 pp

Tissot BP, Espitalie J (1975) L'Evolution thermique de la matiere organique des sediments – application d'une simulation mathematique. Rev Inst Fr Pet 30: 743–777

Tissot BP, Welte DH (1978) Petroleum formation and occurrence. Springer, Berlin Heidelberg New York, 538 pp

Willey LM, Kharaka YK, Presser TS, Rapp JB, Barnes I (1975) Short-chain aliphatic acid in oil-field waters of Kettleman North Dome oil field, California. Geochim Cosmochim Acta 39: 1707–1710

Workman AL, Hanor JS (1985) Evidence for large-scale vertical migration of dissolved fatty acids in Louisiana oil field brines: Iberia field, south-central Louisiana. Gulf Coast Assoc Geol Soc Trans 35: 293–300

Chapter 4   Organic Acids from Petroleum Source Rocks

M.D. Lewan[1] and J.B. Fisher[2]

Summary

The vastness and dynamics of sedimentary basins make it difficult to assess organic acid generation from petroleum source rocks on the basis of the limited subsurface data currently available. An alternative approach involves simulating the natural process in laboratory pyrolysis experiments that maintain a liquid-water phase, utilize whole rock, avoid extreme temperatures, and minimize reactor-wall effects. Appropriately conducted laboratory pyrolysis experiments show that saturated acyclic monocarboxylic acids are the dominant organic acids generated, that C_2-C_5 monocarboxylic acids dominate the aqueous phase assemblages, and that acetic acid (C_2) is typically the dominant aqueous organic acid. Low activation energies derived from laboratory pyrolysis experiments indicate that these organic acids are retained in sedimentary organic matter by weak noncovalent bonds. Thus, the organic acid potential of a source rock is largely dependent on the amount of C_2-C_5 monocarboxylic acids assimilated by noncovalent bonds into sedimentary organic matter during its early development into kerogen. Significant quantities of these acids may be released from petroleum source rocks by diffusion during early diagenesis. However, the sluggishness of this diffusion process is only likely to cause local enhancement of porosity within a source rock and in rocks immediately adjacent to it. Release of the remaining organic acids is most likely to occur during the expulsion of petroleum from a source rock. Dissolved organic acids in the expelled petroleum will redistribute themselves within the associated formation waters as the migrating or entrapped petroleum cools, degasses, or encounters lower salinity waters. As a result, enhanced porosity may occur within carrier beds during secondary petroleum migration or within reservoirs during or after petroleum entrapment.

---

[1] US Geological Survey, Box 25046, MS977 Denver Federal Center, Denver, Colorado 80225, USA

[2] Amoco Production Company, Research Center, P.O. Box 3385, Tulsa, Oklahoma 74102, USA

1 Introduction

Subsurface waters associated with petroleum accumulations have been shown to have organic acid concentrations as high as 5000 mg/l (Carothers and Kharaka 1978). Saturated aliphatic carboxylic acids with two to five carbon atoms are the most commonly observed organic acids in subsurface waters (Fisher 1987). Distributions of these short-chain carboxylic acids are usually dominated by acetic acid (C_2), with propionic (C_3), butyric (C_4), and valeric (C_5) acids being subordinate in varying proportions. Petroleum engineers in the mid- and late-1940s recognized that these short-chain carboxylic acids significantly contributed to the internal corrosion of well-head equipment and production pipes in gas-condensate fields (Menaul 1944; Greco and Griffin 1946; Holmberg 1946; Shock and Hackerman 1948). The serious impact of these corrosive organic acids on the economics of producing fields was sufficient to initiate field and laboratory studies that were collectively published by the Natural Gasoline Association of America (Prange et al. 1953). Geologists and geochemists over the following two decades became more aware of these acids in subsurface waters and their effect on alkalinity measurements (Obukhova and Kutovaya 1968; Shvets and Shilov 1968; Bykova et al. 1971; Willey et al. 1975). However, it was not until the mid-1980s that organic acids were considered responsible for the generation of secondary porosity (Surdam et al. 1984; Surdam and Crossey 1985). The resulting hypothesis proposed that organic acids are released from petroleum source rocks prior to petroleum generation. These expelled acids generate secondary porosity during their migration away from the source rock, thereby enhancing the potential of carrier and reservoir beds for later expelled oil.

This hypothesis continues to be critically questioned (Giles and Marshall, Chap. 14, this Vol.), but it also continues to be used in explaining occurrences of secondary porosity in the subsurface (Hansley and Nuccio 1992; Mazzullo and Harris 1992). A pivotal concern of this hypothesis is whether organic acids can be generated and expelled from petroleum source rocks in sufficient quantities and in the appropriate time frame. Addressing this concern solely from available subsurface data is difficult because of the dynamics of fluid flow within the vastness of sedimentary basins. In addition, subsurface data from a specific petroleum source rock are either limited to a narrow range of thermal maturities or are not comparable over broader ranges of thermal maturities due to regional variations in types and amounts of organic matter. These uncertainties and lack of subsurface data may be alleviated in part by conducting laboratory experiments designed to simulate the natural process. Using results from such experiments within natural constraints defined by available subsurface data, a reasonable evaluation of the hypothesis and related natural processes may be undertaken. This chapter will employ this approach by (1) examining experimental considerations germane to simulating the natural process, (2) reviewing the types and quantities of organic acids recovered from laboratory experiments, and (3) evaluating the generation and expulsion of organic acids in sedimentary basins on the basis of experimental results and natural constraints.

2 Experimental Considerations

The simple heating of a sample to determine its ability to generate organic acids may appear straightforward from a chemistry perspective, but from a geochemistry perspective, there are experimental constraints that must be considered in order to understand the natural process. These experimental considerations include temperature and time conditions, presence of water, type of sample, reactor wall composition, oil/water partitioning, pressure, and water chemistry.

2.1 Time and Temperature Conditions

The time and temperature at which a sample is heated are obvious considerations in any experiment, but they are particularly important in kinetically controlled reactions. Reactions involving organic matter in subsiding sedimentary basins under low to moderate temperatures (i.e., 50 to 150 °C) are typically kinetically controlled. The degree of kinetic control is variable, with some reactions being more time-dependent than others (e.g., petroleum generation relative to vitrinite reflectivity). The problem that this time dependence poses is that natural reactions occurring at low temperatures (<200 °C) over long periods of geologic time (1 to 100 m.y.) cannot be *duplicated* in laboratory experiments. As a result, laboratory experiments must use higher temperatures to achieve measurable amounts of reaction products within reasonable durations (i.e., days to months).

An inherent assumption in this type of substitution is that the mechanisms and their relative rates of occurrence are not significantly altered at the higher experimental temperatures. Assessing the reality of this assumption is difficult and requires a thorough kinetic understanding of the elementary mechanisms making up the overall reaction. Some degree of comfort in this assumption may be obtained from compositional similarities between the products derived experimentally at higher temperatures and those derived naturally at lower temperatures. The reality of this assumption is also more approachable by conducting experiments at the lowest possible temperatures to reduce the possibility of high temperature artifacts and reversals in the relative rates of elementary reactions. In addition, the use of higher temperatures increases the likelihood of product destruction or modification by less time-dependent and more temperature-dependent reactions.

The experimental studies by Kawamura and Ishiwatari (1981, 1985a,b) best illustrate the futility of generating organic acids from sedimentary organic matter at temperatures comparable to those experienced by source rocks in sedimentary basins (i.e., <150 °C). Their experiments involved isothermally heating wet surface sediments from Lake Biwa in borosilicate glass tubes at 65 and 83 °C for 24 to 5952 h (Kawamura and Ishiwatari 1985a), 120 to 198 °C for 48 h (Kawamura and Ishiwatari 1981), and 68 to 325 °C for 24 h (Kawamura and Ishiwatari 1985b). Organic acids with 12 through 32 carbon atoms (i.e., C_{12}–C_{32}) extracted from the wet sediments showed no significant increase in concentration after being heated at 83 °C for 5952 h, 120 °C for 48 h, and 129 °C for 24 h. A small but detectable increase in acid concentrations was first observed after heating at 140 °C for 24 h, and a 50% increase relative to the concentration of the unheated sample required heating temperatures of 300 and 325 °C for 24 h.

Subsequent experimental investigations indicate that optimum conditions for the generation of organic acids from sedimentary organic matter ranges from 200 to 350 °C over durations of a few days to several weeks. Barth et al. (1987, 1989) have shown that concentrations of organic acids generated from Kimmeridgian shales and lower Jurassic coal increased in experiments for 72-h durations from 200 to 350 °C. Similarly, Eglinton et al. (1987) observed the concentration of organic acids generated from organic matter isolated from a Kimmeridgian shale to increase in experiments for 72-h durations from 200 to 330 °C. Cooles et al. (1987) also reported concentrations of organic acids generated from Cretaceous and Kimmeridgian mudstones to increase in experiments for 72-h durations from 300 to 350 °C. No obvious high-temperature artifacts nor peculiar compositional distributions were observed in these experiments. Therefore, experimental conditions of 200 to 350 °C for 72-h durations appear to provide a reasonable *simulation* of organic acid formation at lower temperatures and longer durations in the subsurface of sedimentary basins.

2.2 Presence of Water

Water is ubiquitous in the interstices of subsurface sedimentary rocks. Those interstices include intergranular voids as well as fractures. Although the specific role of water in geologic reactions remains an active area of research, its presence has been shown to be critical to granite melts (Goranson 1932), metamorphism (Rumble et al. 1982), clay-mineral transformations (Whitney 1990), coalification (Schuhmacher et al. 1960), and petroleum generation (Lewan et al. 1979). Eisma and Jurg (1969) showed that the thermal decomposition rate of behenic acid to low-molecular-weight hydrocarbons at 250 °C in the presence of bentonite was 1.5 orders of magnitude higher in the absence of water than in its presence. Kawamura et al. (1986) found that heating kerogen isolated from the Green River Shale at 300 °C for 2 to 100 h generated twice as much organic acid in the presence of water than in its

absence. Mechanisms responsible for these effects of water on organic acids
have not been established, but the ubiquity of water in the subsurface and
its significant enhancement of organic acid preservation or generation make
inclusion of water in laboratory experiments imperative to simulating the
natural system.

In addition to conducting experiments in the presence of water, it is also
important to consider the water phases present in the experiments. With the
exception of localized occurrences of supercritical water fluid in contact
metamorphic regimes and water vapor in shallow hydrothermal vents, liquid
water is the prevailing phase in sedimentary basins. Therefore, maintaining
a liquid-water phase in contact with heated samples is preferred considering
the significantly different properties and reactivities of the other two nonsolid
phases of water. As discussed by Lewan (1993a), the volume of liquid water
(V_l^T) present at a given experimental temperature (T) may be calculated
with the following equation,

$$V_l^T = \frac{(M_w^o \gamma_v^T - [V_r - V_s])\gamma_l^T}{\gamma_v^T - \gamma_l^T}$$

by knowing the reactor volume (V_r), sample volume (V_s), specific volumes
of water liquid (γ_l^T) and vapor (γ_v^T) at the experimental temperature (Meyer
et al. 1983), and mass of liquid water (M_w^o) orginally added to the reactor at
room temperature. The critical aspects of this determination is that sufficient
water is used in an experiment to insure that the heated sample is submerged
in liquid water at the desired experimental temperature. Supercritical conditions should also be considered. Although the supercritical temperature
for pure water is 374 °C, the introduction of CO_2 from a heated sample
may result in supercritical temperatures less than 300 °C (Takenouchi and
Kennedy 1964).

2.3   Type of Sample

Experiments designed to understand the generation of organic acids from
sedimentary organic matter have used either samples of whole rock or
isolated insoluble organic matter (i.e., kerogen). Whole rock samples provide
an overall evaluation of the natural process including organic acid generation
within, migration through, and expulsion from, the rock. Possible catalytic
and absorption effects caused by the rock matrix may also be represented in
the final results because the embedded organic matter thermally matures in
contact with natural mineral assemblages, rock fabrics, and intergranular
porosities. One consideration in this approach is the size of the whole rock
used in the experiments. Oil-prone organic matter in thermally immature
rocks usually has dimensions in the order of tens of micrometers. Therefore,
pulverizing rock samples to grain sizes less than sand size (i.e., $<62\,\mu$m)
risks reducing the amount of natural contact between minerals and organic

matter and increasing the amount of contact between water and organic matter to an unnatural level.

A variation in the use of whole rocks is to preextract soluble organic matter (i.e., bitumen) from a rock with nonpolar organic solvents prior to experimentation (e.g., Lundegard and Senftle 1987). The intent of this treatment is to allow the distinction between preexisting organic acids from those generated during the experiment. Bitumen (i.e., soluble organic matter) in thermally immature rocks typically comprises less than 5 wt% of the sedimentary organic matter. The significance of the removal of this minor amount of organic matter remains to be evaluated, but its removal is clearly a deviation from the natural system.

The use of isolated kerogen (i.e., insoluble organic matter) in experiments provides an evaluation of the amount of organic acids that may be generated without the influence of a rock matrix. Experiments of this type are important in evaluating the effects of rock matrices when compared with whole-rock experiments, but by themselves they are not particularly relevant to evaluating the natural system. Surdam et al. (1984) provide an example of a study utilizing both whole rock and isolated organic matter from the Green River Formation. Although an experimental temperature of only 100 °C was employed for 336 h, the results showed that oxalate concentrations were in trace but measurable quantities in the isolated-kerogen experiments (6 ppm) and below detection limits in the whole-rock experiments.

Consideration should also be given to the methods employed to isolate kerogen from its host rock, and how these methods may affect the ability of a kerogen to generate organic acids. Most isolation procedures remove the mineral matter through a series of HCl and HF treatments (e.g., Lewan 1986) that expose the kerogen to atmospheric oxygen. Whether this exposure influences organic acid generation remains to be determined, but its potential to do so should be considered. Experiments have also been conducted on isolated kerogen that has been mixed with different mineral powders, with the intent of evaluating rock matrix effects (Kawamura et al. 1986). Unfortunately, in the presence of liquid water, these unconsolidated powdered mixtures are not representative of the natural system due to the development of water films, which prevent contact between mineral grains and the organic matter.

2.4 Reactor Wall Composition

Some reactor wall compositions have been shown by Palmer and Drummond (1986) to catalyze the thermal degradation of acetic acid at temperatures ranging from 300 to 400 °C. Their study indicates the catalytic effect to be a heterogeneous interaction between the reactor wall and aqueous acetic acid. This catalytic effect is dependent on the composition of the reactor wall. The rate of thermal degradation is not significantly enhanced by gold surfaces, but is enhanced by an order of magnitude by stainless steel surfaces. The obvious concern that emerges from these experimental results is that the rate at which organic acids are destroyed by surface catalysis may approach the rate at which they are generated. As a result, the true potential of the organic-acid generating capability of sedimentary organic matter could be underestimated. Using whole rock instead of isolated kerogen significantly minimizes the contact between the reactor walls and organic acids generated within the rock matrix. However, as generated organic acids are expelled into the surrounding water phase, their access to reactor walls is greatly enhanced and must be considered.

In order to more fully evaluated this potential experimental problem, a series of experiments were conducted in the authors' laboratory. The experiments involve isothermally heating 300-g aliquots of a thermally immature sample of Monterey Shale (MR-216) with liquid water at 330 °C for 72 h in 1-l reactors with walls of different surface compositions. The six surface compositions used include fresh stainless steel-316, carburized stainless steel-316, gold-plated stainless steel-316, fresh Hastelloy C-276, carburized Hastelloy C-276, and borosilicate glass (Pryex liner). After the experiments cooled to room temperature, the expelled oil floating on the water surface was removed (Lewan 1993a) and the recovered water was decanted from the reactor. The recovered water was immediately filtered through a 0.45-μm filter and analyzed for dissolved inorganic and organic species, which are given in Tables 1 and 2. Recovered waters from these experiments had pH's between 6.7 and 7.1 and Eh's between -402 and -420 mV.

Stainless steel-316 is an iron-based alloy with subordinate amounts of chromium, nickel, and molybdenum. Hastelloy C-276 is a nickel-based alloy with subordinate amounts of molybdenum, chromium, iron, and tungsten. Comparison of the inorganic species dissolved in the waters recovered from the fresh-metal surfaces of these two alloys shows no significant differences (Table 1). However, dissolved organic species (Table 2) show the fresh-metal stainless steel-316 reactor to have significant reductions in total acid, butyric acid, and propionic acid relative to the fresh-metal Hastelloy C-276 reactor. These results indicate that butyric acid is preferentially degraded in fresh-metal stainless steel-316 reactors.

The bright silvery surfaces of the fresh-metal walls of both alloys dull and darken with successive experiments utilizing sedimentary organic matter (Lewan 1993a). This surface seasoning is the result of carbon from hydrocarbon and CO_2 degradation at the metal surface diffusing into the alloy (Stanley 1970). This process is referred to as carburization and typically extends several μm into the reactor wall in these types of experiments where expelled oils and generated gases are produced. Dissolved inorganic species show no significant differences between the fresh-metal and carburized surfaces with the exception of chlorine, which has a significantly higher concentration in the carburized reactors of both alloys. Salts in the original sample are the source of this chlorine, and its increased concentration in the

Table 1. Dissolved inorganic species released from a sample of Monterey Shale (MR-216) into the surrounding water after heating for 72 h at 330 °C in 1-l reactors with different surface compositions. Concentrations are given in mg/l and were determined by AA or ICP-AES

Reactor surface composition	Li	Na	K	Mg	Ca	Sr	B	C	Al	Si	S	Cl	Cr	Mn	Fe	Total dissolved solids
Stainless steel-316																
Fresh metal	2.0	2189	98	41	1199	3.2	13.1	1865	1.0	118	571	3865	0.06	0.06	0.15	10660
Carburized	2.1	2177	93	44	1275	3.0	13.2	1875	1.1	122	586	4163	0.06	0.06	0.14	10780
Gold plated	2.1	2229	85	16	1311	3.1	16.6	2015	1.0	115	563	4095	0.06	0.06	0.19	10530
Hastelloy C-276																
Fresh metal	2.1	2229	88	43	1290	3.2	14.8	1888	1.2	125	545	3922	0.06	0.06	0.15	11430
Carburized	2.0	2203	90	44	1387	3.2	13.6	1982	1.1	121	570	4210	0.06	0.06	0.13	11350
Borosilicate glass	1.6	1979	59	31	813	2.3	515.0	1776	0.7	180	423	3021	0.06	0.15	0.19	10090

Table 2. Dissolved organic species released from 300 g of crushed (0.5–2.0 cm) Monterey Shale (MR-216) into 365 g of water after heating for 72 h at 330°C in 1-l reactors with different surface compositions. Aliphatic acids were determined by ion exclusion chromatography[a] and dissolved organic carbon (Org. C) was determined with a UV photooxidation carbon anayzer[b]

Reactor surface composition	Acetic (mg)	Propionic (mg)	Butyric (mg)	Total (mg)	Org. C (mg)	Normalized percentage			% Org. C as acid C
						Acetic	Propionic	Bytyric	
Stainless Steel-316									
Fresh metal	245	123	201	569	607	43.1	21.6	35.3	44.7
Carburized	235	128	246	609	603	38.6	21.0	40.4	48.8
Gold plated	254	128	255	637	648	39.9	20.1	40.0	47.4
Hastelloy C-276									
Fresh metal	245	128	310	683	613	35.9	18.7	45.4	54.5
Carburized	235	137	301	673	627	34.9	20.4	44.7	52.5
Borosilicate glass[c]	266	91	161	518	611	51.3	17.6	31.1	39.6

[a] Dionex 2120: ion chromatograph with 0.001 M HCl eluant, 0.8 ml/min flow, AS-3 separator, Ag-H+ resin suppressor, and conductivity detector.
[b] Astro resources model 1850.
[c] Full-length Pyrex glass liner with bottom in stainless steel-316 1-l reactor.

carburized reactors suggests that more chloride reacts with the fresh-metal surfaces. Acetic acid concentrations decreased and propionic acid concentrations increased slightly in the carburized reactors of both alloys (Table 2). Butyric acid concentrations increased significantly in the carburized stainless steel-316 reactor, but decreased slightly in the carburized Hastelloy C-276 reactor. Although the differences between the fresh-metal and carburized Hastelloy C-276 reactors may not be significant, it is evident from the data in Table 2 that under either condition this alloy has less effect on the amount and distribution of acids than either fresh-metal or carburized stainless steel-316.

Lewan (1993a) noted that the amount of H_2S present in the gases generated by the experiments conducted in stainless steel-316 reactors was inversely proportional to the amount of pyrrhotite that formed on the reactor walls. Pyrrohite coatings were most apparent on the fresh-metal reactor and least apparent on the carburized reactor. The intermediate degree of pyrhotite coating and notable deterioration of gold plating in the gas-phase portion of the gold-plated reactor suggest the existence of an imperfect gold barrier between the underlying fresh-metal stainless steel-316 and dissolved acids. As a result, the amounts and proportionality of the acids in the gold-plated reactor are intermediate to the Hastelloy C-276 reactors and the carburized stainless steel-316 reactor (Table 2). Significant quantities of gold (i.e., 2538 to 7224 mg/kg H_2O) are soluble in near-neutral, H_2S-enriched waters at temperatures between 300 and 350 °C (Shenberger and Barnes 1989). This high susceptibility to dissolution under commonly employed experimental conditions and the ability of some types of sedimentary organic matter to generate high concentrations of H_2S make the utility of gold-plated stainless steel-316 reactors uncertain and less advantageous than Hastelloy C-276 reactors.

Some investigators have conducted their experiments in borosilicate glass tubes under the presumption of minimizing catalytic surface effects (Tannenbaum and Kaplan 1985). Contrary to this presumption, data on the use of borosilicate glass tubing for reactors are more detrimental to simulating the natural system than the use of carburized metal reactors (Lewan 1993a). Palmer and Drummond (1986) showed that the established first-order rate of acetic degradation is altered to a zero-ordered rate when experiments are conducted in fused quartz or borosilicate glass tubes. In addition, Dawidowicz et al. (1984, 1986) have shown that boron within silica glass leads to the formation of unreactive carbon during the heating of alcohols. The amount of unreactive carbon formed was found to be proportional to the amount of boron in the silica glass. Lewan (1993a) suggested this process to be responsible for the reduced yields of expelled oil obtained in an experiment using a borosilicate glass liner in a metal reactor. The glass liner recovered at the conclusion of this experiment was completely frosted and contained carbon deposits where the expelled-oil layer occurred at the experimental temperatures. Dissolved inorganic species in the waters recovered from this experiment are similar to the waters from the metal reactors, with the exception of anomalously high boron and elevated silicon concentrations (Table 1). However, the dissolved organic species show significant reductions in propionic and butyric acids, and an increase in acetic acid relative to the waters from the metal reactors (Table 2). These data and the data from Palmer and Drummond (1986) clearly discourage the use of borosilicate glass in simulating the natural system at temperatures in excess of 300 °C.

2.5 Partitioning Between Oil and Water

The dichotomous chemical character of most organic acids, resulting from their hydrophobic hydrocarbon group and hydrophillic hetero-atom acid group, allows them to occur in varying degrees as dissolved species in both water and organic liquids. Their distribution as dissolved species in coexisting immiscible organic liquid and water phases is expressed as a dimensionless ratio referred to as the partition coefficient (Leo et al. 1971). This parameter is conventionally defined as the concentration of a solute in an organic liquid divided by its concentration in a coexisting water phase. Partition coefficients for organic acids may vary significantly depending on structural configurations and the relative abundance of hetero-atom acid groups to hydrocarbon groups. Figure 1 shows that as the hydrocarbon group of monocarboxylic acids increase in length with addition of aliphatic carbons, there is an increased preference for their occurrence in organic liquids rather than water. These partition coefficients indicate that C_1 through C_4 monocarboxylic acids will preferentially concentrate in water rather than coexisting oils, benzene, or chloroform. Conversely, monocarboxylic acids with five or more carbon atoms will preferentially concentrate in the organic liquids rather than in water. Organic liquids that are more polar in character have higher partition coefficients for the same acids are shown in Fig. 1 by the greater preference for C_3 and C_4 monocarboxylic acids to occur in octanol than in coexisting water.

Hydrous pyrolysis experiments generate two types of organic liquids: (1) bitumen that is retained within the rock and (2) expelled oil that accumulates on the water surface. The expelled oil is rich in hydrocarbons and is likely to have partition coefficients for organic acids similar to those for benzene, chloroform, or oils, as shown in Fig. 1. Bitumen retained in the rock is more polar in character and is likely to have partition coefficients for organic acids between those for octanol and chloroform (Fig. 1). Most hydrous pyrolysis studies conducted to date only determined the amount of organic acids dissolved in the recovered water at room temperature. Depending on the final ratio of organic liquids (bitumen + oil) to water, the results obtained from this procedure may be misleading. This dependence is illustrated in Fig. 2 for acetic, propionic, and butyric acids. As the mass ratio of organic

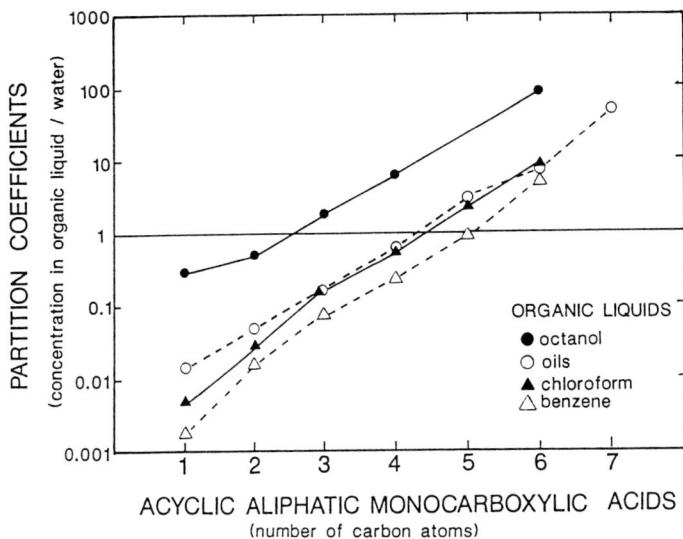

Fig. 1. Partition coefficients of acyclic aliphatic monocarboxylic acids dissolved in organic liquids and coexisting water at 25 °C. Values are from Leo et al. (1971)

liquid to water increases, the percent of the total organic acid dissolved in the water diminishes significantly. This decrease becomes more apparent as the number of carbon atoms in the organic acid increases. At an expelled oil/water ratio of one, only 95% of the acetic acid, 85% of the propionic acid, and 62% of the butyric acid are present in the water at room temperature. This effect becomes more pronounced with bitumen, which would give results somewhere between the octanol and chloroform lines in Fig. 1. It is apparent from Fig. 2 that the amounts as well as the proportionality of the organic acids may be influenced at organic liquid/water ratios greater than 0.1. Therefore, partitioning of dissolved organic acids between organic liquids and water is an important consideration in designing experiments and in interpreting experimental results.

2.6 Pressure and Water Chemistry

Other experimental considerations that may prove to be important in simulating natural organic-acid generation include pressure and water chemistry. Experimental studies have shown pressure (Michels et al. 1992) to have significant effects on petroleum generation. Most experiments concerned with organic-acid generation use pressures generated by the vapor pressure of water at the experimental temperature. An exception is the experimental study by Knauss et al. (1992) in which petroleum source rocks were heated at 330 °C with water in a flexible gold-bag reactor at 30 and 60 MPa for

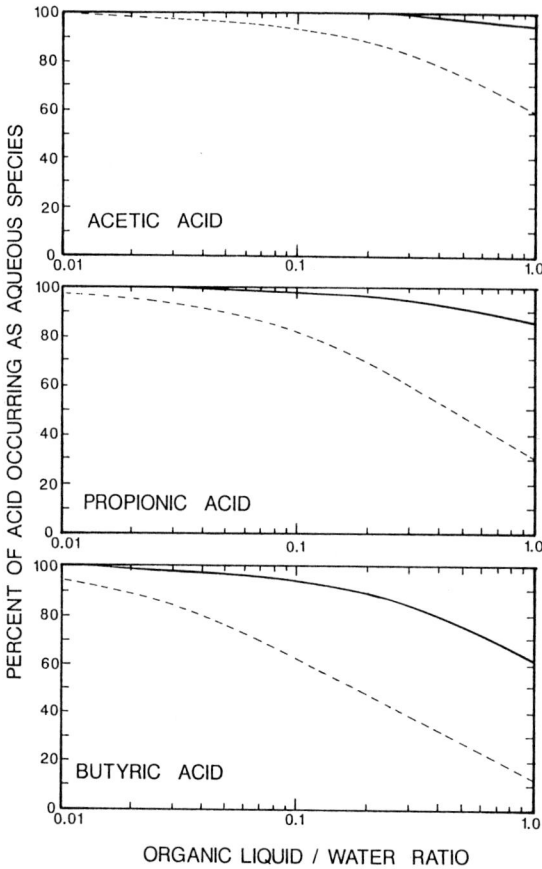

Fig. 2. Graphs showing influence of organic liquid/water ratios on percent of total acetic, propionic, and butyric acids dissolved in water coexisting with oils (*solid curves*) or octanol (*dashed curves*). Curves are calculated from partition coefficients given by Leo et al. (1971)

durations up to 65 days. No significant variations were observed in the concentrations of acetic and propionic acids over this pressure difference. Water chemistry has been shown to significantly enhance petroleum formation (Soldan and Cerqueira 1986), but most experiments concerned with organic-acid generation use distilled or fresh water. An exception is the experimental study by Thornton and Seyfried (1987) in which organic-rich sediment was heated with its original seawater at 350 °C in a flexible titanium reactor with a fixed pressure of 50 MPa. Acetic acid was released under these conditions, but no comparative experiments with different water chemistries were conducted to evaluate their effect. The importance of water chemistry on organic-acid generation remains to be determined.

3   Acid Types and Quantities

Organic acids detected in waters associated with petroleum may be classified as aliphatic carboxylic acids or aromatic carboxylic acids. With respect to generation of these acids from thermally maturing sedimentary organic matter, the aliphatic carboxylic acids have been the most studied. These acids may consist of straight, branched, or cyclic aliphatic groups that are either saturated or unsaturated and contain one or more carboxylic acid groups. Acyclic, saturated, mono- and di-carboxylic acids are the most commonly detected and emphasis will be placed on these acid types.

3.1   Saturated Acyclic Aliphatic Monocarboxylic Acids

These acids are described by their number of carbon atoms and structural configuration (i.e., presence or absence of branch chains). Straight-chain acids with 1 to 32 carbon atoms are the most common acids of this type that are generated from experimentally heated sedimentary organic matter. C_{12} to C_{32} acids of this type in surface sediments of Lake Biwa have been monitored through a series of isothermal heating experiments at temperatures from 68 to 325 °C for 24-h durations (Kawamura and Ishiwatari 1985b). As shown in Fig. 3, there is an initial decrease in their total concentration up to 100 °C, which is attributed to their assimilation into insoluble organic matter during early diagenetic formation of kerogen (Kawamura and Ishiwatari 1985a). At experimental temperatures in excess of 100 °C, these assimilated acids once again become extractable as the kerogen thermally decomposes. The total concentration of these acids continues to increase beyond their original concentration with increasing temperatures until a constant concentration is maintained between 179 and 279 °C. This concentration level is followed by another increase in concentrations at 300 and 325 °C. Assuming a 65 wt% carbon content for the original organic matter, the maximum yield of these acids accounts of 0.83 wt% of the organic matter. Although these long-chain monocarboxylic acids in rock bitumens and crude oils have been used to evaluate sources and transformations of sedimentary organic matter (Parker 1969), they are not likely to be significant in geologic processes mediated by water. Large partition coefficients (>1; Fig. 1), resulting from the large aliphatic groups in these acids, limit their occurrence to source-rock bitumens and expelled oils of a source rock. Concentrations of $C_{14}-C_{30}$ aliphatic monocarboxylic acids have been reported in subsurface waters associated with petroleum accumulations, but their concentrations seldom exceed one μmol/l (Cooper and Bray 1963).

C_2 to C_8 acids of this type are also generated from sedimentary organic matter by hydrous pyrolysis at temperatures from 250 °C for 72 h to 330 °C for 168 h (Eglinton et al. 1987). These experiments used kerogen isolated

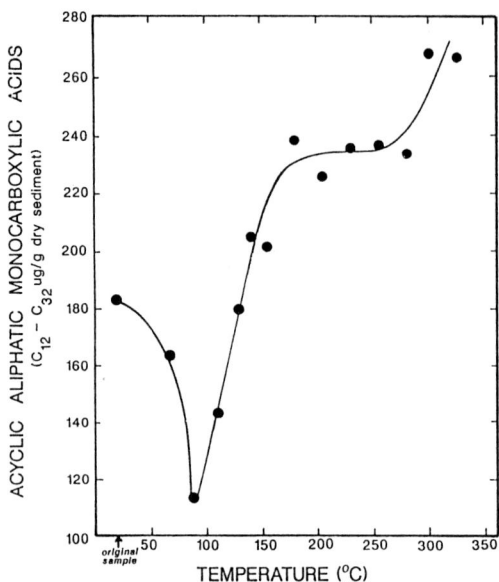

Fig. 3. Concentration of $C_{12}-C_{32}$ acyclic aliphatic carboxylic acids extracted from aliquots of Lake Biwa surface sediments after isothermal heating at temperatures from 68 to 325 °C for 24 h. (Kawamura and Ishiwatari, 1985a)

from an oil shale in the Kimmeridge Clay Formation. A maximum yield of C_2 to C_8 monocarboxylic acids from this kerogen was reached at 330 °C after 168 h. At this maximum, 0.66 wt% of the original kerogen was converted to straight- and branched-aliphatic C_2-C_8 monocarboxylic acids. Straight-chained C_2-C_5 acids comprise nearly 95% of this maximum yield, with branched C_4 and C_5 acids comprising most of the remaining 5%. Small partition coefficients (<1; Fig. 1) for these C_2-C_5 acids insure their occurrence as aqueous species that may be significant in geological processes mediated by water. C_2-C_5 acids are the prevalent aliphatic monocarboxylic acids detected in subsurface waters associated with petroleum accumulations (Carothers and Kharaka 1978; Fisher 1987). Although formic acid (C_1) is also favored to occur as an aqueous species (Fig. 1), it only occurs as a minor component in waters reacted with source rocks during hydrous pyrolysis (Lundegard and Senftle 1987) and in subsurface waters associated with petroleum accumulations (MacGown and Surdam 1988).

Acetic acid (C_2) is typically the most abundant of the monocarboxylic acids and comprises more than 50 mol% of the total acids generated from source rocks during hydrous pyrolysis. Barth et al. (1989) conducted a series of hydrous pyrolysis experiments on aliquots of a powdered shale sample from the Kimmeridge Clay Formation. The aliquots were isothermally heated for 72 h at temperatures ranging from 200 to 350 °C. Acetic acid in the reacted waters increased with increasing temperature. The percent of acetic acid comprising the total acid concentration also increased from 56% at temperatures between 200 and 270 °C to 65% at temperatures between 330 and 350 °C. The maximum amount of kerogen converted to acetic acid was 1.06 wt% at 350 °C after 72 h. Compared with other experimental results (Table 3), maximum conversion of organic matter to acetic acid usually requires experimental temperatures equal to or in excess of 350 °C for 72-h durations. Maximum generation of expelled oil also occurs under these experimental conditions, which will be elaborated on later.

Notable exceptions to the results presented in Table 3 are from hydrous pyrolysis experiments conducted on peats by Pickering and Batts (1992a). Forest and grass peats were heated isothermally for 72 h at temperatures between 95 and 330 °C. Maximum yields of acetic acid occurred at 200 °C with organic matter conversions of 0.15 wt% for the forest peat and 1.18 wt% for the grass peat. In addition to these maximum yields occurring at an anomalously low temperature, the distribution of the C_1–C_5 carboxylic acids was peculiar and not representative of subsurface waters. The weight percentages of the total C_1–C_5 carboxylic acids were 1.1 to 4.9 for formic acid, 9.3 to 12.1 for acetic acid, 32.5 to 64.4 for propionic acid, 6.2 to 24.1 for normal- and iso-butric acids, and 16.2 to 29.2 for normal- and iso-valeric acids. Pickering and Batts (1992b) also reported peculiar distributions with anomalously high percentages of valeric acid from hydrous pyrolysis experiments on Australian coals at 120 °C for 72-h durations. No explanations for these anomalous distributions are given by the investigators. If these peculiar distributions are not a result of the neutralization procedure (i.e., addition of NaOH) imposed on the sample-water system before the experiments, then they may indicate that some coals are not typical sources of these acids in subsurface waters.

The maximum conversions of organic matter to acetic acid in Table 3 are variable. Maximum conversion to acetic acid varies for coals from 0.45 to 2.01 wt% and for marine source rocks from 0.51 to 2.05 wt% (Table 3). All the experiments in Table 3, except for those by Cooles et al. (1987), were conducted in stainless steel-316 reactors, and therefore, the influence of reactor wall composition is not likely to be responsible for these variations. Another factor unlikely to be responsible for this variation is mineral oxides within the rocks. Eglinton et al. (1987) enhanced the generation of acetic acid nearly fourfold by including limonite ($HFeO_2$) with isolated kerogen in a hydrous pyrolysis experiment at 330 °C for 72 h. Kerogen oxidation by oxygen from the reduction of ferric oxide during the experiments is the most likely cause of this enhanced generation of organic acids. The problem with applying this observation to petroleum source rocks like those given in Table 3 is that ferric oxides are not present in the reducing depositional environments under which sedimentary organic matter accumulates. Under these conditions, reduced species like ferrous sulfides (e.g., pyrite) occur rather than ferric oxides. The only exception is if weathered source rock samples containing hematite or jarosite were used in the experiments.

Table 3. Experimental conditions and the weight percent of original organic matter converted to acetic, propionic, butyric, and valeric acids at the maximum for acetic acid generation from organic matter in samples of whole rock (WR), solvent-extracted rock (ER) and isolated kerogen (IK)

Rock unit	Sample type	Kerogen type	Range of conditions	Conditions for maximum acetic acid	Acetic acid (% org. matter)	Propionic acid (% org. matter)	Butyric acid (% org. matter)	Valeric acid (% org. matter)	Total (% org. matter)
Green River Shale[a] (Mahogany Zone)	WR	I	260–361°C/ 72–144 h	361°C/72 h	0.53	0.18	0.21	0.05	0.97
Kimmeridge Clay[b] (Blackstone Band)	IK	II	250–330°C/ 7.2–168 h	330°C/168 h	0.47	0.14	0.05	0.01	0.67
Kimmeridge Clay[a] (Blackstone Band)	WR	II	300–366°C/ 72 h	350°C/72 h	0.51	0.15	0.08	n.d.	0.74
Kimmeridge Clay[c] (Oil Shale)	ER	II	200–350°C/ 72 h	350°C/72 h	1.06	n.r.	n.r.	n.r.	1.06
Kimmeridge Clay[d] (Oil Shale)	WR	II	250–350°C/ 72 h	340°C/72 h	0.73	0.20	n.r.	n.r.	0.93
Phosphoria Fm.[a] (Retort Shale)	WR	II	240–361°C/ 72 h	355°C/72 h	0.51	0.19	0.21	0.03	0.94
Monterey Shale[a] (Phosphatic mb.)	WR	II	240–360°C/ 72 h	355–360°C/ 72 h	1.35	0.31	0.17	0.03	1.86

Kreyenhagen Shale[a] (Siliceous unit)	WR	II	270–365 °C/72 h	360 °C/72 h	2.19	0.38	0.09	0.08	2.74
Non-Marine Shale[e] (TOC = 4.41%)	ER	II	250–350 °C/72 h	350 °C/72 h	0.54	0.16	0.04	n.r.	0.74
Non-Marine Shale[e] (TOC = 7.71%)	ER	II/III	250–350 °C/72 h	350 °C/72 h	0.36	0.09	0.02	n.r.	0.47
Non-Marine Shale[e] (TOC = 4.05%)	ER	III/II	250–350 °C/72–360 h	350 °C/72 h	0.71	0.17	n.d.	n.r.	0.88
Norwegian Coal[f] (Lower Jurassic)	WR	III	290–350 °C/72 h	350 °C/72 h	0.45	n.r.	n.r.	n.r.	0.45
Australian Coal WR[d] (Cretaceous-Paleogene)	WR	III	300–350 °C/72 h	350 °C/72 h	2.01	0.63	n.r.	n.r.	2.64

n.r., not reported; n.d., not detected.
[a] Authors' data.
[b] Eglinton et al. (1987).
[c] Barth et al. (1989).
[d] Cooles et al. (1987).
[e] Lundegard and Senftle (1987).
[f] Barth et al. (1987).

Barth and Bjørlykke (1993) used mulivarient statistics in an attempt to establish correlations between acetic acid yields from hydrous pyrolysis experiments and the organic carbon contents, open-system pyrolysis data (i.e., Rock-Eval parameters), and vitrinite reflectances of the pyrolyzed source rocks. As reported by these investigators, the principal component analysis was inconclusive. Other investigators have suggested a general relationship between the original oxygen content of the organic matter and the amount of acetic acid generated (Cooles et al. 1987; Lundegard and Senftle 1987). Although high acetic acid concentrations are usually generated from oganic matter with high oxygen contents, all organic matter with high oxygen contents does not generate high acetic acid concentrations (Eglinton et al. 1987). Data from hydrous pyrolysis experiments conducted in the authors' laboratory on Phosphoria Retort, Monterey, Kimmeridge Clay, and Kreyenhagen Shales (Table 3) indicate that maximum yields of C_2–C_5 monocarboxylic acids only account for 5 to 10 mol% of the oxygen in the original kerogens. A more direct control on the amount of kerogen converted to acetic acid is the amount of acetic acid originally incorporated into the organic matter during early diagenesis, when insoluble kerogen is forming. As previously discussed, C_{12}–C_{32} monocarboxylic aliphatic acids are initially assimilated into insoluble organic matter (i.e., kerogen) of recent sediments at experimental temperatures below 100°C (Fig. 3). Therefore, the amount of acetic acid available during early diagenesis may dictate the amount of acetic acid that can be generated from a kerogen during the latter stages of diagenesis and petroleum formation. Availability of assimilable acetic acid would depend on the balance between their generation and consumption by microbial activity in the original sediment.

The bonding types responsible for the assimilation of acetic acid into the insoluble kerogen during early diagenesis are not known. However, the low activation energy of 5.4 kcal/mol determined by Barth et al. (1989) for the generation of acetic acid from an oil shale indicates weak noncovalent bonding. It has been well documented that prior to petroleum formation, the oxygen content of a kerogen decreases relative to its carbon content (Tissot et al. 1974). This relationship has been expressed on a van Krevelen diagram in which the atomic H/C ratio of a kerogen is plotted against its atomic O/C ratio (Fig. 4). In addition to loss of CO_2 through decarboxylation, loss of acetic acid through the cleavage of noncovalent bonds may also contribute to the initial decrease in the atomic O/C ratio. As a result, the amount of acetic acid generated by a kerogen in hydrous pyrolysis experiments may depend on its thermal maturity with respect to its atomic O/C ratio prior to experimentation. Figure 4 shows that the rocks with the most thermally immature Type II kerogens (i.e., Monterey Shale and Kreyenhagen Shale), based on high atomic O/C ratios, have generated the most acetic acid during the hydrous pyrolysis experiments (Table 3).

The resulting implication is that the maximum amounts of acetic acid generated from sedimentary organic matter are best assessed in hydrous

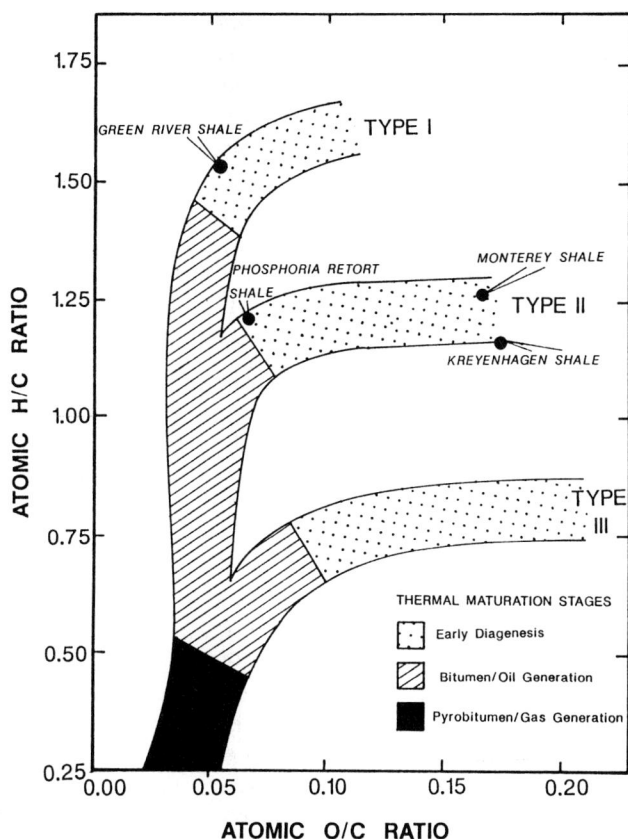

Fig. 4. Modified van Krevelen diagram after Tissot et al. (1974) showing the three major kerogen types based on their atomic hydrogen/carbon and oxygen/carbon ratios. *Labeled points* refer to kerogens isolated from rocks used to study organic acid generation by hydrous pyrolysis in the authors' laboratory

pyrolysis experiments conducted with samples that are thermally immature relative to the oxygen loss during early diagenesis. Samples of Paleozoic age, like the Phosphoria Formation (Permian) in Fig. 4, are immature with respect to petroleum generation (Lewan 1985), but thermally mature with respect to oxygen loss during early diagenesis. It should not be inferred from this implication that all kerogens with high atomic O/C ratios will generate large amounts of acetic acid, but rather these kerogens provide the best assessment of maximum acetic acid generation from maturing source rocks. Accordingly, maxima for acetic acid generation from Type II kerogen are best represented by the experiments on samples from the Monterey Shale and Kreyenhagen Shale, which are 1.35 and 2.17wt% (Table 3), respectively.

Although these conversions of sedimentary organic matter to acetic acid are an acceptable maximum range for Type-II kerogens, sufficient experimental data are not available to assess the maximum range of conversion for Type-III and Type-I kerogens. Elemental analyses (i.e., C, H, O, and N) are not given for the Type-III kerogens of the two coal samples studied (Table 3), but the Australian coal generates the most acetic acid with a conversion of 2.01 wt%. This amount of conversion may not be representative of a maximum for Type-III kerogen, because the vitrinite reflectance of 0.4% Ro given for this coal (Cooles et al. 1987) suggests that it is of sub-bituminous C rank. More representative maximum conversions for Type-III kerogens are most likely to be obtained from future hydrous pyrolysis experiments on lignites, which are more thermally immature with respect to oxygen loss during diagenesis. Similarly, more experimental work is needed to determine the maximum conversions of Type-I kerogens to acetic acid. A complete set of hydrous pyrolysis data is only available on one source rock bearing Type-I kerogen (i.e., Green River Shale), and the original sample is not immature with respect to oxygen loss during early diagenesis (Fig. 4).

Although all of the C_2 through C_5 monocarboxylic acids typically increase with increasing thermal stress (Fig. 5), the amount each acid increases is not always at a constant proportion to the other acids. This variable proportionality among the C_2 through C_5 acids with increasing thermal stress is shown in Fig. 6. Acetic acid becomes more dominant relative to the other acids with increasing thermal stress for the Kreyenhagen Shale. Conversely, butyric acid plus valeric acid becomes more dominant relative to the other acids with increasing thermal stress for the Phosphoria Retort Shale. Intermediate between these two opposite trends are the acids generated from the Monterey Shale, which show no change in proportionality among one another with increasing thermal stress. One explanation for these differences (Fig. 6) may be the preferential loss of acetic acid and to a lesser extent propionic acid during the early diagenetic loss of oxygen, which is at a higher level (i.e., lower atomic O/C ratio) for the original organic matter in the Phosphoria Retort Shale than in the Kreyenhagen Shale (Fig. 4). Implications of this explanation are discussed in the following section concerning generation and expulsion of organic acids.

3.2 Saturated Dicarboxylic Aliphatic Acids

These acids consist of a saturated aliphatic group bearing two carboxylic groups. The saturated aliphatic group may have a straight, branched, or cyclic structural configuration. Straight and to a lesser extent branched configurations are the most commonly reported acids of this type generated in laboratory heating experiments involving sedimentary organic matter. Similar to monocarboxylic aliphatic acids, high-molecular-weight dicarboxylic

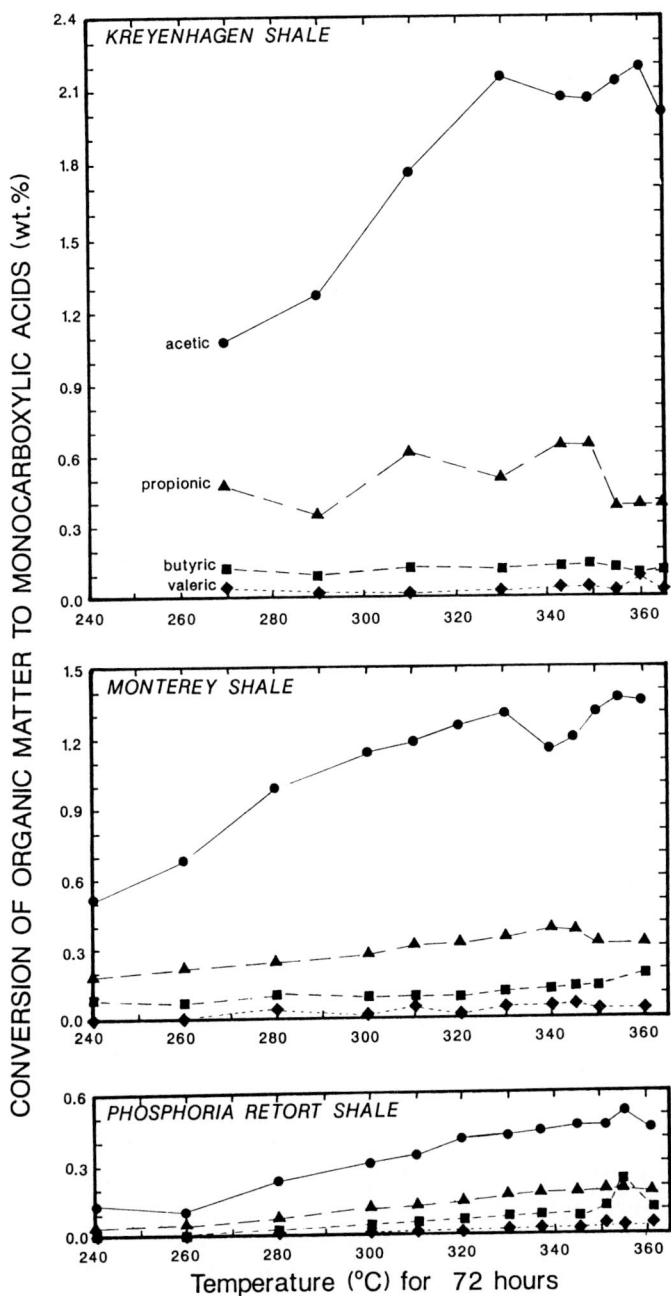

Fig. 5. Weight percent of original organic matter converted to aqueous C_2-C_5 monocarboxylic acids after hydrous pyrolysis of aliquots of Kreyenhagen, Monterey, and Phosphoria Retort shales isothermally heated at temperatures from 240 to 365 °C for 72-h durations. Experiments were conducted in carburized stainless steel-316 reactors in the authors' laboratory

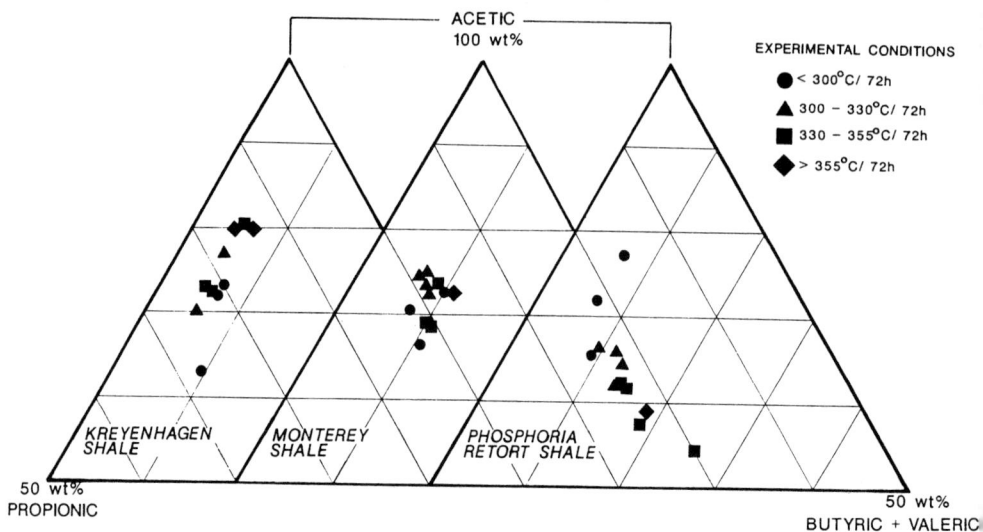

Fig. 6. Ternary diagrams showing the proportionality of C_2–C_5 monocarboxylic acids generated by hydrous pyrolysis of Kreyenhagen, Monterey, and Phosphoria Retort shales as shown in Fig. 5

aliphatic acids with 9 to 30 carbons have been generated from Lake Biwa surface sediments in laboratory heating experiments (Kawamura and Ishiwatari 1985b). Isothermal experiments conducted for 24-h durations generate a steady increase of these acids from 129 °C to a maximum at 279 °C, which is followed by a decrease at 300 and 325 °C. The maximum amount of organic matter converted to C_9–C_{30} dicarboxylic aliphatic acids is 0.11 wt%. It is unlikely that this amount of high-molecular-weight dicarboxylic acids would have a role in geological processes mediated by water because of their positive partition coefficients between organic liquids and water (Leo et al. 1971). Expelled oil and source-rock bitumen would retain most of the C_9 through C_{30} dicarboxylic aliphatic acids with only trace amounts in the associated formation water.

Conversely, the negative partition coefficients for the low-molecular-weight dicarboxylic acids with 2 through 9 carbon atoms makes them more relevant to geological processes mediated by water. Although these C_2 through C_{10} acids have been reported in waters heated with sedimentary organic matter during hydrous pyrolysis experiments (Eglinton et al. 1987), their thermal stability is less than that of the monocarboxylic aliphatic acids (Crossey 1991). Hydrous pyrolysis experiments on Type-II kerogen isolated from the Kimmeridge Clay Formation yielded maximum dicarboxylic acid concentrations at 250 °C after 720 h (Eglinton et al. 1987). Kawamura and Kaplan (1987) also generated dicarboxylic acids with 2 through 10 carbon atoms from sedimentary organic matter in hydrous pyrolysis experiments at

Table 4. Experimental conditions and weight percent of original organic matter converted to oxalic (C_2), succinic (C_4), pyrotartaric (C_5), glutaric (C_5), iso-adipic (C_6), adipic (C_6), pimelic (C_7), suberic (C_8), azelaic (C_9), and sebacic (C_{10}) acids at maximum generation of total dicarboxylic aliphatic acids

Rock unit	Kimmeridge Clay[a]	Monterey Shale[b]	Green River Shale[b]	Tanner Basin Recent Sediment[b]	Recent Bog Sediment[b]
Sample type	Isolated kerogen	Isolated kerogen	Isolated kerogen	Isolated kerogen	Extracted humic acid
Kerogen type	II	II	I	II(?)	III(?)
Range of conditions	250–330 °C/ 7.2–168 h	270 °C/2 h	270 °C/2 h	270 °C/2 h	270 °C/2 h
Conditions at maximum generation	250 °C/720 h	270 °C/2 h	270 °C/2 h	270 °C/2 h	270 °C/2 h
Oxalic acid (% org. matter)	n.r.	0.152	0.238	0.895	0.844
Malonic acid (% org. matter)	n.r.	n.d.	n.d.	n.d.	n.d.
Succinic acid (% org. matter)	0.016	0.032	0.050	0.444	0.256
Pyrotartaric acid (% org. matter)	0.006	0.035	0.057	0.272	0.073
Glutaric acid (% org. matter)	0.014	0.006	0.010	0.055	0.025
Iso-adipic acid[c] (% org. matter)	n.r.	0.007	0.010	0.016	0.005
Adipic acid (% org. matter)	n.d.	0.002	0.004	0.016	0.006
Pimelic acid (% org. matter)	0.001	n.r.	n.r.	n.r.	n.r.
Suberic acid (% org. matter)	0.019	0.004	0.026	0.027	0.022
Azelaic acid (% org. matter)	0.002	n.d.	0.004	0.019	n.d.
Sebacic acid (% org. matter)	tr.	n.d.	0.003	0.008	n.d.
Total dicarboxylic aliphatic acids (% org. matter)	0.058	0.238	0.402	1.752	1.231

n.r., not reported; n.d., not detected
[a] Eglinton et al. (1987).
[b] Kawamura and Kaplan (1987).
[c] 2,3-Dimethyl succinic acid.

270 °C after 2 h. The results of these experiments (Table 4) indicate that oxalic acid accounts for more than half of this type of acid.

Similar to the monocarboxylic aliphatic acids, the most thermally immature samples should provide the best assessment of maximum conversion of organic matter to dicarboxylic aliphatic acids. The assumed Type-II kerogen from recent sediment in the Tanner Basin is the most thermally immature and generates the highest yield. However, some caution is warranted in using this yield as a maximum because the organic matter in this unconsolidated sediment may undergo further microbial alteration before it is buried in the subsurface. In addition, this value and the others given by Kawamura and Kaplan (1987) were determined at only one experimental condition (i.e., 270 °C/2 h), which may not be representative of maximum yields attainable under other experimental conditions. Although concerns over this limited data inhibit establishing a maximum conversion of organic matter to dicarboxylic aliphatic acids, it may be stated that at least 0.40 wt% of Type-I kerogen and 0.24 wt% of Type-II kerogen may be converted to C_2-C_{10} dicarboxylic acids.

Minimum conversions of Type-III kerogen to dicarboxylic acids are more difficult to determine. The high conversion of 1.23 wt% given in Table 4 for a humic acid is not representative because this organic extract may only account for 2 to 42 wt% of Type-III kerogens in peats (Vandenbroucke et al. 1985). Similar to the peculiar results on monocarboxylic acids, the same experimental data reported by Pickering and Batts (1992a) on maximum conversions of peats to dicarboxylic acids also occurred at the anomalously low temperature of 200 °C for a 72-h duration. These maximum conversions of extremely immature Type-III kerogens to C_2-C_4 dicarboxylic acids are 0.49 wt% for forest peat and 2.96 wt% for grass peat. Unlike distributions in subsurface waters, succinic acid (C_4) accounted for more than 62 wt% of these maximum yields. Anomalously high concentrations of succinic acid were also observed in their experiments on Australian coals (Pickering and Batts 1992b). More experimental work is clearly needed to fully appreciate the experimental results reported by these investigators.

3.3   Unsaturated Aliphatic Carboxylic Acids

These acids have an aliphatic group that contains one or more double carbon bonds, which determines the degree of unsaturation. More than half of the aliphatic carboxylic acids in most living organisms are unsaturated (Parker 1969). However, double carbon bonds are highly reactive sites and at burial of less than 1-m unsaturated carboxylic acids are typically absent in recent sediments (Rosenfield 1948; Parker and Leo 1965; Rhead et al. 1971). This loss of unsaturated carboxylic acids at low temperatures is considered to be related to bacterial activity within sediment pore waters (Johnson and Calder 1973). Some of these acids survive early diagenesis as indicated by their occasional detection in some sedimentary rocks (Parker 1969). However, their generation from organic matter or persistence with increasing burial depth into the subsurface is not likely to be significant due to their low thermal stability.

Kawamura and Ishiwatari (1985a,b) showed that unsaturated aliphatic carboxylic acids with 16 and 18 carbon atoms rapidly decreased in concentration from samples of Lake Biwa surface sediments with increasing temperatures from 68 to 279 °C for 24 h. Unsaturated C_{18} carboxylic acids were undetectable at 279 °C after 24 h if they contained one double carbon bond, at 229 °C after 24 h if they contained two double carbon bonds, and at 204 °C after 24 h if they contained three double carbon bonds. This decrease in thermal stability with the increasing number of double carbon bonds in the aliphatic group of carboxylic acids is also supported by laboratory heating experiments conducted on green algae (Abelson 1962). In these experiments, the di- and tri-unsaturated C_{18} carboxylic acids were absent after 18 days at 190 °C, but the mono-unsaturated C_{18} carboxylic acid was present.

Two unsaturated aliphatic carboxylic acids generated from sedimentary organic matter in laboratory heating experiments are fumaric acid and maleic acid (Kawamura and Kaplan 1987). Both of these dicarboxylic acids have four carbon atoms with the former having a *trans* configuration and the latter having a *cis* configuration. Fumaric acid is generated in significantly higher concentrations than maleic acid, and accounts for 0.35 to 0.58 wt% conversion of organic matter from Recent sediments (Tanner Basin) and humic acids, respectively at 270 °C after 2 h (Kawamura and Kaplan 1987). This unsaturated acid and its isomer are essential to the metabolism of most plants and animals, which makes its survival past microbial activity during early diagenesis unlikely. This inference is supported in part by the low conversions of 0.007 to 0.008 wt% of organic matter from indurated rock samples of the Monterey and Green River shales, which have passed through early diagenesis (Kawamura and Kaplan 1987). Some analyses of subsurface waters have reported the occurrence of maleic acid but not fumaric acid (Surdam et al. 1984) despite its higher generation yields from heated sedimentary organic matter. This discrepancy may be related to maleic acid being more soluble in water by two orders of magnitude (March 1985, p. 112) and having a significantly lower partition coefficient than fumaric acid (Leo et al. 1971).

3.4 Aromatic Carboxylic Acids

These acids consist of one or more carboxyl groups bound to an aromatic hydrocarbon. The simplest of these acids is benzoic acid (C_6H_5COOH), which has been reported in the reacted waters from hydrous pyrolysis experiments on sedimentary organic matter. Kawamura et al. (1986) observed that only 0.006 wt% of isolated kerogen from Green River Shale occurred as benzoic acid in the reacted water after heating at 300 °C for 10 h. Similarly, Eglinton et al. (1987) observed that isolated kerogen from the Kimmeridge Clay Formation generated only trace amounts ($\sim$<0.005 wt%) of benzoic acid in reacted waters after 168 and 720 h at 250 °C. At higher experimental temperatures (i.e., 280 and 330 °C) for 7.2 to 720 h, no benzoic acid was detected in the reacted waters. Hydrous pyrolysis of isolated kerogens in the presence of illite, montmorillonite, and calcite showed no appreciable increase in conversion to benzoic acid after 10 h at 300 °C (Kawamura et al. 1986). However, 0.01 to 0.02 wt% of isolated kerogen converted to benzoic acid when heated in contact with limonite after 7.2 to 720 h at 250 to 330 °C (Eglinton et al. 1987). In addition, minor amounts (0.002 to 0.0066 wt%) of the kerogen were also converted to toluic acid [$H_3C(C_5H_4)CO$] under these conditions. As previously stated, the unlikelihood of ferric oxide minerals occurring under reducing depositional conditions makes this source of aromatic carboxylic acids from petroleum source rocks unlikely.

Other aromatic carboxylic acids that have been reported in trace quantities of subsurface waters associated with petroleum accumulations include phthalic and salicylic acids (Fisher and Boles 1990; Harrison and Thyne 1992). Currently, no heating experiments involving sedimentary organic matter have reported these acids in reacted waters. An important consideration in evaluating the potential of these two acids, as well as benzoic acids, is that their partition coefficients at 25 °C indicate their occurrence is favored in an organic liquid rather than in water (Leo et al. 1971).

4   Organic Acid Generation and Expulsion

Organic acid generation within and expulsion from a petroleum source rock are two additional factors that must be evaluated when considering the effectiveness of these acids as agents of secondary porosity formation. The timing of acid generation and expulsion relative to petroleum generation are of particular importance. If generation of organic acids and petroleum involves independent processes, the experimental observations may not reflect their relative relationships at the higher temperatures and shorter time conditions employed in hydrous pyrolysis experiments. This variation is especially true if the overall reaction rate constants significantly diverge or converge with one another as temperature increases. It is therefore important to understand the relationships between organic acid and petroleum formation under the experimental conditions, and to evaluate these relationships through extrapolations to subsurface conditions in sedimentary basins. Before embarking on this type of evaluation, it is first important to review the current understanding of petroleum formation (i.e., generation and expulsion) as simulated in hydrous pyrolysis experiments.

4.1 Petroleum Formation

An important attribute of hydrous pyrolysis that distinguishes it from an-hydrous pyrolysis (i.e., absence of liquid water) is its ability to generate an immiscible oil phase that is expelled from a source rock in a manner considered operative in subsiding sedimentary basins (Lewan 1987, 1992a,b). The expelled oil accumulates on the overlying water surface within the reactor where it may be quantitatively collected at the end of an experi-ment (Fig. 7). Petroleum formation may be described by this experimental approach in terms of four stages: (1) pre-generation; (2) bitumen gener-ation; (3) oil generation; and (4) pyrobitumen generation (Fig. 8). Organic matter of a source rock in the pre-generation stage consists predominantly (i.e., >95 wt%) of an aggregate of macromolecular organic compounds that are insoluble in organic solvents (e.g., benzene, dichloromethane, and chloroform). This insoluble fraction is referred to as kerogen and the re-maining (i.e., <5 wt%) solvent-soluble fraction is referred to as bitumen. As the thermal stress experienced by a source rock increases, the bitumen-

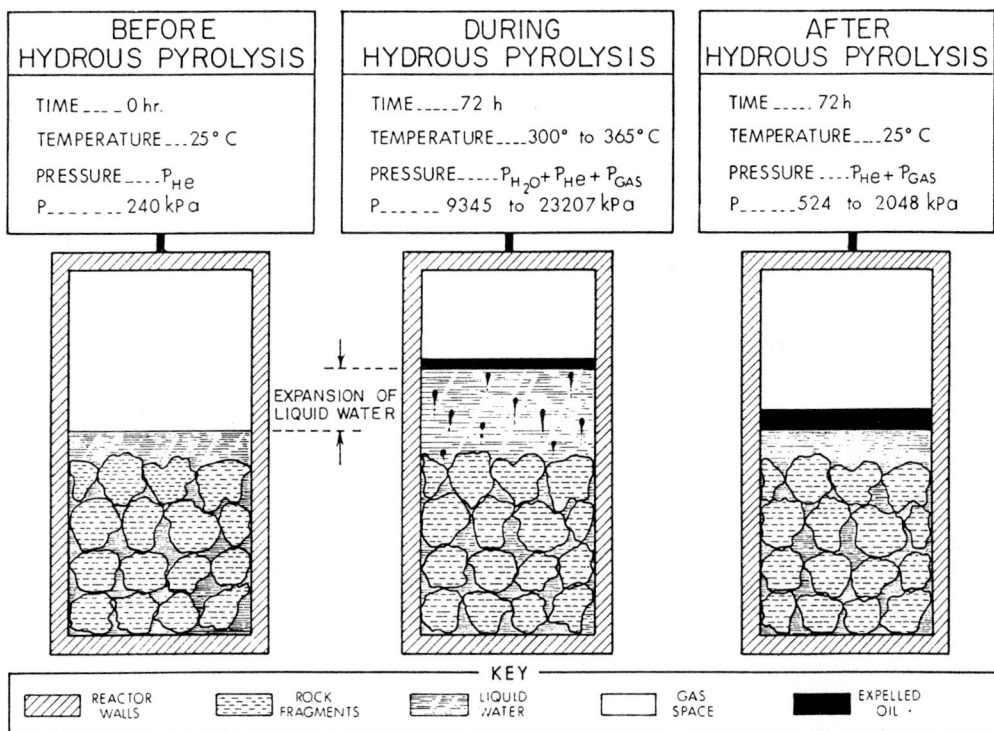

BEFORE HYDROUS PYROLYSIS	DURING HYDROUS PYROLYSIS	AFTER HYDROUS PYROLYSIS
TIME ____ 0 hr.	TIME ____ 72 h	TIME ____ 72 h
TEMPERATURE ... 25° C	TEMPERATURE ____ 300° to 365° C	TEMPERATURE ... 25° C
PRESSURE .... P_{He}	PRESSURE ..... $P_{H_2O} + P_{He} + P_{GAS}$	PRESSURE .... $P_{He} + P_{GAS}$
P ____ 240 kPa	P ____ 9345 to 23207 kPa	P ____ 524 to 2048 kPa

EXPANSION OF LIQUID WATER

KEY — REACTOR WALLS — ROCK FRAGMENTS — LIQUID WATER — GAS SPACE — EXPELLED OIL

Fig. 7. Diagramatic cross section of physical conditions before, during, and after hydrous pyrolysis of a petroleum source rock (Lewan 1992a). Specifics on experimental procedures are given by Lewan (1993a)

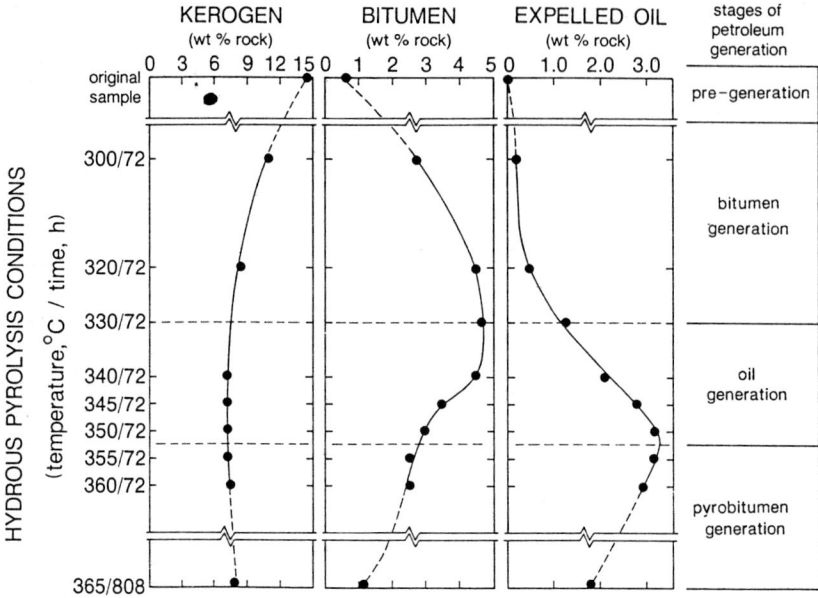

Fig. 8. Changes in kerogen, bitumen, and expelled oil through an isothermal heating series of hydrous pyrolysis experiments conducted on aliquots of Woodford Shale

generation stage commences with the partial decomposition of kerogen to bitumen (Fig. 8). The generated bitumen consists predominantly of high-molecular-weight organics that were originally held in the kerogen by weak noncovalent bonds (e.g., hydrogen bonds, chemical sorption, and electron donor-acceptor complexes). A net volume increase accompanying this over-all reaction causes the generated bitumen to expand into the micropores and along bedding-plane partings to form a continuous bitumen network within the source rock. In addition to displacing some of the interstitial water within the rock, several mole percent of the interstitial water is dissolved in the bitumen network (Lewan 1992a).

The amount of interstitial water that may be dissolved in bitumen in-creases with increasing temperature and is two orders of magnitude greater than the amount of hydrocarbon that may be dissolved in water (Lewan 1992a). A water-saturated bitumen network within a source rock is main-tained during hydrous pyrolysis by the excess water surrounding the rock sample within the reactor. This excess water in the natural system occurs in fractures and faults that dissect source rocks, and in porous strata adjacent to source rocks. As discussed by Lewan (1992a,b), dissolved water is chemically and physically important to the oil-generation stage, which involves partial decomposition of the bitumen network to an expelled immiscible oil (Fig. 8). Data suggest that the dissolved water acts as a hydrogen donor to terminate free-radical sites and as an oxygen donor, to generate excess carbon dioxide (Lewan 1992a). Termination of free-radical sites with water derived hydrogen retards thermal cracking, carbon-carbon bond cross-linking, and aromatization, which reduces pyrobitumen generation (Lewan and Winters 1991). Physically, the dissolved water in the bitumen allows an immiscible oil phase to separate from the water-saturated bitumen network. Under anhydrous conditions, the lack of dissolved water in the bitumen network does not allow formation of an immiscible oil phase.

The cleavage of covalent bonds in the partial decomposition of water-saturated bitumen to an immiscible oil is accompanied by a net volume increase in the resulting organic products (Lewan 1987). If the bitumen network fills the preexisting porosity of a source rock, the volume increase caused by the generation of immiscible oil will not be accommodated within the confining rock matrix. As a result, generated immiscible oil is expelled from the bitumen impregnated rock into the surrounding water and buoys to the water surface within the reactor where it accumulates. It is envisaged in the natural system that the immiscible oil is expelled into preexisting regional fractures that typically dissect source rock. These types of fractures typically occur in orthogonal sets that are parallel to the long and short axes of the basins in which they developed (Nelson 1985). Their ubiquitous character in near-horizontal strata, early development after induration of sediment to rock, and persistence with burial depth (Lorenz et al. 1991) makes regional fracture systems excellent water reservoirs for maintaining water-saturated bitumen networks in maturing source rocks. In addition, these water-filled fractures may serve as conduits for migration of expelled oil by buoyancy through a source rock or into adjacent permeable carrier beds.

The pyrobitumen-generation stage follows the oil-generation stage as thermal stress increases (Fig. 8). Immiscible oil and remaining bitumen from the oil-generation stage thermally decompose to pyrobitumen and gas during this stage. Pyrobitumen is defined as organic matter that has become insoluble in organic solvents as a result of thermal stress. This insolubility is caused by polymerization through carbon-carbon bond cross-linking and subsequent aromatization through disproportionation of cyclic structures. It is not feasible to separate pyrobitumen from matured kerogen, and as a result, there is an apparent increase in kerogen during this stage (Fig. 8). Lewan (1992a,b) has shown that in the absence of liquid water, the oil-generation stage is reduced and the pyrobitumen-generation stage is enhanced.

4.2 Kinetics and Bonding

Acetic acid is the most abundant and studied organic acid generated from source rocks, and therefore it will be the focus of this and subsequent discussions. Although this acid may have several different precursory link-

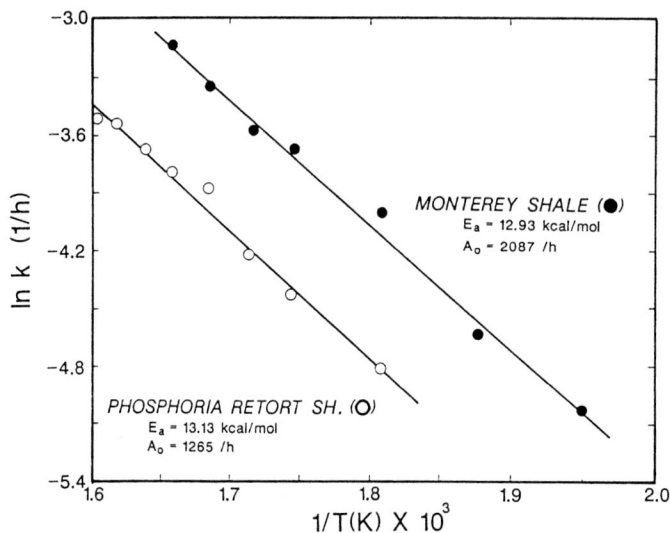

Fig. 9. Arrhenius plot for acetic acid generation from hydrous pyrolysis experiments conducted on the Monterey and Phosphoria Retort shale (Fig. 5). Calculated activation energies (E_a) and frequency factors (A_o) assume a first-order rate law

ages in sedimentary organic matter, its generation in hydrous pyrolysis experiments may initially be treated as one overall reaction. A first approximation of this type is shown by the Arrhenius plots for acetic acid generation from the Monterey and Phosphoria Retort shales in Fig. 9. First-order reaction rates are assumed and calculated on the basis of maximum yields observed in Fig. 5. The resulting linear expressions give similar activation energies (E_a) and frequency factors (A_o) for acetic acid generation from both rocks. Activation energies of 13 kcal/mol are too low to represent cleavage of covalent bonds, which typically have activation energies in excess of 40 kcal/mol for oil generation in hydrous pyrolysis experiments (Lewan 1985; Hunt et al. 1991). It should be noted that these low activation energies for acetic acid generation include energy consumed in transporting acetic acid through the bitumen network of the rock into the surrounding water. Therefore, the activation energies responsible for cleavage of these weak noncovalent bonds are likely to be lower and more in the range of hydrogen bonding for dimeric carboxylic acids (6 to 9 kcal mol^{-1} bond^{-1}; Lewis and Randall 1960). This type of bonding is supported by the low activation energy of 5.4 kcal/mol determined for acetic acid generation from hydrous pyrolysis of Kimmeridgian oil shale by Barth et al. (1989). An integral method that did not require a maximum yield for determining first-order rate laws was used to derive this activation energy. This approach avoids the problem of underestimating maximum yield due to acid destruction, which results in higher activation energies.

(a)

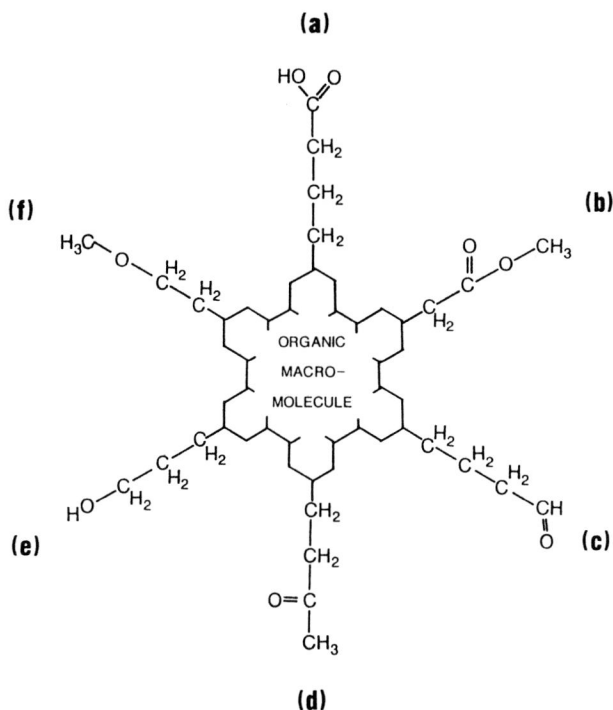

Fig. 10. Diagramatic representation of oxygen-functionalized aliphatic chains covalently bound to an organic macromolecule in bitumen or kerogen: **a** aliphatic carboxylic acid; **b** ester; **c** ketone; **d** aldehyde; **e** alcohol; and **f** ether

An important implication of these low activation energies is that carboxylic aliphatic chains covalently bound to carbon constituents in sedimentary organic matter (Fig. 10a) are not likely to be a source of acetic acid. This unlikelihood is also apparent because the covalent bond between the carboxylic carbon and the adjacent aliphatic carbon is weaker than the covalent bond between aliphatic carbon-carbon bonds in the hydrocarbon chain. Therefore, decarboxylation of the carboxylic group is more likely to occur than the cleavage of aliphatic carbon-carbon bonds. Aliphatic carboxylic-acid side chains may also result from the cleavage of a primary carbon ($\cdot CH_3$) at the end of an ester side chain (Fig. 10b). However, decarboxylation of this carboxylic group to form CO_2 is again more likely than cleavage of the stronger bond between the secondary aliphatic carbons ($\cdot CH_2 \cdot$) to form acetic acid.

Another consideration is the working hypothesis that water dissolved in the bitumen network of a source rock oxidizes carbonyl groups to form carboxylic groups, which decarboxylate with increasing thermal stress to form excess CO_2 (Lewan 1992b). Oxidation of carbonyl groups in aldehyde side chains (Fig. 10c) by water would only result in the formation of a carboxylic aliphatic chain. Similar to preexisting carboxylic-aliphatic and ester side chains, the aldehyde-derived carboxylic aliphatic chains are likely to be only a source of CO_2 and not an aliphtic carboxylic acid. However, oxidation of carbonyl groups in ketone side chains (Fig. 10d) by water may result in the formation of acetic acid. This reaction would occur due to oxidative cleavage of the weaker bond between the carbonyl carbon and their secondary aliphatic carbon ($\cdot CH_2 \cdot$), rather than the stronger bond between the carbonyl carbon and the primary aliphatic carbon ($\cdot CH_3$). Although this form of oxidation by water may source some acetic acid, the lack of an increase in the slope of the Arrhenius plots (Fig. 9) at higher temperatures suggests it is not significant.

It appears that acetic acid, as well as the other short-chain (C_3 to C_5) carboxylic acids, represent initial components formed during early diagenesis that are not covalently bound into condensing higher-molecular-weight organic components. As condensation of macromolecules proceeds to form insoluble organic matter, more favorable sites for hydrogen bonding of the unassimilated short-chain carboxylic acids become available. Covalently bound high-molecular-weight organic components with carboxylic acid groups make the most favorable sites for hydrogen bonding with the short-chain carboxylic acids, but other oxygen-bearing functional groups covalently bound into the organic matter (Fig. 10b–f) may also act temporarily as hydrogen-bonding sites. The amount of available carboxylic acid sites for hydrogen bonding is more than sufficient when one considers that Type-II kerogens have 15 to 19% of their oxygen as undifferentiated carboxylic acid groups (Tissot and Welte 1984, p. 154; Behar and Vandenbroucke 1987) and only 5 to 10% of their oxygen occurs as short-chain carboxylic acids after hydrous pyrolysis experimentation. Hydrogen bonding reactions are reversible, and the making and breaking of these bonds over time are likely to result in most of the short-chain carboxylic acids being associated with the more stable covalently bound carboxylic acid groups. Another implication of the reversibility in making and breaking of these weak noncovalent bonds is that no transport of short-chain carboxylic acids out of the sedimentary organic matter is likely without the occurrence of a diffusion gradient within or fluid expulsion from a source rock.

4.3 Expulsion Timing and Mechanisms

Although organic acids generated from source rocks have been considered to be responsible for secondary porosity, research and discussion have not focused on mechanisms by which these acids are expelled from a source rock. Three periods for expulsion of organic acids are envisaged. The first period relies on diffusion prior to bitumen generation, the second period relies on diffusion after a bitumen network has been generated, and the third period relies on bulk flow during oil expulsion. The first period of expulsion occurs after lithification of the sediment and prior to the thermal decomposition of kerogen to bitumen (i.e., early diagenesis). Organic matter in a source rock during this period consists predominantly (>95 wt%) of insoluble kerogen typically dispersed as insular domains within a mineral matrix having a water-filled pore system. Periodic cleavage of hydrogen bonds between short-chain carboxylic acids and kerogen allows the released acid to diffuse into the water-filled pore system and eventually through water-filled fractures that typically dissect source rocks. The chemical potential necessary to drive this diffusion would be the higher affinity of short-chain carboxylic acids to reside in a water phase than an organic phase, as demonstrated by their low partition coefficients (Fig. 1).

Sufficient data are not available to conduct a rigorous evaluation on the significance of this period of early diffusion. However, a general assessment may be made assuming one-dimensional diffusion in a semi-infinite homogenous pore water system, a constant concentration of acids at the kerogen-water interface, and no acids in the pore water at the start of the diffusion process. Under these assumptions, the distance (x) and time (t) required for diffusing acids to reach a given concentration (C) may be calculated using the expression

$$C = C_o erfc\left\{\frac{x}{2(Dt)^{1/2}}\right\},$$

which is derived by a Laplace transform of Fick's second law (Crank 1979, p. 20–21). D is the diffusion coefficient and C_o is the initial concentration of a diffusing acid at a given temperature. Considering the time (t) and distance (x) at which a diffusing acid front would reach half of its initial concentration (i.e., $C/C_o = 0.50$) gives the expression

$$erfc\left\{\frac{x}{2(Dt)^{1/2}}\right\} = 0.50.$$

Based on tabulated error-function complements (erfc; Crank 1979, p. 375), the bracketed term must equal 0.4772 in this expression, which reduces to the expression

$$x = 0.9544(Dt)^{1/2}.$$

This reduced expression allows one to calculate the distance at which diffusing species reach half of their initial concentration in a specified amount of time (t) by knowing their diffusion coefficients (D). Assuming effective diffusion coefficients of 10^{-12} to 10^{-11} m^2 s^{-1} for dissolved hydrocarbon gas diffusion through source rocks (Krooss 1987) are similar in magnitude to short-chain carboxylic acids, the significance of early diffusion prior to bitumen generation may be evaluated. For example, half of the initial organic-acid concentration will occur only 93 to 293 m from the kerogen-water interface after 300 million years at 25 °C. This approximation represents maximum distances because the calculation does not include the rate of hydrogen-bond cleavage, which may reduce the overall rate.

Although the sluggishness of this process is not likely to generate timely nor significant concentrations of organic acids to generate appreciable secondary porosity, it may account for the lower yields of short-chain carboxylic acids generated by hydrous pyrolysis of the older Phosphoria Retort Shale (Permian) relative to the younger Monterey (Miocene) and Kreyenhagen (Oligocene-Eocene) shales (Table 3). In addition, the flux of acetic acid diffusing out of a source rock is likely to be greater than that of the C_3 through C_5 carboxylic acids, because of its greater affinity for water (Fig. 1) and higher diffusion coefficient (Oelkers 1991). As a result, one would expect preferential depletion of acetic acid in a rock with increasing geologic time. This preferential depletion may explain in part the significantly lower absolute (Table 3) and relative (Fig. 6) concentrations of acetic acid generated from the Phosphoria Retort Shale of Paleozoic Age relative to the Monterey and Kreyenhagen shales of Tertiary age. Other variations in the amount of generated acetic acid may be attributed to interactions with minerals along its diffusion pathways within a source rock as suggested by Giles and Marshall (1986). It should be noted that this early diagenetic process is not simulated by hydrous pyrolysis.

The second period of expulsion is after the development of a bitumen network and before the bitumen partially decomposes to an immiscible oil. Expulsion of short-chain carboxylic acids dispersed in the bitumen network is envisaged to occur by their diffusion into water-filled fractures dissecting a source rock and adjacent rock units. Diffusion of the acids from the bitumen network at fracture surfaces into water-filled fractures is expected to be faster than diffusion during the first expulsion period because of higher effective diffusion coefficients resulting from a decrease in tortuosity. Although a two-order of magnitude increase in the diffusion coefficient may be expected (10^{-10} to $10^{-9}\,m^2\,s^{-1}$; Thomas 1989), this increase suggests that half of the initial organic acid concentration in the bitumen network will only diffuse 54 to 534 m from the interface in 10 m.y. Once again, this approximation represents maximum distances under the previously stated assumptions. In addition, the higher solubility of short-chain carboxylic acids in water than in an organic liquid (Fig. 1) indicates that the flux of these diffusing acids through a bitumen network to a fracture surface may control and lower the flux of these diffusing acids through a water-filled fracture. More experimental work is needed to fully assess this period of expulsion, but first approximations suggest that timely and sufficient concentrations of short-chain carboxylic acids from this expulsion period are only likely to be agents of secondary porosity on a localized scale in rocks immediately adjacent to maturing source rocks.

The third period of acid expulsion is envisaged to coincide with the expulsion of immiscible oil as it is generated from the partial decomposition of the bitumen network. As shown in Fig. 11, the amount of aliphatic

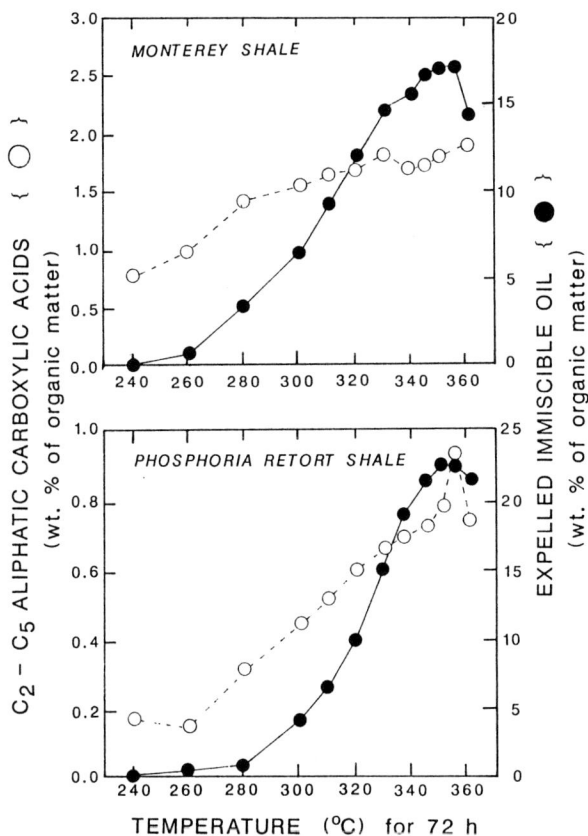

Fig. 11. Plots of expelled immiscible oil and aqueous C_2–C_5 monocarboxylic acids generated through a series of isothermal hydrous pyrolysis experiments conducted on aliquots of Monterey and Phosphoria Retort shales

carboxylic acids in the recovered water of hydrous pyrolysis experiments increases with the amount of expelled immiscible oil. Comparison of the two source rocks shows that there is no direct nor constant relationship between amounts of expelled immiscible oil and organic acids. Additional research is needed to elucidate the factors responsible for these relationships, but the amount of organic acids lost prior to this period of expulsion and the effects of mineral and organic matter composition on partitioning of organic acids between bitumen and immiscible oil are likely factors worth considering. Despite this uncertainty, mass flow of expelled immiscible oil bearing dissolved organic acids from a source rock may be a viable expulsion mechanism. Organic acids dissolved in expelled oil may repartition into formation waters during secondary migration of oil through carrier beds or during entrapment of oil in reservoir rocks as pressures, temperatures, or water salinities decrease.

The effects of temperature, pressure, and water salinity on partitioning of aliphatic carboxylic acids between oil and water phases are not currently known. However, some qualitative insights on the effects of these variables on partition coefficients may be evaluated from experiments conducted by Knaepen et al. (1990). Their experiments determined the partition coefficients for pentanol-1 and ethylacetate in isooctane/water or North Sea oil/water systems at different temperatures (22 to 150 °C), water salinities (0 to 10 wt% NaCl), and gas/oil ratios (0 to 230 v/v). An increase in temperature from 22 to 150 °C increased the partition coefficient for pentanol-1 in the isooctane/water system sevenfold. A water salinity of 10 wt% NaCl at 22 °C has a partition coefficient for pentanol-1 that is almost threefold greater than pure water. The amount of dissolved methane in North Sea oil as determined by gas/oil ratios has only a minor effect on the partition coefficient of ethyl acetate at room temperature, but at 96 °C the partition coefficient increases twofold.

Assuming similar changes in partition coefficients for short-chain carboxylic acids, these acids may be released from the source rocks during hydrous pyrolysis experiments as dissolved species in the expelled immiscible oils at the experimental temperatures (240 to 365 °C). As the experiments cool to room temperature after the 72-h isothermal heating, these acids may diffuse into the water phase as their partition coefficients decrease as a result of the falling temperature. Condensation of H_2O vapor to reduce the water salinity and the release of dissolved methane from the expelled oil would accompany this fall in temperature and further augment the decrease in the partition coefficients. Similarly, in subsiding sedimentary basins, short-chain carboxylic acids not lost to diffusion prior to oil expulsion may be released from source rocks as dissolved species in expelled oils. As these oils cool, degas, or encounter low salinity waters during migration and entrapment, the dissolved short-chain carboxylic acids may be redistributed in adjacent formation waters.

An obvious implication of this third period of expulsion is that porosity enhancement by organic acids is most likely to occur during or after oil expulsion rather than before it. As a result, oil migration pathways may be lined with enhanced porosity because of the release of organic acids into adjacent formation waters. Similarly, enhanced porosity may occur in reservoirs as a result of the release of organic acids from entrapped oil into peripheral formation waters. Amounts of organic acids within formation waters associated with oil reservoirs will be determined in part by the amount of water the reservoired oil encountered during migration and the differential in temperature, dissolved gas content, and water salinity between the source rock and the reservoir. An additional control is the amount of organic acids lost from a source rock by the first and second periods of diffusion prior to oil generation and expulsion.

4.4 Kinetic Modeling

Kinetic modeling based on laboratory experiments has become a common practice in trying to predict the occurrence of kinetically controlled reactions in sedimentary basins (Whitney and Lewan 1992). For example, kinetic parameters determined for oil formation (i.e., generation and expulsion) from hydrous pyrolysis experiments have been shown to extrapolate well to lower temperatures and longer geologic times encountered by source rocks during their subsidence in sedimentary basins (Hunt et al. 1991; Hunt and Hennet 1992). The success of these extrapolations is attributed to similar reaction mechanisms and pathways over a wide temperature range (100 to 350 °C) and to the expulsion of oil from a source rock being a direct consequence of oil generation within a source rock (Lewan 1993b). The similarity of organic acid compositions between recovered hydrous pyrolysis waters and natural subsurface waters suggests that the reaction mechanisms and pathways responsible for their generation within source rocks are also similar over a wide temperature range. However, the processes responsible for their expulsion from a source rock (i.e., diffusion and oil expulsion) are independent of organic acid generation. As a result, kinetic parameters determined from hydrous pyrolysis experiments cannot be used without taking into account these expulsion processes. This problem may be illustrated by the kinetic parameters derived from hydrous pyrolysis experiments on the Monterey Shale (Fig. 9). A simple calculation using these kinetic parameters in the Arrhenius equation for a first-order rate law indicates that essentially all of the acetic acid is generated and expelled from the Monterey Shale after 1000 years at 27 °C. Paradoxically, this sample resided at this or slightly higher temperatures for more than 5 million years before the experiments were conducted. Kinetic parameters for organic acid formation that avoid this paradox have been derived by assuming a distribution of activation energies and a single frequency factor (Braum et al. 1992; Knauss et al. 1992; Barth and Nielsen 1993). These kinetic parameters are commonly similar to kinetic parameters derived in the same manner for oil formation (Knauss et al. 1992) and may be significantly different for the same type of organic matter in different samples from the same locality and rock unit (Barth and Nielsen 1993). Although these curve-fitting methods avoid the time paradox without considering expulsion processes, their scientific reality remains questionable (Lakshmanan et al. 1991).

5 Conclusions

Laboratory pyrolysis experiments are the most effective means of assessing the amounts and types of organic acids derived from petroleum source rocks. This approach requires careful consideration of experimental conditions within the natural constraints of subsiding sedimentary basins. The role of water and its chemistry in organic acid generation is not completely understood, but experiments conducted without water result in reduced concentrations and deviate from the water-saturated state of sedimentary basins. Kinetic controls on organic acid generation require experiments to use higher temperatures for shorter durations to compensate for the lower temperatures and longer durations experienced in sedimentary basins. This substitution of temperature for time appears to be appropriate between temperatures of 200 and 360 °C for 72- to 360-h durations. Within these experimental conditions, no high-temperature artifacts are apparent and generated organic acids are similar in type and distribution to those in natural subsurface waters.

Experimentally simulating natural processes responsible for organic acid generation may best be approached with whole-rock samples that are larger than 62 μm in size and not preextracted with organic solvents. This type of sample insures natural contacts between minerals, organic matter, and pore waters, which are not simulated by experiments using isolated kerogens or unconsolidated artificial mixtures of isolated kerogens and powdered minerals. Differences in the partitioning of organic acids between water and organic liquids (i.e., bitumen and oil) require consideration of organic liquid to water ratios and product collection procedures. Another experimental consideration is the composition of reactor wall surfaces. Borosilicate glass has been shown to cause more alteration to organic acid concentrations and distributions than fresh metal surfaces of stainless steel-316. Carburized reactor walls reduce the effects of stainless steel-316 but carburized Hastelloy-C276 appears to impose the least effect on the overall water chemistry of an experiment. Experimental fluid pressures do not appear to influence organic acid generation below 60 MPa. No data are currently available on the influence of water chemistry, particularly salinity, on organic acid generation.

Acyclic aliphatic carboxylic acids are the most prevalent organic acids generated from petroleum source rocks. These acids may contain as many as 32 carbon atoms, but partition coefficients indicate that only those with fewer than five carbon atoms are likely to be important in geological processes mediated by water. Similar to organic acid concentrations in natural subsurface waters, acetic acid (C_2) usually dominates among these C_1–C_5 acids, and formic acid (C_1) usually occurs below detection limits. Changes in concentration and distributions of these generated C_2–C_5 acids with increasing experimental temperatures are variable for different rock units bearing the same kerogen types. These variations may be attributed in part to differences in the level of early diagenesis samples experienced before being subject to pyrolysis experiments. Rocks at higher levels of early diagenesis (i.e., low atomic O/C ratios of kerogen) may have preferentially lost more of their acetic acid and to a lesser extent propionic acid through diffusion than rocks at lower levels of early diagenesis (i.e., high atomic O/C ratios of kerogen). Therefore, petroleum source rocks at the lowest level of early diagenesis following the terminus of microbial activity provide the best assessment of maximum acid-generation potential. Maximum yields of these monocarboxylic aliphatic acids typically occur at 350 °C after 72-h durations, which are similar conditions for maximum oil generation. Available experimental data indicate that maximum conversions of organic matter to acetic acid are 2.17 wt% for Type-II kerogen, 2.01 wt% for Type-III kerogen, and 0.53 wt% for Type-I kerogen. Sufficient data are available to accept the maximum conversion for Type-II kerogen, but more experimental work is needed to confirm the maximum conversions for Types-I and -III kerogens. Dicarboxylic aliphatic acids are not as thermally stable as the monocarboxylic aliphatic acids, and from limited data reach maximum yields at experimental temperatures below 300 °C for durations of less than 72 h. Oxalic acid is the most dominant of the dicarboxylic acids generated. More experimental data are needed to determine maximum conversions of organic matter to these acids, but available data indicate at least 0.15 wt% of Type-II kerogen and 0.24 wt% of Type-I kerogen convert to oxalic acid. Significantly higher yields of mono- and dicarboxylic acids have been generated in experiments utilizing an iron oxide with isolated kerogen. However, it is unlikely that iron oxides would coexist with organic matter under the reducing conditions associated with depositional environments under which petroleum source rocks accumulate.

Low activation energies (<13 kcal/mol) determined for the overall generation of acetic acid indicate that it is retained in sedimentary organic matter by weak noncovalent bonds. Hydrogen bonding of carboxylic oxygens in acetic acid with other oxygen functional groups covalently bound in the sedimentary organic matter is one possibility that may also apply to the other short-chain carboxylic acids. This lack of covalent-bond cleavage in their generation implies that carboxylic aliphatic chains covalently bound to macromolecules in sedimentary organic matter are not likely sources of these acids. Instead, these carboxylic aliphatic chains are more likely to decarboxylate to form carbon dioxide than cleave from a macromolecule to form a short-chain carboxylic acid. Therefore, the potential for a source rock to generate short-chain carboxylic acids may be controlled by the amount of short-chain carboxylic acids assimilated through weak noncovalent bonding into sedimentary organic matter during its early development into kerogen.

Three expulsion periods are envisaged for the release of short-chain carboxylic acids from source rocks. The first period occurs during early diagenesis before thermal stress is sufficient to generate a bitumen network within a source rock. The higher affinity of short-chain carboxylic acids for water establishes a diffusion gradient from kerogen domains through pore-water systems within the rock matrix. This sluggish diffusion process is not likely to expel sufficient quantities of these acids in a timely manner to generate secondary porosity. However, the effectiveness of this diffusion process over long periods of geologic time may explain the low yield of these acids from pyrolysis experiments conducted on older rocks near the end of early diagenesis. The second period is also diffusion controlled and occurs after a bitumen network has been established within a rock matrix. Again, the higher affinity of short-chain carboxylic acids for water establishes a diffusion gradient from the bitumen network through water-filled fractures dissecting the source rock. This period of expulsion is also sluggish and may only enhance porosity locally in adjacent rocks.

The third expulsion period occurs during the expulsion of oil from a source rock. Short-chain carboxylic acids are released from a source rock as dissolved species in an expelled oil. As the expelled oil cools, degasses, or encounters lower salinity waters along its migration pathway or within its reservoir, these dissolved acids regain their affinity to diffuse into adjacent formation waters. If the concentrations of these diffusing acids are sufficient, secondary porosity may occur along migration pathways and near oil-water interfaces in reservoirs. Secondary porosity development envisaged for this expulsion period occurs during or after petroleum formation and not before it. Additional research addressing the effects that temperature, pressure, water salinity, and oil composition have on the partition coefficients of organic acids is needed.

References

Abelson PH (1962) Thermal stability of algae. Carn Inst Wash Year Book 61: 179–181

Barth T, Bjørlykke K (1993) Organic acids from source rock maturation: generation potentials, transport mechanisms, and relevance for mineral diagenesis. Appl Geochem 8: 650–660

Barth T, Nielsen SB (1993) Estimating kinetic parameters for generation of petroleum and single compounds from hydrous pyrolysis of source rocks. Energy Fuels 7: 101–110

Barth T, Borgund AE, Hopeland AL, Grave A (1987) Volatile organic acids produced during kerogen maturation – amounts, composition and role in migration of oil. Org Geochem 13: 461–465

Barth T, Borgund AE, Hopeland AL (1989) Generation of organic compounds by hydrous pyrolysis of Kimmeridge oil shale – bulk results and activation energy calculations. Org Geochem 14: 69–76

Behar F, Vandenbroucke M (1987) Chemical modelling of kerogens. Org Geochem 11: 15–24

Braum RL, Burnham AK, Reynolds JG (1992) Oil and gas evolution kinetics for oil shale and petroleum source rocks determined from pyrolysis TQMS data at two heating rates. Energy Fuels 6: 468–474

Bykova EL, Mel'Kanovitskaya SG, Shvets VM (1971) Distribution of organic acids in formation waters. Sov Geol 14: 135–142

Carothers WW, Kharaka YK (1978) Aliphatic acid anions in oil-field waters – implications for the origin of natural gas. Am Assoc Pet Geol Bull 62: 2441–2453

Cooles GP, Mackenzie AS, Parkes RJ (1987) Non-hydrocarbons of significance in petroleum exploration: volatile fatty acids and non-hydrocarbon gases. Mineral Mag 51: 483–493

Cooper JE, Bray EE (1963) A postulated role of fatty acids in petroleum formation. Geochim Cosmochim Acta 27: 1113–1127

Crank J (1979) The mathematics of diffusion. Clarendon Press, Oxford, 414 pp

Crossey LJ (1991) Thermal degradation of aqueous oxalate species. Geochim Cosmochim Acta 55: 1515–1527

Dawidowicz AL, Nazimek D, Pikus S, Skubiszewska J (1984) The influence of boron atoms on the surface of controlled porous glasses on the properties of the carbon deposit obtained by pyrolysis of alcohol. J Anal Appl Pyrolysis 7: 53–63

Dawidowicz AL, Pikus S, Nazimek D (1986) Properties of the material surfaces obtained by pyrolysis of alkanols on boron-enriched controlled porous glass. J Anal Appl Pyrolysis 10: 59–69

Eglinton TI, Curtis CD, Rowland SJ (1987) Generation of water-soluble organic acids from kerogen during hydrous pyrolysis: implications for porosity development. Mineral Mag 51: 495–503

Eisma E, Jurg JW (1969) Fundamental aspects of the generation of petroleum. In: Eglinton G, Murphy M T J (eds) Organic geochemistry. Methods and results. Springer, Berlin Heidelberg New York, pp 676–698

Fisher JB (1987) Distribution and occurrence of aliphatic acid anions in deep subsurface waters. Geochim Cosmochim Acta 51: 2459–2468

Fisher JB, Boles JR (1990) Water-rock interaction in Tertiary sandstones, San Joaquin Basin, California, USA: diagenetic controls on water composition. Chem Geol 82: 83–101

Giles MR, Marshall JD (1986) Constraints on the development of secondary porosity in the subsurface: re-evaluation of processes. Mar Pet Geol 3: 243–255

Goranson RW (1932) Some notes on the melting of granite. Am J Sci 23: 227–236

Greco EC, Griffin HT (1946) Laboratory studies for determination of organic acids as related to internal corrosion of high pressure condensate wells. Corrosion 2: 138–152

Hansley PL, Nuccio VF (1992) Upper Cretaceous Shannon Sandstone Reservoirs, Powder River Basin, Wyoming: evidence for organic acid diagenesis. Am Assoc Pet Geol Bull 76: 781–791

Harrison WJ, Thyne GD (1992) Predictions of diagenetic reactions in the presence of organic acids. Geochim Cosmochim Acta 56: 565–586

Holmberg M E (1946) Some metallurigical observations with respect to corrosion in distillate wells. Corrosion 2: 278–285

Hunt JM, Hennet RJ-C (1992) Modeling petroleum generation in sedimentary basins. In: Whelan J K, Farrington J W (eds) Organic matter: productivity, accumulation, and preservation in Recent and ancient sediments. Columbia University Press, New York, pp 20–52

Hunt JM, Lewan MD, Hennet RJ-C (1991) Modeling oil generation with time-temperature index graphs based on the Arrhenius equation. Am Assoc Pet Geol Bull 75: 795–807

Johnson RW, Calder JA (1973) Early diagenesis of fatty acids and hydrocarbons in a salt marsh environment. Geochim Cosmochim Acta 37: 1943–1955

Kawamura K, Ishiwatari R (1981) Experimental diagenesis of fatty acids in a sediment: changes in their existence forms upon heating. Geochem J 15: 1–8

Kawamura K, Ishiwatari R (1985a) Conversion of sedimentary fatty acids from extractable (unbound + bound) to tightly bound form during mild heating. Org Geochem 8: 197–201

Kawamura K, Ishiwatari R (1985b) Behavior of lipid compounds on laboratory heating of a Recent sediment. Geochem J 19: 113–126

Kawamura K, Kaplan IR (1987) Dicarboxylic acids generated by thermal alteration of kerogen and humic acids. Geochim Cosmochim Acta 51: 3201–3207

Kawamura K, Tannenbaum E, Huizinga BJ, Kaplan IR (1986) Volatile organic acids generated from kerogen during laboratory heating. Geochem J 20: 51–59

Knaepen WAI, Tijssen R, van den Bergen EA (1990) Experimental aspects of partitioning tracer tests for residual oil saturation determination with FIA-based laboratory equipment. Soc Pet Eng Reservoir Eng 5: 239–244

Knauss KG, Copenhaver SA, Braum RL, Burnham AK (1992) Hydrous pyrolysis of New Albany and Phosphoria Shales: effects of temperature and pressure on the kinetics of production of carboxylic acids and light hydrocarbons. Am Chem Soc Division Fuel Chem Preprints 37: 1621–1627

Krooss BM (1987) Experimental investigation of the molecular migration of C_1–C_6 hydrocarbons: kinetics of hydrocarbon release from source rocks. Org Geochem 13: 513–523

Lakshmanan CC, Bennett ML, White N (1991) Implications of multiplicity in kinetic parameters to petroleum exploration: distributed activation energy models. Energy Fuels 5: 110–117

Leo A, Hansch C, Elins D (1971) Partition coefficients and their uses. Chem Rev 71: 525–616

Lewan MD (1985) Evaluation of petroleum generation by hydrous pyrolysis experimentation. Philos Trans R Soc Lond 315A: 123–134

Lewan MD (1986) Stable carbon isotopes of amorphous kerogens from Phanerozoic sedimentary rocks. Geochim Cosmochim Acta 50: 1583–1591

Lewan MD (1987) Petrographic study of primary petroleum migration in the Woodford Shale and related rock units. In: Doligez B (ed) Migration of hydrocarbons in sedimentary basins. Technip, Paris, pp 113–130

Lewan MD (1992a) Primary oil migration and expulsion as determined by hydrous pyrolysis. Proc 13th World Petroleum Congr 1991, vol 2. John Wiley, Chichester, pp 215–223

Lewan MD (1992b) Water as a source of hydrogen and oxygen in petroleum formation. Am Chem Soc Fuel Chem Division Preprints 37: 1643–1649

Lewan MD (1993a) Laboratory simulation of petroleum formation: hydrous pyrolysis. In: Engel M H, Macko S A (eds) Organic geochemistry. Plenum Press, New York, pp 419–442

Lewan MD (1993b) Assessing natural oil expulsion from source rocks by laboratory pyrolysis. In: Magoon L, Dow W (eds) The petroleum system – from source to trap. Am Assoc Pet Geol Mem 60: 200–219

Lewan MD, Winters JC (1991) Retardation of the thermal decomposition of organic matter in shales under hydrous conditions. Geol Soc Am Abstr Programs 23: 24

Lewan MD, Winters JC, McDonald JH (1979) Generation of oil-like pyrolyzates from organic-rich shales. Science 203: 897–899

Lewis GN, Randall M (1960) Thermodynamic properties of hydrogen bonds. In: Pimentel GC, McClellan AL (eds) The hydrogen bond. Freeman, San Francisco, pp 206–225

Lorenz JC, Teufel LW, Warpinski NR (1991) Regional fractures I: a mechanism for the formation of regional fractures at depth in flat-lying reservoirs. Am Assoc Pet Geol Bull 75: 1714–1737

Lundegard PD, Senftle JT (1987) Hydrous pyrolysis: a tool for the study of organic acid synthesis. Appl Geochem 2: 605–612

MacGowan DB, Surdam RC (1988) Difunctional carboxylic acid anions in oilfield waters. Org Geochem 12: 245–259

March J (1985) Advanced organic chemistry, 3rd edn. Wiley-Interscience, New York, 1346 pp

Mazzullo SJ, Harris PM (1992) Mesogenetic dissolution: its role in porosity development in carbonate reservoirs. Am Assoc Pet Geol Bull 76: 607–620

Menaul PL (1944) Causative agents of corrosion in distillate field. Oil Gas J, Nov 11: 80–81.

Meyer CA, McClintock RB, Silvestri GJ, Spencer RC Jr (1983) ASME steam tables, 5th edn. The American Society of Mechanical Engineers, New York, 332 pp

Michels R, Landais P, Elie M, Gerard L, Mansuy L (1992) Evaluation of factors influencing the thermal maturation of organic matter during confined pyrolysis experiments. Am Chem Soc Division Fuel Chem Preprints 37: 1588–1594

Nelson RA (1985) Geologic analysis of naturally fractured reservoirs. Gulf, Houston, 320 pp

Obukhova ZP, Kutovaya AA (1968) Distribution of organic acids in condensate waters along the area of gas-condensate deposit. Korroziyi i Zashchita v Neftedobyvayushchei Promyshlennost Nauch-Fekh 4: 16–19

Oelkers EH (1991) Calculation of diffusion coefficients for aqueous organic species at temperatures from 0 to 350 °C. Geochim Cosmochim Acta 55: 3515–3529

Palmer DA, Drummond SE (1986) Thermal decarboxylation of acetate. Part I. The kinetics and mechanism of reaction in aqueous solution. Geochim Cosmochim Acta 50: 813–823

Parker PL (1969) Fatty acids and alchohols. In: Eglinton G, Murphy MTJ (eds) Organic geochemistry. Methods and results. Springer, Berlin Heidelberg New York, pp 357–373

Parker PL, Leo RF (1965) Fatty acids in blue-green algal mat communities. Science 148: 373–374

Pickering SA, Batts BD (1992a) Patterns of mono- and difunctional carboxylic acids in hydrothermal leachates of humified soil organic matter and peat: geological and environmental implications. Org Geochem 18: 683–693

Pickering SA, Batts BD (1992b) Persistence patterns of aqueous leachable carboxylic acids in immature to peak mature coals: implications for carboxylic acid patterns in hydrocarbon reservoir waters. Org Geochem 18: 695–700

Prange FA, Edwards WH, Greco EC, Griffith TE, Grimshaw JA, Nathan CC, Shock DA (1953) Condensate well corrosion. Natural Gasoline Association of America, Tulsa, 203 pp

Rhead MM, Eglinton G, Draffan GH, England PJ (1971) Conversion of oleic acid to saturated fatty acids in Severn estuary sediments. Nature 232: 327–330

Rosenfield WD (1948) Fatty acid transformations by anaerobic bacteria. Arch Biochem Biophys 16: 263–273

Rumble D III, Ferry JM, Hoering TC, Boucot AJ (1982) Fluid flow during metamorphism at the Beaver Brook fossil locality, New Hampshire. Am J Sci 282: 886–919

Schuhmacher JP, Huntjens FJ, van Krevelen DW (1960) Chemical structure and properties of coal XXVI. Studies on artificial coalification. Fuel 39: 223–234

Shenberger DM, Barnes HL (1989) Solubility of gold in aqueous sulfide solutions from 150 to 350 °C. Geochim Cosmochim Acta 53: 269–278

Shock DA, Hackerman N (1948) Extraction of polar constituents from hydrocarbon solutions. Ind Eng Chem 40: 2169–2172

Shvets VM, Shilov IK (1968) Organic matter in subterranean waters of the southwestern Azov-Kulan Artesian Basin. Geol Nefti I Gaza 12: 46–49

Soldan AL, Cerqueira JR (1986) Effects of thermal maturation on geochemical parameters obtained by simulated generation of hydrocarbons. Org Geochem 10: 339–345

Stanley JK (1970) The carburization of four austenitic stainless steels. J Mater 5: 957–971

Surdam RC, Crossey LJ (1985) Organic-inorganic reactions during progressive burial: key to porosity and permeability enhancement and preservation. Philos Trans R Soc Lond 315A: 135–156

Surdam RC, Boese SW, Crossey LJ (1984) The chemistry of secondary porosity. In: McDonald DA, Surdam RC (eds) Clastic diagenesis. Am Assoc Pet Geol Mem 37: 127–149

Takenouchi S, Kennedy GC (1964) The binary system H_2O-CO_2 at high temperatures and pressures. Am J Sci 262: 1055–1074

Tannenbaum E, Kaplan IR (1985) Role of minerals in the thermal alteration of organic matter. I. Generation of gases and condensates under dry conditions. Geochim Cosmochim Acta 49: 2589–2604

Thomas MM (1989) Comments on calculation of diffusion coefficients from hydrocarbon concentration profiles in rocks. Am Assoc Pet Geol Bull 73: 787–791

Thornton EC, Seyfried WE Jr (1987) Reactivity of organic-rich sediment in seawater at 350 °C, 500 bars: experimental and theoretical constraints and implications for the Guaymas Basin hydrothermal system. Geochim Cosmochim Acta 51: 1997–2010

Tissot BP, Welte DH (1984) Petroleum formation and occurrence. Springer, Berlin Heidelberg New York, 699 pp

Tissot B, Durand B, Espitalié J, Combaz A (1974) Influence of nature and diagenesis of organic matter in formation of petroleum. Am Assoc Pet Geol Bull 58: 499–506

Vandenbroucke M, Pelet R, Debyser V (1985) Geochemistry of humic substances in marine sediments. In: Aiken G R, McKnight D M, Wershaw R L, MacCarthy P (eds) Humic substances in soils, sediments, and water. Wiley, New York, pp 249–273

Whitney G (1990) Role of water in the smectite-to-illite reaction. Clays Clay Minerals 38: 343–350

Whitney G, Lewan D (1992) Diagenesis in a bottle – experimental strategies for studying thermal maturity in clays and organic matter. US Geol Surv Circ 1074: 81

Willey LM, Kharaka YK, Presser TS, Rapp JB, Barnes I (1975) Short chain aliphatic anions in oil field waters and their contribution to the measured alkalinity. Geochim Cosmochim Acta 39: 1707–1711

Chapter 5   Material Balance Considerations for the Generation of Secondary Porosity by Organic Acids and Carbonic Acid Derived from Kerogen, Denver Basin, Colorado, USA

Edward D. Pittman[1] and Lori A. Hathon[2]

Summary

Data on source and reservoir rocks in the Wattenberg area of the Denver Basin were used to evaluate the hypothesis that carboxylic acids and carbon dioxide generated from the thermal maturation of organic matter create secondary porosity in sandstone reservoirs. For the Lower Cretaceous source rocks, information on the type of kerogen, richness, maturity, and thickness was collected for each formation. We used subsidence modeling to determine the temperature and burial history of the source rocks. For the reservoir and potential reservoir sandstones (Fox Hills, Larimer-Rocky Ridge, Terry, Hygiene, Codell, Dakota J, and Lyons) overlying the deepest part of the basin, the composition, total macroporosity, and secondary porosity were determined using 115 thin sections cut from cores from 18 wells. Data on the stratigraphic thickness needed to calculate volume (for $1\,km^2$) of sandstones came from geophysical logs. Porosity in nonreservoir sandstones throughout the stratigraphic column was evaluated using geophysical logs and thin sections of drill cuttings. These nonreservoir sandstones contain only trace amounts of porosity and were not included in the calculations. Also, three carbonate formations were examined in thin section, but disregarded for this study because of lack of macroporosity.

Two types of calculations were made. In the first type, the amount of plagioclase and calcite that could be dissolved by carboxylic and carbonic acids, respectively, generated by decarboxylation of organic matter, is complementary and additive. A second type of calculation is additive, but in our opinion speculative, and involves the generation of organic acids from redox reactions. We used data from hydrous pyrolysis experiments using shale source rocks to establish the conversion efficiency of kerogen oxygen to mono- and dicarboxylic oxygen. To maximize the calculations we considered the ratio of dicarboxylic to total carboxylic to be 0.52, which is a high value

Amoco Production Co., Research Center, Tulsa, Oklahoma, USA
[1] Present address: Consultant, Sedona, Arizona, USA
[2] Present address: Amoco Production Co., Houston, Texas, USA

reported for the San Joaquin Basin. For these optimum conditions, we
calculate that 1.72% secondary porosity can form in the sandstones. By
point count, using a conservative approach in identifying secondary porosity,
the secondary porosity is 2.49%. The secondary porosity is probably higher
than this, but it is impossible to accurately document. These calculations
suggest that the supply of acids may be insufficient to explain the observed
secondary porosity. The source rocks in the Denver Basin contain almost
exclusively Type-II kerogen. If this kerogen had been Type III, which is more
oxygen-rich, then presumably more carboxylic acids and carbon dioxide
could have been generated to create greater porosity.

This modeling approach to evaluate the material balance considerations
for the generation of acids from source rocks and the creation of secondary
porosity is different from previous attempts because we use data from a
specific basin rather than taking a general model using published average
data for kerogen.

1 Introduction

Ever since Hayes (1979) and Schmidt and McDonald (1979) showed that
secondary porosity was common in sandstones of all ages worldwide,
scientists have been concerned with how this porosity formed. Curtis (1978)
and Schmidt and McDonald (1979) attributed the porosity-creating process
to the formation of carbon dioxide generated by the decarboxylation of
organic matter. This became a popular hypothesis because: (1) the role of
carbon dioxide in calcite solubility has been known for over 200 years; and
(2) significant amounts of carbon dioxide are generated in the laboratory
when organic matter is heated.

Work by Bjørlykke (1983, 1984) and Lundegard and Land (1986) casts
doubt regarding the adequacy of the kerogen as a source of carbon dioxide
to create secondary porosity. These studies, which approached the problem
differently, reached a similar conclusion: carbon dioxide can account for an
insignificant amount (less than one-fifth) of what is considered to be porosity
of secondary origin in Frio sandstones of the Gulf Coast. The techniques
used by Bjørlykke (1983) and Lundegard and Land (1986) to model these
processes are discussed later.

As interest in carbon dioxide for the origin of secondary porosity waned,
Surdam et al. (1984) called attention to the possibility that organic acids
generated by thermal maturation of kerogen had the potential to create
significant amounts of secondary porosity.

The purpose of this chapter is to present material balance calculations to
evaluate whether there were suffecent acidic solutions available to explain
observed amounts of secondary porosity in sandstones overlying the deepest
portion of the Denver Basin (Colorado, USA), which has thermally mature

shales and limestones considered to be the source of hydrocarbons and acidic solutions.

1.1 Previous Work on Material Balance Calculations

Bjørlykke (1983), whose work was based on data published on kerogen by Tissot and Welte (1978) and Hunt (1979), calculated the amount of calcite in a sandstone that could be dissolved through carbon dioxide generated by decarboxylation of organic matter in an underlying shale. He used a shale/sand ratio of 10 and an average kerogen content of 1%. If this kerogen is humic, it will yield about 10% carbon dioxide, whereas a sapropelic kerogen will yield only about 1% carbon dioxide. Bjørlykke (1983) calculated that a humic kerogen will produce sufficient carbon dioxide to create about 2% secondary porosity, whereas a sapropelic kerogen will create only about 0.2% secondary porosity. These figures are too low to explain the high percentages (up to 11% in the Gulf Coast Basin) of secondary porosity that commonly have been reported in the literature.

Lundegard and Land (1986) made material balance calculations for the Paleogene of the Gulf Coast and evaluated the observed secondary porosity versus the volume of minerals that could be dissolved by carbon dioxide generated by decarboxylation of organic matter. The amount of secondary porosity had been established by an extensive petrological data base (Loucks et al. 1984), which indicated that the Frio and Wilcox sandstones have 11 and 7% secondary porosity, respectively. Lundegard and Land (1986) pointed out that there is at least 5% secondary porosity that is attributable to dissolution of K-feldspar.

The procedure that Lundegard and Land (1986) used is described briefly below. Based on equations for the dissolution of K-feldspar and plagioclase, at least 1 mol of protons for each mole of feldspar dissolved was required. The amount of carbon dioxide that a given shale can produce by decarboxylation of organic matter depends on the abundance and type of organic matter. Therefore, they assumed the shale source beds had 0.5 wt% Type-III kerogen. Lundegard and Land (1986) used a value of 6.3 wt% for the amount of oxygen in the kerogen that could be converted to carbon dioxide by decarboxylation. This is based on 25% oxygen, of which 25% is considered to be in carboxylic groups (Tissot and Welte 1978). Based on these assumptions, Lundegard and Land determined the moles of carbon dioxide per gram of shale produced by decarboxylation. Next, they established the properties of a shale for purposes of calculations (10% porosity and a solid density of 2.6 g/cm) and how many moles of carbon dioxide a cubic centimeter of shale could produce by decarboxylation. Finally, they needed to know the ratio of thermally mature shale to sandstone, which was established from geophysical logs based on work by Galloway et al. (1982). This ratio ranged from 0 to 13 and averaged approximately 3. They calculated that areas with abundant shale (high ratio) could yield enough acidic solutions to produce about 1% secondary porosity by dissolution of calcite or about 3% secondary porosity by dissolution of feldspar. Areas with low ratios would generate lesser amounts of secondary porosity. Therefore, their calculations suggest that acids generated from decarboxylation are inadequate to explain the amount of secondary porosity believed to be present in the Paleogene sandstones of the Gulf Coast.

Giles et al. (Chap. 14, this Vol.) considered the mass-balance problem by evaluating the relative thickness of source rock required to generate enough organic acid to create 3 vol% of secondary porosity through dissolution of feldspar. Their calculations were based on the acid-generating capacity from source rocks as determined by Lundegard and Kharaka (1990) in hydrous pyrolysis experiments. Giles et al. (Chap. 14, this Vol.) concluded that unrealistically high ratios of source rock to sandstone are required to account for 3% secondary porosity.

Lundegard et al. (1984) and Bjørlykke (1984) have argued that there is insufficient oxygen in kerogen to produce the required quantities of acids needed to account for the secondary porosity interpreted to occur in some subsurface formations. MacGowan and Surdam (1988) pointed out that these calculations are based on elemental analyses of kerogen isolated from source rocks. The kerogen is isolated by soxhlet extraction of bitumen followed by decomposition of inorganic material by HCl and HF acids. MacGowan and Surdam (1988) have called attention to the work of Vandergrift et al. (1980) which showed that a significant amount of organic oxygen was lost during the extraction process. The modeling presented in this chapter is based on hydrous pyrolysis of untreated shale source rocks, which eliminates this criticism.

2 Collection of Pertinent Data for the Denver Basin

The Denver Basin is asymmetric with a gently dipping eastern flank and a sharply upturned western flank (Fig. 1). This basin was selected for study because it is a relatively small basin with hydrocarbon production from predominantly sandstone reservoirs, which have been extensively cored. Also, considerable data are available on the source rocks in this basin.

Data were collected for source rocks and sandstones to provide the basis for material balance calculations. Pertinent source rock data were compiled (published and new data) for the Lower Cretaceous shales and limestones (see columnar section, Fig. 2) in the Wattenberg area, which is the deepest and hottest portion of the basin. Petrologic data were derived from the study of 115 thin sections of sandstones throughout the stratigraphic column. In the Wattenberg area, the Upper Cretaceous Terry and Hygiene sandstones produce oil at Spindle field and the Lower Cretaceous Niobrara

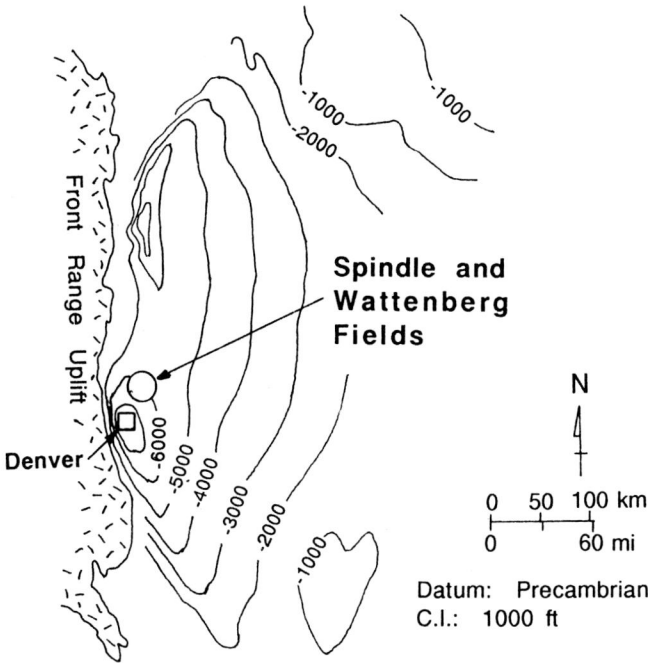

Modified from Moredock and Williams (1976)

Fig. 1. May showing the asymmetric configuration of the Denver Basin and the area of study, Spindle and Wattenberg fields, which overlies the deep portion of the basin

Formation, Codell Sandstone, and Dakota J Sandstone yield gas at the underlying Wattenberg field.

Subsidence modeling was used to evaluate the thermal maturity profile of the source beds. The subsidence model developed for this area is presented by Pittman (1988). A 107-m core in the Niobrara Formation and Carlile Shale provided consistent vitrinite reflectance data to calibrate the subsidence curves. An R_o of 1.41% corresponds to the base of the Niobrara Formation. This value was used as a control on the development of the subsidence profile; that is, a match of 1.41% was obtained for the base of the Niobrara.

2.1 Source Rocks

Oil in the small Lyons Sandstone reservoirs along the northwest margin of the Denver Basin has distinctly different characteristics than oil in the Cretaceous reservoirs. The source for the Lyons oil is unknown. The source for the oil in the Lower and Upper Cretaceous reservoirs has been established as Cretaceous shales and limestones (Clayton and Swetland 1980) by oil-to-rock correlation techniques. The Carlile Shale, Greenhorn Limestone,

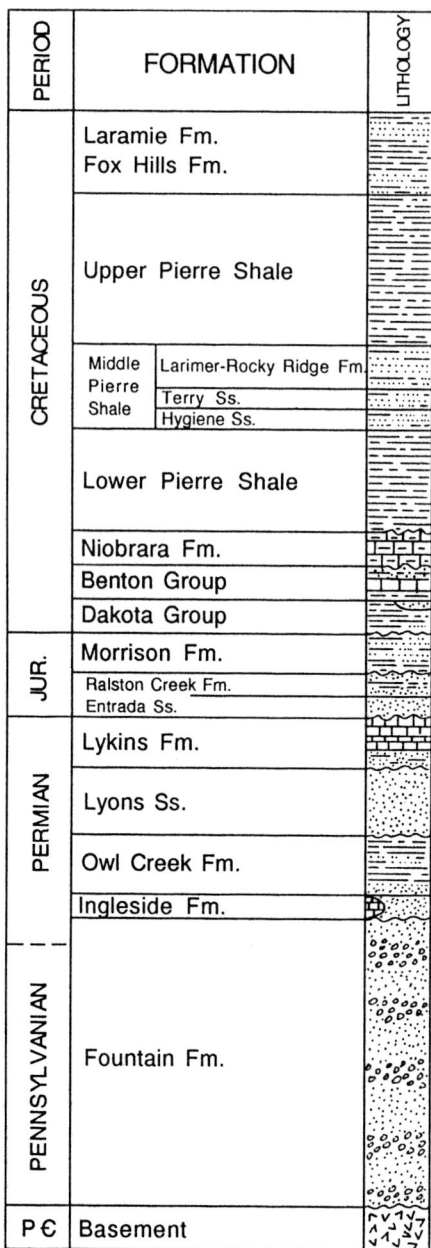

Fig. 2. Columnar section for the Wattenberg area of the Denver Basin. The primary source rocks are the shales and limestones of the Dakota Group, Benton Group, and Niobrara Formation. The most important hydrocarbon reservoirs are the Dakota J Sandstone and the Hygiene and Terry sandstones of the Middle Pierre Shale. The Lyons Sandstone, Codell Sandstone of the Benton Group, and limestone of the Niobrara Formation also produce hydrocarbons

and Graneros Shale of the Benton Group and Mowry Shale of the Dakota Group are the most probable sources of oil produced from Cretaceous reservoirs. The Skull Creek Shale of the Dakota Group, although leaner in organic matter, is also a possible source. Clayton and Swetland (1980) lacked data for the Lower Pierre Shale, but pointed out that this interval is a less likely source rock; however, the Lower Pierre cannot be totally discounted as a potential source for oil trapped in the Terry and Hygiene reservoirs in Spindle field. They also lacked data on the Niobrara in the deep part of the basin. Paleozoic source rocks have not been identified in the Wattenberg area.

Source rock analyses were made for the Niobrara Formation and the Carlile Shale because of the paucity of data available in public domain literature. The Niobrara in the Wattenberg area is a black shaley limestone and is the richest source rock in the basin. Source rock data used in calculations presented later in this chapter are shown in Table 1. Only analyses made on core samples were used except for the Pierre Shale where a lack of cores necessitated the use of data derived from outcrops and drill cuttings.

The Niobrara Formation yields thermochemically generated gas condensate from fractured reservoirs in the deep part of the basin. The Niobrara probably serves as an indigenous source in this situation. In contrast, the light-colored Niobrara chalk on the shallow eastern flank of the basin produces biogenic methane as shown by the isotopically light carbon (δ^{13}C, ranging from -55 to $-65‰$; Rice 1984). Cretaceous source rocks are only thermally mature in the deep portion of the basin near the axis (Clayton and Swetland 1980; Tainter 1984). This means that updip, lateral migration of oil

Table 1. Source rock data for Wattenberg area, Denver Basin, used in material balance calculations. Kerogen in these source rocks is almost exclusively Type II

Formation (No. Samples)	TOC (wt%)	Thickness[e] (m)	Volume source rock[f] (km^3)
M. & U. Pierre Sh. (10)[a,d]	1.8	1036.3	1.04
L. Pierre Sh. (0)[c]	1.8	487.7	4.88×10^{-1}
Niobrara Fm. (6)[b]	3.7	91.1	9.1×10^{-2}
Carlile Sh. (16)[b]	2.0	42.7	4.27×10^{-2}
Greenhorn Ls. (1)[a]	1.5	31.7	3.17×10^{-2}
Graneros Sh. (5)	2.6	52.1	5.21×10^{-2}
Sh. in Dakota Gp. (19)[a]	1.8	50.3	5.03×10^{-2}

[a] Clayton and Swetland (1980).
[b] This chapter.
[c] No source rock analyses; TOC was assumed to be the same as M. & U. Pierre Shale.
[d] Includes data from outcrops and drill cuttings.
[e] Based on 13 wells in Wattenberg field.
[f] Based on area of 1 km^2.

is required on the eastern flank to fill reservoirs that are up to 160 km from the source. Vertical migration through about 488 m of Lower Pierre Shale is required to charge the Terry and Hygiene reservoirs overlying the axis of the basin.

The Cretaceous shales and limestones are also sources for carbon dioxide and organic acids. The richness and volume of the source rocks, as well as the type of kerogen, are determinants of the volume and type of acids. Humic-rich Type-III kerogen is more oxygen-rich and therefore capable of generating more carbon dioxide and organic acids (Hunt 1979). Cretaceous source rocks in the Denver Basin, however, contain Type-II kerogen almost exclusively.

2.2 Petrology of Sandstones

2.2.1 Procedure

One of our goals was to sample cores from all major sandstone formations present in the Wattenberg area in order to determine the type and amount of macroporosity. Sandstone stratigraphic units and the number of wells (in parentheses) utilized are Fox Hills (1); Larimer-Rocky Ridge (2); Terry (3); Hygiene (3); Codell (2); Dakota J (6); and Lyons (1). No cores were available from the Fountain Formation; however, based on Walker's 1984 study, the deeply buried Fountain Formation sandstones have undergone extensive diagenesis and lack macroporosity. We were unable to evaluate sandstones in the Permian and Jurassic because of lack of cores; however, based on log evaluation and drill cuttings these sandstones appear to have essentially no porosity and were not considered in the calculations. We studied thin sections of drill cuttings through the Upper Pierre Shale in the Amoco No. 1 Bacon (SW NW Section 3-2N-68W) to evaluate secondary porosity in sandy-silty beds known to be present. Porosity, including secondary porosity, generally was limited to trace amounts in these beds and was not considered in any calculations. We also sampled three cores from carbonate formations to evaluate the amount and type of porosity. None of the carbonate units have macroporosity and were excluded from calculations.

We briefly described the sandstone cores and sampled each lithology type. For each sample, routine X-ray diffraction analyses and thin sections were made. Thin sections were prepared from rock containing pressure impregnated, blue-dyed, epoxy resin so that pore space would be readily discernable. Thin sections were stained to facilitate recognition of carbonates and K-feldspar using the techniques of Dickson (1966) and Bailey and Stevens (1960), respectively. Point-count analyses were made using a voice recognition system described by Dunn et al. (1985). A minimum of 300 points was counted per thin section.

2.2.2 Composition

With increasing age, there is a progressive shift toward greater mineralogical maturity in sandstones of the Cretaceous strata (Fig. 3). Fox Hills, the youngest formation sampled, contains about equal proportions of quartz, feldspar, and lithic fragments and is an arkosic or lithic arenite in a modified Dott scheme (Pettijohn et al. 1987). Sandstones in the Middle Pierre Shale (Larimer-Rocky Ridge, Terry, and Hygiene) are more quartz-rich and predominantly subarkose or sublitharenite. The Codell Sandstone (of the Benton Group), based on only eight samples, has a more diverse composition and varies from lithic arenite to quartz arenite. The Codell has compositional attributes that bridge the gap between sandstones of the Middle Pierre Shale and the Dakota Group. The Dakota J Sandstone has only minor feldspar and can be classified as quartz arenite or sublitharenite.

The total feldspar content (Fig. 4) decreases downward in the Cretaceous strata and ranges from a mean high in the Fox Hills of 11.4 vol% through 8.7 vol% for sandstones in the Middle Pierre Shale, 4.3 vol% for the Codell Sandstone to 0.5 vol% for the Dakota J Sandstone. The total feldspar content rises again to 5.9 vol% in the Permian Lyons Sandstone.

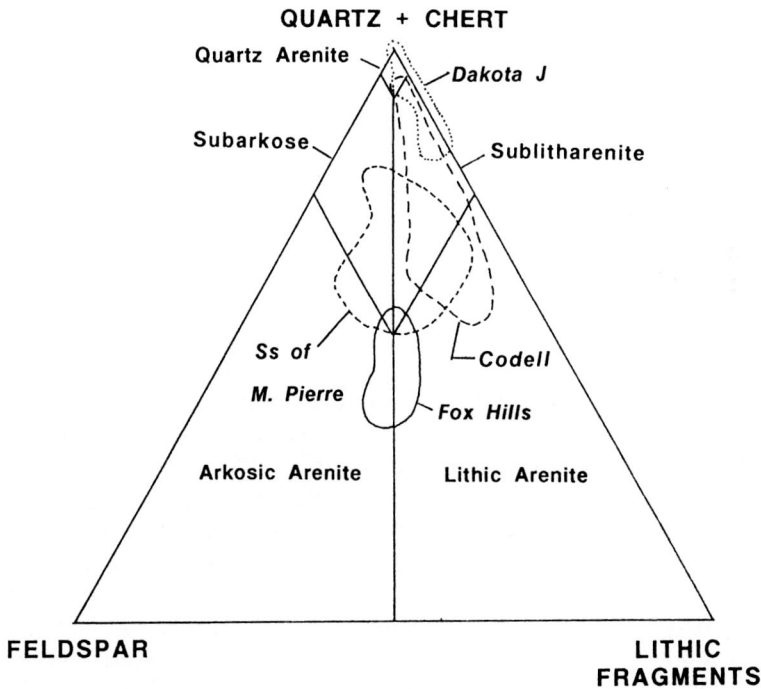

Fig. 3. Compositional ranges for Cretaceous sandstones in the Wattenberg area of the Denver Basin plotted on a modified (Pettijohn et al. 1987) Dott classification triangle

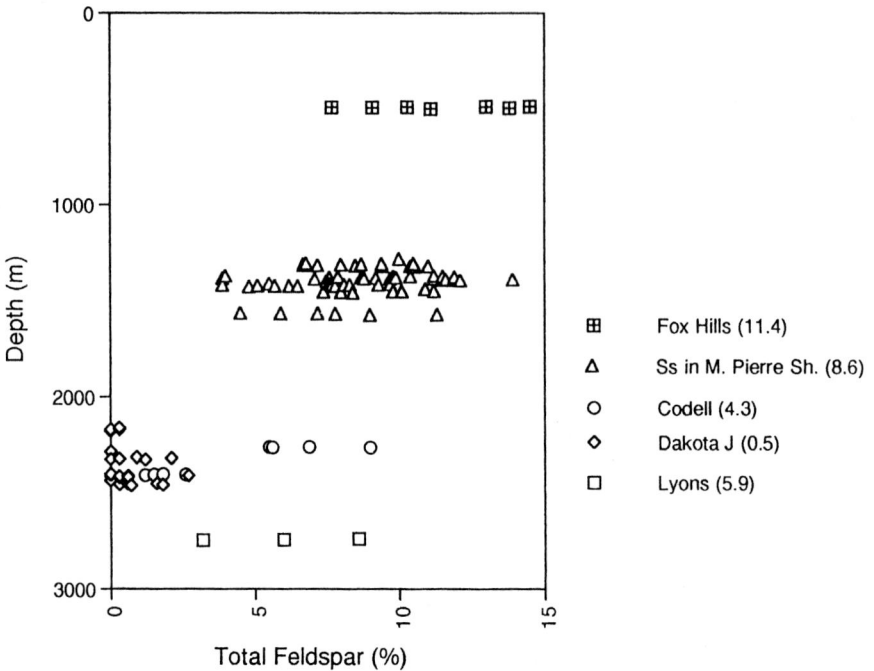

Fig. 4. Relationship between depth and volume percentage of total feldspar for sandstones in the Wattenberg area. The *numbers in parentheses* following the stratigraphic names are the mean values for total feldspar

The proportion of feldspar types changes along with the decrease in total feldspar content in the Cretaceous (Fig. 5). K-feldspar averages 5.4 vol% in the Fox Hills, drops to 1.8 vol% in sandstones of the Middle Pierre Shale, and is absent in the Codell and Dakota J sandstones. This decrease does not appear to be due to diagenesis. K-feldspar sometimes contains intragranular pores, but appears to be less commonly dissolved than plagioclase. Also, there is no evidence of progressive loss of K-feldspar by replacement. K-feldspar appears again in the Permian Lyons Sandstone where it averages 5.5 vol%. Plagioclase is most abundant in the Fox Hills and sandstones of the Middle Pierre Shale (6.0 and 6.8 vol%, respectively), drops to 4.3 vol% in the Codell Sandstone, and to trace amounts in the Dakota J Sandstone (0.5 vol%) and the Lyons Sandstone (0.4 vol%). Plagioclase has undergone albitization and most of the plagioclase is albitized at a depth of 1600 m (Pittman 1988). The decrease in plagioclase with depth does not appear to be due to diagenesis, although feldspar certainly has been dissolved. Neither the amount of plagioclase nor K-feldspar appears to correlate with the amount of secondary porosity in any formation.

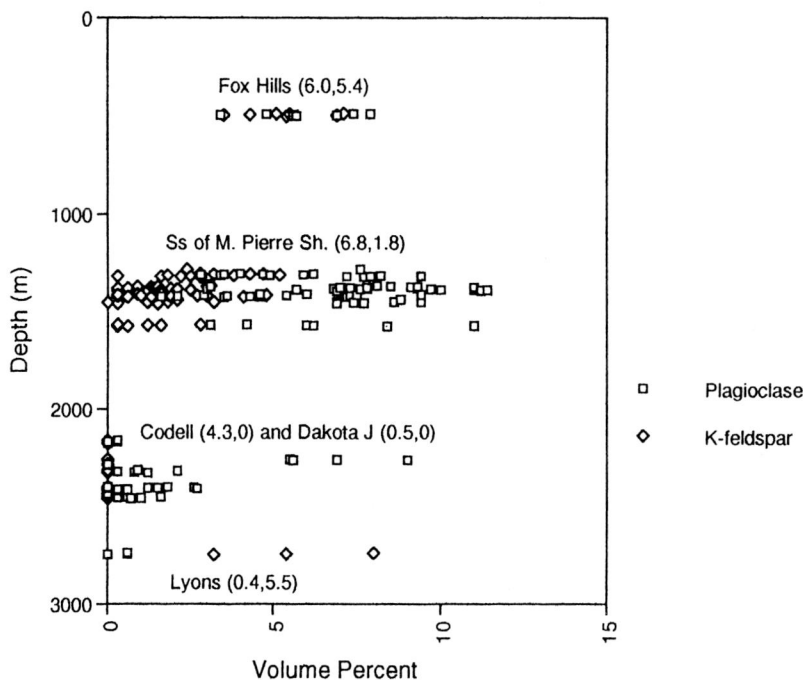

Fig. 5. Cross-plot of depth versus plagioclase and K-feldspar. The *numbers in parentheses* are the mean values for plagioclase followed by K-feldspar for each stratigraphic unit. Only in the Lyons Sandstone is K-feldspar more abundant than plagioclase

2.2.3 Porosity

Figure 6 shows the volume percent of macroporsity based on point-count analyses for the Fox Hills through to the Lyons at a depth of approximately 2740 m. The total effective porosity, that is, the porosity measured by routine core analysis, would be considerably higher for many of the Cretaceous sandstone samples because of microporosity. Generally, the macroporosity decreases with depth and increasing age. The shallowest formation, the Fox Hills, has a mean macroporosity of 13.1 vol%. Sandstones of the Middle Pierre Shale have 5.6 vol% mean macroporosity, whereas the Codell Sandstone has less than 1 vol%, the Dakota J Sandstone 3.9 vol%, and the Lyons Sandstone 0.9 vol% porosity.

The recognition of secondary intragranular porosity associated with feldspar grains is not a problem, but any identification of intergranular pores as being of secondary origin is interpretative and subjective, and is based on textural criteria established by Schmidt and McDonald (1979). Calcite cement may be dissolved to create secondary intergranular porosity that

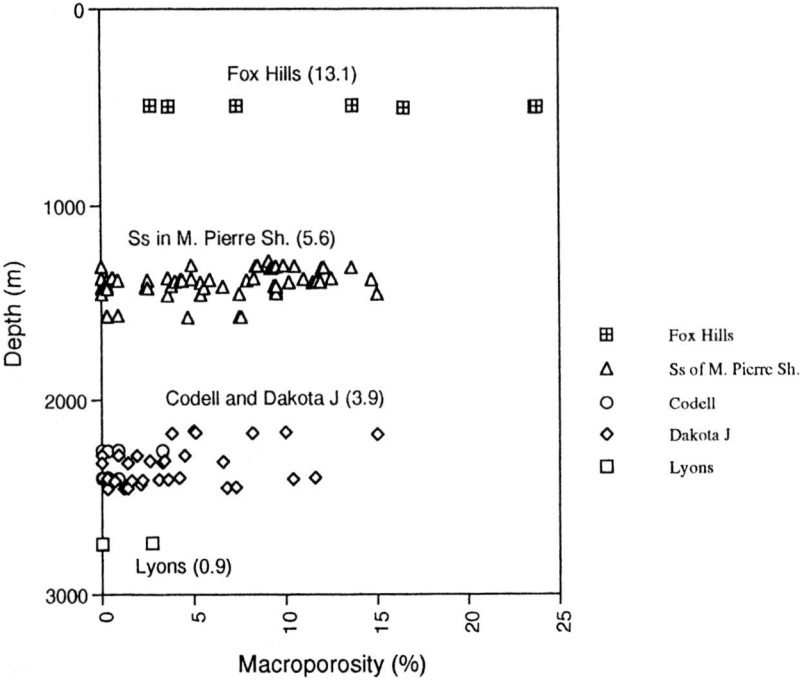

Fig. 6. Relationship between depth and volume percentage of macroporosity for sandstones in the Wattenberg area. The *numbers in parentheses* are the mean values for each stratigraphic interval

closely mimics primary intergranular porosity. The observation that calcite cement has been dissolved in one area of a thin section is often used as the basis for an inference that all intergranular pores visible in the thin section are secondary in origin. We have avoided making that inference. Instead, we have taken a conservative approach: only an intergranular pore adjacent to calcite cement with corroded crystals was considered to be secondary. We did not extend the interpretation to include any other intergranular pores.

Even using the conservative approach discussed above, the majority of the porosity in sandstones in this portion of the Denver Basin is secondary in origin (Fig. 7). Mean values for the percentage of secondary porosity grouped stratigraphically are: Fox Hills, 74%; sandstones in the Middle Pierre Shale, 78%; Codell, 100%; Dakota J, 53%; and Lyons, 56%.

The most common origin of secondary porosity in sandstones studied is due to dissolution of plagioclase or else dissolution of carbonate that replaced plagioclase. Replacement of silicate minerals by calcite is not uncommon, and much of the plagioclase dissolution porosity may actually be replacement-calcite dissolution porosity. We know of no way to resolve this problem. For purposes of calculations, we assumed that plagioclase was dissolved.

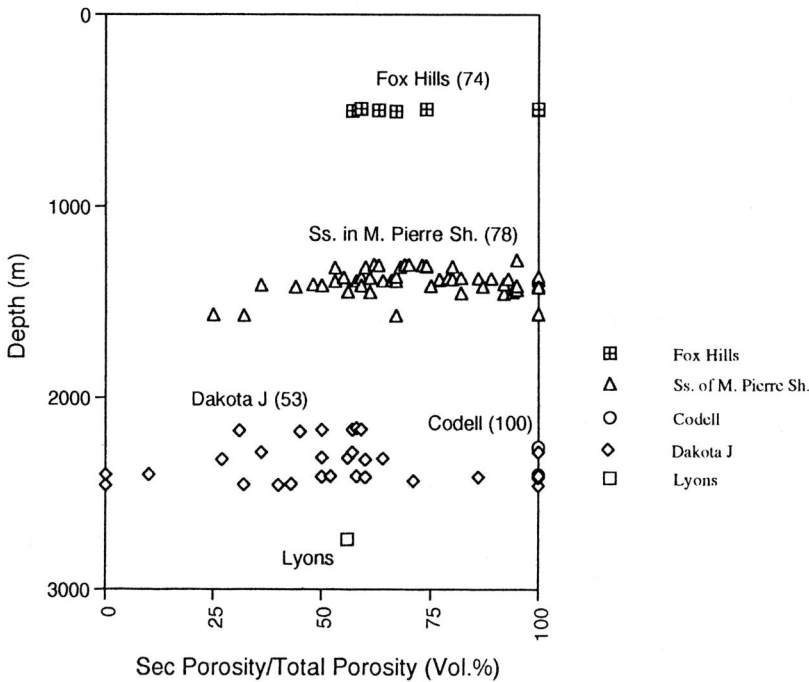

Fig. 7. Depth versus the percentage of secondary porosity in the total porosity for sandstones in the Wattenberg area. The *numbers in parentheses* are the mean values for each stratigraphic interval

Not only is the shallowest formation (Fox Hills) plagioclase-rich, but it also has the most porosity attributable to plagioclase dissolution. The Codell Sandstone is highly argillaceous with low permeability, which has retarded the dissolution of framework grains. In all sandstones, secondary porosity attributable to dissolution of K-feldspar is limited to trace amounts. The third form of secondary porosity, which is dissolution of calcite cement and replacement, is sparse (maximum of 1.5 vol%) and absent in most samples. Trace amounts of secondary porosity also formed by the dissolution of lithic fragments.

Cement, particularly carbonate, and argillaceous matrix are effective at reducing macroporosity in the sandstones studied. As is usually the case, the amounts of matrix and cement in the sandstones vary inversely. For all sandstones, argillaceous matrix is detrimental to macroporosity. In some samples, macroporosity is fairly good (>10 vol%) despite considerable matrix (15 to 20 vol%). This is possible because bioturbation has produced patches of interspersed porous and clay-filled sandstone.

2.2.4 Carbonate Cement

Carbonate cement/replacement in sandstones of the Denver Basin is most commonly ferroan calcite based on staining techniques. Dolomite may occur, but is volumetrically unimportant except in the Fox Hills Formation where dolomite (mean of 6.7 vol%) is more common than ferroan calcite (mean of 4.5 vol%). Ferroan calcite averages 2.8 vol% for sandstones of the Middle Pierre Shale compared with less than 1 vol% for dolomite. Sandstones in the two Codell cores are quite different in terms of the amount of ferroan calcite cement, but average 2.1 vol% with no dolomite. The Dakota J Sandstone has only trace amounts of ferroan calcite and dolomite cement. The Lyons Sandstone contains trace amounts of calcite and no dolomite.

3   Material Balance Considerations

Two types of calculations were made. In the first type, the amount of plagioclase and calcite that could be dissolved by organic acids and carbon dioxide generated by decarboxylation of organic matter is complementary and additive. The second type of calculation also is additive, but speculative in our opinion, and involves the generation of organic acids from redox reactions (Surdam and Crossey 1985). We have used data from hydrous pyrolysis experiments using shale source rocks with Type-II kerogen to establish the conversion efficiency of kerogen oxygen to mono- and dicarboxylic oxygen. A 1-km^2 reference area was used in calculating volumes. Data on thickness of source rocks and sandstones were derived from averaging penetration thickness for 13 wells in the Wattenberg area. The total volume of sandstone in the study area is 0.1645 km^3. The sandstones in the Wattenburg area have 2.2% secondary porosity (0.0036 km^3) based on point-count data. The thickness and volume of source rocks used in the calculations are given in Table 1.

3.1   Dissolution of Plagioclase by Carboxylic Acids

For purposes of calculations, a plagioclase of composition An_{20} was used. Dissolution of plagioclase may be aided by the presence of aluminum-complexing, organic agents. Acetate and malonate are used as representative mono- and dicarboxylic acid anions to complex aluminum. A hypothetical expression for the dissolution of plagioclase in the presence of acetic acid is:

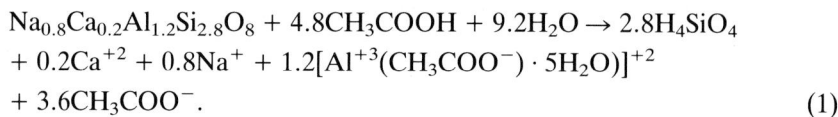

$$Na_{0.8}Ca_{0.2}Al_{1.2}Si_{2.8}O_8 + 4.8CH_3COOH + 9.2H_2O \rightarrow 2.8H_4SiO_4$$
$$+ 0.2Ca^{+2} + 0.8Na^+ + 1.2[Al^{+3}(CH_3COO^-) \cdot 5H_2O)]^{+2}$$
$$+ 3.6CH_3COO^-. \qquad (1)$$

Martell and Smith (1977) proposed that aluminum monoacetate is a stable complex and this gives a maximum value for secondary porosity development by acetic acid. The reaction is written using CH_3COO^- anions for charge balance rather than H^+ as a reactant, in order to show porosity generating potential only from organic acids with no external acid source required. This approach was suggested by I.D. Meshri (pers. comm. 1985).

The dissolution of plagioclase in the presence of malonic acid may be expressed as:

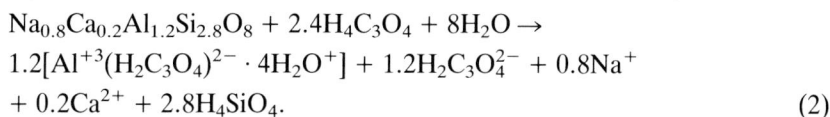

$$Na_{0.8}Ca_{0.2}Al_{1.2}Si_{2.8}O_8 + 2.4H_4C_3O_4 + 8H_2O \rightarrow$$
$$1.2[Al^{+3}(H_2C_3O_4)^{2-} \cdot 4H_2O^+] + 1.2H_2C_3O_4^{2-} + 0.8Na^+$$
$$+ 0.2Ca^{2+} + 2.8H_4SiO_4. \qquad (2)$$

We used 1 tonne (10^6 g) as the unit mass of source material. The mass ratio of carbon in the total kerogen and the percent of oxygen in the kerogen were assumed to be 0.70 and 10%, respectively, based on hydrous pyrolysis work involving shale source rocks (M.D. Lewan, pers. comm. 1985). The number of moles of acetic and malonic acid generated per tonne of source rock was calculated taking into account the maximum experimental (M.D. Lewan, pers. comm. 1985) conversion efficiency of kerogen oxygen to carboxylic oxygen of 10% for acetic acid. A malonic to total carboxylic acid ratio of 0.52 was based on the maximum reported concentration of malonic acid in formation waters (MacGowan and Surdam 1988). Then, the volume of plagioclase that could be dissolved per tonne of source rock for acetic and malonic acids was calculated. The next step was to use the data from the Denver Basin (Table 1) and calculate the volume of plagioclase that could be dissolved for each stratigraphic source rock unit. Based on the above procedures, 7.35×10^{-4} km^3/total source of plagioclase was dissolved by carboxylic acids, which accounts for 0.45% secondary porosity.

Assuming that plagioclase is more unstable than K-feldspar and therefore more likely to be dissolved, any carboxylic acids remaining after plagioclase dissolution may be used for K-feldspar dissolution. The carboxylic acids produced are insufficient to account for all of the porosity generated by plagioclase dissolution. Therefore, no calculations for K-feldspar were made.

3.2 Dissolution of Calcite

For the dissolution of calcite, we assumed an initial kerogen oxygen content of 10%. After removal of oxygen as mono- and dicarboxylic acids, 8.8 wt% oxygen will remain. M.D. Lewan (pers. comm. 1985) suggested that 5.5 wt% oxygen is the minimum residual oxygen content for Type-II kerogen based on hydrous pyrolysis experiments. Thus, 3.3 wt% oxygen may be removed as carbon dioxide. Following the procedure outlined above for organic acids, the moles of carbon dioxide produced per unit mass of source rock, the potential volume of calcite dissolved (8.88×10^{-4} km^3/total source), and the secondary porosity due to calcite dissolution (0.54%) by carbonic acid were calculated. Based on point-count data, the actual secondary porosity due to calcite dissolution is 0.29%.

3.3 Organic Acids Derived from Redox Reactions

Surdam and Crossey (1985) suggested that iron released during the smectite to illite transformation acts as a mineral oxidant with organic matter, forming more oxygen-bearing functional groups in the kerogen and thereby enhancing the overall yield of carboxylic acids. Eslinger et al. (1979) suggested that the iron may be reduced within the clay mineral structure. In either case, the reduction of mineral oxidants and simultaneous oxidation of kerogen may be a possible mechanism for generating organic acids. Surdam and Crossey (1985) calculated the amount of kerogen converted to specific organic products (e.g., acetate and oxalate) for various concentrations and types of kerogen. We utilized their work to estimate the amount of acetate and oxalate derived from Type-II kerogen during the smectite to illite transformation. Subsidence modeling indicated that only the Carlile and Graneros formations of the Benton Group and the Dakota Group were hot enough for the smectite to illite conversion to occur in the source rocks. Therefore, shallower source rocks were not included in these calculations. We considered, for purposes of these calculations, that malonic acid and oxalic acid (used by Surdam and Crossey 1985), both dicarboxylic acids, were comparable. Based on the above approach, the volume of plagioclase dissolved by organic acids generated by the redox mechanism is 0.34% by acetic acid and 0.39% for malonic acid (based on a malonic acid to total carboxylic acid ratio of 0.52).

3.4 Discussion

Calculated secondary porosities [from organic acids ($\Phi_{\text{Ac and Mal}}$), carbon dioxide (Φ_{CO_2}), and the redox mechanism ($\Phi_{\text{Ac}}^{\text{Redox}} + \Phi_{\text{Mal}}^{\text{Redox}}$)] based on the maximum ratio (0.52) of dicarboxylic acids to total carboxylic acids reported in the literature (MacGowan and Surdam 1988) would yield:

$$\Phi_{\text{Ac and Mal}} + \Phi_{\text{CO}_2} + \Phi_{\text{Ac}}^{\text{Redox}} + \Phi_{\text{Mal}}^{\text{Redox}} = 0.45 + 0.54 + 0.34 + 0.39$$
$$= 1.72\% \text{ secondary}$$

porosity. The actual secondary porosity determined by point count was 2.49% (2.2% plagioclase and 0.29% calcite). Thus, the modeling predicts that a capability to generate organic acids and carbon dioxide accounts for approximately 69% of the secondary porosity documented in thin section. Keep in mind that this is a conservative approach and that almost certainly there is more secondary porosity than identified in the point-count analyses. If the source rocks were exceptionally rich (e.g., 16% TOC), such as immature Woodford or Phosphoria Formation, then it would be possible to generate 2.85% secondary porosity. Also, if the kerogen in the source rocks of the Denver Basin had been predominantly Type III instead of Type II, then presumably more oxygen would have been available for conversion to organic acids and carbon dioxide (Hunt 1979) and more secondary porosity could have formed. However, based on experimental work by Eglinton et al. (1987), organic matter with a high oxygen content does not always generate high carboxylic acid concentrations.

We consider the redox mechanism of generating carboxylic acids to be speculative because it seems likely that any oxygen in the subsurface would be used up early in oxidizing organic matter, which is why the organic matter is preserved. Where does the free oxygen come from to make Fe_2O_3 from the Fe^{+3} released by the smectite to illite transformation? Also, the volume of solids estimated to be dissolved using the redox mechanism represents the maximum possible because: (1) the equations used by Surdam and Crossey (1985) have only organic acids as an organic reaction product, whereas it would be possible to have other materials produced; (2) complete dissociation of the acids is assumed; and (3) the equations used for the dissolution of plagioclase form an organo-aluminum complex in solution with no precipitation of reaction products. If the equations for the dissolution of plagioclase are written so that kaolinite is formed, then there is less porosity created.

4 Conclusions

The material balance study to estimate whether the acids generated by thermal maturation of shales could create the secondary porosity in sandsones in the Denver Basin was based on the following: (1) determination of the volume and type of organic matter in each source rock unit; (2) point-count analyses to determine the type and amount of macroporosity in each sandstone unit in the stratigraphic column; (3) use of geophysical logs to determine the thickness of stratigraphic units enabling calculation of volumes based on $1 \, km^2$ overlying the deepest part of the basin; and (4) use of hydrous pyrolysis experimental data on untreated shale source rocks to establish the conversion efficiency of kerogen oxygen to mono- and dicarboxylic oxygen.

Calculations were maximized by using a ratio of dicarboxylic to total carboxylic of 0.52, the highest value cited in the literature and by using a conservative approach to identify secondary porosity in thin section. We calculated that 1.72% secondary porosity could form in the sandstones from all sources (CO_2 and carboxylic acids), whereas by point-count analyses there is 2.49% secondary porosity; that is, 69% of the secondary porosity is explainable by the material balance calculations.

We recognize that modeling approaches are quite simplified compared to the complicated processes of nature. The modeling presented in this chapter certainly can be improved with additional data and a better understanding of the processes involved in creating secondary porosity in sandstones. Nevertheless, this study, which uses a conservative approach to identify secondary porosity, suggests that insignificant amounts of secondary porosity can be generated from the maturation of organic matter.

Acknowledgments. We thank M.D. Lewan for hydrous pyrolysis data on shale source rocks used as a basis for the calculations. Also, we thank J.B. Fisher and I.D. Meshri for advice on various chemical aspects of this chapter. J.B. Fisher, I.D. Meshri, and R.R. Thompson read portions of the manuscript and provided helpful comments. We thank Amoco Production Co. for granting permission to publish this chapter, which is based on Research Dept. Report F86-G-3.

References

Bailey EH, Stevens RE (1960) Selective staining of K-feldspar and plagioclase on rock slabs and thin sections. Am Mineral 45: 1020–1025

Bjørlykke K (1983) Diagenetic reactions in sandstones. In: Parker A, Sellwood, BW (eds) Sediment diagenesis. Reidel, Dordrecht, pp 169–213

Bjørlykke K (1984) Formation of secondary porosity: how important is it? In: McDonald DA, Surdam RC (eds) Clastic diagenesis. Am Assoc Pet Geol Mem 37: 277–286

Clayton JL, Swetland PJ (1980) Petroleum generation and migration in Denver Basin. Am Assoc Pet Geol Bull 64: 1613–1633

Curtis CD (1978) Possible links between sandstone diagenesis and depth-related geochemical reactions occurring in enclosing mudstones. J Geol Soc (Lond) 135: 107–117

Dickson JAD (1966) Carbonate identification and genesis as revealed by staining. J Sediment Petrol 36: 491–505

Dunn TL, Hessing RB, Sandkuhl DL (1985) Application of voice-recognition computer-assisted point counting. J Sediment Petrol 55: 602–603

Eglinton TI, Curtis CD, Rowland SJ (1987) Generation of water-soluble organic acids from kerogen during hydrous pyrolysis: implications for porosity development. Mineral Mag 51: 495–503

Eslinger E, Highsmith P, Albers D, DeMayo B (1979) Role of iron reduction in the conversion of smectite to illite in bentonites in the disturbed belt, Montana. Clays Clay Minerals 27: 327–338

Galloway WE, Hobday DK, Magara K (1982) Frio Formation of Texas Gulf coastal plain: depositional systems, structural framework, and hydrocarbon distribution. Am Assoc Pet Geol Bull 66: 649–688

Hayes JB (1979) Sandstone diagenesis – the hole truth. In: Scholle PA, Schluger PR (eds) Aspects of diagenesis. Soc Econ Paleontol Mineral Spec Publ 26, Tulsa, OK, pp 127–139

Hunt JM (1979) Petroleum geochemistry and geology. Freeman, San Francisco, 617 pp

Loucks RG, Dodge MM, Galloway WE (1984) Regional controls on digenesis and reservoir quality in Lower Tertiary sandstones along the Texas Gulf Coast. In: McDonald DA, Surdam RC (eds) Clastic diagenesis. Am Assoc Pet Geol Mem 37: 15–45

Lundegard PD, Kharaka YK (1990) Geochemistry of organic acids in subsurface waters. In: Melchior DC, Bassett RL (eds) Chemical modeling of aqueous systems II. Am Chem Soc Symp Ser 146, Washington DC, pp 169–189

Lundegard PD, Land LS (1986) Carbon dioxide and organic acids: their role in porosity enhancement and cementation, Paleogene of the Texas Gulf Coast. In: Gautier DL (ed) Soc Econ Paleontol Mineral Spec Publ 38, Tulsa, OK, pp 129–146

Lundegard PD, Land LS, Galloway WE (1984) Problem of secondary porosity: Frio Formation (Oligocene), Texas Gulf Coast. Geology 12: 399–402

MacGowan DB, Surdam RC (1988) Difunctional carboxylic acid anions in oilfield waters. Org Geochem 12: 245–259

Martell AE, Smith RM (1977) Critical stability constants, vol 3. Other ligands. Plenum Press, New York, 495 pp

Moredock DE, Williams SJ (1976) Upper Cretaceous Terry and Hygiene Sandstones – Singletree, Spindle and Surrey fields, Weld County, Colorado. In: Epis RC, Weimer RJ (eds) Colorado School Mines Prof Contrib 8, Golden, CO, pp 264–274

Pettijohn FJ, Potter PD, Siever R (1987) Sand and sandstone. 2nd edn. Springer, Berlin Heidelberg New York, 553 pp

Pittman ED (1988) Diagenesis of Terry Sandstone (Upper Cretaceous), Spindle Field, Colorado. J Sediment Petrol 58: 785–800

Rice DD (1984) Relationship of hydrocarbon occurrence to thermal maturity of organic matter in the Upper Cretaceous Niobrara Formation, eastern Denver Basin: evidence of biologic versus thermogenic origin of hydrocarbons. In: Woodward J, Meissner FF, Clayton JL (eds) Hydrocarbon source rocks of the Greater Rocky Mountain Region. Rocky Mountain Assoc Geol, Denver, CO, pp 365–368

Schmidt V, McDonald DA (1979) The role of secondary porosity in the course of sandstone diagenesis. In: Scholle P A, Schluger PR (eds) Aspects of diagenesis. Soc Econ Paleontol Mineral Spec Publ 26, Tulsa, OK, pp 175–207

Surdam RC, Crossey LJ (1985) Mechanism of organic/inorganic interactions in sandstone/shale sequences: relationship of organic matter and mineral diagenesis. In: Gautier DL, Kkaraka YK, Surdam RC (eds) Relationship of organic matter and mineral diagenesis. Soc Econ Paleontol Mineral Short Course 17, pp 177–232

Surdam RC, Boese SW, Crossey LJ (1984) The chemistry of secondary porosity. In: McDonald DA, Surdam RC (eds) Clastic diagenesis. Am Assoc Pet Geol Mem 37: 127–149

Tainter PA (1984) Stratigraphic and paleostructural controls on hydrocarbon migration in Cretaceous D and J Sandstones of the Denver Basin. In: Woodward J, Meissner FF, Clayton JL (eds) Hydrocarbon source rocks of the Greater Rocky Mountain Region. Rocky Mountain Assoc Geol, Denver, CO, pp 339–354

Tissot BP, Welte DH (1978) Petroleum formation and occurrence. Springer, Berlin Heidelberg New York, 538 pp

Vandergrift GF, Winans RE, Scott RG, Horwitz EP (1980) Quantitative study of the carboxylic acids in Green River oil shale bitumen. Fuel 59: 627–633

Walker TR (1984) Diagenetic albitization of potassium feldspar in arkosic sandstones. J Sediment Petrol 54: 3–16

Appendix

Symbols Used

Cr   = % carbon in source rock.
Ok   = % oxygen in kerogen.
Ct   = grams of organic carbon per tonne source rock.
Mk   = grams kerogen per tonne source.
gp   = grams plagioclase dissolved.
MWp = molecular weight plagioclase.
Vp   = volume plagioclase dissolved.
SGp = specific gravity plagioclase.
gc   = grams calcite dissolved.
MWc = molecular weight calcite.
Vc   = volume calcite dissolved by carbonic acid.
SGc = specific gravity calcite.

Assumptions

The following calculations assume:

1. Unit mass of source material $= 10^6$ g.
2. % Carbon in the source rock (%Cr), and % oxygen in the kerogen (%Ok) are known.
3. Mass ratio of carbon in the total kerogen is 0.7 (70%); from analyses of immature kerogens (M.D. Lewan, pers. comm. 1985).
4. Conversion efficiency of kerogen oxygen to carboxylic oxygen is 10%; maximum experimental conversion efficiency of M.D. Lewan (pers. comm. 1985).
5. The ratio of dicarboxyl to monocarboxyl acids produced (in this case, malonic:acetic acid ratio) is 1.08:1 (or 0.052:0.048 for a total conversion efficiency of 0.1; based on the maximum observed ratio of dicarboxyl to monocarboxylic acid anions in oil-field waters reported by MacGowan and Surdam (1988).
6. Carboxylic acids generated by source rocks are strongly dissociated.

Acetic Acid

Given the above, the total grams of organic carbon per tonne source rock is:

$$Ct = (\%Cr/100) \times 10^6$$

and the total organic matter (grams kerogen per tonne source) is:

$$Mk = Ct/0.7.$$

The mass of oxygen (g) per tonne of source is:

$Ok = (Mk) \times (\%Ok/100)$

and moles of oxygen are given by:

mol $Ok = Ok/16$.

From the moles of oxygen in the kerogen, the number of moles of acetic acid that may be generated is given by:

$mCH_3COOH = (mol\ Ok/2) \times 0.048$ (see assumptions above).

Combining all constants, the number of moles of acetic acid generated is:

$mCH_3COOH = 0.214 \times \%Cr \times \%Ok$.

Assuming a plagioclase composition of An_{20} and the chemical reaction shown in Eq. (1) (text), the number of grams of plagioclase potentially dissolved by acetic acid is given by:

$gp = MWp \times (mCH_3COOH/4.8)$.

Then the potential volume of plagioclase dissolved by acetic acid (in $km^3/$ tonne source) is given by:

$Vp = (gp/SGp) \times 10^{-15}$,

where SGp is the specific gravity of plagioclase. Combining all constants

$Vp = 4.5 \times 10^{-15} \times \%Cr \times \%Ok$

will give the potential volume of plagioclase dissolvable by acetic acid produced by each source horizon.

Malonic Acid

Similarly, the number of moles of malonic acid that may be generated is given by:

$mC_3H_4O_4 = (mol\ Ok/4) \times 0.052$.

Combining all constants (as outlined above), the number of moles of malonic acid generated is given by:

$mC_3H_4O_4 = 0.116 \times \%Cr \times \%Ok$.

The number of grams of plagioclase dissolved per tonne of source rock using Eq. (2) in the text is given by:

$gp = MWp \times (mC_3H_4O_4/2.4)$,

and the potential volume of plagioclase dissolved by malonic acid is given by:

$$Vp = 4.8 \times 10^{-15} \times \%Cr \times \%Ok.$$

The total potential volume of plagioclase feldspar dissolved is the sum of Vp acetic and Vp malonic.

Carbon Dioxide

After removal of 10% of the total kerogen oxygen as mono- and dicarboxylic acids, the remaining oxygen will be given off as CO_2 until a base level oxygen content of approximately 5.5 wt% is reached. M.D. Lewan (pers. comm. 1985) suggests that this percentage is the minimum residual oxygen content of a mature Type-II kerogen. As described above, the number of moles of oxygen per tonne of source rock is given by:

$$mol\ Ok = [(\%Cr/100)/0.7] \times 10^6 \times [(\%Ok^*/100)/16],$$

where $\%Ok^*$ is residual oxygen after production of aliphatic acids. The number of moles of CO_2 generated under these conditions is given by:

$$mCO_2 = (mol\ Ok/2) \times (z),$$

where z is a conversion efficiency factor dependent upon the original oxygen content of the kerogen. Combining all constants, the number of moles of CO_2 produced from a given source rock is given by:

$$mCO_2 = 4.45 \times (z) \times \%Cr \times \%Ok.$$

The number of grams of calcite that may be dissolved per tonne of source rock is given by:

$$gc = MWc \times mCO_2,$$

and the potential volume of calcite (km^3) dissolved by carbonic acid per tonne of source rock is given by:

$$Vc = (gc/SGc) \times 10^{-15}.$$

Combining all constants, the total potential volume of calcite dissolved is given by:

$$Vc = 1.65 \times 10^{-13} \times (z) \times \%Cr \times \%Ok.$$

Discussion

Because our calculations involve a unit mass of source rock, we must convert to a unit volume of source rock. This value is then multiplied by a $1\ km \times 1\ km \times h\ km$ volume term where h is the actual thickness of a given source horizon as interpreted from well logs. In this way, the plagioclase dissolution potential for each source rock in the Wattenberg area can be calculated, and the total plagioclase dissolution potential for the system is the sum of the potentials for each source horizon. This value may be converted to percent porosity generated in associated sandstones using a sand/shale ratio of 0.1645:1, also determined from well logs.

For the redox method, a slightly different technique was applied. The mass ratio of carbon in the total kerogen was assumed to be 0.75. An average shale density of $2.3 \, g/cm^3$ was used for converting mass of source material to volume of source material. The moles of H^+ generated and the potential volume of plagioclase dissolved were calculated from the source rock data using the following equation:

$$\text{Mol } H^+ = (\text{wt\% TOC}/75) \times [\text{wt\% converted to carbox. acid}/(100 \times \text{mol. wt. carbox. acid})] \times 2.3 \, g/cm^3.$$

The values used for the wt% conversion term in the above equation were estimated from the calculated curves of Surdam and Crossey (1985). The redox calculation was made for acetic and malonic acids, assuming a conversion efficiency for malonic acid equivalent to that reported for the production of oxalic acid. This calculation was applied only to source rocks undergoing the smectite-illite transformation. The potential volume of plagioclase dissolved was calculated using the relationship:

$$Vp = (\text{total moles } H^+/A) \times \text{mol. vol. plag} \times 10^{-15},$$

where A is dictated by the stoichiometry of the dissolution reactions given in Eqs. (1) and (2) in the text.

Chapter 6   Role of Soil Organic Acids in Mineral Weathering Processes

James I. Drever[1] and George F. Vance[2]

Summary

The soluble organic acids in soils consist largely of complex mixtures of polymeric compounds referred to collectively as fulvic and humic acids. These compounds are relatively refractory, and are broken down only slowly by microorganisms. Low-molecular-mass acids (e.g., acetic, oxalic, formic) exist in dynamic balance: they are rapidly produced and consumed by microorganisms. Their concentrations may be quite high where rapid decomposition of plant material is taking place, and in microenvironments adjacent to roots and fungal hyphae, but are generally low (typically less than 1 mM) in bulk soil solution. Concentrations of organic acids are generally highest in the organic layer at the top of the soil profile and decrease with depth. Particularly high concentrations of dissolved organic acids occur in peatlands and waterlogged soils.

Organic acids affect the mineralogy of soils mostly through their ability to complex and transport iron and aluminum, resulting in a characteristic profile development of humid regions. There is considerable controversy as to whether organic acids at natural concentrations significantly accelerate the rate of dissolution of primary silicate minerals. Proposed mechanisms for increasing rates involve formation of "surface complexes" with ions on the surface of the mineral, and lowering of the pH of the soil solution. Neither mechanism is likely to have a significant effect on the dissolution rates of primary minerals of granitic rocks, but may have an effect on more mafic rock types. There will also be significant effects in the microenvironments around rootlets or fungal hyphae, where organic acid concentrations may be much higher than in bulk soil solution.

[1] Department of Geology and Geophysics, University of Wyoming, Laramie, Wyoming, USA
[2] Department of Plant, Soil and Insect Sciences, University of Wyoming, Laramie, Wyoming, USA

1 Introduction

It has been known for a long time that organic acids, particularly humic and fulvic acids, are present in soil solutions, and that these acids have a major influence on the translocation of iron and aluminum in soil profiles (McKeague et al. 1986; McColl et al. 1990). Organic acids may also accelerate the weathering of primary silicate minerals (Tan 1986). In this chapter, we shall summarize the occurrence and distribution of soluble organic acids in soils, and shall review critically the role of organic acids in mineral weathering.

2 Source, Nature, and Distribution of Dissolved Organic Carbon in Soils

Solid and dissolved organic matter in soils is a complex mixture of organic molecules that has defied complete characterization (Aiken et al. 1985; Hayes et al. 1989). A major portion of the solid and dissolved organics in soils is composed of humic substances, which are defined (Aiken et al. 1985) as "A general category of naturally occurring, biogenic, heterogeneous organic substances that can generally be characterized as being yellow to black in color, of high molecular weight, and refractory." Identifiable specific compounds such as low-molecular-mass organic acids, proteins, polysaccharides, and lipids also make up a portion of soil organic matter (Schnitzer 1978; Stevenson 1982; Thurman 1985).

2.1 Humic Substances

Soil humic substances are classified on the basis of solubility differences into a humin fraction (insoluble in alkaline and acidic solutions), a humic acid fraction (soluble in alkaline solution but precipitates upon acidification), and a fulvic acid fraction (soluble in alkaline and acidic solutions). The insoluble fractions are generally separated from the soluble material by centrifugation, or by passing through a $0.45-\mu m$ membrane filter; the distinction between "soluble" and "insoluble" is, in fact, operationally defined by the ability to pass through a $0.45-\mu m$ filter. Humic and fulvic acid materials are described as a complex mixture of polyelectrolytes with highly variable molecular weights; the nature of the humin fraction is relatively unknown (Hatcher et al. 1985). Fulvic acids are lower in molecular weight than humic acids and are the predominant form of dissolved organic carbon (DOC) in soil solutions and surface and groundwaters (Thurman 1985). Fulvic acids have been identified in virtually every soil solution, surface water, and groundwater tested.

Fulvic acids typically have a molecular mass range of about 500–5000 daltons, and are highly variable in composition. Those found in lakes and streams appear to be more aliphatic in nature than those of terrestrial soils (Malcolm 1985). Humic acids, on the other hand, have higher molecular masses (typically 2000 to 50000 daltons). Soil humic acids are generally older and higher in molecular mass than are those found in surface waters. The basic structure of humic acid consists of aromatic rings of the di- or trihydroxyphenyl type bridged by $-CH_2-$, $-O-$, $-NH-$, $-N=$, $-S-$, and other groups. In addition, the humic acid structure may contain several types of $-OH$ groups and quinone linkages (Stevenson 1982). Theoretical structures for humic and fulvic acids have been proposed by several researchers; Fig. 1 represents proposed humic acid (Stevenson 1982) and fulvic acid (Buffle 1977) structures. We should stress, however, that fulvic and humic acids represent complex mixtures, and cannot be represented by any single formula.

All humic substances contain $-COOH$ groups (as well as $-OH$), which vary in acidity and position in the organic molecule. In the natural state,

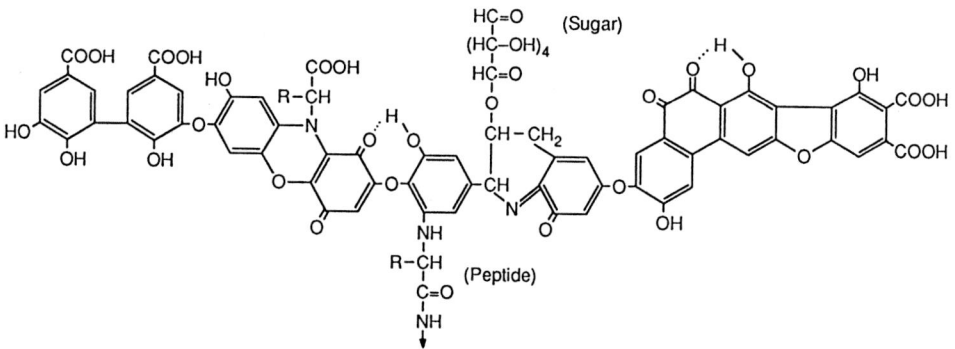

(1) Humic Acid

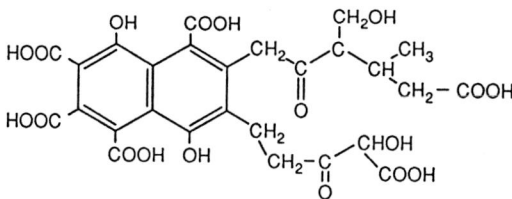

(2) Fulvic Acid

Fig. 1. Proposed structures for humic (*1*) and fulvic (*2*) acids. (After Stevenson 1982, Buffle 1977)

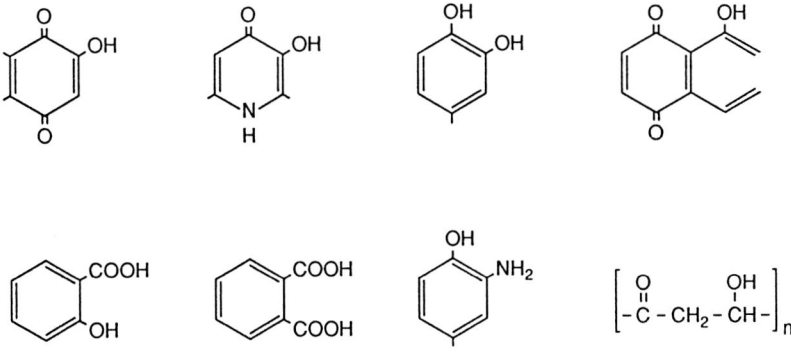

Fig. 2. Potential binding sites on humic and fulvic acids

humic substances contain attached proteinaceous and carbohydrate residues, which are capable of forming complexes with polyvalent cations (Stevenson and Fitch 1986). Some potential binding sites in humic substances are shown in Fig. 2.

The ability of humic substances to form stable complexes with polyvalent cations is due to their high contents of oxygen-containing functional groups, which include $-COOH$, phenolic$-OH$, enolic$-OH$, alcoholic$-OH$, and $>C=O$. The estimated average contents (and ranges) of functional groups in humic and fulvic acids, as reported by Stevenson and Vance (1989), are respectively in mEq g^{-1}, 7.2 ± 1.6 and 10.3 ± 3.9 for total acidity, 3.6 ± 2.1 and 8.2 ± 3.0 for $-COOH$ groups, 3.9 ± 1.8 and 3.0 ± 2.7 for acidic$-OH$ and, 2.6 ± 2.4 and 6.0 ± 3.4 for weakly acidic$-OH$. The acidic nature of humic substances is due primarily to $-COOH$ and acidic$-OH$ groups with $-COOH$ being the most important.

Metal ion binding by humic substances has been suggested to occur at salicylic ($-COOH$ ortho to a phenolic$-OH$) and phthalic (two adjacent $-COOH$ on an aromatic ring) sites (Schnitzer and Khan 1972). Vance et al. (1986) and Sikora and McBride (1990) concluded that organic substances containing catechol (ortho phenolic$-OH$) sites may be significant in the complexation, mobilization, and translocation of Fe and Al during podzolization. Due to their ubiquitous nature, humic substances may play a greater role than low-molecular-mass organic acids in modifying the reactivity of metals on oxide surfaces and complexation of metals in most soils and natural waters (Tan 1986; Stevenson and Vance 1989).

The concentrations of humic and fulvic acids vary considerably from soil to soil, but are closely related to the total organic matter content in the soil. Soil organic matter in well-drained soils ranges from as little as 1.0 wt% or less in coarse-textured soils (Psamments) to more than 6.5 wt% in prairie grassland soils (e.g., Hapludolls) (Soil Survey Staff 1975). Poorly drained mineral soils (aquic suborders), including hydric soils in wetlands, often have organic matter contents approaching 20 wt%; many Histosols (peat) have very high organic matter contents (>30 wt%). Soil organic matter of histic and mollic epipedons (organic-rich surface horizons) is characterized by a high proportion of humic acid, whereas that of forest soils (Alfisols, Spodosols, and Ultisols) contain high proportions of fulvic acid (Stevenson 1985). The humic acid to fulvic acid ratio usually decreases with depth in the soil profile.

2.2 Organic Ligands of Known Composition

Soil organic substances are also comprised of biochemical compounds, such as simple aliphatic and aromatic acids, phenols and phenolic acids, hydroxamate siderophores, sugar acids, and complex polymeric phenols, which are synthesized by living organisms. Biochemical compounds are common to all soils and natural waters and would be expected to play a dominant role in zones where microbial activity is intense. However, biochemical chelating agents normally have only a transitory existence in the pedosphere, and the amounts found at any one time represent a balance between synthesis and destruction by microorganisms (Stevenson 1967). The amount of potential chelate formers present in soils is normally low and variable (Stevenson and Fitch 1986). Table 1 lists examples of different environments and the various types of biochemical compounds present within them.

Organic acids of low-molecular mass are of particular interest due to their ubiquitous nature and because many of them form stable chelate complexes

Table 1. Location and nature of biochemical compounds that exist in the biosphere. Abundant amounts of any one of the compounds listed may accelerate mineral weathering

Environment	Products
Plant canopy	LMM[a] aliphatic acids, phenolic acids and aldehydes, polymeric phenols (flavanoids)
Forest litter	LMM aliphatic and aromatic acids, phenolic acids and aldehydes, polymeric phenols (flavanoids)
Surface horizons	LMM aliphatic and aromatic acids, phenolic acids, polymeric phenols, sugar acids, amino acids
Subsurface horizons	LMM aliphatic and aromatic acids, phenolic acids, sugar acids, amino acids
Soil solutions	LMM aliphatic and aromatic acids, phenolic acids and aldehydes, polymeric phenols, sugar acids, amino acids
Rhizosphere	LMM aliphatic acids and aldehydes, phenolic acids, sugar acids, amino acids, hydroxamate siderophores
Rock surfaces	LMM aliphatic acids, polymeric phenols (lichen acids)

[a] LMM = low molecular mass.

with polyvalent cations. Studies on the isolation and identification of low-molecular-mass aliphatic acids have indicated that the following acids are present in soil: formic, acetic, propionic, butyric, valeric, hexanoic, lactic, oxalic, succinic, fumaric, malic, tartaric, and citric. Hydroxy acids, such as citric, generally form stronger complexes than those acids containing a single —COOH group. Aromatic low-molecular-mass acids found in soil include: benzoic, phenylacetic, and cinnamic acids. Although the concentration of low-molecular-mass organic acids in the soil solution is normally low (usually less than 1 mM), substantially higher amounts can be found in the rhizosphere (environment surrounding plant roots) (Fox and Comerford 1990).

Phenolic compounds present in soils are synthesized by a variety of microorganisms, found in root exudates, added to soils in forest canopy leachates, and derived from decomposing plant and animal remains (Stevenson 1982). The concentration of phenolic acids in the soil solution has been estimated at 0.05 to 0.3 mM. Phenolic compounds released during the decay of plant material include: 4-hydroxybenzoic, protocatechuic, gallic, vanillic, orsellinic, syringic, 4-hydroxyphenylpropionic, 3,4-dihydroxy-phenylpropionic, 4-hydroxycinnamic, caffeic, ferulic, and sinapic acids and their aldehydes. The phenolic acids, 4-hydroxybenzoic, vanillic, 4-hydroxycinnamic, and ferulic acids were reported by Whitehead et al. (1982) to be widely distributed in surface horizons. Protocatechuic acid was identified as the dominant phenolic acid in spodic subsurface horizons (Vance et al. 1986). Aldehyde and methoxy derivatives, 4-hydroxy-benzaldehyde, vanillin, and syringaldehyde have also been detected in surface and subsurface horizons but at lower concentrations. Phenolic acids that contain adjacent OH groups, such as protocatechuic, gallic, and caffeic acids, are believed to be of considerable importance in the complexation and translocation of Al and Fe in soils due to their ability to form chelates with polyvalent cations.

Polymeric phenols refer to substances containing more than one aromatic ring along with one or more phenolic—OH group. They include the flavanoids of plant origin and comprise one of the largest and most widespread groups of secondary plant products. Flavanoids have been identified in aqueous extracts of the leaves and needles of several plants, including the catechin isomers (D- and epi-catechin), quercetin, and the two isomers of gallocatechin. Coulson et al. (1960) concluded that catechin isomers were of considerable importance in soil formation processes due to their ability to form highly stable complexes with Fe and Al.

Lichens are known to dissolve mineral substances during the weathering of rocks and minerals through the synthesis of a variety of complex phenolic compounds that form highly stable complexes with metal ions. Iskandar and Syers (1972) demonstrated the complexing ability of six lichen acids (salazinic, stictic, evernic, lecanoric, roccellic, and atranorin) in weathering silicate minerals and rocks (biotite, granite, and basalt).

Sugar acids are also important natural chelators that can enhance the weathering of soil minerals. Gluconic, glucuronic, and galacturonic acids are all common metabolites of microorganisms. Environments rich in organic matter have been found to contain a large number of microorganisms that synthesize 2-ketogluconic acid. A high proportion of the bacteria in soil, as well as those living on rock surfaces, can produce 2-ketogluconic acid (Stevenson 1967).

Siderophores produced by plants and soil microorganisms may play an important role in the complexation and weathering of Fe-bearing minerals. Several bacteria and fungi produce siderophores in order to enhance Fe solubility. Certain grasses also excrete phytosiderophores under Fe-deficient conditions; the affinity of phytosiderophores for Fe (III) is less than that of microbial siderophores. All siderophores contain the anionic reactive group $(R-CO-NO-)$, which enables these compounds to be extremely effective chelators of Fe.

Carbohydrates and proteins may also play an important role in weathering of soils. As much as 30% of the DOC in soil solutions can occur as saccharides, although only a small portion has been accounted for as poly-saccharides. Polysaccharides extracted from soil usually contain Al, Fe, and Si; an indication of their ability to complex polyvalent ions. Amino acids, peptides, and proteins are also capable of forming complexes with metal ions. These substances have a strong affinity for binding to silicate minerals and are able to perturb the hydrolytic reactions of Al.

3 Adsorption and Precipitation of Organic Acids

Spodosol formation is an example of how organic acids participate in the weathering and eluviation of Fe and Al during soil formation (Vance et al. 1985; Dahlgren and Ugolini 1989). These soils developed under climatic and biological conditions that resulted in the eluviation of considerable quantities of Al, Fe, and organic matter to lower depths.

Morphological characteristics of Spodosols are clear manifestations of the processes that are involved in their formation. At the soil surface is an organic-rich layer (O and A horizons) that consists of decomposition products of the forest litter. Underlying the organic-rich layer is a light-colored eluvial horizon (E), which has lost substantial amounts of Al and Fe relative to Si. Below the E horizon lies a dark-colored illuvial horizon (B) in which the major accumulation products are Fe and Al (Bs), and organic matter (Bh), alone or in combination (Bhs). Other soils with an E horizon also show evidence of transport of Fe and Al in association with organic matter (Soil Survey Staff 1975).

Although Fe, Al, and organic matter constitute the major amorphous products in spodic horizons, the mechanism(s) responsible for translocation

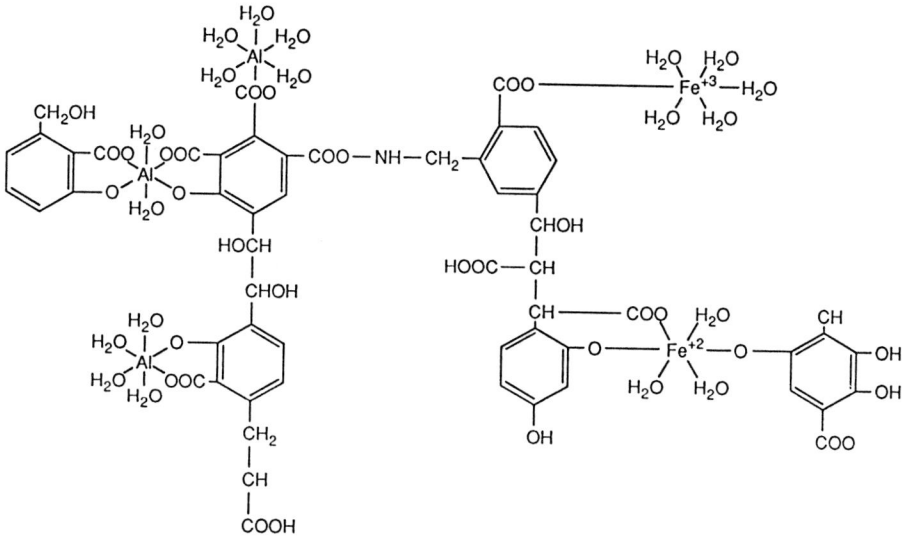

Fig. 3. Example of a soluble hypothetical metal-organic complex. (After DeConinck 1980)

of these materials is not clearly understood. One popular theory is that Fe and Al are complexed by organic acids in the eluvial horizons (O, A, and E) and translocated downward as soluble complexes (Fig. 3). The precipitation of the metal-organic complexes at lower depths is due to metal saturation, organic polymerization, or changes in ionic strength or pH levels (DeConinck 1980). In another theory, Farmer (1982) suggests that Al is transported as a hydroxy-Al silicate product (imogolite and allophane); soluble organic colloids migrating downward are then precipitated on the previously deposited imogolite and allophane. Both of the processes mentioned above for the translocation of Al, organometallic complexes, and hydroxy-Al silicates probably occur to some degree; the relative importance of each is likely dependent on the environmental conditions (Ugolini and Dahlgren 1988).

3.1 Adsorption

Adsorption of organics to mineral surfaces can occur by several mechanisms including: water bridging, electrostatic (coulombic) attraction, coordinate linkage with a single donor group, and chelation (Fig. 4). The stronger interactions, coordinate linkages, and chelation would be expected to represent the predominant forms associated with metal ions.

Soil organics form both soluble and insoluble complexes with metal ions; solubility depends on pH, presence of salt or electrolytes (i.e., ionic strength effect), and degree of saturation of organic binding sites (Dempsey et al.

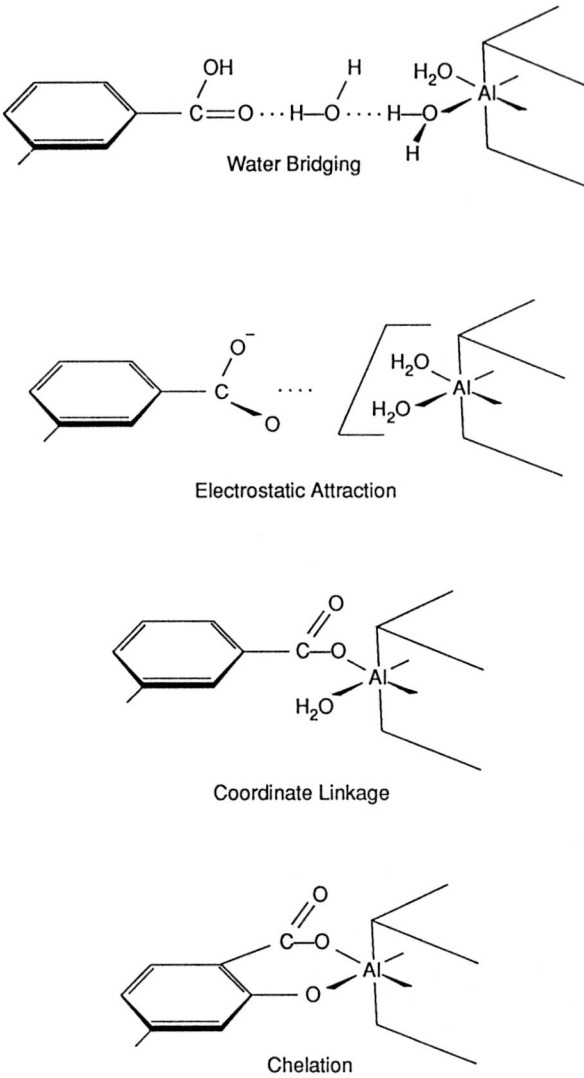

Fig. 4. Mechanisms involved in the adsorption of soil organic compounds to mineral surfaces

1984; Stevenson and Vance 1989). Other factors affecting complexation and coagulation include the concentration and source of the organic substance. In mineral soils, most of the humified organic matter is bound to mineral surfaces through linkages with polyvalent cations (Stevenson and Vance 1989).

The adsorption of humic and fulvic acids at oxide surfaces is believed to occur primarily through Fe, Al, and other polyvalent cation linkages. Both

Fe and Al can also bind low-molecular-mass organic acids to clay and oxide surfaces. Several mechanisms have been proposed that account for the retention/solubility of organic substances in soil including: (1) protonation and subsequent reduction of charges on the humic ploymer causing pre-cipitation (i.e., molecule becomes more hydrophobic); (2) formation of chain-like structures through cation bridges; (3) attachment to clay particles and oxide surfaces, such as through an Fe or Al linkage (Dempsey et al. 1984) and; (4) formation of metal-hydroxyl complexes at high pH values.

The cross-linking of soil organics to clay particles and oxide surfaces through Fe, Al, and other cations is regarded as important in the aggregation of soil minerals (Emerson et al. 1986). Another consequence of adsorption is found in Spodosols, which exhibit an illuvial spodic horizon composed of Fe- and Al-fulvate complexes that coat sand particles (Lowe 1980). The high organic matter contents of allophanic tropical soils have also been attributed to adsorption and stabilization of humic matter through linkages with Fe, Al, and possibly other metal ions (Martin and Haider 1986).

Fulvic acids are the primary type of humic substances that bind metal ions since they contain many oxygen-containing functional groups. Many of the low-molecular-mass ligands, however, can form more stable complexes with polyvalent cations than do fulvic acids.

4 Rhizosphere and Mycosphere Microenvironments

Various biochemicals of the types mentioned in previous sections are pro-duced periodically in soil through the activities of microorganisms; others are found in excretions from plant roots and in leachates of plant residues, including forest floor litter (Robert and Berthelin 1986). Measurable quantities can usually be found in zones favorable for the proliferation of microorganisms, such as in the rhizosphere and in organic-rich soils.

Organic acid-producing microorganisms are typically numerous in the rhizosphere environment. Aliphatic organic acids synthesized by rhizosphere microorganisms include: formic, acetic, propionic, butyric, oxalic, fumaric, glycolic, succinic, tartaric, citric, and 2-ketogluconic acid (Rovira and McDougall 1967; Rovira and Davey 1974). Many fungi are prolific producers of oxalic acid, including the vesicular-arbuscular mycorrhizal fungi, where calcium oxalate crystals can form at the soil-hypha interface (Graustein et al. 1977). Some species of *Pseudomonas* are capable of thriving under strongly acid conditions and are able to produce significant amounts of chelating substances (Robert and Berthelin 1986).

The amounts of hydroxamate siderophores contained in the rhizosphere of plants can be 10 to 50% higher than in the bulk soil. Hydroxamate siderophores have also been shown to be produced by soil fungi, including the ectomycorrhizal fungi, which associate closely with fine roots of many plants. Root exudates also contain phenolic acids in addition to the aliphatic acids discussed above.

5 Effects on Dissolution Rates of Silicate Minerals

The mechanisms that control the rate of dissolution of silicate minerals have been the subject of controversy for many years (for a summary, see Drever 1988). In the late 1970s and early 1980s it was established that the dissolution rate for most minerals was controlled by the detachment of atoms or ions from the surface of the solid rather than by diffusion through some sort of altered layer (Petrovic et al. 1976; Berner and Holdren 1977, 1979; Holdren and Berner 1979; Schott et al. 1981; Schott and Berner 1983). Subsequent work has focused on the detachment mechanism and how it is affected by pH and the presence of organic ligands.

Organic acids may affect the dissolution of minerals in several ways: they may lower the pH, they may form "surface complexes" (see below) which accelerate the detachment of ions from the crystal, they may affect the speciation of ions in solution (notably Al) that themselves affect dissolution rate (Chou and Wollast 1985; Amrhein and Suarez 1992), and they may allow more extensive dissolution before the solution approaches saturation with respect to the primary mineral. When a solution approaches saturation with respect to a dissolving mineral, the "back reaction" – reprecipitation of the primary mineral – becomes significant so that the net rate of dissolution decreases. Most laboratory experiments are conducted far from equilibrium and back reactions are ignored.

5.1 The Surface Complexation/Transition State Theory Model

The currently accepted mechanistic model for the kinetics of mineral dissolution is an analogy to the transition state model for homogeneous gas-phase reactions (Eyring 1935). According to this theory, the rate-limiting step in the dissolution of a mineral is the detachment of an "activated complex" from the surface of the mineral (for a more complete discussion, see Stumm and Wieland 1990). The precursor to the activated complex is a specific surface species (for example, a protonated surface-OH group), or a surface complex involving a metal at the surface of the solid and a ligand from solution. The dissolution rate is proportional to the concentration of the activated complex, which is in turn related to the concentration of the precursor species. The precursor species may be a protonated surface OH group $(-OH_2^+)$ associated with a specific metal, a deprotonated surface OH group $(-O^-)$, or a metal-ligand complex at the surface of the solid.

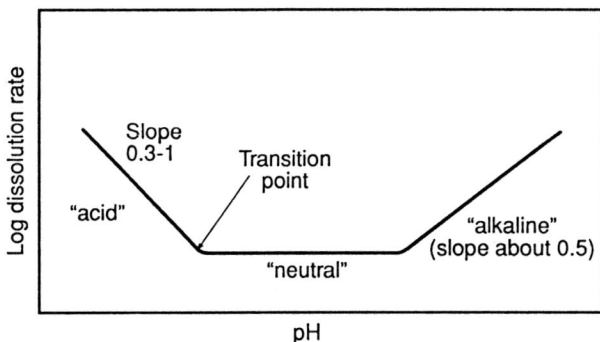

Fig. 5. Schematic relationship between silicate mineral dissolution rate and pH. The pH of the transition points between pH-dependent and pH-independent behavior varies from mineral to mineral

In the absence of organic ligands, the rate of dissolution of most silicate minerals depends on pH as shown in Fig. 5. In the acid region, we have proton-promoted dissolution:

$$(\text{Rate})_H = k_H[H^+]^n, \tag{1}$$

where k represents a rate constant, [ ] is the concentration of H^+ in solution, and n is a numerical value particular to the mineral studied. The n value varies from mineral to mineral; for silicates it typically ranges from 0.3 to 1 ("slope in acid region" in Table 2). In the circumneutral region, the rate is independent of pH

$$(\text{Rate})_{neutral} = k_n, \tag{2}$$

and in the alkaline region, the rate increases with increasing pH

$$(\text{Rate})_{OH} = k_{OH}[OH^-]^m, \tag{3}$$

where m typically has a value of about 0.3 to 0.5 (e.g., Chou and Wollast 1985; Blum and Lasaga 1988, 1991; Brady and Walther 1989). Adsorbing ligands may also accelerate dissolution. The ligand-promoted rate is proportional to the concentration of adsorbed ligand, which is related to the concentration of the ligand in solution

$$(\text{Rate})_{ligand} = k_L[\text{ligand}]^p. \tag{4}$$

The total rate of dissolution will represent the sum of the rates of the individual processes

$$(\text{Rate})_{total} = k_H[H^+]^n + k_n + k_{OH}[OH^-]^m + k_L[\text{ligand}]^p. \tag{5}$$

The exponents n, m, and p reflect the stoichiometry of the surface complex and the fact that the concentration of a species adsorbed on a solid is not necessarily linearly related to its concentration in solution. The effect of a change in pH or ligand concentration on the overall dissolution rate will depend on the relative magnitudes of the individual terms in Eq. (5). If the term $k_L[\text{ligand}]^p$ is small compared to the other terms, the ligand will have no appreciable effect on the overall dissolution rate; if it is the dominant term, the dissolution rate will increase with increasing concentration of the ligand.

Organic acids in soils may accelerate mineral dissolution rates through their effect on pH or through ligand effects.

Effect of pH. Most minerals, including carbonate minerals (Wollast 1990), exhibit a transition from pH-independent to pH-dependent dissolution (Fig. 5). The critical question is whether the pH of the transition is above or below the pH of the soil solution. The transition points for a number of minerals are shown in Table 2. If alkali feldspars and intermediate plagioclases are the most important sources of cations in weathering (Garrels and Mackenzie 1967), then the transition pH will be about 4.5, and the weathering rate will increase as $[H^+]^{0.5}$. What is a reasonable pH range for soil solutions? If pH were controlled by the CO_2 system, the pH (at a given temperature) would be determined by the P_{CO_2} and alkalinity (or base cation concentration in the absence of other anions) of the soil solution. pH values below 4.5 would occur only at very high P_{CO_2} values (around 0.1 atm)

Table 2. pH of the transition points (see Fig. 5) and slope of log (rate) vs. pH curve in the acid region for silicate minerals

Mineral	Transition pH (acid)	Slope in acid region	Source
Albite	4.5	−0.5	Sverdrup (1990)[a]
Oligoclase	4.5	−0.5	Oxburgh (1991)
Andesine	4.5	−0.5	Oxburgh (1991)
Bytownite	5	−0.75	Oxburgh (1991)
Anorthite	4.5	−3	Amrhein and Suarez (1988)
K-feldspar	5	−0.5	Schweda (1989)
Forsterite	7[b]	−0.9	Sverdrup (1990)[a]
Forsterite	4.5 (?)	−0.6	Blum and Lasaga (1988)
Forsterite	?	−0.54	Wogelius and Walther (1991)
Garnet	5.5	−0.9	Sverdrup (1990)[a]
Amphiboles	5.5	−0.8	Sverdrup (1990)[a]
Amphibole	<3		Mast and Drever (1987)
Diopside, augite	ca. 6	−0.7 to −0.9	Sverdrup (1990)[a]
Phlogopite, Biotite		−0.4	Sverdrup (1990)[a]
Kaolinite	4	−0.4	Wieland and Stumm (1992)

[a] Sverdrup (1990) represents both original work and a compilation of data from the literature.
[b] No pH-independent region.

and low cation concentrations. Lower pH values occur in soils as a result of organic acids (and atmospheric deposition of strong mineral acids). If organic acids lower the soil solution pH to 3.5, the increase in weathering rate would be a factor of about 3 over the rate at near-neutral pH.

Effect of Organic Ligands. The literature on the effect of organic ligands on mineral weathering is confusing. The confusion stems from several sources: in some studies pH was not buffered, so the ligand effect could not be unambiguously separated from the pH effect (e.g., Huang and Keller 1970; Manley and Evans 1986); in some studies it is not possible to distinguish between the effect on the dissolution of the primary mineral and the effect on inhibition of precipitation of a secondary mineral, and in most studies the concentration of organic ligand was higher than would reasonably be expected in soil solutions. As one would expect, the higher the concentration of the ligand, the more effect it has on the dissolution rate. The most-studied ligand in mineral dissolution experiments is oxalate, in part because it forms strong complexes with Al and is among the most effective in accelerating silicate dissolution (Stumm and Wieland 1990). Some data on the effect of oxalate on silicate dissolution rates are shown in Table 3.

At a concentration of 1 mM and moderately acidic conditions, oxalate has no effect on the dissolution rate of oligoclase feldspar, but a significant effect on more calcic feldspars (Mast and Drever 1987; Amrhein and Suarez 1988; Oxburgh 1991). Swoboda-Colberg and Drever (unpubl.) measured the effect of 1 mM oxalate on the dissolution rate at pH 4 and 4.5 of the 75–150 μm size fraction of natural soil minerals from a Spodosol in Maine.

Table 3. Effect of oxalate on silicate mineral dissolution rates

Mineral	Conc. (m)	pH	Dissolution rate[a]	Source
Oligoclase	10^{-3}	4–9	1	Mast and Drever (1987)
Andesine	10^{-3}	4 and 5	2	Oxburgh (1991)
Bytownite	10^{-3}	4	1	Oxburgh (1991)
Bytownite		4.5	1.7	Oxburgh (1991)
Bytownite		5	3	Oxburgh (1991)
Anorthite	10^{-4}	4	2.2	Amrhein and Suarez (1988)
K-feldspar[b]	2×10^{-2}	3.6	1.6	Bevan and Savage (1989)
Olivine	10^{-3}	4.5	23	Grandstaff (1986)
Amphibole	10^{-3}	3–9	1	Mast and Drever (1987)
Kaolinite	10^{-3}	4	1.4	Stumm and Wieland (1990)
Kaolinite	10^{-2}	4	2.3	Stumm and Wieland (1990)
75–150 μm fraction of soil	10^{-3}	4, 4.5	1.5	Swoboda-Colberg and Drever (unpubl.)

[a] Rate with ligand divided by rate without ligand.
[b] At 70 °C.

They observed that the dissolution rate increased by a factor of 1.5, which is probably a reasonable estimate of the overall effect of 1 mM oxalate on weathering rates.

How does 1 mM oxalate compare to natural dissolved organic carbon (DOC) levels? For comparison, DOC from soil solutions in the Adirondacks has an estimated 6.5 mEq carboxylate functional groups per gram C (Cronan and Aiken 1985). Thus, the DOC concentration required to provide the same carboxylate concentration as 1 mM oxalate would be about 300 mg C/l. Thurman (1985) gives 2 to 30 mg C/l as the typical range for soil solutions, which is more than an order of magnitude lower. Also, all the carboxylate groups in oxalate are optimally located in the molecule to cause chelation, whereas the distribution of carboxylate groups in natural DOC is presumably more random, with some fractions not optimally located for chelation. Some high-molecular-weight carboxylic acids may actually decrease the rate of feldspar weathering (Welch and Ullman 1992). Thus, even on an equivalent carboxylate basis, natural DOC is likely to be a less effective chelator than oxalate, and less effective in accelerating the dissolution of silicate minerals. Lundström and Öhman (1990) measured the effect of mor and peat extracts and natural high-DOC waters on the dissolution rate of feldspars and the silt fraction of a soil at pH 5.1. DOC concentrations ranged from 14 to 200 mg C/l. Their results are hard to interpret. In some instance they observed enhancements in dissolution rate with a factor of 2–3, in other instances (at similar DOC levels), they observed decreases in dissolution rate; in the data set as a whole, there was no correlation between DOC concentration and enhancement of dissolution rate. In summary, chelation by natural DOC should have only a minimal effect on the weathering rates of granitic rocks (alkali and intermediate feldspars, amphiboles), but it might have a significant effect on ultramafic and basaltic rocks; weathering of anorthite and olivine is much more strongly affected by oxalate than weathering of alkali feldspars.

In the microenvironments associated with fungal hyphae and roots, however, concentrations of organic acids may be many times higher than in the bulk soil solution (see discussion above). In such microenvironments, significant acceleration of silicate dissolution is to be expected. Such acceleration has been clearly demonstrated in microbial in microbial cultures (e.g., Eckhardt 1979). An apparent effect of lichens in accelerating basalt weathering in Hawaii was described by Jackson and Keller (1970), however, their interpretation has recently been questioned (Cochran and Berner 1992). It is hard to quantify the overall effect of microenvironments on mineral weathering rates.

The above discussion is based on the assumption that the soil solution remains far from saturation with respect to the primary minerals undergoing weathering regardless of whether organic ligands are present. Velbel (1989) has addressed this question in detail: as equilibrium is approched the net reaction rates must decrease as the rate of the reverse reaction increases, but the decrease is significant only when the solution is fairly close to equilibrium. Most soil solutions on silicate rocks are sufficiently far from equilibrium for the back reaction to be negligible. The only plausible exceptions among primary silicate minerals are K-feldspar and muscovite; even if their dissolution rates decreased, it would not affect the overall weathering rate much, because potassium makes up a relatively small fraction of the cations released by weathering. The situation is very different with carbonate rocks, where solutions commonly approach equilibrium with calcite. The assumption that solutions are far from equilibrium is probably reasonable for the soil environment, in strong contrast to the diagenetic environment. There, long contact times and higher temperatures will result in solutions closer to equilibrium, and the effect of organic acids on equilibrium solubility of minerals may be more important than their effects on dissolution kinetics.

6   Effects on Transport and Deposition of Iron and Aluminum

6.1   Complexation of Iron and Aluminum

Virtually every facet of Fe and Al chemistry in soils and soil solutions is subjected to the influence of reactions involving organic substances (see Sposito 1989 for a comprehensive review of the environmental chemistry of Al). The properties of many present-day soils, notably those of humid and semihumid environments, are directly related to the effect of organic substances on Fe and Al transformations during pedogenesis (Peterson 1976). In many soils, interactions between organic substances and mineral matter occur through metal ion bridging (Emerson et al. 1986). For a complete understanding of organic acid reactions with soil minerals, information is required on: (1) thermodynamic stability constants of metal-organic complexes; (2) binding capacity of the soil organic acids as a function of pH and ionic strength; (3) dissociation rate or lability; and (4) kinetics of complex formation (Tuschall and Brezonik 1983). Stability constants for metal-organic complexes are required in computer codes (e.g., GEOCHEM, MINTEQ, MICROQL, etc.) designed for predicting the speciation of metal ions in soil solutions or natural waters as a function of mineral solubility.

Because soil organic substances are extremely heterogeneous, stability constants for Fe and Al are most often reported as apparent or conditional stability constants that are functions of pH, ionic strength, and concentrations of metal and organic material (Weber 1988; Stevenson and Vance 1989). Our understanding of complexation is limited not only because of the complicated structure of humic substances, but also because they contain multiple complexing sites.

6.2   Neogenesis of Iron- and Aluminum-Bearing Minerals

Organic compounds produced in the biosphere enhance the weathering of
Fe- and Al-bearing rocks and minerals, and are involved in the transport of
Fe and Al during pedogenesis (McKeague et al. 1986; Tan 1986). The initial
stage of soil formation is characterized by the colonization of rock surfaces
by algae, lichens, and fungi, all of which produce chelating agents that
solubilize metal cations from minerals such as silicates, sulfides, and oxides
(Robert and Berthelin 1986). By forming metal-organic complexes, organic
acids and other complexing agents not only accelerate the dissolution of Fe-
and Al-containing minerals but also facilitate the movement of Fe and Al to
lower horizons in the soil profile (McKeague et al. 1986).

Orgnic acids can also retard or enhance the formation of secondary Fe
and Al minerals (Huang and Violante 1986). A wide variety of organic
ligands (e.g., citrate, oxalate, salicylate, catechol) can hinder the precipitation of solid phase Al minerals (Huang and Violante 1986). The
mechanism postulated for this effect is through a restraint on subsequent
hydrolysis reactions imposed by the ligand through occupation of coordination sites of Al-hydroxides (Fig. 6).

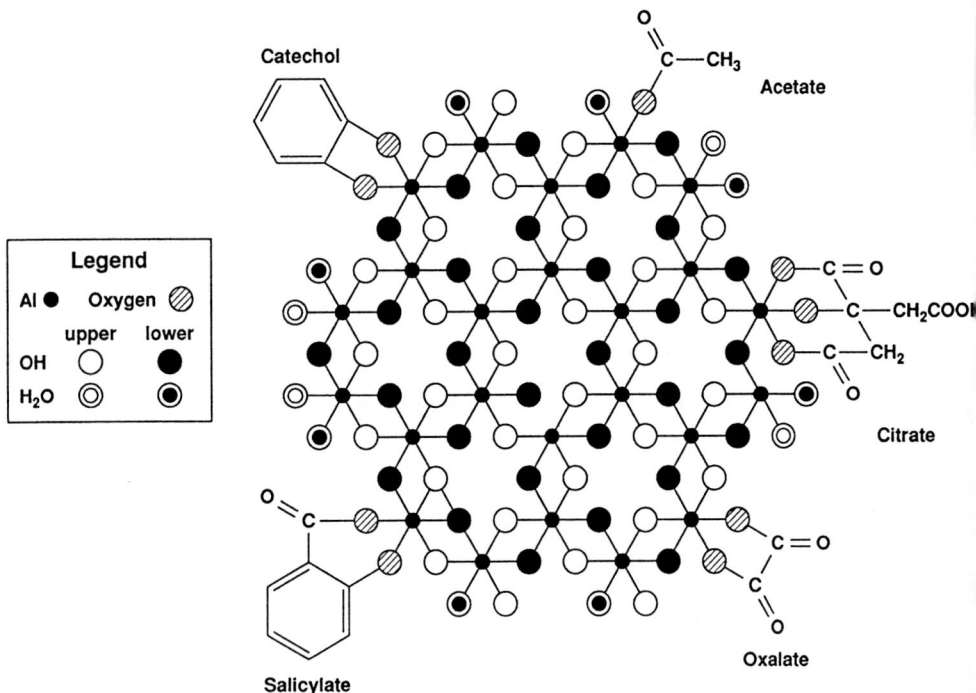

Fig. 6. Example of low-molecular-mass organic complexing ligands interacting with an Al
hydroxide polymer. Hydrolysis of Al may be retarded by soil organic acids

Organic chelating agents also are able to distort the arrangement of the unit sheets normally found in crystalline Al-hydroxides, thus leading to the formation of short-range, ordered Al precipitation products (Kwong and Huang 1979). Citric acid has also been found to perturb hydroxy-Al interlayering in montmorillonite. Both the type and formation of crystalline Al-hydroxides in soils and sediments can be influenced in environments where organic acids tend to accumulate (especially those that are acidic) (Huang and Violante 1986).

7 Significance of Organic Acids in Pedogenesis

The mobilization and transport of Fe, Al, and other polyvalent cations in leached soils are facilitated through formation of complexes with organic substances (Antweiler and Drever 1983; Weber 1988). Speciation of Fe and Al in the soil solution depends on such factors as pH, content of dissolved organic substances, and types of competing inorganic ligands (Stevenson and Vance 1989). Several studies (David and Driscoll 1984; Driscoll et al. 1985; Ares 1986) indicate that organic ligands play a dynamic and important role in defining the speciation of Fe and Al in forest soil solutions. At the pH levels found in most humid region soils, hydrolysis reactions may produce insoluble Al-hydroxides; when complexed by organic ligands, Al can be maintained in solution and transported into lower horizons of the soil profile (B and C horizons) or to lakes and streams (Driscoll and Schecher 1988).

Plants and microorganisms have long been regarded as vital factors in weathering processes and soil genesis (Stotzky 1986). Water-soluble organic substances produced in the biosphere are capable of complexing and mobilizing metal ions. Many processes are involved in the mobilization and transport of Fe and Al in the pedosphere, including: disintegration of Fe- and Al-bearing rocks and minerals, eluviation of Fe and Al to lower soil horizons during soil formation, uptake by plant roots and translocation into leaf tissue, incorporation into the humus layer of the soil, and transport to natural waters as soluble metal-organic complexes (Driscoll 1989; Jeffries and Hendershot 1989).

David and Driscoll (1984) found that one of the dominant forms of Al in throughfall and leachates of the O, E, and B horizons of Spodosols (Typic Haplorthod) in the Adirondack mountains of New York is organically complexed Al. Positive correlations have also been found between DOC levels and soluble Fe and Al concentrations in natural waters (Perdue et al. 1976).

7.1 Soil Mineralogy/Profile Development

Soils are formed through weathering of parent materials by chemical and physical processes. Soluble humic and biochemical compounds are believed to accelerate weathering of primary and secondary soil minerals by forming soluble metal-organic complexes. These complexes enhance the downward movement and separation of soil chemical constituents, which facilitate horizon differentiation. Soil organic substances can also influence the formation or crystallization of secondary minerals by either promoting or inhibiting their development.

Soils of the Alfisol, Histosol, Mollisol, and Spodosol orders, as well as others, are influenced to a large degree by organic compounds. Histosols are soils that have formed from the accumulation of organic matter, generally in areas of poor drainage. Alfisols, Mollisols, and Spodosols are mineral soils which have been weathered, in part, by organic acids and humic substances. Organic matter in these soils may be important for either stabilizing mineral matter near the soil surface (mollic epipedon) or solubilizing and trans-locating metals and possibly clays to lower horizons. For a more complete description of the role of organic substances in relation to soil pedological processes, see McKeague et al. (1986).

Soil organics are also considered to be essential to the formation of stable aggregates in a wide range of soils. Physical properties of soil, such as structure and bulk density, are influence by the quantity and nature of complexes that form between organic and mineral matter. The physical characteristics of a soil are important to plant growth due to their effects on aeration, water penetration and retention, and mechanical impedance to

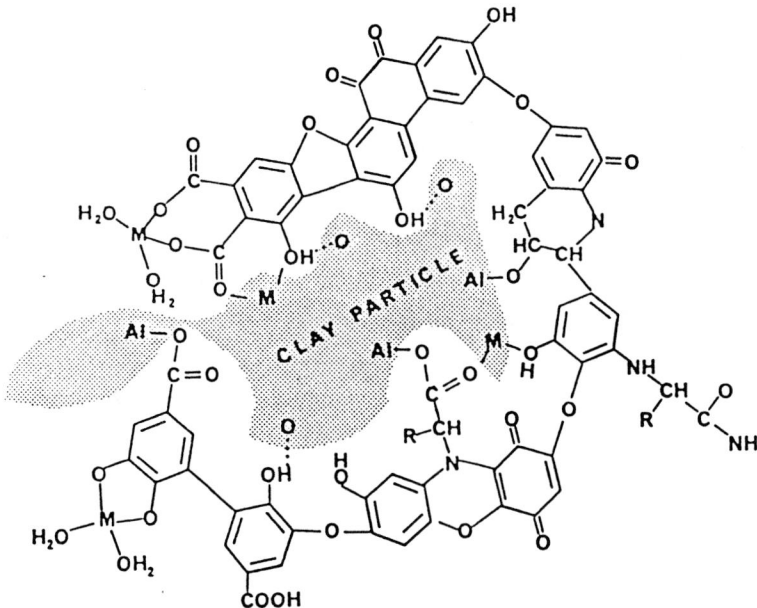

Fig. 7. Bridging of clay and soil organic constituents through direct interactions or by metal ions to form larger aggregates. (Koskinen and Harper 1990)

roots (Russell 1977). An important physical factor is the cohesion of clay and/or clay-humus complexes into larger particles through bridging with polyvalent cations (Fig. 7; Emerson et al. 1986).

8 Conclusions

1. Soluble organic acids are produced in the soil by microbial activity. Their concentrations are highest in the organic layer at the top of the soil profile and decrease with depth as a result of bacterial decomposition, adsorption, and coprecipitation with hydrated oxides of Fe and Al.

2. Fulvic acid, a general term for a complex mixture of refractory polymeric compounds, is generally the dominant organic acid in soil solutions. A wide range of low-molecular-mass organic acids, most notably acetic, oxalic, and formic, are also present in soil solutions. The concentrations of these acids represent a dynamic balance between production and consumption by microbial processes.

3. Both fulvic acid and various low-molecular-mass compounds are very effective in complexing Al and Fe, which results in the vertical translocation of these elements in the soil profile.

4. Organic acids, at the concentrations common in bulk soil solutions, have probably only a minor effect on the weathering rates of silicate minerals. Concentrations are much higher in the microenvironments surrounding the rhizosphere and the mycosphere. Weathering rates may be greatly accelerated in these microenvironments.

Acknowledgments. The authors wish to express their appreciation to Larry C. Munn, Stephen E. Williams, and Michael J. Blaylock for their critical review of this chapter, and to Robert A. Berner for stimulating discussions. We are also grateful to the book editors for their helpful comments and suggestions. Partial funding was provided by the Environmental Protection Agency Watershed Manipulation Project. Although this study was supported by the US. Environmental Protection Agency, it has not been subjected to the agency's review and therefore does not necessarily reflect the views of the agency, and no official endorsement should be inferred.

References

Aiken GR, McKnight DM, Wershaw RL, MacCarthy P (eds) (1985) Humic substances in soil, sediment, and water. Wiley-Interscience, New York, 692 pp
Amrhein C, Suarez DL (1988) The use of a surface complexation model to describe the kinetics of ligand-promoted dissolution of anorthite. Geochim Cosmochim Acta 52: 2785–2793

Amrhein C, Suarez DL (1992) Some factors affecting the dissolution kinetics of anorthite
  at 25 °C. Geochim Cosmochim Acta 56: 1815–1826
Antweiler RC, Drever JI (1983) The weathering of a late Tertiary volcanic ash: importance
  of organic solutes. Geochim Cosmochim Acta 47: 623–629
Ares J (1986) Identification of Al species in acid forest soil solution on the basis of AL:F
  reaction kinetics. 1. Reaction path in pure solutions. Soil Sci 141: 399–407
Berner RA, Holdren GR Jr (1977) Mechanism of feldspar weathering: some observational
  evidence. Geology 5: 369–372
Berner RA, Holdren GR Jr (1979) Mechanism of feldspar weathering. II. Observations
  of feldspars from soils. Geochim Cosmochim Acta 43: 1173–1186
Bevan J, Savage D (1989) The effect of organic acids on the dissolution of K-feldspar
  under conditions relevant to burial diagenesis. Mineral Mag 53: 415–425
Blum AE, Lasaga AC (1988) The role of surface speciation in the low-temperature
  dissolution of minerals. Nature 4: 431–433
Blum AE, Lasaga AC (1991) The role of surface speciation in the dissolution of albite.
  Geochim Cosmochim Acta 55: 2193–2201
Brady PV, Walther JV (1989) Controls on silicate dissolution rates in neutral and basic
  pH solutions at 25 °C. Geochim Cosmochim Acta 53: 2823–2830
Buffle J (1977) Les substances humiques et leurs interactions avec les ions minéraux. Conf
  Proc Commission d'Hydrologie Appliquée de l'AGHTM L'Université d'Orsay, pp
  3–10
Chou L, Wollast R (1985) Steady-state kinetics and dissolution mechanisms of albite. Am
  J Sci 285: 963–993
Cochran MF, Berner RA (1992) The quantitative role of plants in weathering. In:
  Kharaka YK, Maest AS (eds) Water-rock interaction, WRI-7, vol 1. Balkema,
  Rotterdam, pp 473–476
Coulson CB, Davis RI, Lewis DA (1960) Polyphenols in plant, humus and soil. II.
  Reduction and transport by polyphenols of iron in model soil columns. J Soil Sci 11:
  30–44
Cronan CS, Aiken GR (1985) Chemistry and transport of soluble humic substances in
  forested watersheds of the Adirondack Park, New York. Geochim Cosmochim Acta
  49: 1697–1705
Dahlgren RA, Ugolini FC (1989) Aluminum fractionation of soil solutions from
  unperturbed and tephra-treated Spodosols, Cascade Range, Washington, USA. Soil
  Sci Soc Am J 53: 559–566
David MB, Driscoll CT (1984) Aluminum speciation and equilibria in soil solutions of a
  Haplorthod in the Adirondack mountains (New York, USA). Geoderma 33: 297–318
DeConinck F (1980) Major mechanisms in formation of spodic horizons. Geoderma 24:
  101–128
Dempsey BA, Ganho RM, O'Melia CR (1984) The coagulation of humic substances by
  means of aluminum salts. J Am Water Works Assoc 76: 141–150
Drever JI (1988) The geochemistry of natural waters, 2nd edn. Prentice-Hall, Englewood
  Cliffs, 437 pp
Driscoll CT (1989) The chemistry of aluminum in surface waters. In: Sposito G (ed)
  The environmental chemistry of aluminum. CRC Press, Boca Raton, pp 241–
  278
Driscoll CT, Schecher WD (1988) Aluminum in the environment. In: Sigel H (ed) Metal
  ions in biological systems. Dekker, New York, pp 59–122
Driscoll CT, van Breemen N, Mulder J (1985) Aluminum chemistry in a forested
  Spodosol. Soil Sci Soc Am J 49: 437–443
Eckhardt FEW (1979) Über die Einwirkung heterotropher Microorganismen auf die
  Zersetzung silikatischer Minerale. Z Pflanzernaehr Bodenkd 142: 434–445
Emerson WW, Foster RC, Oades JM (1986) Organo-mineral complexes in relation to soil
  aggregation and structure. In: Huang PM, Schnitzer M (eds) Interactions of soil

minerals with natural organics and microbes. Soil Sci Soc Am Spec Publ 17, pp 521–548

Eyring H (1935) The activated complex in chemical reactions. J Chem Phys 3: 107–115

Farmer VC (1982) Significance of the presence of allophane and imogolite in Podzol Bs horizons for podzolization mechanisms: a review. Soil Sci Plant Nutr 28: 571–578

Fox TR, Comerford NB (1990) Low-molecular-weight organic acids in selected forest soils of the southeastern USA. Soil Sci Soc Am J 54: 1139–1144

Garrels RM, Mackenzie FT (1967) Origin of the chemical composition of some springs and lakes. Equilibrium concepts in natural water systems. Am Chem Soc Adv Chem Ser 67: 222–242

Grandstaff DE (1986) The dissolution rate of forsteritic olivine from Hawalian beach sand. In: Colman SM, Dethier DP (eds) Rates of chemical weathering of rocks and minerals. Academic Press, Orlando, pp 41–59

Graustein WC, Cromack K Jr, Sollins P (1977) Calcium oxalate: occurrence in soils and effect on nutrient and geochemical cycles. Science 198: 1252–1254

Hatcher PG, Greger IA, Maciel GE, Szeverenyi NM (1985) Geochemistry of humin. In: Aiken GR, McKnight DM, Wershaw RL, MacCarthy P (eds) Humic substances In soil, sediment, and water. Wiley-Interscience, New York, pp 275–302

Hayes MHB, MacCarthy P, Malcolm RL, Swift RS (eds) (1989) Humic substances II. Search of structure. Wiley, New York, 764 pp

Holdren GR Jr, Berner RA (1979) Mechanism of feldspar weathering. I. Experimental studies. Geochim Cosmochim Acta 43: 1161–1171

Huang WH, Keller WD (1970) Dissolution of rock-forming silicate minerals in organic acids: simulated first-stage weathering of fresh mineral surfaces. Am Mineral 55: 2076–2094

Huang PM, Violante AO (1986) Influence of organic acids on crystallization and surface properties of precipitation products of aluminum. In: Huang PM, Schnitzer M (eds) Interactions of soil minerals with natural organics and microbes. Soil Sci Soc Am Spec Publ 17, pp 159–222

Iskandar IK, Syers JK (1972) Metal-complex formation by lichen compounds. J Soil Sci 23: 255–265

Jackson TA, Keller WD (1970) Comparative study of the role of lichens and inorganic processes in the chemical weathering of recent Hawaiian lava flows. Am J Sci 269: 446–466

Jeffries DS, Hendershot WH (1989) Aluminum geochemistry at the catchment scale in watersheds influenced by acidic precipitation. In: Sposito G (ed) The environmental chemistry of aluminum. CRC Press, Boca Raton, pp 279–302

Koskinen WC, Harper SS (1990) The retention process. Mechanisms. In: Cheng HH (ed) Pesticides in the soil environment: processes, impacts, and modeling. Soil Sci Soc Am Book Ser 2, Madison WI, pp 51–77

Kwong NKKF, Huang PM (1979) The relative influence of low-molecular-weight complexing organic acids and the hydrolysis and precipitation of aluminum. Soil Sci 128: 337–342

Lowe LE (1980) Humus fraction ratios as a means of discriminating between horizon types. Can J Soil Sci 60: 219–229

Lundström U, Öhman L-O (1990) Dissolution of feldspars in the presence of natural organic solutes. J Soil Sci 41: 359–369

Malcolm RL (1985) Geochemistry of stream fulvic and humic substances. In: Aiken GR, McKnight DM, Wershaw RL, MacCarthy P (eds) Humic substances in soil, sediment, and water. Wiley-Interscience, New York, pp 181–210

Manley EP, Evans LJ (1986) Dissolution of feldspars by low-molecular-weight aliphatic and aromatic acids. Soil Sci 141: 106–112

Martin JP, Haider K (1986) Influence of mineral colloids on turnover rates of soil organic carbon. In: Huang PM, Schnitzer M (eds) Interactions of soil minerals with natural organics and microbes. Soil Sci Soc Am Spec Publ 17, pp 283–304

Mast MA, Drever JI (1987) The effect of oxalate on the dissolution rates of oligoclase and tremolite. Geochim Cosmochim Acta 51: 2559–2568

McColl JG, Pohlman AA, Jersak JM, Tam SC, Northup RR, (1990) Organics and metal solubility in California forest soils. North America Forest Soils Conf Proc. University of British Columbia, Vancouver, pp 178–195

McKeague JA, Cheshire MV, Andreux F, Berthelin J (1986) Organo-mineral complexes in relation to pedogenesis. In: Huang PM, Schnitzer M (eds) Interactions of soil minerals with natural organics and microbes. Soil Sci Soc Am Spec Publ 17, pp 549–592

Oxburgh R (1991) The effect of pH, oxalate ion and mineral composition on the dissolution rates of plagioclase feldspars. MS Thesis, University of Wyoming, Laramie, 69 pp

Perdue EM, Beck KC, Reuter JH (1976) Organic complexes of iron and aluminium in natural waters. Nature 260: 418–420

Peterson L (1976) Podzols and podzolization. PhD Dissertation, Royal Veterinary and Agricultural University, Copenhagen, 293 pp

Petrovic R, Berner RA, Goldhaber MB (1976) Rate control in dissolution of alkali feldspars. I. Study of residual feldspar grains by X-ray photoelectron spectroscopy. Geochim Cosmochim Acta 40: 537–548

Robert M, Berthelin J (1986) Role of biological and biochemical factors in soil mineral weathering. In: Huang PM, Schnitzer M (eds) Interactions of soil minerals with natural organics and microbes. Soil Sci Soc Am Spec Publ 17, pp 453–496

Rovira AD, Davey CB (1974) Biology of the rhizosphere. In: Carson EW (ed) The plant root and its environment. University Press of Virginia, Charlottesville, pp 153–204

Rovira AD, McDougall BM (1967) Microbiological and biochemical aspects of the rhizosphere. In: McLaren AD, Peterson GH (eds) Soil biochemistry, vol 1. Dekker, New York, pp 417–463

Russell RS (1977) Plant root systems: their function and interaction with the soil. McGraw-Hill, London, 298 pp

Schnitzer M (1978) Humic substances: Chemistry and reactions. In: Schnitzer M, Khan SU (eds) Soil organic matter. Elsevier, New York, pp 1–64

Schnitzer M, Khan SU (1972) Humic substances in the environment. Dekker, New York, 327 pp

Schott J, Berner RA (1983) X-ray photoelectron studies of the mechanism of iron silicate dissolution during weathering. Geochim Cosmochim Acta 47: 2233–2240

Schott J, Berner RA, Sjøberg EL (1981) Mechanism of pyroxene and amphibole weathering – I. Experimental studies of iron-free minerals. Geochim Cosmochim Acta 45: 2123–2135

Schweda P (1989) Kinetics of alkali feldspar dissolution at low temperature. In: Miles DL (ed) Water-rock interaction, WRI-6. Balkema, Rotterdam, pp 609–612

Sikora FJ, McBride MB (1990) Aluminum complexation by protocatechuic and caffeic acids as determined by ultraviolet spectroscopy. Soil Sci Soc Am J 54: 78–86

Soil Survey Staff (1975) Soil taxonomy: a basic system of soil classification for making and interpreting soil surveys. Washington, US Dep Agriculture, Soil Conservation Service, Agricultural Handbook 436. US Govt Printing Office, Washington DC, 754 pp

Sposito G (ed) (1989) The environmental chemistry of aluminum. CRC Press, Boca Raton, 317 pp

Stevenson FJ (1967) Organic acids in soil. In: McLaren AD, Peterson GH (eds) Soil biochemistry, vol 1. Dekker, New York, pp 119–146

Stevenson FJ (1982) Humus chemistry; genesis, composition, reactions. Wiley-Inter-science, New York, 443 pp

Stevenson FJ (1985) Geochemistry of soil humic substances. In: Aiken GR, McKnight DM, Wershaw RL, MacCarthy P (eds) Humic substances in soil, sediment, and water. Wiley-Interscience, New York, pp 275–302

Stevenson FJ, Fitch A (1986) Chemistry of complexation of metal ions with soil solution organics. In: Huang PM, Schnitzer M (eds) Interactions of soil minerals with natural organics and microbes. Soil Sci Soc Am Spec Publ 17, pp 29–58

Stevenson FJ, Vance GF (1989) Naturally occurring aluminum-organic complexes. In: Sposito G (ed) The environmental chemistry of aluminum. CRC Press, Boca Raton, pp 117–146

Stotzky G (1986) Influence of soil mineral colloids on metabolic processes, growth, adhesion and ecology of microbes and viruses. In: Huang PM, Schnitzer M (eds) Interactions of soil minerals with natural organics and microbes. Soil Sci Soc Am Spec Publ 17, pp 305–428

Stumm W, Wieland E (1990) Dissolution of oxide and silicate minerals: rates depend on surface speciation. In: Stumm W (ed) Aquatic chemical kinetics. Wiley-Interscience, New York, pp 367–400

Sverdrup HU (1990) The kinetics of base cation release due to chemical weathering. Lund University Press, Lund, 246 pp

Tan KH (1986) Degradation of soil minerals by organic acids. In: Huang PM, Schnitzer M (eds) Interactions of soil minerals with natural organics and microbes. Soil Sci Soc Am Spec Publ 17, pp 1–28

Thurman EM (1985) Organic geochemistry of natural waters. Nijhoff/Junk, Dordrecht, 497 pp

Tuschall JR, Brezonik PL (1983) Complexation of heavy metals by aquatic humus: a comparative study of five analytical methods. In: Christman RF, Gjessing ET (eds) Aquatic and terrestrial humic materials. Ann Arbor Science, Ann Arbor, pp 275–294

Ugolini FC, Dahlgren R (1988) Formation and role of imogolite in Spodosols. Am Soc Agron Abstr, Madison, WI, p 206

Vance GF, Boyd SA, Mokma DL (1985) Extraction of phenolic compounds from a Spodosol profile: an evaluation of three extractants. Soil Sci 140: 412–420

Vance GF, Mokma DL, Boyd SA (1986) Phenolic compounds in soils of hydrosequences and developmental sequences of Spodosols. Soil Sci Soc Am J 50: 992–996

Velbel MA (1989) Effect of chemical affinity on feldspar hydrolysis rates in two natural weathering systems. Chem Geol 78: 245–253

Weber JH (1988) Binding and transport of metals by humic materials. In: Frimmel FH, Christman RF (eds) Humic substances and their role in the environment. Wiley, New York, pp 165–178

Welch SA, Ullman WJ (1992) Microbially produced compounds and feldspar dissolution. VM Goldschmidt Conf, Program and Abstracts, Reston, VA, pp A-119–A-120

Whitehead DC, Dibb H, Hartley RD (1982) Phenolic compounds in soil as influenced by the growth of different plant species. J Appl Ecol 19; 579–588

Wieland E, Stumm W (1992) Dissolution kinetics of kaolinite in acid aqueous solutions at 25 °C. Geochim Cosmochim Acta 56: 3339–3355

Wogelius RA, Walther JV (1991) Olivine dissolution at 25 °C: effects of pH, CO_2, and organic acids. Geochim Cosmochim Acta 55: 943–954

Wollast R (1990) Rate and mechanism of dissolution of carbonates in the system $CaCO_3$-$MgCO_3$. In: Stumm W (ed) Aquatic chemical kinetics. Wiley-Interscience, New York, pp 431–445

Chapter 7 Chemistry and Mechanisms of Low-Temperature Dissolution of Silicates by Organic Acids

Philip C. Bennett[1] and William Casey[2]

Summary

Organic acids are important diagenetic agents in near-surface weathering processes. The mechanism of interaction, however, is complex, and their significance in a particular environment will depend on a variety of geo-chemical parameters. Organic acids in low-temperature aqueous systems can complex metals and metalloids in solution, thereby increasing their solubility and mobility. Organic acids are also implicated in mineral-surface interactions where they can act in ligand exchange reactions to increase the rate of mineral dissolution independent of solution concentration constraints. The manner in which organic acids accelerate dissolution rate can be examined by looking at how organic acids complex metals in solution, and by examining the fundamental mechanisms of silicate dissolution.

In the case of aluminosilicates, organic acids can complex aluminum, and to a lesser degree silica, in solution, thereby changing the total solubility of the mineral. Solution characteristics have a large influence on this interaction, in particular pH, ionic strength, the ligand concentration, and the stability (structure) of the resulting chelate. At neutral pH, aluminum complexes are less favored, while complexes of silica may be favored. At weakly acidic pH, in contrast, silica is not affected by the presence of organic acids, whereas aluminum complexes are stable. Aluminum is chelated by a variety of organic ligands, probably through a dissociative ligand exchange reaction (S_N1 reaction), whereas silica is chelated in an associative (S_N2 reaction). A system where aluminosilicates are dissolving and aluminum is chelated in solution by organic acids, for example, would increase the total aluminum solubility, thus shifting the equilibria with respect to secondary minerals such as gibbsite.

At the silicate surface, organic acids interact with both the silicon and aluminum metal centers in an associative ligand exchange reaction. This

---

[1] Department of Geological Sciences, University of Texas at Austin, Austin, Texas 78712, USA
[2] Department of Land, Air and Water Resources, University of California, Davis, California 95616, USA

reaction acts to polarize and weaken framework crystal bonds, thus changing the energetics of the rate-limiting step of silicate dissolution. The result is an increase in dissolution rate independent of solubility constraints. Both surface and solution characteristics will directly influence this reaction by changing the speciation of the organic acid ligand, the speciation of the surface, and the distribution and characteristics of the surface metal-oxide sites.

The outcome of these interactions can be seen in field settings, especially where microbial activity in the subsurface produces high concentrations of organic acids. In an oil-contaminated aquifer, microbial degradation of the petroleum in a groundwater at neutral pH produces a variety of reactive organic compounds that mobilizes silica from quartz and feldspars, even in waters greatly supersaturated with respect to quartz. In this system, the quartz and feldspar dissolution rate is greatly increased, but only silica is mobilized, as aluminum is conserved in a solid phase. In a peat bog setting, in contrast, a low system pH and a high organic acid concentration result in enhanced aluminosilicate dissolution and both aluminum and silica transport. In both cases, the presence of organic acids dominates the control of silicate weathering in the subsurface.

1 Introduction

Previous chapters of this volume have described the origin and distribution of organic acids in natural systems, and the role of organic acids in soil development. From these reports it is clear that organic acids are ubiquitous, reactive, and an important aspect of mineral weathering in the subsurface. This chapter will focus specifically on the mechanisms by which organic acids enhance the dissolution of silicates at low temperature, and suggest several possible reaction pathways. High-temperature interactions are discussed elsewhere in this volume (Surdam and Yin, Chap. 13; Hajash, Chap. 8, this Vol.).

After a review of the history of speculation on the interaction of organic substances with silicates, we briefly summarize mechanisms of silicate dissolution in *inorganic* aqueous systems. From this base, we suggest mechanisms of interaction between organic acids and silicate surfaces, first examining complexation/ligand exchange reactions of metals and metalloids in solution. Then, using the solution model as a basis of understanding, the analogous surface complexation/ligand exchange reactions are evaluated for their role in silicate dissolution kinetics. The potential role of subsurface microbes in creating organic-rich microenvironments on silicate surfaces is also discussed. Finally, two field examples of enhanced silicate dissolution in natural waters are reviewed.

1.1 Historical Perspective

There is a rich history of debate over the role of organic acids in weathering silicate minerals. Julien (1879), for example, discussed the influence of "humus" on silicate dissolution, concluding that these organic acids are important agents in the chemical weathering of quartz and feldspars (unfortunately, he also concluded that organic acids are the catalyst for marble formation from limestone!). Clarke (1911), in the second edition of *The Data of Geochemistry*, concluded that Julien (1879) lacked reliable experimental evidence, and discounted any effect of organic acids on mineral decomposition. Later, Gruner (1922) concluded that organic acids do in fact dissolve silicate minerals. Moore and Maynard (1929) looked at the association of organic-rich waters with iron and silica transport, whereas Graham (1941) examined the importance of colloidal organic substances on anorthite weathering. Duff et al. (1963) looked specifically at microbial systems and concluded that 2-ketogluconic acid is an effective weathering agent, but even as late as 1967 researchers questioned the significance of organic acids in mediating the rate of silicate dissolution (i.e., Krauskopf 1967; see Baker 1972, for a summary). This history of reversing conclusions would suggest that the interaction of organic substances with silicates is somewhat difficult to characterize.

Since 1970 there have been numerous studies verifying the influence of organic acids on aluminosilicates. Schnitzer (1971) and Baker (1972) discussed the role of humic acids in mineral weathering in soil environments (see Drever and Vance, Chap. 6, this Vol.). Huang and Keller (1970) examined the stoichiometry of silicate dissolution and the dissolution kinetics of aluminosilicates in organic acid solutions. Ong et al. (1970), Harter (1977), and Tan (1980) considered the weathering of primary minerals, whereas Boyle et al. (1974) and Kodama et al. (1983) investigated the weathering of biotite and chlorite by organic acids. Jorgensen (1976) studied silicate dissolution kinetics in catechol solutions, an important and reactive class of organic chelates. Antweiler and Drever (1983) and Grandstaff (1986) showed unequivocally that organic acids accelerate the dissolution of aluminosilicates at low temperature, and that the interaction is dependent on organic acid species, solution conditions, and ligand concentration. These investigations focused primarily on the role of organic acids as proton donors and as aluminum chelates, and outline the importance of organic electrolytes in silicate diagenesis (e.g., Surdam et al. 1984).

Direct evidence for silica – organic acid interactions and enhanced quartz dissolution in aqueous organic acid solutions is sparse. Evans (1964) reviewed some of the early work in determining the actions of nucleic acids, blood serum, and ATP on quartz dissolution. Duff et al. (1963) suggested that organic acids may enhance quartz dissolution. Crook (1968) and Cleary and Conolly (1972) found chemically weathered quartz grains in soil root zones. Siever (1960) investigated the solubility of silica polymorphs in a variety of water types, including "peat water"; he concluded, however, that the humic acids in this water actually decreased the solubility of amorphous silica at acidic pH. Others have reported evidence of accelerated quartz dissolution and increased apparent solubility in water rich in organic acids (e.g., Beckwith and Reeve 1964). Jorgensen (1976) confirmed that catechol, a hydroxy-phenol, accelerates the dissolution of quartz at alkaline pH.

From this historical review, it is clear that the effects of organic acids on mineral stability are complex and multifaceted. The evidence suggests that organic acids influence silica and aluminum mobility, quartz and aluminosilicate dissolution kinetics, and the solubility of silicates in natural waters. This makes it difficult to distinguish the *mechanism* by which organic acids interact with silicate minerals and, thus, difficult to predict the final outcome. In natural waters, organic acids increase acidity, accelerating the hydrolysis of aluminosilicates and increasing the solubility of dissolved inorganic hydroxy complexes of aluminum. This would have little effect on quartz dissolution, however, as quartz solubility is insensitive to changes in pH below pH 9, and the kinetics of quartz dissolution decreases only slightly with decreasing pH below pH 8 (Knauss and Wolery 1988; Bennett 1991). Organic acid chelates of aluminum decrease the activity of monomeric aluminum species in solution, thus increasing total solubility, and changing equilibria with respect to secondary clay minerals. Solution complexes, however, could only influence the kinetics of concentration-dependent, first-order rate mechanisms, and the silicate dissolution rate is usually considered to be surface controlled and dependent on surface speciation, not solution concentration (e.g., Stumm 1990). This is also the case for quartz, where dissolution is zero-order, far from equilibrium, and solution complexes of silica would have little effect on dissolution kinetics. In order for the rate of silicate dissolution to be accelerated, coordination of organic ligands onto silicate surfaces must also be recognized as a fundamental interaction and distinguished from solution complexes as influencing kinetics without necessarily changing the solubility (e.g., Furrer and Stumm 1986; Stumm and Furrer 1987; Wieland et al. 1988). Surface effects, however, will still be strongly influenced by the solution condition, especially pH, and by the structure and chemistry of the silicate surface. The purpose of this chapter is to distinguish between these mechanisms in their effects on silicate dissolution.

2 Mechanisms of Mineral Dissolution

Organic acids enhance the dissolution of silicates in aqueous systems by influencing the fundamental rate-controlling steps of *inorganic* aqueous dissolution mechanisms. A basic understanding of these fundamental reactions is needed before addressing the additional influence of organic acids. This overview is directed toward understanding the fundamental reactions influenced by organic acids, rather than a comprehensive discussion of mineral dissolution kinetics. The interested reader is directed to Lasaga (1984) or Stumm (1990), for example, for a more comprehensive treatment of mineral dissolution kinetics.

2.1 Transport and Surface Control of Mineral Dissolution Rates

At the simplest level, there are three sequential steps to removing a metal from a dissolving mineral surface. First, reactive solutes (e.g., hydrogen ions or organic ligands) must migrate through the aqueous solution to the mineral surface. Second, the solutes combine with metals, metalloids, or oxygen on the mineral surface to form a reactive complex, which then decays and the products desorb. Third, the dissolved products diffuse away from the mineral surface along concentration gradients. The slowest of these three sequential processes controls the rate of the overall dissolution reaction.

Under *experimental* conditions, most rock-forming silicate minerals are sufficiently nonreactive that transport of solutes toward, or away from, the dissolving surface does not control the rate. The rate of the overall reaction is controlled by reactions at the interfaces, that is, the second sequential step. In soil, solute transport is sufficiently rapid that silicate weathering is controlled by the reaction kinetics at the mineral-solution interface (Berner 1978).

The range of surface-controlled mineral dissolution rates is quite large. Rates vary: (1) with solution composition; (2) for minerals of similar structure but varying composition (e.g., Co_2SiO_4, Mg_2SiO_4, Fe_2SiO_4); and (3) for minerals of similar composition but varying structure (e.g., Mg_2SiO_4; $MgSiO_3$; $Mg_6(Si_8O_{20})(OH)_4$). Even the dissolution rates of simple divalent oxide minerals (e.g., MgO, BeO) vary by a factor of 10^8 under otherwise identical conditions (Casey 1991).

Attempts to systematize these variations must begin at a fine scale. Large-scale features of the mineral structure, such as lattice dislocations, have surprisingly little effect on the dissolution rates (e.g., Casey et al. 1988a; Blum and Lasaga 1990). Under some conditions, complicated mineral structures dissolve nonuniformly (Murata 1943), and reaction of plagioclase in acid, for example, causes the aluminum, calcium, and sodium to leach away from the near-surface region, leaving a silica-rich surface (Casey et al. 1988b). Dissolution in this case proceeds at the scale of individual metal-oxygen bonds at the mineral surface.

2.2 Dissolution as a Ligand-Exchange Reaction

An important working hypothesis is that the reactions, which remove a metal from a mineral surface, have much in common with the reactions that replace ligands around a dissolved metal. In both cases, bonds between the metal and coordinated oxygens are broken and reformed. A typical ligand-exchange reaction would be the replacement of a hydration water in the inner coordination sphere of dissolved ferric iron $[Fe(H_2O)_6^{+3}]$ with a water molecule from the solvent. In this example, these reacting bonds are between water molecules and iron in the complex, $Fe(H_2O)_6^{+3}$. Extrapolated to hematite, the bonds could exist between ferric iron and bridging oxygens at the surface. Continued replacement of surface-bridging oxygens with coordinated water molecules releases the $Fe(H_2O)_6^{+3}$ into solution, and dissolution occurs.

The number of ligands coordinated to the metal ion is conserved during ligand exchange and dissolution. Ferric iron in the above example is co-ordinated to six oxygens in the aquated ion and in hematite. Silicon is tetrahedrally coordinated to four oxygens in silicic acid and in quartz. Furthermore, the oxygen coordination chemistry of metals in a mineral is not too different from the ion in many inorganic aqueous solutions. For example, magnesium is coordinated to six oxygens in periclase (MgO), forsterite (Mg_2SiO_4), and the hydrated ion $[Mg(H_2O)_6^{+2}]$. The Mg—O bond length in periclase (2.11 Å) and forsterite (2.101–2.126 Å) (Table 1) com-pares well to that in the hydrated ion (2.10 Å). In general, the metal-oxygen bond distances in common oxide and silicate minerals are within a few hundredths of an angstrom of the value for the aquated ion (Table 1).

Large variations between solution and surface properties are only ob-served for cases where the coordination number in the mineral is different

Table 1. Metal-oxygen bond distances (Å) in hydrated ions (Burgess 1988), oxide minerals (see compilation in Casey 1991), and orthosilicate minerals (Brown 1980; Bish and Burnham 1984) Divalent cations in the minerals are hexacoordinated to oxygens except for ZnO, Zn_2SiO_4, BeO, and Be_2SiO_4, which have tetrahedral coordination. Zinc has a different M-O distance because the coordination number is from six in the aquated ion and four in the minerals

Ion	M—O distance	Oxide	M—O distance	Silicate	M—O distance
$Ca(H_2O)_6^{+2}$	2.39–2.46	CaO	2.38	Ca_2SiO_4	2.346–2.392
$Mg(H_2O)_6^{+2}$	2.10	MgO	2.11	Mg_2SiO_4	2.101–2.126
$Be(H_2O)_4^{+2}$	–	BeO	1.65	Be_2SiO_4	1.645
$Zn(H_2O)_6^{+2}$	2.08–2.17	ZnO	1.92	Zn_2SiO_4	1.92
$Mn(H_2O)_6^{+2}$	2.18–2.20	MnO	2.22	Mn_2SiO_4	2.185–2.227
$Fe(H_2O)_6^{+2}$	2.10–2.12	FeO	2.13	Fe_2SiO_4	2.157–2.182
$Co(H_2O)_6^{+2}$	2.05–2.08	CoO	2.13	Co_2SiO_4	2.119–2.142
$Ni(H_2O)_6^{+2}$	2.04–2.10	NiO	2.095	Ni_2SiO_4	2.076–2.102

from that in the dissolved complex. Such changes in coordination number in inorganic systems are generally uncommon, but do exist for geologically important elements (e.g., Zn^{+2}, Al^{+3}, B^{+3}, Si^{+4}). Silicon, for example, is tetrahedrally coordinated as silicic acid and in virtually all rock-forming silicates except the high-pressure mineral stishovite, but is octahedrally coordinated in aqueous solutions by F (SiF_6^{-2}) and by organic chelates (discussed further in later sections).

The coordination chemistry of the chelated metal is an important consideration. Some pathways for ligand exchange (and, by extrapolation, dissolution) are sensitive to ligand and ion size. There must be sufficient room in the coordination complex to configure the metal and the incoming and outgoing ligands into an activated complex. Changes in coordination number might increase (or decrease) steric hindrance of the initial nucleophilic attack of a ligand (i.e., ligand exchange reaction). Alternatively, the ability of a metal to form a stable hypervalent species with a higher coordination number would increase the stability of the activated complex (i.e., penta and hexa coordination of silicon and aluminum).

2.3 Associative/Dissociative Mechanisms of Ligand Exchange

There are two extreme or end-member mechanisms of ligand exchange to distinguish (e.g., Burgess 1988). In *dissociative* exchange, there is a strong element of bond destruction in the activated equilibrium (Fig. 1). This bond destruction is manifested in a positive activation volume for the reaction and a positive activation entropy. In a simple scenario, the metal-ligand bond lengthens to the point of dissociation before entry of the new ligand. The

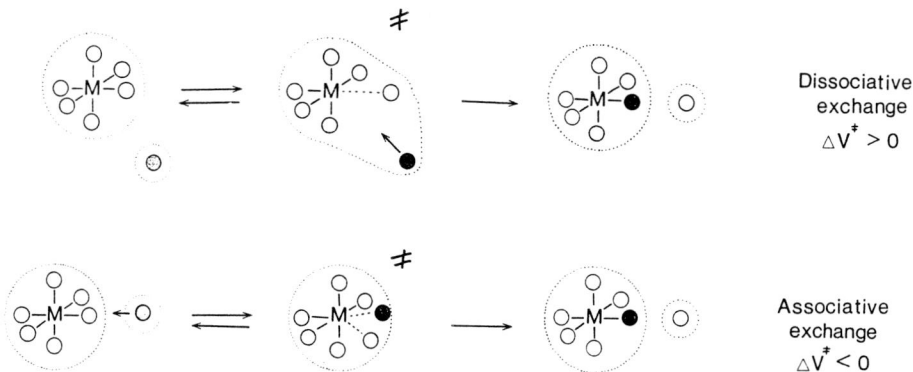

Fig. 1. An illustration of dissociative (*above*) and associative (*below*) pathways for ligand exchange. A similar, albeit more complicated, figure could be created for destruction of a bridging oxygen at a mineral surface. In this case, a ligand is added to coordinate with one of the two previously bridged metals

rate of dissociative exchange is not significantly affected by the character of the incoming ligand because the metal-ligand coordination number decreases in the activated state (Fig. 1). The final metal-ligand *stability* is a direct function of the "softness" of the ligand (Martell and Motekaitis 1989). An ion-pair dissociative mechanism is probably involved in inorganic and organic complexation of aquo-aluminum in solution (Eigen 1961; Secco and Venturini 1974).

In *associative* exchange, there is bond creation, and the metal center assumes an increased coordination number in the activated state (Fig. 1). Rates of associative exchange are significantly affected by the character of the incoming ligand since this ligand must have sufficient electron density to intrude into, and distort, the inner coordination sphere of the metal. These reactions have a negative activation volume and negative activation entropies. Reactions involving silicic acid and organic chelates probably involve an associative S_N2 mechanism and a penta-coordinate transition state (Bennett 1991).

Ion size and bond directionality are important in associative exchange because the rate-controlling step requires an increase in coordination number in the activated state. Large metal ions accommodate the extra ligand in associative exchange more easily than smaller ions. Likewise, the d-electron properties of first-row transition metals, which are all approximately the same ion size, are important because of the considerable directionality of bonds to ligands (e.g., Basolo and Pearson 1958; Burns 1970). Even the unoccupied d-orbitals in silicon and aluminum can be important in controlling the geometry of the associative hypervalent transition state. These mechanisms should be thought of only as extremes in a continuum and the associative/dissociative character of a reaction is primarily relevant in comparisons.

2.4   Dissolution Rates and Solution Composition

Sorption reactions, which introduce potentially reactive ligands to a metal at the mineral surface, or modify a preexisting ligand, are important to the dissolution rate (Westall 1988). The adsorption reactions, however, are very rapid whereas dissolution is slow. The discrepancy in reaction time suggests that the adsorbates indirectly influence but do not control dissolution rates. To understand the influence of solution composition on dissolution rates, it is useful to examine the simplest possible analog for mineral dissolution: dissociation of a dissolved metal dimer.

A dimer consists of two metal ions that are associated by way of a common oxide or hydroxide bridge and are surrounded by waters of hydration (think of this dimer as a two-metal mineral). A simple example might be the $Fe_2(OH)_2(H_2O)_8^{+4}$ species (Fig. 2). This dimer dissociates in water to form two iron monomers, either as hydroxide species, $Fe(H_2O)_5(OH)^{+2}$,

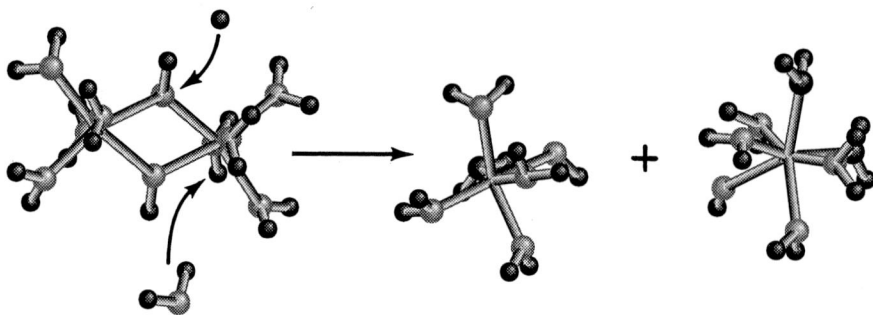

Fig. 2. Ball and stick model for the proton-promoted pathway for the dissociation of the iron dimer. Protonation of a bridging hydroxyl promotes the exchange of water for a bridging bond, resulting in dissociation while conserving the original coordination state

which can react further with the aqueous solution, or as fully aquated ions, $Fe(H_2O)_6^{+3}$ (Fig. 2). Each iron ion in the dimer and monomers is surrounded by six oxide ligands (in this case either hydroxyl ions or water molecules). Because the number of ligands around each iron ion remains the same before and after reaction, this may be considered a ligand-exchange reaction.

The dissociation rate of the iron dimer is very sensitive to the solution pH. In fact, the first-order rate constant for dissociation can be resolved into two terms:

$$k = k_1 + k_2(H^+). \tag{1}$$

One term (k_1, M^{-1} s^{-1}) is for the zero-order rate of dissociation into monohydroxide species. The other term (k_2, s^{-1}) is first-order and proportional to the hydrogen ion concentration, and corresponds to a pathway where a hydrogen ion associates with, and weakens, a bridging hydroxide ion (Wilkens 1974, p. 81) prior to dissociation. The presence of the associated hydrogen ion causes the underlying metal-oxygen bond to take on a more ionic character. In this manner, a rapid acid-base reaction introduces a distinct pH dependence to a slow reaction rate.

From the case of a proton-enhanced rate, one can imagine the case where other ligands affect the dissociation rate. Cyanide ion, for example, could penetrate the inner coordination sphere of an iron in the dimer, thereby displacing a preexisting ligand and inducing dissociation (Fig. 3). Such a reaction would introduce yet a third pathway for dissociation of the dimer:

$$k = k_1 + k_2(H^+) + k_3(CN^-) \tag{2}$$

and would create one iron-cyanide complex as a dissociation product. Note the fundamental difference in the way that hydrogen ions and the incoming

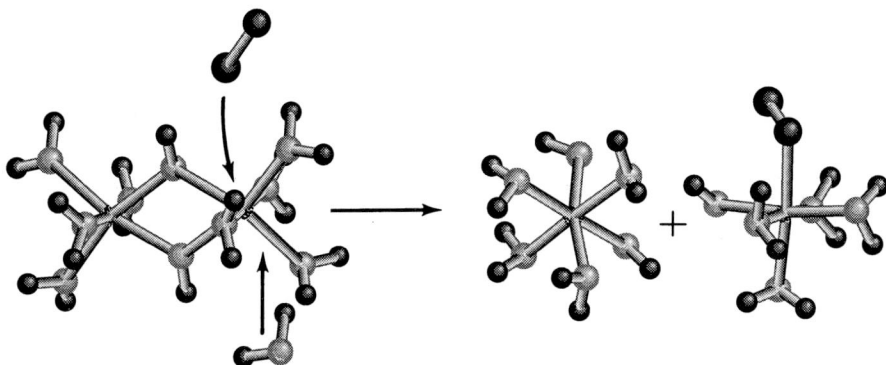

Fig. 3. Ball and stick model for the ligand-promoted pathway for the dissociation of the iron dimer. Cyanate ion coordinates with the iron metal, exchanging with the hydroxyl bridging bond, rupturing these bonds to produce two monomers

cyanide ligand affect the reaction. The hydrogen ion modifies a preexisting ligand coordinated to the metal. The cyanide ion, however, directly associates with the metal and is a wholly new ligand (Fig. 3). Whereas the presence of either proton or cyanide results in a faster rate of dissociation, the fundamental mechanisms are quite different. Thus, there exist *proton*-promoted and *ligand*-promoted pathways for dissociation of the dimer. (Note that neither the hydrogen ion nor the cyanide ions is a true catalyst because each remains associated with the metal after dissociation.)

Metals and metalloids on the surface of silicate minerals are also connected by oxide or hydroxide ion bridges, which undergo acid-base reactions with the adjacent aqueous solution. The results of these acid-base reactions are measurable as surface charge, which varies in concentration with solution pH. (The situation is a little more complicated than this, especially for minerals that have a structural charge due to uncompensated cation substitutions. The reader is directed to, for example, Schindler and Stumm 1988.) The enhancement of dissolution rates via adsorption of hydrogen ions is referred to as the *proton-promoted* pathway for dissolution (Furrer and Stumm 1986) and is analogous to the proton-promoted pathway for the dimer dissociation discussed above.

The acid-base chemistry of a particular mineral surface is characteristic of its composition and structure, just as the hydrolysis chemistry of a particularly dissolved metal is commonly distinct. Figure 4 shows modeled changes in a concentration of positively charged groups ($X-OH_2^+$) and negatively charged groups ($X-O^-$) for the mineral forsterite (Mg_2SiO_4). These surface groups are generic in the sense that they cannot be assigned a true stoichiometry, but are calculated from the variation in total surface charge concentration with solution pH (and magnesium concentration in this case). Note that the concentrations have distinct slopes at pH < 4 and pH >

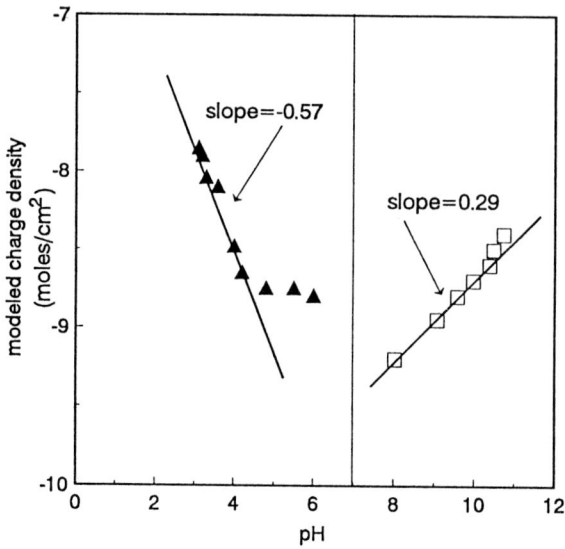

Fig. 4. Variations in modeled concentrations of positively and negatively charged groups on the surface of forsterite as a function of solution pH (Blum and Lasaga 1988)

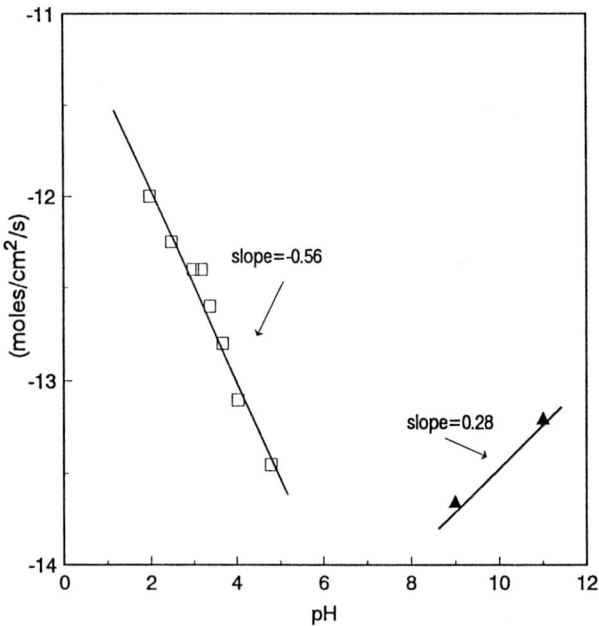

Fig. 5. Rates of dissolution of γ-Al_2O_3 vary with the surface concentration of certain organic ligands (Furrer and Stumm 1986). In the pH range 5–6, small variations in pH do not affect the surface concentration of the ligand or the dissolution rates

8. Dissolution rates vary in the same way as these modeled concentrations (Fig. 5). The slow dissolution kinetics is enhanced by rapid adsorption reactions, and the extent of adsorption varies with pH, just as in the case of the iron dimer. Similar trends are found for other silicates, and a strong case can be made for surface-speciation control on inorganic silicate dissolution kinetics (Schott and Petit 1988; Brady and Walther 1989; Stumm and Wieland 1990).

A further effect of surface speciation is to control ion exchange and ion adsorption at the surface. A quartz surface is largely neutral at acidic pH, but is 50% anionic at neutral pH. Alumina surface (γ-Al$_2$O$_3$), however, is neutral at pH 9.1, whereas Al(OH)$_3$ is neutral at pH 5 (Stumm and Morgan 1981). As the anionic character of a surface increases, cation sorption is selected over anion sorption. Because the first step in dissolution is the sorption of the reacting ligand to the surface, solution pH influences ligand-promoted pathways by influencing the ability or extent to which charged ligands sorb to the charged surface, and even directing the ligand to a specific functional group (i.e., alumina vs. silica groups on the aluminosilicate surface).

2.5 Reactivity Trends

Apart from the variations in rate with solution composition, some metal-oxygen bonds in silicates are inherently more reactive than others. A good way to systematize these inherent reactivities is, again, to examine the reactivity trends for an analogous process in an aqueous solution, such as the replacement of hydration waters around a metal in solution. As discussed earlier (Table 1), the metal-oxygen coordination chemistry of an aquated metal ion is not significantly different from that ion in some simple oxide and silicate minerals. Thus, exchange of waters around an aquated metal requires destruction of metal-oxygen bonds, which are similar to those in a silicate or oxide mineral.

From this gross similarity, one might expect a correlation between mineral dissolution rates under well-defined conditions and rates of water exchange about the corresponding dissolved cation. Well-defined conditions mean: (1) compared minerals have similar structures but variable compositions; (2) solution compositions are identical; (3) conditions are excluded whereby the valence state of metals at the mineral surface is altered before dissolution; and (4) the metal-oxygen coordination chemistry of metals in the mineral somewhat resembles that in the cation in solution. Such a correlation is shown in Fig. 4 for both simple oxide minerals containing divalent cations and end-member orthosilicate minerals at pH 2 and 25 °C. As one can see, there is considerable similarity between rates of oxide and orthosilicate dissolution rates (Fig. 5) and rates of water exchange around the corresponding aquated cation (Fig. 4).

3   Organic Acids and Silicate Dissolution

Organic acids influence the dissolution of silicate minerals in much the same manner as inorganic ligands. The two mechanisms that seem most important are: (1) increase in solubility due to the complexing of metals and metalloids in solution; and (2) surface complexation followed by rupturing of framework metal-oxygen bonds, resulting in an increase in dissolution rate. Either or both of these pathways may be important in a specific natural system.

3.1   Solution Complexation by Organic Acids

Ligand-exchange reactions involving organic acids and organic acid anions are analogous to the inorganic systems discussed above. Organic anions are reasonably hard ligands and act as strong electron donors that can participate in hydrogen-bonded (or electron-donor-acceptor) structures with water and with each other, and form coordinative covalent bonds with metals and metalloids. The resulting complex may be a weak electron donor-acceptor complex, or a uni- or multi dentate chelate (Eigen 1961; Buffle 1990).

Stability constants for simple organic monoacid and diacid complexes with metals in aqueous systems are well documented (Smith and Martell 1989). The stability of a particular aqueous coordination complex is influenced by several variables, notably by organic-acid species and pH. pH is particularly important because the properties of both the organic acid and the aquated metal ion are sensitive to hydrogen ion activity. Aluminum speciation is especially sensitive to proton activity, and aluminum-organic acid complexes are strongly pH-dependent (Martell and Motekaitis 1989).

For many organic – ligand-metal pairs, 1:1, 2:1, and 3:1 complexes are possible, as dictated by the ligand-metal activity ratio. In general, ligand-exchange reactions that form multidentate complexes are favored over monodentate complexes, and multiprotic acids such as oxalic, citric, and phthallic acids are usually much better complexing agents than the monoprotic acids such as acetic and benzoic acids (Smith and Martell 1989). Multifunctional α-hydroxy acids, such as tartaric and citric acids, and the very weakly acidic dihydroxy aromatic compounds such as catechol are very effective in ligand exchange reactions, even though the hydroxyl groups easily protonate at even very low proton activity.

3.1.1   Organic–Aluminum Complexes

Organic – acid-aluminum complexes have been well characterized in part because of their importance in buffering monomeric aluminum activity in water impacted by acidic precipitation, and the fundamental role of aluminum speciation in mineral weathering (Driscoll et al. 1985; Stumm and Furrer 1987; Nordstrom and May 1989).

Aluminum is a hard trivalent ion, and is usually found as an aquo- or hydroxy-complex or hydrous oxide solid (Martell and Motekaitis 1989). Aluminum can act as a complexing agent of other metals, or as a metal center to other inorganic and organic ligands (Buffle 1990). As the hexa-coordinated hydrolyzed cation, aluminum is octahedrally coordinated by six waters (Fig. 6), and as such acts as a multiprotic acid (Stumm and Morgan 1981).

In organic systems aluminum is presumed to be complexed primarily by difunctional organic acids through a bidentate chelate (Martell and Motekaitis 1989), forming a ring structure that incorporates two $Al-O-C$ bonds (Fig. 7). An often reported order of complexing ability for aliphatic organic acids (e.g. Stumm and Furrer 1987; Nordstrom and May 1989) is:

Citric > Oxalic > Malonic > Succinic ≫ Acetic

Greatest ⎯⎯⎯→ Stability ⎯⎯⎯→ Least,

whereas for the aromatic chelates:

Catechol > Salicylic > Phthalic ≫ Benzoic.

The order of stability of these aqueous complexes closely follows the order of greatest to least promotion of dissolution of γ-alumina (Furrer and Stumm 1986; Schindler and Stumm 1988).

The mechanism of complex formation for organic-aluminum chelates probably involves a dissociative mechanism, as the aquo-aluminum aqueous complex involves a full octahedral coordination shell (Hewkin and Prince

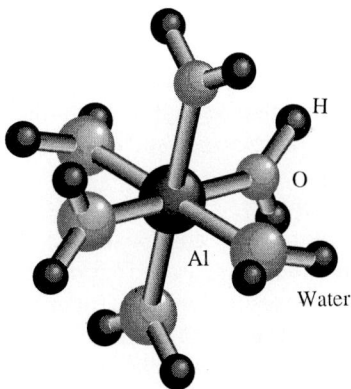

Fig. 6. Ball and stick model of the fully coordinated aquo-aluminum complex in an octahedral geometry

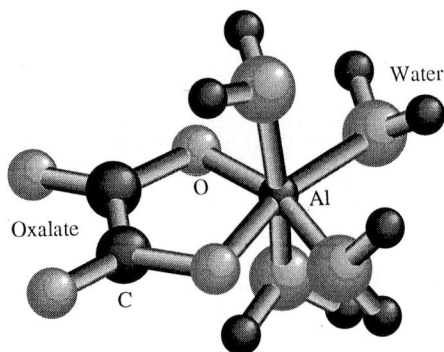

Fig. 7. Ball and stick model of a bidentate aluminum-oxalate 1:1 complex with coordinating water in an octahedral geometry

1970). An associative pathway would involve a hepta-coordinate transition state complex, which is unlikely if only based on steric criteria. In a study of salicylate-aluminum complexes at 25°C, Secco and Venturini (1974) proposed that the first step is a fast ion-pair association of the anion and the aquo-aluminum coordination complex, i.e.:

$$(H_2O)_6Al^{3+} + HA^- <\!-\!> [(H_2O)_5\,Al(H_2O):HA]^{2+},\tag{3}$$

where they found that the rate of ligand penetration to be in good agreement with the rate of water exchange on aluminum, as discussed earlier. This is followed by the rate-limiting departure of one water molecule from the aquo-aluminum coordination sphere:

$$[(H_2O)_5Al(H_2O):HA]^{2+} <\!-\!> (H_2O)_5Al\text{-}AH^{2+} + H_2O,\tag{4}$$

followed by the loss of a hydronium ion with subsequent ring closure:

$$(H_2O)_5Al\text{-}AH^{2+} <\!-\!> (H_2O)_4Al\text{-}A + H_3O^+,\tag{5}$$

resulting in a 1:1 bidentate chelate. This might be followed by additional reaction to produce 2:1 and 3:1 salicylate-aluminum complexes. Similar reaction sequences probably occur for other organic acids, such as oxalic acid (Fig. 7).

From studies of aluminum coordination chemistry, it is clear that an important factor in the stability of organic-aluminum complexes is the geometry of the final bidentate chelate. Five-member rings were found to be most stable, such as the oxalate or catechol complexes, whereas six-member ring chelates, such as malonate or salicylate complexes of aluminum, were less stable (Furrer and Stumm 1986; Martell and Motekaitis 1989). From a simple standpoint, therefore, stability decreases with increasing ring size over five. This relationship requires that effective organic ligands have adjacent oxygen-containing functional groups, such as with oxalic acid, catechol, and citric acid, and allows at least a simplistic basis for predicting the ability of an organic compound to mobilize aluminum by complexation (e.g., Ohman and Sjoberg 1983).

The stability of the organic-aluminum complexes in varying solution types has become an important problem, in part due to recent debate on the mobility of aluminum in subsurface brines (see Surdam and Yin, Chap. 13, this Vol.). Much of the work detailing aluminum complexes with natural organic acids uses chromatographic separation methods (e.g., Stevenson and Vance 1989), titration methods (e.g., Ohman and Sjoberg 1983), or mass transfer dissolution methods (e.g., Furrer and Stumm 1986) to determine stability constants. Direct characterization of the stability of these complexes under different environmental conditions, however, is difficult. Organic and inorganic complexes have been characterized using an indirect fluorescence method in several solution types, and catechol complexes of aluminum have been characterized using UV-difference methods (Sikora and McBride 1989). Aluminum complexes with acetate were characterized using CIRIR methods (Aines and Jackson 1990). Recently, Tait et al. (1991) examined the stability of aluminum-oxalate complexes as a function of pH and temperature using Raman and FTIR spectroscopy. Both the 1:1 (Fig. 7) and 3:1 complexes were tentatively identified based on vibrational spectra and found to be stable in aqueous solutions up to 140 °C. On the basis of preliminary peak-height comparison, it also appears that the stability of the 3:1 complex is greater at higher temperature, arguing for a strong coordinative interaction rather than the temperature-labile electron donor-acceptor complex.

3.1.2 Organic-Silica Complexes

Organic-silica complexation in natural waters has been hypothesized but not well documented. Coordination of silicon by organic ligands via Si—C and Si—O—C linkages are well characterized in nonaqueous systems (see reviews by Carlstrom 1977; Chvalovsky and Bellama 1984), though the existence of hydrogen-bonded complexes is recognized in biochemical systems (Schwarz 1973; Weiss and Herzog 1977; Birchall and Espie 1986; Carlisle 1986; Sullivan 1986).

The aqueous speciation of Si^{+4} is fundamentally different from that of Al^{+3} in water at low temperature. Whereas aluminum is octahedrally coordinated with water and/or hydroxide, silicon is tetrahedrally coordinated by hydroxide only, with no coordinating water. In contrast to alumina $Al(OH)_n(H_2O)_3^{3-n}$, the hydroxyls on $Si(OH)_4$ (Fig. 8) participate in hydrogen-bonding interactions with water and with each other, creating a complex structure in solution and making it impossible to isolate pure monomeric $Si(OH)_4$ (Iler 1979). Also, the silica hydroxyl hydrogens are much less acidic than the hydrogens associated with coordinated water on

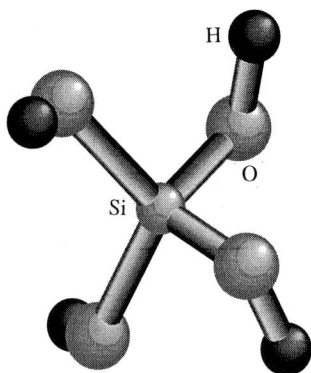

Fig. 8. Ball and stick model of silicic acid [Si(OH)$_4$] in a tetrahedral geometry

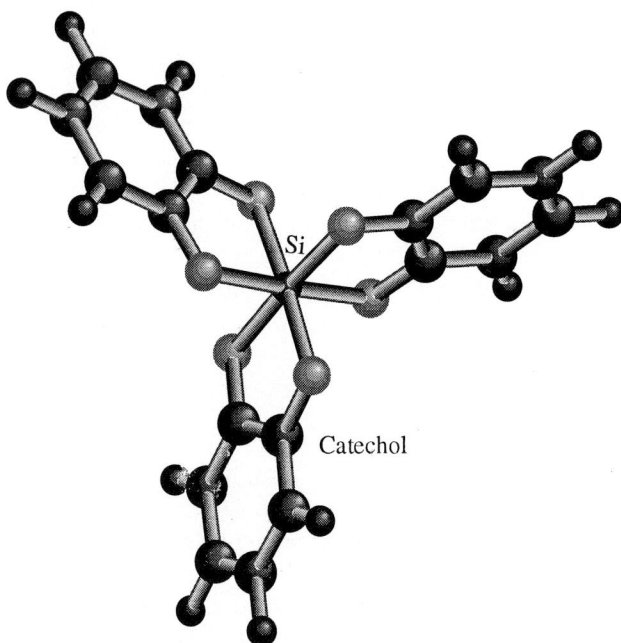

Fig. 9. Silica-catechol 3:1 complex, showing the five-membered ring with the silicon in an octahedral geometry

aluminum. So while aluminum speciation will change with pH, and aluminum is effectively complexed by many inorganic ligands, monomeric silicic-acid is stable and will remain largely uncomplexed in water.

The differences between alumina and silica in the inorganic system suggest that organic-silica complexes will differ significantly from alumina. In aqueous systems, only catechol-silica complexes are well established (Iler 1977, 1979), and 1:1, 2:1, and 3:1 chelates have been documented (Fig. 9).

Other complexes involving nitrogen-containing organic bases have also been characterized (Iler 1977). Hydrogen-bonded complexes involving an =Si—O—H ... O—C – R bridge may also occur in aqueous systems and on silica surfaces (e.g., Iler 1977), and have been implicated in some soil zone interactions.

Dissolution experiments using quartz in organic-acid solutions provide an indirect evaluation of the ability of organic-acid anions to complex silica in solution. Assuming that the fundamental thermodynamic equilibrium relationship between solubility constant and monomeric silicic-acid activity is unchanged, a change in solubility as measured by *total silica concentration* versus the solubility constant should provide a measure of the amount of complexed silica in solution, regardless of the form of the complex. The results of such experiments (Bennett et al. 1988; Bennett 1991) showed that citrate increases quartz solubility as a function of citrate concentration at near-neutral pH, although the stability constant is small. In a 0.020 molal solution of citrate, the solubility of quartz only increases by about 100 micromolal (Fig. 10). The increase in solubility decreases with decreasing pH, and is not significant at pH 5. Only citrate and oxalate were found to

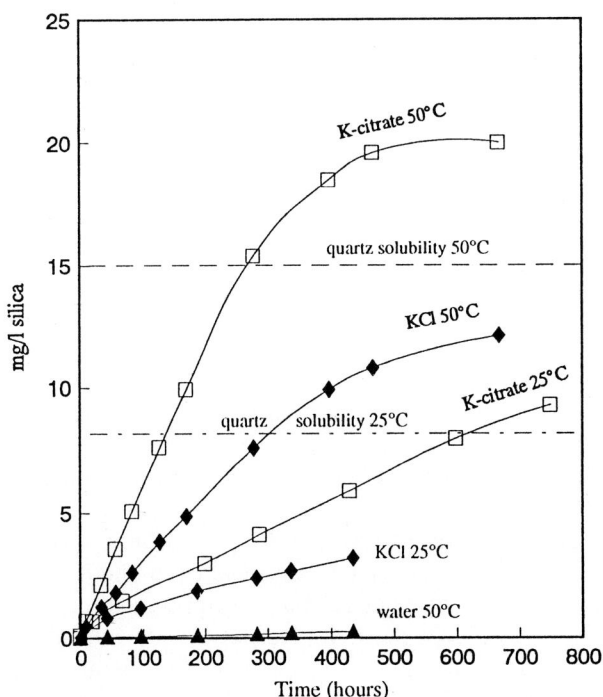

Fig. 10. Dissolution of quartz plotted as silica concentration over time for 20 mmol citrate solutions at pH 7, at 25 and 50 °C

clearly increase the solubility of quartz. and oxalate only at high concentration. High concentrations (>10 mmolal) of salicylate and humic substances also may increase quartz solubility, but the evidence is not as strong. This is in contrast to alumina, where salicylate is an effective chelating agent.

With the exception of the catechol-silica complex, aqueous organic-silica complexation is extremely difficult to *directly* characterize. UV-visible spectroscopic methods (Bennett et al. 1988) suggest that citrate and oxalate interact with silica via an electron donor-acceptor interaction at near-neutral pH. Further investigations using Raman and FTIR vibrational spectroscopic methods, although successful for aluminum complexes (Tait et al. 1991), have yielded only ambiguous (Bennett 1991) or even suspect results (Marley et al. 1989). Ongoing experiments using enriched ^{29}Si NMR are examining the organic-silica interactions but to date the results are masked by the extreme peak-broadening effects of the silica-water and silica-carboxylate hydrogen bonding. Thus, direct evidence of a silica-organic – acid anion chelate beyond hydrogen bonding remains poor.

As with the coordination chemistry of aquo-aluminum, the critical factor in the formation of aqueous chelates of silica may be the formation of a five-member ring complex between a difunctional organic electron donor and the silicon metal center through a pair of linkages. Only chelates with that structure have been clearly identified, whereas only weak evidence for six-member rings and no evidence for seven-member rings exist. While the five-member ring can be rationalized with aluminum, given the octahedral geometry of the aquo-complex, the five-member ring is problematic with the silicon-oxygen tetrahedral geometry. Theoretically, a six-member ring in a slightly twisted conformation would be expected to have a lower angle-strain component.

The 3:1 catecholate-silica complex offers insight into this problem, where silicon is octahedrally coordinated by three bidentate ligands, forming a hypervalent silicon chelate. Expanded valence-shell geometries have been reported for other silicon-organic complexes, where in addition to the tetra-coordinated, sp^3 chelates, hypervalent penta- and hexa-coordinated silicon species involving unoccupied d-orbitals are found (Sommer 1965; Kwart and King 1977; Cella et al. 1980; Corriu et al. 1984; Tandura et al. 1986). The expanded valence shell structures of silicon were also found to occur with either F (i.e., SiF_6^{2-}) or electron-rich oxygen functional groups, such as organic-acid anions.

Observational and modeling evidence suggests that the d-orbital polarization of organic-silicon coordination geometry is a critical component of the bonding reaction sequence, and is a link between the inorganic and organic-silica systems. Ab initio calculations of the hydrolysis of Si−O−Si bridging bonds (Lasaga and Gibbs 1990) suggest that this reaction forms an intermediate in which silicon is penta-coordinated by the addition of nucleophilic water, followed by the rupturing of the Si−O bond (*associative* ligand exchange). This involves a distinct shift in silicon geometry from a tetrahe-

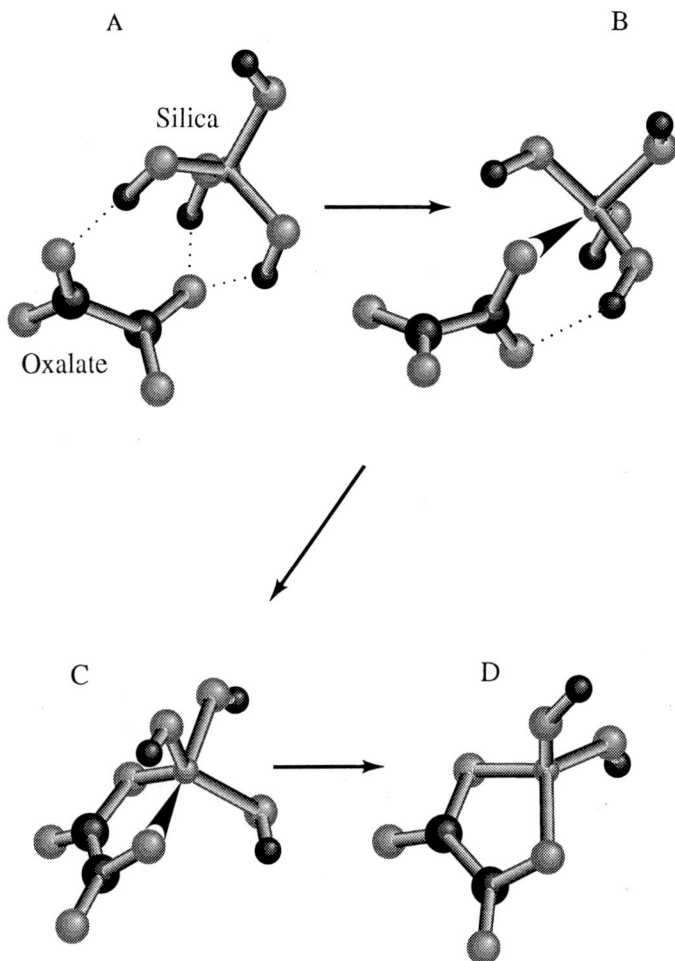

Fig. 11A–D. Modeled reaction sequences of the oxalate-silica complex. **A** Oriented hydrogen-bonded complex (probably the most common species). **B** Transition state to the monodentate complex via S$_N$2-type ligand exchange. Geometry is a penta-coordinated triganol bipyramid. **C** Monodentate complex proceeding into the transition state to the bidentate via the same reaction as **B**. **D** The bidentate silica-oxalate 1:1 complex

dral to a trigonal bipyramid form. Organic-acid anions in water are strong nucleophiles, and molecular orbital calculations using the frontier molecular orbital approach suggest that silica undergoes a nucleophilic S$_N$2 reaction by catechol and specific organic acid anions similar to that proposed for water. Here, the intermediate structure forms a penta- or hexa-coordinated inter-mediate stabilized by the utilization of low lying d-orbitals. In the hexa-coordinated structure, an octahedral geometry is the lowest energy confor-mation, with 90° bond angles between ligands. This intermediate structure decays with the release of a hydroxyl ion as the leaving group (Bennett 1991; Fig. 11). The final geometry is still influenced by the now unoccupied d-orbitals, where polarization of the sp^3 orbitals results in an $O-Si-O$ bond angle stabilized at 90° to 94°, with five-member ring chelates being the most stable (Fig. 11D).

3.2   Surface Interactions By Organic Acids

Extrapolating the solution-model of organic-metal interactions to the silicate surface in a molecular, coordination chemistry approach (e.g., Amrhein and Suarez 1988; Wieland et al. 1988) suggests several possible interactions and outcomes. In making this extrapolation, however, the influences of surface charge and speciation, and the energetics of sorption must be taken into account. Also, the coordination of surface aluminum is problematic, and for feldspars may be intermediate between the solid-state tetrahedral coordination and the octahedral coordination. Therefore, it is not clear whether the surface association will be associative or dissociative.

3.2.1   Sorption and Ligand Exchange Reactions

As discussed above, the first step of a pathway where organic solutes enhance silicate dissolution must be the association of the organic solute with the aqueous surface of the silicate (i.e., adsorption). Sorption of an organic solute to a silicate surface may be generalized to occur by one of three mechanisms (e.g., Westall 1988). Hydrophobic bonding occurs when nonpolar organic solutes partition to the solid surface as an inverse function of aqueous solubility (e.g., Karickhoff et al. 1979). This interaction will not be considered further here. A second interaction is coulombic in nature, involving electrostatic interaction and exchange of ions according to the specific selectivity coefficients (Westall 1988). Whereas this interaction itself is probably not significant in the dissolution process, it does serve to bring an organic anion present at low aqueous concentration, for example, into proximity with the silicate surface. The third mechanism, sometimes called chemisorption, is where an organic solute forms specific chemical bonds with silicate surface groups. In this case, because the aquated silicate surface is fully coordinated by water or hydroxyls, the interaction is actually a ligand exchange reaction analogous to the solution exchange reaction discussed above (Kummert and Stumm 1980). This interaction, where organic-acid anions coordinate directly with a silicate metal center, results in enhanced dissolution.

The importance of solution pH becomes clear when considering the initial sorption step of organic-enhanced silicate dissolution. Organic anions, as the stronger nucleophile, are probably the most important reactants, and most carboxyl groups deprotonate above pH 4–5, whereas the organic hydroxyl hydrogens have a higher pK. At high pH, however, the quartz surface, for example, is weakly anionic. At the ZPC (zero point of charge) of quartz, pH ~2, the overall surface charge is neutral, and the concentration of cationic sites equals the concentration of anionic sites. At the pK_o of quartz (a measure of the acidity of the surface silanol groups), pH ~6.8, 50% of the surface sites are anionic, or negatively charged, with the rest dominantly neutral. Above the pK_o the surface is dominantly anionic. The silicate surface, in contrast, will have a heterogeneous distribution of anionic and neutral silanol groups with cationic and neutral alumina groups (feldspar ZPC is about 2.4, Stumm and Morgan 1981). As pH increases, the overall anionic character of the silicate surface dominates, selecting against an initial electrostatic interaction with anionic organic solutes of like charge. This creates an adsorption "window" (Buffle 1990) where pH control of solute and surface speciation directly influences anion adsorption. For silicates, the anion adsorption window is roughly between pH 5 and 7.5 in natural waters.

3.2.2 Ligand-Exchange Interactions by Organic Acids

Because of the ligand-exchange adsorption reactions at the silicate surface, certain organic ligands affect silicate dissolution rates by introducing *ligand-promoted* pathways for dissolution (e.g., Kummert and Stumm 1980). The variation in the dissolution rate of γ-Al_2O_3 as a function of the surface concentration of two organic acids is shown in Fig. 12. As in the case of proton-promoted dissolution, the effect correlates with surface concentration of the ligands, not dissolved concentration. Rates of dissolution vary linearly with the concentration of adsorbed salicylate and citrate, which form a bidentate complex with surface metals to form a ring chelate on the mineral surface. As was found in the solution model, the effect of any particular organic ligand on the dissolution rates varies with the size of the resulting ring. Oxalic acid, which forms a five-membered ring, has a much stronger effect on oxide dissolution rates than succinic acid, which forms a seven-member ring (Furrer and Stumm 1986), whereas monodentate ligands do not appreciably affect dissolution rates. Rate coefficients for ligand-promoted dissolution of γ-Al_2O_3 decrease as pH decreases below 5 (Fig. 13), where high proton concentrations cause the ring compounds to protonate and open, thereby forming an inert, monodentate surface ligand. At the same time, as pH decreases, the proton-promoted dissolution rate increases.

The coordinating organic ligand polarizes framework metal linkages, forcing the bond to rupture faster than for the equivalent inorganic rate where water is the coordinating ligand (Bennett 1991; Casey and Westrich 1992). For aluminum, this may be either a dissociative or associative mechanism, depending on the surface coordination state, whereas with silicon

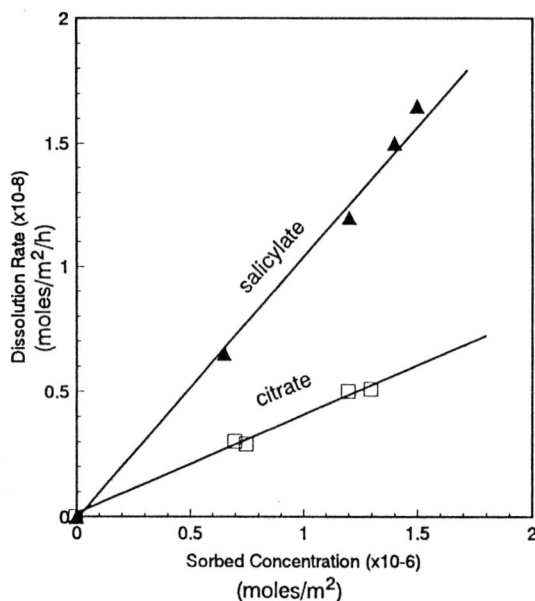

Fig. 12. Rate of feldspar dissolution versus adsorbed concentration of salicylate and citrate

Table 2. Rate coefficients for ligand-promoted dissolution of γ-Al$_2$O$_3$ (Furrer and Stumm 1986). Ligand-promoted dissolution exhibits a first-order dependence on the surface concentration of ligand; the rate coefficient has units of h^{-1}

	Ligand	$k\ (h^{-1})$
Five-membered rings:	Oxalate	$10.8 \times 10^{-3}\,h^{-1}$
Six-membered rings:	Malonate	$6.9 \times 10^{-3}\,h^{-1}$
	Salicylate	$12.5 \times 10^{-3}\,h^{-1}$
Seven-membered rings:	Succinate	$2.4 \times 10^{-3}\,h^{-1}$
	Phthalate	$3.0 \times 10^{-3}\,h^{-1}$
Monodentate:	Benzoate	$\ll 10^{-3}\,h^{-1}$

the interaction is associative. For example, the rate coefficients for ligand-promoted dissolution of γ-Al$_2$O$_3$ are compiled in Table 2. As one can see (Table 2), the rate coefficients vary by a factor of 2–6 for different ligands in a systematic manner; that is, the rate depends (albeit weakly) on the character of the adsorbed ligand.

Similar, though less dramatic, results are obtained for the dissolution rate of quartz. At low temperature, citrate accelerates the dissolution of quartz between pH 5 and 7 and increases the solubility (Bennett 1991). Dissolution

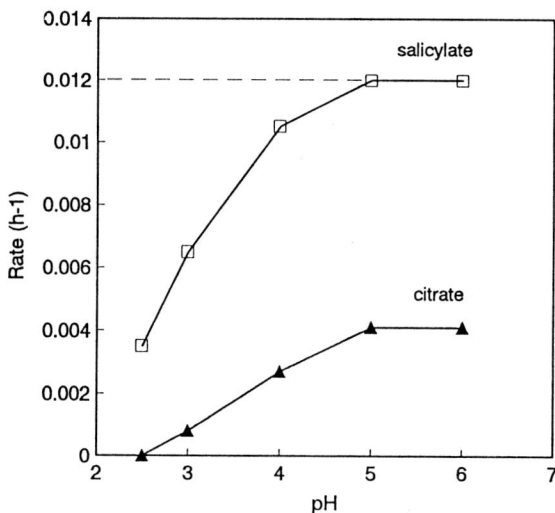

Fig. 13. Ligand-promoted rate plotted against pH. As pH decreases the enhancement due to the organic ligand decreases

rate increases with increasing citrate concentration between 0.001 and 0.020 molal, as does measured solubility (Table 3; Fig. 10). Oxalate also increases the dissolution rate, but not as strongly. Monoprotic acids, and acids that form six-member or larger rings, have little or no effect on dissolution rate. As the dissolution of quartz is zero-order when far from equilibrium, the increase in dissolution rate by organic acids is a surface phenomenon not directly related to the increased solubility due to solution complexation.

Analysis of dissolution rate as a function of temperature shows that the difference in quartz dissolution rate is greatest at low temperature, and decreases with increasing temperature. Application of the Arrhenius equation:

$$k = Ae^{-Ea/RT},$$

where k is the rate constant, A a preexponential function, R the universal gas constant, T temperature in Kelvin, and Ea the activation energy, suggests that the increase in dissolution rate results from a decrease in the activation energy of dissolution (Table 3). At temperatures above 70 °C, however, there is no detectable influence of citrate on quartz dissolution rate. In the case of silica mobility, therefore, citrate influence on dissolution and solubility is a low-temperature phenomenon.

A model for quartz dissolution can be proposed from the experimental data by extrapolation of the results of ab initio and semiempirical molecular orbital calculations from the static single molecule to the aquated quartz surface (Fig. 14; Bennett 1991). The rate-limiting step of quartz dissolution

Table 3. Influence of electrolyte type and solution concentration on quartz dissolution kinetics. Experiments were conducted at pH 7. Electrolyte type gives the anion and counter ion present. Concentration refers to the anion, and is expressed in mmol/l; rate is in $mol\,cm^{-2}s^{-1}$; Ea is activation energy in kcal/mol

Electrolyte	mmol/l	log Rate	Ea
Water		−16.9	17.9
NaCl	5	−16.7	17.9
	10	−16.4	
	20	−16.2	
	50	−15.9	
Na-Citrate	5	−15.8	13.9
	10	−15.6	
	20	−15.4	
Na-Oxalate	2	−16.0	15.0
	20	−15.6	

is the breaking of framework Si—O bonds during hydrolysis, a reaction that is preceded by the formation of a penta-coordinated [≡Si—OH]-H_2O surface species (Lasaga and Gibbs 1990). An initial reaction involving the adsorption of an organic-acid anion, followed by chelation of a surface silanol group and polarization of framework bonds, could result in a significantly lower energy barrier for the subsequent hydrolysis reaction. This model fits the finding of a decrease in the activation energy of quartz dissolution in dilute solutions of some organic electrolytes, and is supported by molecular orbital calculations.

3.2.3 Silicate Solubility and Phase Equilibria

Organic acids influence silicate solubility primarily by complexing dissolved aluminum and silica and decreasing their solution activities. In the case of quartz, high concentrations of citrate or catechol result in a significant increase in solubility at low temperature and neutral (for citrate) or basic (for catechol) pH (Iler 1979; Bennett 1991). At lower pH (pH 5–6) organic acids are effective complexing agents of aluminum (Martell and Motekaitis 1989; Nordstrom and May 1989), and during feldspar dissolution, high concentrations of organic acids might increase the mobility of aluminum (e.g., Surdam and Yin, Chap. 13, this Vol.). In such a system, the precipitation of a secondary solid phase may be indirectly controlled by the presence and subsequent degradation of organic acids.

Solution pH may also have an indirect control on silicate solubility above the direct control offered by proton concentration in the proton-enhanced

Oxalate

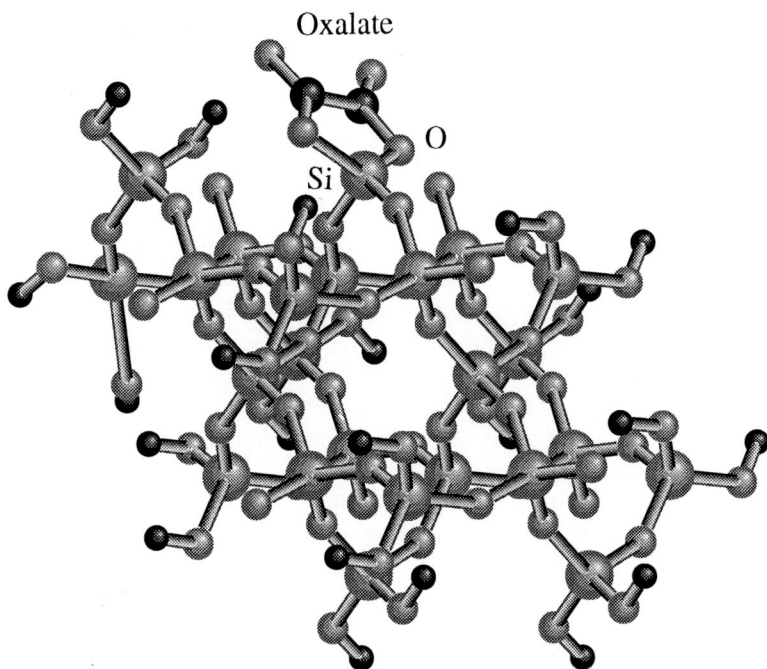

Fig. 14. Ball and stick model of a silica-oxalate chelate on a quartz surface. The coordination of some organic acids on the surface would both alter the geometry at the chelated silicon center, and polarize framework bonds, thereby decreasing the energy barrier associated with hydrolysis and increasing dissolution rate

hydrolysis. At acidic pH (pH < 4.5), organic acids are more likely to be protonated, and in the protonated form the carboxyl acid is a weaker complexing agent of metals (Buffle 1990); thus, the proton-enhanced pathways of dissolution are favored. At mildly acidic pH, aluminum is readily complexed by organic acids in solution and on the silicate surface, and aluminum ligand-exchange – enhanced pathways of dissolution are favored over proton-enhanced. Aluminum is mobilized as organic chelates, and the equilibria with respect to secondary aluminum solid phases are perturbed. Organic complexes of silica, however, are less stable, and silica is mobilized simply as a result of the rupturing of the silicate framework.

At near-neutral pH, aluminum is preferentially complexed by hydroxide, total aluminum solubility is very low, and organic-aluminum complexes are not favored. At this pH, however, organic-silica complexes are more favored, silica is mobilized, and silicate and quartz dissolution rate is accelerated. Organic-acid concentration, however, must be much higher than for the equivalent mobilization of aluminum because of the weaker silica – organic-acid interaction. In a near-neutral pH, organic-rich system, silica is mobilized from the silicate surface with the released aluminum unstable with respect to solid $Al(OH)_3$ or clay phases. In this scenario, the observer might find silicates rapidly dissolving and silica concentrations very high, without a significant increase in dissolved aluminum.

3.3   Microbial Influence on Silicate Dissolution

The relationship between microorganisms and mineral weathering has long been recognized. Webley et al. (1963) and Duff et al. (1963) showed that the organic acids produced by bacteria enhance mineral weathering in soils (see Drever and Vance, Chap. 6, this Vol.). Several researchers have found that both bacteria and fungi preferentially mobilize silica during mineral weathering reactions (Kutuzova 1969; Avakyan et al. 1981), and may utilize silica supplied as quartz or silicate mineral in metabolic processes (e.g., Heinen 1967, 1968; Krumbein and Werner 1983).

Microbes enhance the dissolution of aluminosilicates by a variety of mechanisms (e.g., Price and Morel 1990). Soluble polysaccharides containing xylose and glucuronic acids are associated with algae and bacteria and may complex metals. Polysaccharides are also a major component of external mucus formation during bacterial attachment and contain available carboxyl sites. Bacteria also produce a variety of extracellular chelating substances involved in the transport of hydrophilic micronutrients across the cell membrane (e.g., Duff et al. 1963; Webley et al. 1963; Price and Morel 1990). Citric, oxalic, and 2-ketogluconic acids have been identified in soils and interstitial waters, and it is expected that high concentrations of these organic acids may build up around bacteria adhering to a mineral surface. Enhanced weathering of quartz has been identified in fungal systems and metabolically produced citric acid was found to be the reactive component (Aristovakaya and Kutuzova 1968; Silverman and Munoz 1970), illustrated by the weathering action of lichens (Silverman 1979; Jones and Wilson 1985).

Recent studies by Bennett et al. (1991) and Hiebert and Bennett (1992) showed that native bacteria in an oil-contaminated aquifer rapidly colonize fresh mineral surfaces, where the contaminating petroleum provides a rich carbon substrate for metabolic processes. Quartz and feldspar grains placed in the contaminated portion of the aquifer were deeply weathered after only 14 months, with the feldspar grains showing deep prismatic etch pits on the exposed surface. The sample collection procedure preserved the adhering bacteria, and they were shown to be active populations, with the greatest weathering associated with the highest bacterial density. The findings from this field site are discussed further below.

4   Case Studies

To illustrate the principles discussed above, two case examples are examined to determine the controlling mechanisms of silicate dissolution, and the final outcome. In the first case study, organic acids preferentially mobilize silica in a microbially active oil-contaminated aquifer, whereas the second study shows a transition between enhanced aluminum mobility at acidic pH and enhanced silica mobility at neutral pH in a peat bog.

4.1   Silicate Stability in a Petroleum-Contaminated Groundwater

In a petroleum-contaminated sand and gravel aquifer near Bemidji, Minnesota, native microorganisms actively degrade the oil, producing carbon dioxide in oxidizing waters, and methane and hydrophilic and hydrophobic organic acids in anaerobic waters (Cozzarelli et al. 1990). In the highly reducing waters beneath the oil, dissolved organic carbon is as much as 5 mmol/l C, with much of the carbon as hydrophilic and hydrophobic organic acids in addition to soluble alkyl aromatics (Eganhouse et al. 1993). The groundwater pH decreases from 7.8 in the background area to 6.5 beneath the oil, dissolved iron increases from near zero to 1 mmolal, and dissolved silica increases from 0.3 mmol/l to over 1.0 mmol/l (Bennett et al. 1993). Dissolved aluminum is low, increasing from a background of less than 0.05 μmol/l to ~0.2 μmol/l (Fig. 15).

Microscope examination of sand grains in the aquifer shows that quartz, feldspars, and iron silicates are unaltered in uncontaminated areas of the aquifer (Fig. 16A,C), but chemically weathered in the anaerobic portion of the aquifer where organic acids are present (Fig. 16B,D). Quartz grain surfaces displayed triangular etch pits diagnostic of chemical etching (Bennett and Siegel 1987; Fig. 16B), whereas feldspars were heavily weathered with adhering secondary minerals (Fig. 16D). Subsequent experiments using *in situ* microcosms of clean crushed crystals of quartz and feldspar showed that native microorganisms rapidly colonize the silicate surfaces, with clear evidence of chemical weathering obvious after 14 months (Hiebert and Bennett 1992). Etching of fresh surfaces occurred only near developing bacterial colonies, while precipitation of secondary clay minerals occurred on uncolonized surfaces.

In this example, organic acids are produced at near-neutral pH during the transformation of petroleum by microorganisms. Silica is mobilized from both aluminosilicates and quartz even though the groundwater is greatly supersaturated with respect to quartz, and aluminum is conserved in a solid phase. The rate of the weathering is rapid, with clear etching visible after only 14 months. Organic acids produced by adhering on silicate surfaces will be present at very high concentrations near and under the microbial colony

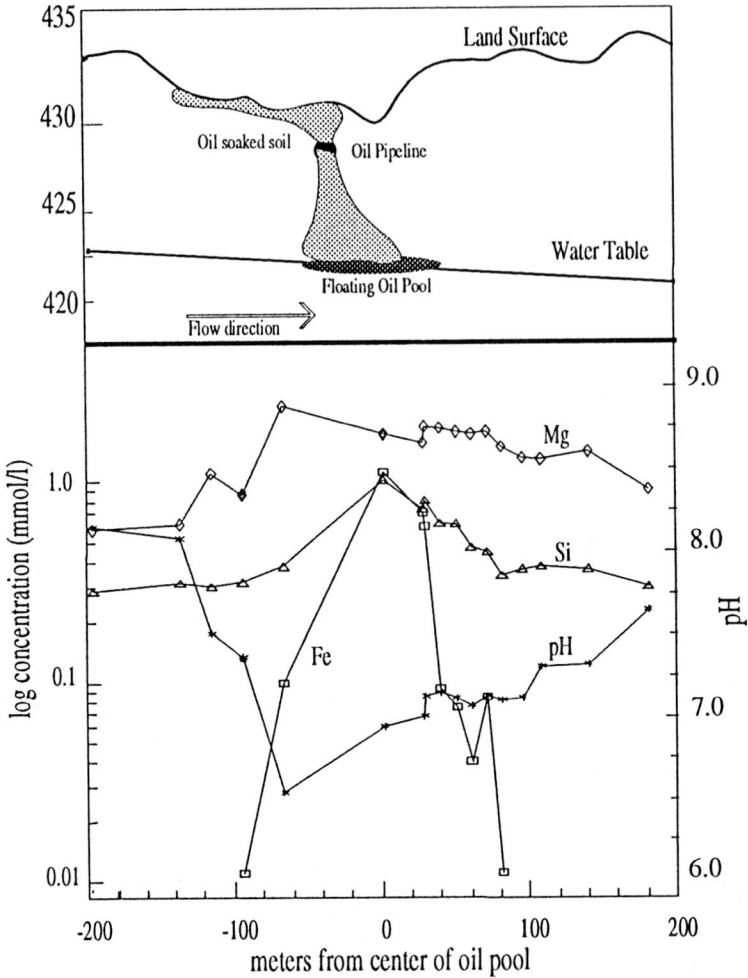

Fig. 15. Cross-section diagram showing physical and chemical features at the Bemidji field site. *Upper section* shows the ground surface and water table elevation in meters above sea level. Trends in geochemical signature are shown on a log scale in mmol/l along the primary flow path through the contaminated area

and will interact with silica both on the surface and in solution, resulting in an increase in quartz solubility and silicate dissolution rate. Down-gradient, where oxygen is again present, the organic complexes are apparently oxidized, and silica precipitates onto otherwise unaltered quartz grains.

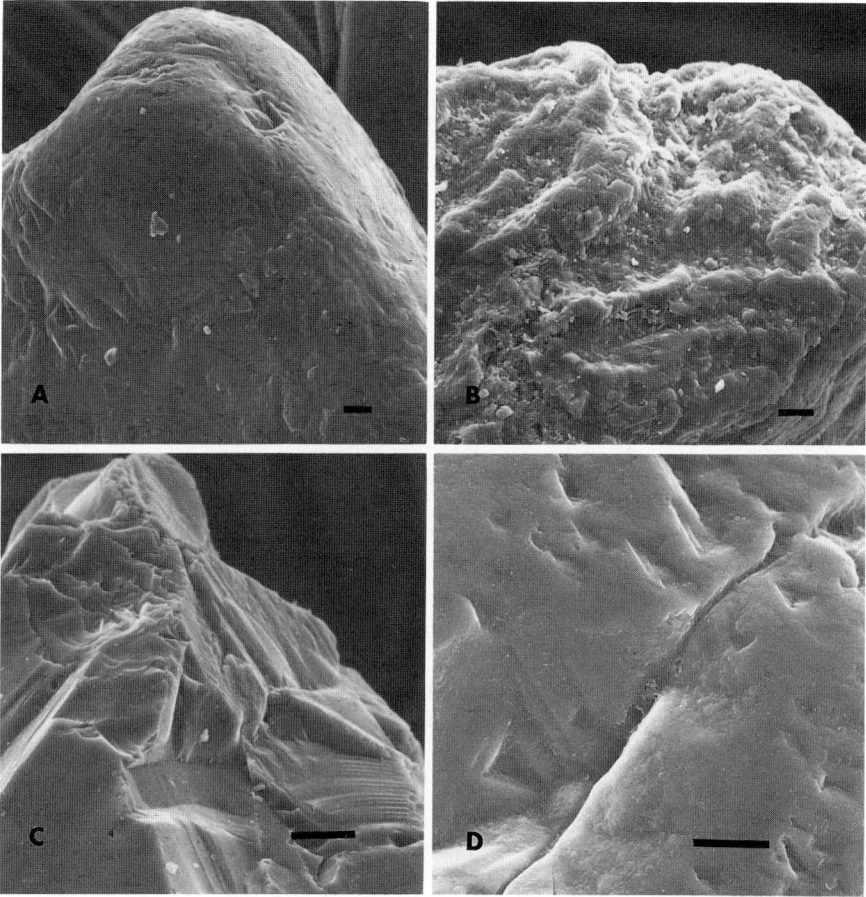

Fig. 16A–D. SEM micrographs of sand grains from the Bemidji field site. Scale *bar* = 10 μm. **A** Clean, unetched quartz collected from background locations; **B** etched quartz collected from the contaminated zone; **C** unetched plagioclase from the background location; **D** heavily weathered plagioclase with adhering secondary minerals from the contaminated zone

4.2 Peat Bogs, Silicates, and Underclays

Many wetlands, and in particular peat bogs and fens, produce waters with extremely high concentrations of organic acids as humic and fulvic acids (Thurman 1979). A peat bog in northern Minnesota was examined to determine the effects of these organic-rich waters on silicate weathering and silica mobility (Bennett et al. 1991). Continuous peat cores and pore waters were collected from a raised bog and a rich spring fen, and hydraulic head was measured at each location. Groundwater generally discharges year-round

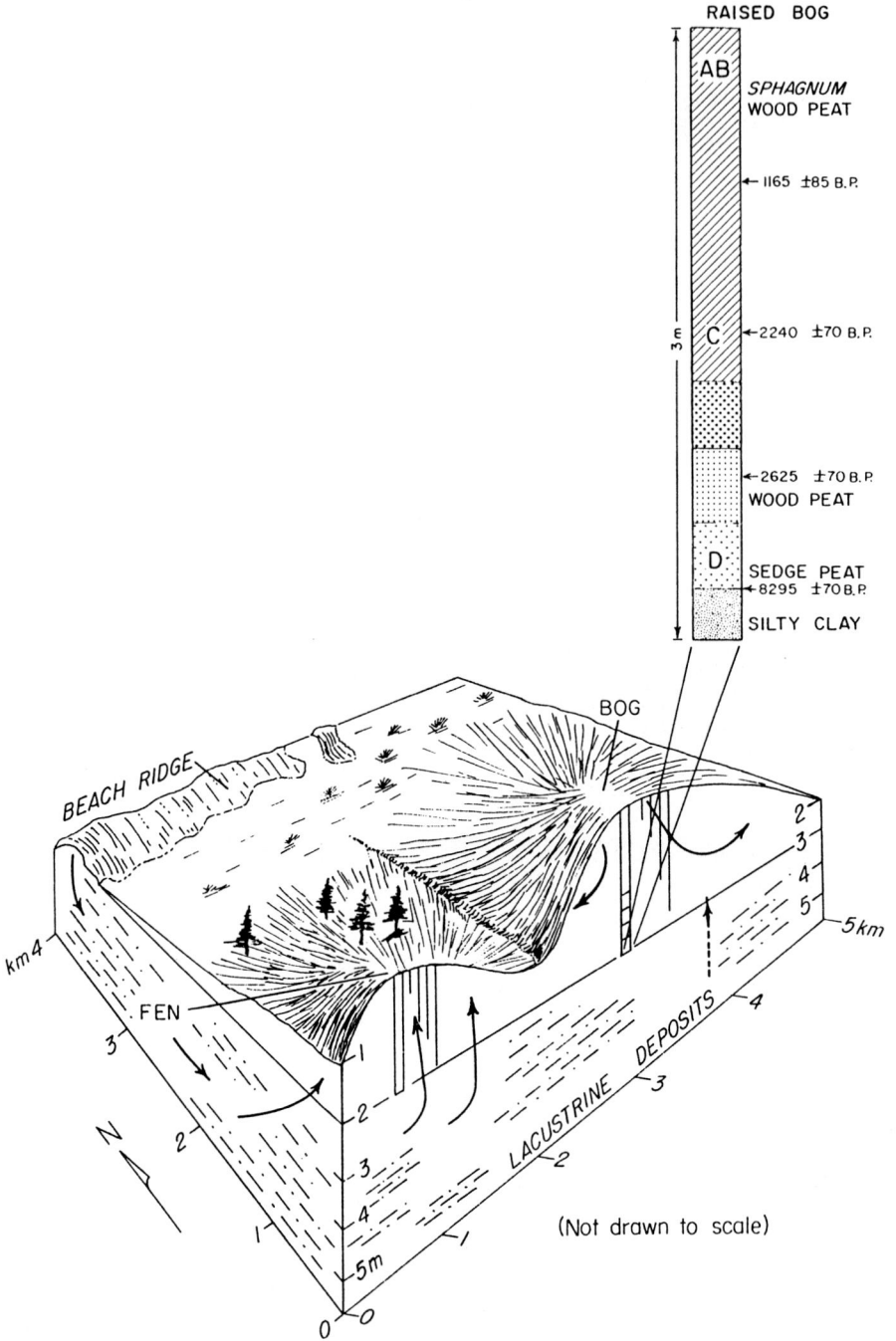

Fig. 17. General diagram of the peat-bog study site. *Arrows* show groundwater flow lines. Blow up of peat core stratigraphy is attached

from the fen, with water visibly flowing at the surface. In the raised bog, however, seasonal head reversals occur, with both recharging and discharging conditions present at different times. In this situation, a hydrologically stagnant zone is created in the interior of the bog (Fig. 17).

The chemistry of the discharging fen water is characteristic of the groundwater extracted from the underlying glacial aquifer: pH and dissolved constituents vary little with depth, and only at the surface does the pH decrease substantially (Fig. 18). These waters are oxidizing, with relatively low concentrations of dissolved organic carbon because of the rapid discharge rate. Silicate silt grains, eolian derived, were extracted from the peat and examined by SEM and found to have little evidence of enhanced chemical weathering (Fig. 19).

At the surface of the raised bog, the pore waters are derived primarily from recharging precipitation, and are high in dissolved organic acids from the oxidation of plant material. pH is low (<5), and dissolved aluminum is high, whereas dissolved silicon and alkaline earth cation concentrations are low. Examination of silt grains from the peat show that aluminosilicate minerals are chemically weathered but quartz is not. In this type of water, aluminum is mobilized as organic-aluminum complexes at acidic pH, and

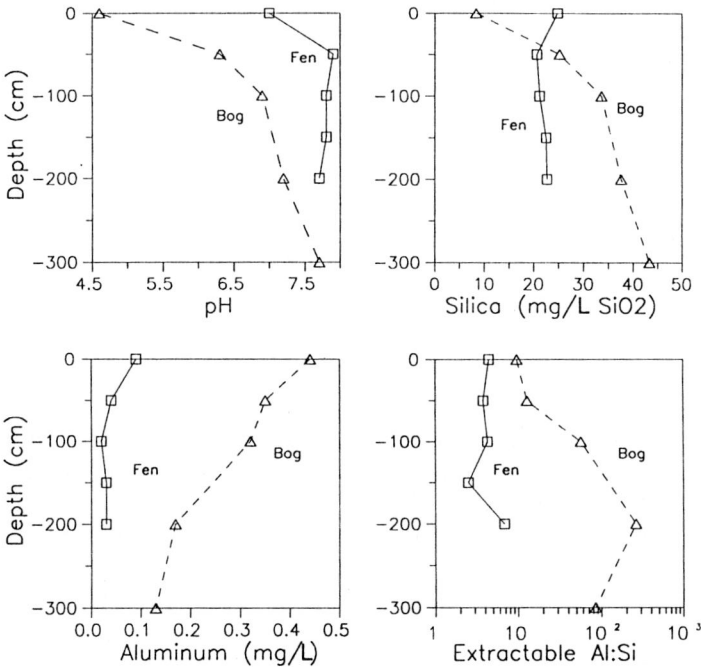

Fig. 18. pH, silica, and aluminum concentration in peat waters, and the residual Al:Si ratio

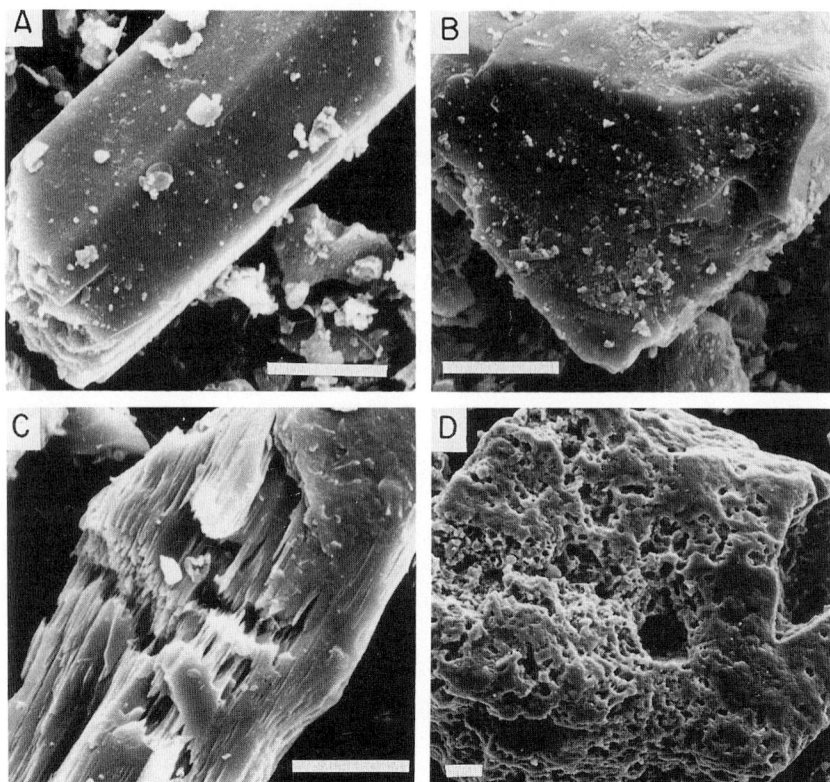

Fig. 19A–D. SEM micrographs of silt grains from the peat bog. Scale *bar* = 10 μm. **A** Unaltered hornblende from the surficial area of the bog; **B** Unaltered quartz grain from the surficial area; **C** weathered hornblende; **D** weathered quartz from the deeper peat sections. (Bennett et al. 1991)

silicates are weathered via both proton-promoted and ligand exchange-promoted pathways.

Approximately 1 m below the surface in the raised bog, the pH increases to near-neutral, the dissolved organic carbon concentration increases, and the waters are extremely reducing. Here, the concentration of alkaline earth cations approaches the concentration found in the underlying groundwater, showing the influence of the mixing of groundwater during seasonal head reversals. In this zone, however, the dissolved silicon concentration increases with depth, exceeding the background groundwater by a factor of 2, while aluminum concentration decreases with depth (Fig. 18). SEM examination of silicate silt grains shows continued weathering of aluminosilicates, as well as clear evidence of quartz weathering (Fig. 19). In this geochemical environment, organic complexes are unstable (high pH), and silica-organic complexes are stable. Quartz and aluminosilicates dissolve due to ligand exchange-promoted pathways at the silicon-containing groups, and the solubility of quartz increases due probably both to solution complexes of silicic acid and microbial activity at the quartz surface.

At the peat-mineral interface, unaltered aluminosilicates are difficult or impossible to find. The extracted aluminum-silicon ratio increases with depth, due to the preferential mobility of silica over aluminum, suggesting that aluminum is conserved in a solid phase (Fig. 18). The aluminum-silicon ratio is highest at the peat-mineral interface, where an aluminiferous clay mineral begins to dominate the mineralogy. It is possible that the preferential mobilization of silica in organic-rich, neutral pH, reducing waters in bogs leaches silica from the sediments, leaving behind an aluminum-rich underclay.

5   Conclusion

Organic acids chelate aluminum, and to a lesser extent, silica, in solution by way of ligand-exchange mechanisms. These chelates are pH-sensitive, with aluminum most effectively chelated at acidic pH, whereas silica is complexed at near-neutral pH. The stability of the organic-metal coordination complex is closely related to the organic acid species, with multifunctional diacids, a-hydroxy aliphatic acids, and hydroxy-aromatic species being the most effective coordinating ligands. These solution complexes can increase the solubility of quartz and aluminosilicates by decreasing the activity of dissolved monomeric silica or aluminum, resulting in the increased mobility of these species as organic chelates.

The chemistry of solution complexation can be applied to silicate surfaces, where the coordination of organic ligands results in accelerated framework bond rupture. The ligand-promoted dissolution is in many cases much faster than the nonorganic system, and results in rapid weathering of primary silicate minerals. Microbes may be an important component of the organic-enhanced weathering scenario, as they may produce steep chemical gradients near microbial colonies and high concentrations of dissolved organic acids. The results of these interactions can be seen in some low-temperature organic-rich natural waters.

Acknowledgments. W.H. Casey acknowledges support from the National Science Foundation via grant #EAR-9105318. P.C. Bennett acknowledges support from the National Science Foundation via grant #EAR-9105778, the ACS Petroleum Research Foundation via grant #22679-G2, and Los Alamos National Laboratory, INC-7&11.

References

Aines RD, Jackson KJ (1990) Direct observation of aqueous organic speciation and measurement of aqueous organic-metal complexation constants. Geol Soc Am Abstr Programs 22: A59

Amrhein C, Suarez DL (1988) The use of a surface complexation model to describe the kinetics of ligand promoted dissolution of anorthite. Geochim Cosmochim Acta 52: 2785–2793

Antweiler RC, Drever JI (1983) The weathering of a late Tertiary ash: importance of organic solutes. Geochim Cosmochim Acta 47: 623–629

Aristovakaya TV, Kutuzova RS (1968) Microbiological factors in the mobilization of silicon from poorly soluble natural compounds. Pochvovedenie 12: 59–66

Avakyan ZA, Karavalko GI, Mel'Nikova EO, Drutsko VS, Ostroushko YuI (1981) Role of microscopic fungi in weathering of rocks and minerals from a pegmatite deposit. Mikrobiologiya 50: 156–162

Baker WE (1972) The role of humic acids from Tasmanian podzolic soils in mineral degradation and metal mobilization. Geochim Cosmochim Acta 37: 269–281

Basolo F, Pearson RG (1958) Mechanisms of inorganic reactions. Wiley, New York, 426 pp

Beckwith RS, Reeve R (1964) Studies on soluble silica in soils. II. The release of monosilicic acid from soils. Aust J Soil Res 1: 157–168

Bennett PC (1991) The dissolution of quartz in organic-rich aqueous systems. Geochim Cosmochim Acta 55: 1781–1797

Bennett PC, Siegel DI (1987) Increased solubility of quartz in water due to complexation by dissolved organic compounds. Nature 326: 684–687

Bennett PC, Melcer DI, Siegel DI, Hassett JP (1988) The dissolution of quartz in dilute aqueous solutions of organic acids at 25 °C. Geochim Cosmochim Acta 52: 1521–1530

Bennett PC, Siegel DI, Hill BM, Glaser PH (1991) Fate of silicate minerals in a peat bog. Geology 19: 328–331

Bennett PC, Siegel DI, Maedecker MJ, Hult MF (1993) Crude oil in a shallow sand and gravel aquifer I. Hydrogeology and inorganic geochemistry. Appl Geochem (in press)

Berner RA (1978) Rate control of mineral dissolution under Earth surface conditions. Am J Sci 278: 1235–1252

Birchall JD, Espie AW (1986) Biological implications of the interaction (via silanol groups) of silicon with metal ions. In: Evered D, O'Connor M (eds) Silicon biochemistry. Ciba Foundation Symposium 121. Wiley, New York, pp 140–159

Bish D, Burnham CW (1984) Structure energy calculations on optimum distance model structures: application to the silicate olivines. Am Mineral 69: 1102–1109

Blum A, Lasaga AC (1988) Role of surface speciation in the low-temperature dissolution of minerals. Nature 331: 431–433

Blum A, Lasaga AC (1990) The effect of dislocation density on the dissolution rate of quartz. Geochim Cosmochim Acta 54: 283–297

Boyle FR, Voigt GK, Sawhney BL (1974) Chemical weathering of biotite by organic acids. Soil Sci 117: 42–45

Brady PV, Walther JV (1989) Controls on silicate dissolution rates in neutral and basic pH solutions at 25 °C. Geochim Cosmochim Acta 53: 2823–2830

Brown GE (1980) Olivines and silicate spinels. In: Ribbe PH (ed) Reviews in mineralogy 5. Ortho-silicates. Mineralogical Society of America, Washington DC, pp 275–333

Buffle J (1990) Complexation reactions in aquatic systems. Ellis Horwood, New York, 692 pp

Burgess J (1988) Ions in solution: basic principles of chemical interactions. Ellis-Norwood, Chichester, 191 pp

Burns RG (1970) Mineralogical applications of crystal field theory. Cambridge University Press, London, 224 pp

Carlisle EM (1986) Silicon as an essential trace element in animal nutrition. In: Evered D, O'Connor M (eds) Silicon biochemistry. Ciba Foundation Symposium 121. Wiley, New York, pp 123–140

Carlstrom D (1977) Structural aspects on organosilicon compounds. In: Bendz G, Lindqvist I (eds) Biochemistry of silicon and related problems. Plenum Press, New York, pp 523–534

Casey WH (1991) On the relative dissolution rates of some oxide and orthosilicate minerals. J Colloid Interface Sci 146: 586–589

Casey WH, Westrich HR (1992) Control of dissolution rates of orthosilicate minerals by local metal-oxygen bonds. Nature 355: 157–159

Casey WH, Carr MJ, Graham RA (1988a) Crystal defects and the dissolution kinetics of rutile. Geochim Cosmochim Acta, 52: 1545–1556

Casey WH, Westrich HR, Arnold GW (1988b) The surface chemistry of labradorite feldspar reacted with aqueous solutions at pH = 2, 3 and 12. Geochim Cosmochim Acta 52: 2795–2807

Cella JA, Cargioli JD, Williams EA (1980) ^{29}Si NMR of five- and six-coordinate organosilicon complexes. J Organometall Chem 186: 13–17

Chvalovsky V, Bellama JM (1984) Carbon-functional organosilicon compounds. Plenum Press New York, 303 pp

Clarke FW (1911) The data of geochemistry. US Geol Surv Bull 491. Govt Printing Office, Washington DC, 782 pp

Cleary WJ, Conolly JR (1972) Embayed quartz grains in soils and their significance. J Sediment Pet 42: 899–904

Corriu RJP, Christian G, Moreau JJE (1984) Stereochemistry at silicon. Top Stereochem 43:43–198

Cozzarelli IM, Eganhouse RP, Baedecker MJ (1990) Transformation of monoaromatic hydrocarbons to organic acids in anoxic groundwater environments. Environ Geol Water Sci 16: 135–141

Crook KAW (1968) Weathering and roundness of quartz sand grains. Sedimentology 11: 171–182

Discoll CT, van Breemen N, Mulder J (1985) Aluminum chemistry in a forested spodosol. Soil Sci Soc Am J 49: 437–444

Duff RB, Webley DM, Scott RD (1963) Solubilization of minerals and related materials by 2-Ketogluconic acid-producing bacteria. Soil Sci 95: 105–114

Eganhouse RP, Baedecker MJ, Cozzarelli IM, Aiken GR, Thorn KA, Dorsey TF (1993) Crude oil in a shallow sand and gravel aquifer II. Organic geochemistry. Appl Geochem (in press)

Eigen M (1961) Advances in chemistry of coordination compounds. In: Kirschner S (ed) Advances in chemistry of coordination compounds. Macmillan, New York, pp 371–379

Evans WD (1964) The organic solubilization of minerals in sediments. In: Colombo U, Hobson GD (eds) Advances in organic geochemistry. Macmillan, New York, pp 263–270

Furrer G, Stumm W (1986) The coordination chemistry of weathering. I. Dissolution kinetics of a-Al_2O_3 and BeO. Geochim Cosmochim Acta 50: 1847–1860

Graham ER (1941) Colloidal organic acids as factors in the weathering of anorthite. Soil Sci 52: 291–295

Grandstaff DE (1986) The dissolution rate of forsteritic olivine from Hawaiian beach sand. In: Colman SM, Detheir DP (eds) Rates of chemical weathering of rocks and minerals. Academic Press, Orlando, pp 41–60

Gruner JW (1922) The origin of sedimentary iron formations: the Biwabik Formation of the Mesabi range. Econ Geol 17:407–460

Harter RD (1977) Reactions of minerals with organic compounds in the soil. In: Dinauer RC (ed) Minerals in soil environments. Soil Science Society of America, Madison, pp 709–740

Heinen W (1967) Ion accumulation in bactrial systems. I. Isolation of two particulate fractions participating in silicon metabolism, from *Proteus mirabilis* cell-free extracts. Arch Biochem Biophys 120: 86–92

Heinen W (1968) The distribution and some properties of accumulated silicate in cell-free bacterial extracts. Acta Bot Neerl 17: 105–113

Hewkin DJ, Prince RH (1970) The mechanism of octahedral complex formation by labile metal ions. Coord Chem Rev 5: 45–73

Hiebert FK, Bennett PC (1992) Microbial control of silicate weathering in organic rich ground water. Science 258: 278–281

Huang WH, Keller WD (1970) Dissolution of rock-forming silicate minerals in organic acids: simulated first-stage weathering of fresh mineral surfaces. Am Mineral 55: 2076–2094

Iler RK (1977) Hydrogen-bonded complexes of silica with organic compounds. In: Bendz G, Lindqvist I (eds) Biochemistry of silicon and related problems. Plenum Press, New York, pp 53–76

Iler RK (1979) Chemistry of silica – solubility, polymerization, colloid and surface properties and biochemistry. Wiley Interscience, New York, 866 pp

Jones D, Wilson MJ (1985) Chemical activity of lichens on mineral surfaces. Int Biodeterior 21: 99–104

Jorgensen SS (1976) Dissolution kinetics of silicate minerals in aqueous catechol solutions. J Soil Sci 27: 183–195

Julien AA (1879) On the geological action of the humus acids. Am Assoc Adv Sci Proc 28: 311–410

Karickhoff SW, Brown DS, Scott TA (1979) Sorption of hydrophobic pollutants on natural sediments. Water Res 13: 241–248

Knauss KG, Wolery TJ (1988) The dissolution kinetics of quartz as a function of pH and time at 70 °C. Geochim Cosmochim Acta 52: 43–53

Kodama H, Schnitzer M, Jaakkimainen M (1983) Chlorite and biotite weathering by fulvic acid solutions in closed and open systems. Can J Soil Sci 63: 619–629

Krauskopf KB (1967) Introduction to geochemistry. McGraw-Hill, New York, 617 pp

Krumbein WE, Werner D (1983) The microbial silica cycle. In: Krumbein W E (ed) Microbial geochemistry. Blackwell Scientific, Oxford, pp 132–143

Kummert R, Stumm W (1980) The surface complexation of organic acids on hydrous a-Al_2O_3. J Colloid Interface Sci 75: 373–385

Kutuzova RS (1969) Release of silica from minerals as a result of microbial activity. Mikrobiologiya 38: 714–721

Kwart H, King K (1977) d-Orbitals in the chemistry of silicon, phosphorus, and sulfur. Springer, Berlin Heidelberg New York, 220 pp

Lasaga AC (1984) Chemical kinetics of water-rock interactions. J Geophys Res 89: 4009–4025

Lasaga AC, Gibbs GV (1990) Ab initio quantum-methanical calculations of surface reactions – a new era? In: Stumm W (ed) Aquatic chemical kinetics. Wiley-Interscience, New York, pp 259–291

Marley NA, Bennett P, Janecky DR, Gaffney JS (1989) Spectroscopic evidence for organic diacid complexation with dissolved silica in aqueous systems. I. Oxalic acid. Org Geochem 14: 525–528

Martell AE, Motekaitis RJ (1989) Coordination chemistry and speciation of Al(III) in aqueous solution. In: Lewis TE (ed) Environmental chemistry and toxicology of aluminum. Lewis, Chelsea, pp 3–18

Moore ES, Maynard JE (1929) Solution transportation and precipitation of iron and silica. Econ Geol 24: 272–303

Murata KJ (1943) Internal structure of silicate minerals that gelatinize with acid. Am Mineral 28: 545–562

Nordstrom DK, May HM (1989) Aqueous equilibrium data for mononuclear aluminum species. In: Sposito G (ed) The environmental chemistry of aluminum. CRC Press, Boca Raton, pp 29–54

Ohman LO, Sjoberg S (1983) Equilibrium and structural studies of silicon(IV) and aluminum(III) in aqueous solution, Part 9. A potentiometric study of mono- and polynuclear aluminum(III) citrates. J Chem Soc Dalton Trans 12: 2513–2517

Ong LH, Swanson VE, Bisque RE (1970) Natural organic acids as agents of chemical weathering. U S Geol Surv Prof Pap 700-C: C130–C137

Price NM, Morel MM (1990) Role of extracellular enzymatic reactions in natural waters. In: Stumm W (ed) Aquatic chemical kinetics. Wiley-Interscience, New York, pp 225–258

Schindler PW, Stumm W (1988) The surface chemistry of oxides, hydroxides, and oxide minerals. In: Stumm W (ed) Aquatic surface chemistry. Wiley-Interscience, New York, pp 83–110

Schnitzer M (1971) Metal-organic matter interactions in soils and waters. In: Faust SD, Hunter JV (eds) Organic compounds in aquatic environments. Dekker, New York, pp 297–315

Schott J, Petit JC (1988) New evidence for the mechanisms of dissolution of silicate minerals. In: Stumm W (ed) Aquatic surface chemistry. Wiley-Interscience, New York, pp 293–318

Schwarz K (1973) A bound form of silicon in glycosaminoglycans and polyuronides. Proc Nat Acad Sci USA 70: 1608–1612

Secco F, Venturini M (1974) Mechanism of complex formation. Reaction between aluminum and salicylate ions. J Inorg Chem 14: 1978–1981

Siever R (1960) Silica solubility, 0 °C–200 °C, and the diagenesis of siliceous sediments. J Geol 70: 127–150

Sikora FJ, McBride MB (1989) Aluminum complexation by catechol as determined by ultraviolet spectrophotometry. Environ Sci Tech 23: 349–356

Silverman MP (1979) Biological and organic chemical decomposition of silicates. In: Trudinger PA, Swain DJ (eds) Biogeochemical cycling of mineral-forming elements. Elsevier, Amsterdam, pp 445–466

Silverman MP, Munoz EF (1970) Fungal attack on rock: solubilization and altered infrared spectra. Science 169: 985–987

Smith RM, Martell AE (1989) Critical stability constants, vol 6. Plenum Press, New York, 822 pp

Sommer LH (1965) Stereochemistry, mechanism, and silicon. McGraw-Hill, New York, 189 pp

Stevenson FJ, Vance GF (1989) Naturally occurring aluminum-organic complexes. In: Sposito G (ed) The environmental chemistry of aluminum. CRC Press, Boca Raton, pp 117–145

Stumm WS (1990) Aquatic chemical kinetics. Wiley-Interscience, New York, 545 pp

Stumm W, Furrer G (1987) The dissolution of oxides and aluminum silicates; examples of surface coordination-controlled kinetics. In: Stumm W (ed) Aquatic surface chemistry. Wiley-Interscience, New York, pp 197–220

Stumm W, Morgan JJ (1981) Aquatic chemistry, 2nd edn. Wiley-Interscience, New York, 780 pp

Stumm W, Wieland E (1990) Dissolution of oxide and silicate minerals: rates depend on surface speciation. In: Stumm W (ed) Aquatic chemical kinetics. Wiley-Interscience, New York, pp 367–400

Sullivan CW (1986) Silicification by diatoms. In: Evered D, O'Connor M (eds) Silicon Biochemistry. Ciba Fundation Symposium 121. Wiley, Chichester, pp 59–83

Surdam RC, Boese SW, Crossey LJ (1984) The chemistry of secondary porosity. In: McDonald DA, Surdam RC (eds) Clastic diagenesis. Am Assoc Pet Geol Mem 37, pp 127–150

Tait CD, Janecky DR, Clark DL, Bennett PC (1991) Oxalate complexation with aluminum(III) and iron(III) at moderately elevated temperatures. In: Kharaka Y K, Maest A S (eds) Water-rock interaction. Balkema, Rotterdam, pp 349–353

Tan KH (1980) The release of silicon, aluminum, and potassium during decomposition of soil minerals by humic acid. Soil Sci 129: 5–11

Tandura SN, Voronkov MG, Alekseev NV (1986) Molecular and electronic structure of penta- and hexacoordinate silicon compounds. Top Curr Chem 131: 99–189

Thurman EM (1979) Organic geochemistry of natural waters. Nijhoff, Dordrecht, 497 pp

Webley DM, Henderson MEF, Taylor EF (1963) The microbiology of rocks and weathered stones. J Soil Sci 14: 102–112

Weiss A, Herzog A (1977) Isolation and characterization of a silicon-organic complex from plants. In: Bendz G, Lindqvist I (eds) Biochemistry of silicon and related problems. Plenum Press, New York, pp 109–128

Westall JC (1988) Adsorption mechanisms in aquatic surface chemistry. In: Stumm W (ed) Aquatic surface chemistry. Wiley-Interscience, New York, pp 3–32

Wieland E, Wehrili B, Stumm W (1988) The coordination chemistry of weathering. III. A generalization on the dissolution rates of minerals. Geochim Cosmochim Acta 52, pp 1969–1981

Wilkens RG (1974) The study of kinetics and mechanisms of reactions of transition metal complexes. Allyn and Bacon, Boston, 652 pp

Chapter 8   Comparison and Evaluation of Experimental Studies on Dissolution of Minerals by Organic Acids

Andrew Hajash, Jr.[1]

Summary

Numerous experiments have been conducted to study the interactions of organic acids and minerals in weathering and diagenetic processes. Most work has focused on feldspar dissolution by water-soluble organic acids. It has been proposed that simple carboxylic acids (e.g., acetic and oxalic) may complex with aluminum and increase the solubility and dissolution kinetics of feldspar. This would provide a mechanism for mobilizing aluminum and creating secondary porosity. In addition, dissolved organic species may act as proton donors and pore-fluid pH buffers.

Feldspar and clay dissolution experiments conducted at elevated temperatures (70–100 °C) are reviewed in this chapter. The data indicate that the aluminum content of carboxylic acid solutions is a complex function of temperature, pH, and concentration and type of organic acid anion. There is considerable scatter in the data reflecting different initial solids, fluids, experimental systems, and procedures. However, most of the data indicate that carboxylic acids do enhance the solubility of feldspar, especially at moderately acidic conditions (pH $\approx$ 4–5). Solutions containing difunctional acids (e.g., oxalic) are more effective at dissolving aluminum than solutions with monofunctional acids (e.g., acetic) and, in general, the aluminum content of the reacted solutions varies inversely with pH and directly with the concentration of the organic acid anion. For example, at 100 °C with buffered pH $\approx$ 4.5–4.7, aluminum reached steady-state concentrations as high as 80 ppm in oxalic acid ($\approx$1000 ppm oxalate) and 14 ppm in acetic acid ($\approx$4000 ppm acetate).

Experimental data also indicate that feldspar dissolution rates increase significantly in oxalate and acetate solutions over that observed for water or CO_2-water solutions. In general, dissolution rates vary inversely with pH and directly with oxalate concentration.

The combination of enhanced solubility and increased dissolution rates indicates that acetate and oxalate may play major roles in mineral dissolu-

---

[1] Department of Geology, Texas A&M University, College Station, Texas 77843-3115, USA

tion and secondary porosity development in sandstones. In addition, the
acetate-acetic acid system appears capable of effectively buffering fluid pH
over a time scale of months, even at high mineral surface area/fluid volume.
The effectiveness of the buffering and dissolution processes will depend on
many factors including type and availability of organic acids, pH, temperature, flow rate, mineralogic composition, and associated reaction kinetics.

1 Introduction

The main objective of this chapter is to review experimental work on
organic acids and mineral dissolution under diagenetic conditions. Consequently, emphasis is placed on experiments conducted at elevated temperatures (70–100 °C). The primary focus is on the effects of carboxylic acids
on feldspar solubility and dissolution kinetics as they relate to potential
aluminum mobility and creation of secondary porosity.

Feldspar dissolution is widely recognized as an important process for the
production of secondary porosity in some sandstone reservoirs (Heald and
Larese 1973; Schmidt and McDonald 1979), expecially those with low intergranular porosity and high feldspar content. Estimates of secondary porosity
in Gulf Coast sandstones range up to 15% of total rock volume (Loucks et
al. 1984; McMahon 1989) and may constitute up to 100% of the total
porosity in some sandstone reservoirs (Franks and Forester 1984; Siebert et
al. 1984).

Secondary porosity development through feldspar dissolution, without
subsequent precipitation of authigenic clays, requires that aluminum be
mobilized and maintained in solution. It has been postulated that simple
organic acids derived during maturation of kerogen may promote secondary
porosity development by increasing feldspar solubility and dissolution rate
(Siebert et al. 1984; Surdam et al. 1984). In particular, anions of carboxylic
acids, such as acetate and oxalate, have been proposed as important complexing agents which may significantly enhance aluminum mobility (Surdam
et al. 1984). In addition, dissolved organic species may act as proton donors
and pore-fluid pH buffers (Willey et al. 1975; Carothers and Kharaka 1978;
Surdam et al. 1984; Kharaka et al. 1986; Surdam and MacGowan 1987).

Water-soluble carboxylic acids have been found in oil-field formation
waters from around the world (Carothers and Kharaka 1978; Surdam et al.
1984; Hanor and Workman 1986; Fisher 1987; Means and Hubbard 1987;
MacGowan and Surdam 1988, 1990; Barth et al. 1990; Fisher and Boles
1990; Lundegard and Kharaka, Chap. 3, this Vol.) with total concentrations
reaching up to 10000 ppm (Surdam et al. 1984). Total concentrations and
abundances of individual species vary widely and only a few of the simple monofunctional and difunctional groups have been identified in each
study. Typically, monofunctional carboxylic acids (formic, acetic, propionic,

butyric, and valeric) reach 5000 ppm, while difunctional acids (oxalic, malonic, and succinic) range from 0 to 300 ppm, although difunctional types have been observed as high as 2640 ppm (MacGowan and Surdam 1988).

Carboxylic acids also account for 50–100% of the alkalinity in some formation waters (Carothers and Kharaka 1978; Surdam et al. 1989). The pH of subsurface waters measured at the well head generally range from 5.5–7.5, but reservoir pH values are generally lower (Kharaka et al. 1986). Calculated in situ pH values for well-characterized waters from the northern Gulf of Mexico basin range from 4.0–5.5 (Kharaka et al. 1986).

Carboxylic acid anions and phenolic compounds are generated by thermogenic degradation of kerogen, especially in the presence of mineral oxidants that serve as catalysts (Surdam and Crossey 1985; Crossey et al. 1986). Hydrous pyrolysis experiments confirm that thermal maturation of kerogen liberates oxygen in the form of water, carbon dioxide, organic acids, and other compounds, at thermal ranks lower than or equal to those of liquid hydrocarbon generation (Rouxhet et al. 1980; Lundegard and Senftle 1987). Carboxylic acid functional groups may be liberated directly from the kerogen structure or may be formed by hydrolysis or oxidation reactions involving other functional groups. Crossey (1985) suggests that hydrolysis of esters and oxidation of aldehydes, alcohols, and alkenes may result in the formation of carboxylic acid functional groups, which may then be liberated from the kerogen structure. After formation, carboxylic acid anions may be sufficiently stable (Carothers and Kharaka 1978; Drummond and Palmer 1986; Palmer and Drummond 1986; Crossey 1991) to be important and active components in water/rock systems undergoing diagenesis.

In summary, a significant amount of secondary porosity appears to be due to dissolution of feldspar and silicate rock fragments; this apparently requires transport of aluminum in solution at relatively high concentrations. Carboxylic acids, which are known to form strong complexes with metal cations (Lind and Hem 1975) and are abundant in oil-field waters, may enhance the solubility and dissolution kinetics of silicates, mobilize aluminum and other cations, and modify the porosity of sandstones undergoing diagenesis.

To evaluate this hypothesis, experimentalists have been trying to provide data on mineral solubility and dissolution kinetics as a function of temperature, pH, and fluid composition. Experimental techniques and results of published studies are summarized below along with new data from our laboratory, including the results of pH-buffered, flow-through experiments that track variations in pore-fluid chemistry through time (Reed 1990; Reed and Hajash 1990, 1992; Franklin 1991; Franklin et al. 1990, 1991).

2 Results of Specific Studies

2.1 Selected Low-Temperature Experiments

The effects of organic acids in weathering environments are well documented and described in detail elsewhere (e.g., Stumm et al. 1985; Grandstaff 1986; Mast and Drever 1987; Amrhein and Suarez 1988; Bennett et al.
1988; Bennett 1991; Chin and Mills 1991; also Drever and Vance, Chap. 6,
this Vol.; Bennett and Casey, Chap. 7, this Vol.). In general, experimental
work at the low temperatures indicates that organic acids may increase both
the solubility and dissolution kinetics of silicates.

Recently, numerous studies have focused on the effects of organic acids
on feldspar dissolution rates. Manley and Evans (1986) conducted feldspar
dissolution experiments at 13 °C using organic acids. Although observing an
increase in dissolution rate with the addition of organic acids, they concluded
that the increase in dissolution kinetics was due to the strength of the acids
rather than complexing.

Mast and Drever (1987) investigated the effects of 0.5 and 1 mM oxalate
solutions ($\approx$50 and 100 ppm, respectively) on the dissolution rate of oligoclase at 22 °C and 1 bar. They used pH-buffered solutions (pH 4, 5, 7, 9) in a
fluidized bed flow-through reactor (modified after Chou and Wollast 1984)
at a water/rock ratio (W/R) of 6.25. The composition of the reactor was not
given. They found that steady-state release rates for aluminum from oligoclase were elevated slightly by oxalate at pH 5 and 7 but were not affected
by oxalate at pH 4 and 9. When oxalate was added to the flow-through
system, large transient "spikes" of aluminum were observed without concomitant increases in silicon; the size of the spike decreased with increasing
pH and was not apparent at pH 9. In contrast, when oxalate was removed
from the input solution, a large silicon spike was observed without a sig-
cificant increase in aluminum. They attributed the aluminum spikes "to
formation and subsequent destruction of a leached zone around dissolution
sites, or to adsorption-desorption reactions on feldspar surfaces away from
sites of dissolution." Oxalate concentrations in the reacted solutions were
not reported.

Amrhein and Suarez (1988) studied dissolution kinetics of anorthite as a
function of pH in solutions containing up to 1 mM oxalate. They found that,
at room conditions, the presence of complex-forming ligands (oxalate and
fluoride) increased the dissolution rate. For example, the dissolution rate in
an 0.1 mM oxalate solution at pH 4 was 221% greater than that in the
absence of oxalate. In addition, the dissolution rate increased linearly with
decreasing pH. They suggested that the reactivity of plagioclase feldspars in
organic acids is directly proportional to aluminum content.

Stillings et al. (1990) also demonstrated experimentally that dissolution
rates for plagioclase feldspar in oxalate solutions increase with increasing

aluminum content. In addition, they showed that the presence of NaCl in solution decreased dissolution rates in oxalate by about a factor of two. In a similar study, Oxburgh and Drever (1990) showed a linear decrease in dissolution rate with increasing pH over the range 3–5 in oxalate solutions.

Chin and Mills (1991) found that the dissolution kinetics of kaolinite also were increased by organic ligands such as oxalate; dissolution rates increased with increasing concentration of organic ligand. They also found that, in the presence of oxalate, Al dissolved more rapidly than Si, consistent with the earlier observations of Huang and Kiang (1972).

2.2 Elevated Temperature Experiments

Although considerable work has been done at the low temperatures of the weathering environment, it was the work of Surdam et al. (1984) that first focused on the possible importance of organic acids at the higher temperatures of diagenesis, especially with regard to enhancing porosity and permeability. Their study included dissolution experiments in which various minerals (andesine, albite, labradorite, microcline) were reacted with pH-buffered and unbuffered organic acids. Experiments were conducted in Teflon-lined steel pressure vessels at 100 °C and fluid vapor pressure for 14 days at a W/R of 100.

Unbuffered solutions used by Surdam et al. (1984) included acetic acid (10000 ppm acetate) and oxalic acid (10, 100, 1000, 10000 ppm oxalate). In the acetic acid experiments (final pH 3.15–3.40), they found high concentrations of aluminum in solution (14–72 ppm). Aluminum was lowest with albite and highest with labradorite; experiments with andesine and microcline resulted in solutions with similar aluminum concentrations (12–24 ppm). In the unbuffered oxalic acid experiments, aluminum correlated directly with oxalate and inversely with pH, reaching concentrations as high as 1400 ppm in a solution with 10000 ppm oxalate (pH 2.7). Increases in aluminum were apparently not just due to increased acidity because solutions with oxalate had higher aluminum than comparable experiments without oxalate at the same pH.

They also reacted andesine with 1000 ppm oxalate solutions (buffered with sodium acetate + acetic acid to pH 3.2, 4, 5, and 6). Andesine experiments with 1000 ppm oxalate (10000 ppm acetate, various pH) showed that the amount of Al and SiO_2 in the reacted solutions varied inversely with pH at essentially constant oxalate. Similar pH-buffered experiments were conducted with laumontite ($\pm$ calcite) to determine the effects of calcite on the aluminum content of the fluid. In the absence of calcite, aluminum concentrations in solution were high at low pH (up to 26 ppm at pH 5) but $\leqslant$ 1 ppm at pH $\geqslant$ 8; however, when calcite was added to the system, oxalate was removed during the reaction, especially at low pH, and aluminum was below the detection limit.

Surdam et al. (1984) concluded from the results summarized above that, depending on the pH, carboxylic acids could significantly increase the solubility, and probably the mobility of aluminum. Acetic and oxalic acids increased the total amount of aluminum in solution by one and three orders of magnitude, respectively, compared to gibbsite solubility in the absence of organic acids. However, the presence of calcite, especially at low to neutral pH, could neutralize the ability of carboxylic acids to enhance the solubility of aluminosilicates.

Surdam and MacGowan (1987), using the same experimental procedures and conditions as Surdam et al. (1984), reacted andesine and a core from the arkosic Stevens Sandstone with solutions containing 20000 ppm NaCl, 50 ppm salicylate, 10000 ppm acetate, and 2500 ppm malonate. Initial pH was 5 and final pH was ≈5.2. Aluminum concentrations in reacted solutions ranged from 9 to 43 ppm, depending on the initial fluid composition. Fluids with NaCl, + acetate had the lowest aluminum, whereas fluids with NaCl + acetate + salicylate + malonate had the highest aluminum. They suggested that "much of the feldspar dissolution observed in sandstones is the result of Al^{3+} complexation due to the presence of difunctional carboxylic acid anions." In addition to Al, they also reported increased concentrations for Si, Fe, and Ca in solutions containing organic acids, especially difunctional types. MacGowan and Surdam (1988) described similar 100 °C experiments in which the starting solids included crushed rock from the Stevens Sandstone, calcite, kaolinite, and andesine. Fluids used contained 20000 ppm NaCl 20000 ppm NaCl + 10000 ppm acetate, 20000 ppm NaCl + 10000 ppm acetate + 2500 ppm malonate, and San Joaquin Basin water with ≈2500 ppm malonate + 2140 ppm acetate. The fluids had an initial pH of 5 and final pH values of 5.0–5.2. Reported aluminum concentrations in the reacted solutions were 0 in NaCl, 15 ppm in the NaCl-acetate solution, 39 ppm in the NaCl-acetate-malonate solution, and 32.7 ppm in the reacted formation water. These aluminum concentrations are several orders of magnitude higher than kaolinite solubility in the absence of organics.

Although the experimental results of Surdam and MacGowan (1987) and MacGowan and Surdam (1988) demonstrated the ability of diverse concentrations and types of organic acids to dissolve aluminum, the initial solids and fluids were very complex so that it is difficult to isolate the effects of individual components. Nevertheless, their data suggest that solutions containing carboxylic acids (including some natural oil-field waters) at pH ≈ 5.2 can maintain high concentrations of aluminum (5–40 ppm), at least over the time span of the experiments.

Hajash et al. (1989) studied the reaction of carboxylic acids with feldspar under experimental conditions similar to those used by Surdam et al. (1984), except that gold capsules were used (no Teflon) and the W/R (≈2–3) was considerably lower. They reacted articially produced granitic sand (0.18–0.25 mm) with acetic (1000, 5000, and 10000 ppm total acetate) and oxalic acids (500, 1000, 2500, 5000 ppm total oxalate) at 100 °C and 345 bar for

12–60 days. Fluids were unbuffered with respect to pH and initial pH values were very acidic (1.4 to 2.4 for the oxalic acid and 2 to 4 for the acetic acid). They found that aluminum concentrations in solution were enhanced substantially by the organic acids and that aluminum concentrations varied directly with the initial concentration of organic acid anion. For example, aluminum concentrations exceeded 200 ppm in oxalic acid (5000 ppm oxalate), similar in magnitude to those reported by Surdam et al. (1984). Acetic acid also resulted in high aluminum concentrations (10–20 ppm Al with 10000 ppm acetate) that were essentially identical to those reported by Surdam et al. (1984). Although these experiments demonstrate the potential of organic acids to mobilize aluminum and create porosity, the fact that the solutions were very acidic and not buffered made it difficult to separate the effects of pH from those of complexing by organic acid anions. Interpretation was also complicated by the precipitation of Fe- and Ca-oxalate salts (humboltine and whewellite), which removed oxalate from solution during the reaction.

Bevan and Savage (1989) conducted experiments to study the dissolution kinetics of feldspar at elevated temperatures in pH-buffered solutions containing oxalate. Dissolution rates were measured for K-feldspar at 70 and 90 °C in solutions with and without oxalate at pH 1, 3.6, and 9. Initial oxalate concentration was 0.02 M ($\approx$1800 ppm) oxalate. Experiments were conducted in direct-sampling batch reactors (Seyfried et al. 1979, 1987) from which samples were withdrawn for periods up to 1 month; the initial W/R was 100 and decreased as fluid was removed for analysis. The composition of the reaction cell (Teflon, gold, or titanium) was not given. Three experiments were conducted in a modified system (Dibble and Potter 1982; Potter et al. 1987) that allowed continuous flow through a small (5 mm diameter × 7.5 cm long) gold tube at flow rates of 0.72 ml/h. Starting solid was crushed orthoclase (125–250 μm) with 17% albite. The pH buffers were 0.2 M solutions of HCl/NaCl (pH 1), Na acetate/acetic acid (pH 3.6), and boric acid/NaCl/NaOH (pH 9).

Bevan and Savage (1989) found that when oxalate was added to the buffer solutions, the dissolution rate increased at pH 3.6 and 9, but remained the same or decreased slightly at pH 1. Oxalate had the greatest effect at the intermediate pH (3.6) and at the highest temperature (95 °C). Under these conditions, the dissolution rate increased by a factor of $\approx$4 compared to similar oxalate-free conditions. However, acetate was also present in the pH buffer at concentrations of 0.2 M. Rate constants determined from both the batch and flow-through experiments agreed closely. Reported rate constants calculated from Si, Al, and K data were similar, indicating that the dissolution process was stoichiometric. They concluded that organic acids affect the "overall dissolution mechanism" rather than selectively complexing aluminum. Exceptions were the early portions of the flow-through experiments in which the concentration of aluminum in solution was "high" early in the experiment and decreased with time, similar to the "spikes" described by Mast and Drever (1987). This was interpreted by Bevan and Savage

(1989) as a preferential reaction of disturbed surfaces. However, only Si and Al showed this behavior, K showed steady-state behavior throughout the experiment. Aluminum concentration data are markedly sparse in this study; they give aluminum concentrations vs. time for only 70 h of one flow-through experiment. No steady-state or equilibrium values for aluminum are given for the batch experiments, although their experiments went more than twice as long as those reported in Surdam et al. (1984) and Hajash et al. (1989) and were at the same water-rock mass ratio as Surdam and others (1984). Apparently, equilibrium or near-equilibrium conditions were not reached. No data on oxalate concentrations in reacted solutions were given. Bevan and Savage (1989) concluded that oxalic acid does not preferentially complex aluminum but does enhance the dissolution rate of orthoclase, especially at pH $\approx$ 4. They also suggested that this "optimum pH regime" coincides with that of a naturally buffered organic acid system, which may result in the production of secondary porosity in sandstone reservoirs.

Stoessell and Pittman (1990) recently reported on experiments designed to study the dissolution of alkali feldspar by carboxylic acids and anions. They reacted crushed microcline (1–2 mm diameter) at 100 °C, 300 bar for 14 days with aqueous solutions containing NaCl (0.15 and 0.075 molal), sodium acetate (0.15 molal), sodium oxalate (0.075 molal), sodium propionate (0.15 and 0.075 molal), and sodium malonate (0.075 molal); they also used acetic acid (0.15 molal) and oxalic acid (0.075 molal). The sodium salt solutions containing organic anions had initial pH values of 6.8–8.3, whereas the pH of the organic acids was much lower (1.5–2.8). Their experiments were conducted in titanium-lined pressure vessels with an initial W/R of 100. They stated that they chose reaction time and temperature "to facilitate comparison with experimental results on organic acids reported by Surdam and others (1984)," however, the starting solids, type of reaction vessels, and especially the pH were very different making such comparisons problematic. Differences in experimental systems and interpretations will be discussed in detail later.

Stoessell and Pittman (1990) found that, under high pH conditions (6.8–8.3), solutions containing high concentrations of organic acid anions (e.g., 8750 ppm acetate, 6600 ppm oxalate) had very low aluminum concentrations (generally <1 ppm). However, under strongly acidic conditions, aluminum concentrations reached 3 ppm in 0.15 molal acetic acid (8780 ppm acetate) and 67 ppm in 0.075 molal oxalic acid (6560 oxalate), with initial pH values of 2.8 and 1.5, respectively. In the oxalic acid experiment, dissolved titanium reached almost 500 ppm due to "leaching of the titanium reaction vessel" and the microcline was coated with "amorphous titanium oxide." These data indicate that the titanium-lined vessel was a major reactant in the low pH, oxalic acid experiment; feldspar dissolution and dissolved aluminum may have been greater if the vessel had been less reactive. They also reported that over half the oxalate was "destroyed" over a period of 170 h in the oxalic acid experiment.

Stoessell and Pittman (1990) concluded that, under reservoir conditions, aluminum-acetate complexes would be insignificant but that aluminum-oxalate and aluminum-malonate complexes may be significant. They also suggested that high aluminum concentrations measured in oxalic acid may be due to enhanced dissolution kinetics at low pH, not the "presence of organic anions."

Recent experimental studies of Fein (1991a,b) attempted to determine the stoichiometry and thermodynamic stability of aluminum-acetate and aluminum-oxalate complexes. Fein (1991a) reacted gibbsite, portlandite, and brucite with sodium acetate/acetic acid solutions of various molalities at pH 3.1–3.5 and 80 °C, 1 bar. One to 5 g of solid and ≈ 100 ml of fluid were sealed in polyethylene bottles and continuously shaken; the initial W/R was ≈ 20–100 and decreased through time as fluid was removed for analysis. Experiments ran for 30–36 days. Aluminum concentrations varied directly with the acetate concentration in solution. For example, solutions reacted with gibbsite contained ≈ 2 ppm Al with 3400 ppm acetate and ≈ 8 ppm Al with 10 700 ppm acetate. Fein (1991a) found that aluminum-acetate complexes ($AlAc^{2+}$ and $AlAc_2^+$) were significantly more stable than Mg- and Ca-acetate complexes; however, because of the relatively low concentrations of aluminum in the reacted solution, he concluded that acetate in sedimentary basin fluids cannot "markedly enhance rock porosity through mineral dissolution."

In contrast, Fein (1991b) concluded that aluminum-oxalate complexes may "play an important role in creating secondary porosity in reservoir rocks." Following the same experimental procedures as in Fein (1991a), gibbsite solubility was determined in sodium oxalate-oxalic acid solutions as a function of total oxalate concentration at a calculated pH of 4.9–5.1. As in the acetate experiments, steady-state concentrations were approached from undersaturation and concentrations of dissolved aluminum in reacted solutions varied directly with initial oxalate values. Final oxalate concentrations were within 1% of the initial values. Solutions reacted with gibbsite for 41 days at 80 °C contained ≈ 34 ppm Al with 570 ppm oxalate and ≈ 51 ppm Al with 900 ppm oxalate. Aluminum molalities were 2.3 to 3.1 orders of magnitude higher than those calculated for solutions without oxalate at the same pH. He concluded that aluminum-oxalate complexing is much more important than aluminum-acetate complexing, but existing data are insufficient to determine the exact nature of the complex and its thermodynamic properties.

Except for the kinetics studies of Bevan and Savage (1989), all of the elevated temperature work described above was conducted in closed systems or batch reactors and/or at relatively high W/R (≈ 100). In our laboratory, we have been investigating feldspar dissolution by carboxylic acids in semistatic, flow-through reactors (Hajash 1986; Hajash and Bloom 1991) that allow pore-fluid chemistry to be monitored through time. These systems allow simulation of many diagenetic conditions including elevated temperatures and pressures, low W/R (<0.3), and high ratios of mineral surface area (A) to fluid volume (V). The primary component of the experimental system is an externally heated pressure vessel (Fig. 1) which contains a reaction cell with independently controlled confining pressure and pore-fluid pressure. The reaction cell consists of a cylindrical Teflon jacket (10 cm long $\times$ 4.4 cm outer diameter) sealed at the ends to titanium closures connected to an Inconel pore-fluid entrance and exit lines. A manually operated back-pressure system is used to extract reacted pore fluid at constant pressure from one end of the sample while unreacted fluid is simultaneously added at the other end by a microprocessor-controlled syringe pump. Average pore-fluid flow rates are controlled by the volume of fluid removed and the time interval between samples. Pore fluid is static except during sampling. Because of the high A/V used in these systems, pore fluids rapidly approach equilibrium compositions, typically in less than 100 h. Low flow rates (0.1–1 ml/h) were used to determine equilibrium solubilities; equilibrium is assumed if steady-state fluid compositions do not vary with flow rate. Dissolution kinetics are studied either by initial concentration increases vs. time or by steady-state fluid chemistry as a function of flow rate.

These flow-through systems were used to investigate the dissolution of granitic sand (0.25–0.5 mm) in pH-buffered oxalic and acetic acids at 100 °C and 345 bar (Reed 1990; Reed and Hajash 1990, 1992). The reaction cell contained $\approx$180 g of sand with an initial porosity of 46%. Pore-fluid chemistry was monitored through time to evaluate the potential ability of these fluids to dissolve aluminum and create secondary porosity. Experiments also examined whether organic components in solution could buffer the pH in systems with high surface area/fluid mass or if silicate hydrolysis reactions would dominate.

Pore fluids used in this study included distilled water, an acetate pH buffer, and 0.01 M oxalic acid with the acetate pH buffer. The buffer (4135 ppm total acetate) had an initial pH at 25 °C (pH_i) of 4.7. The buffered oxalic acid (1000 ppm oxalate, 4135 ppm acetate) had $pH_i = 4.2$. Ammonium acetate/acetic acid (CH_3COONH_4/CH_3COOH) was chosen as the pH buffer because acetic acid commonly occurs in formation waters in conjunction with difunctional carboxylic acids. The concentration of total acetate in the buffer was chosen to be in the range of that found in natural waters. In addition, the initial pH values (4.2–4.7) were chosen to be in the range of calculated values (4.0–5.5) for reservoir waters corrected for known amounts of degassing of CO_2 and H_2S (Kharaka et al. 1986).

Reed (1990) found that, at essentially constant pH, the concentrations of Al, SiO_2, and other components were enhanced in acetate- and oxalate-bearing pore fluids. Aluminum concentrations were exceedingly high in the pH-buffered oxalate solution at all flow rates. Aluminum concentrations ranged from 2.19 (59 ppm) to 2.98 mmolal (80 ppm) and varied inversely with flow rate. Representative data at low volumetric flow rates (0.13–0.23 ml/h) are shown in Fig. 2 along with data from similar experiments (Franklin

Pore-fluid exit line (Inconel™)

Confining-pressure line

Pore-fluid entry line (Inconel™)

Head bolt

Washer

Head

Vessel seal

Titanium closure

Jacket-seal area

Compression ring

Titanium filter

Sample jacket (Teflon™)

Solid + pore fluid

Titanium filter

Titanium closure

Pore-fluid line

50mm

1"

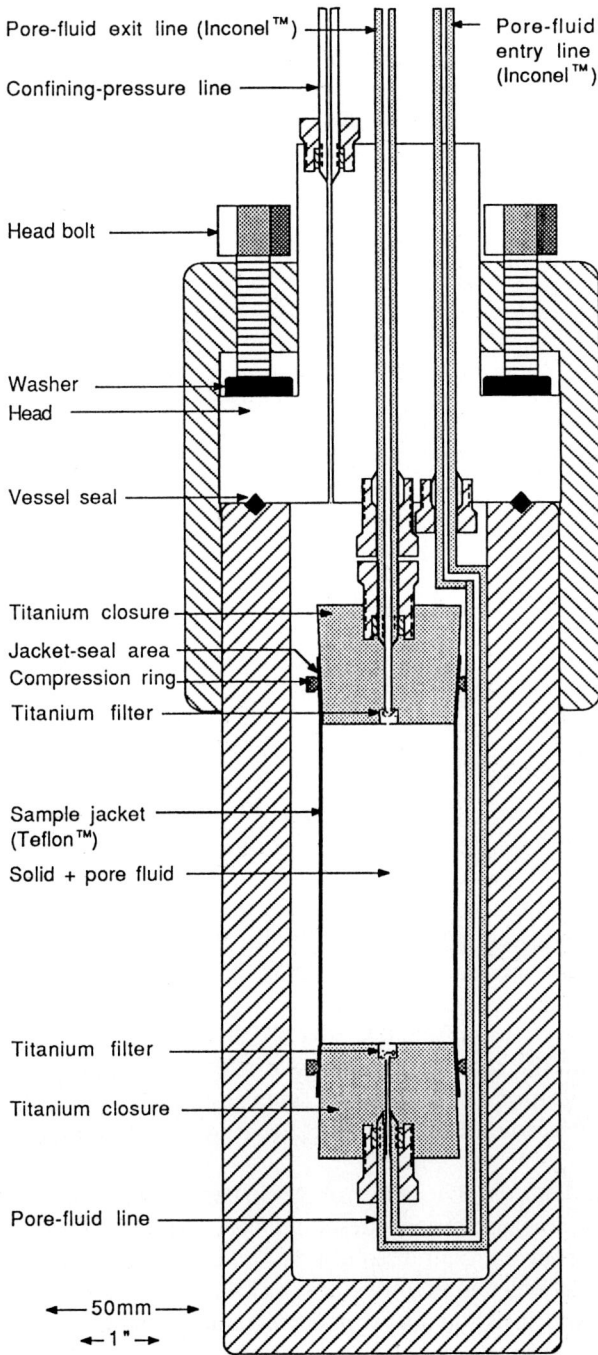

Fig. 1. Schematic diagram of the pressure vessel with internal reaction cell used in the semistatic, flow-through experiments of Reed (1990), Reed and Hajash (1992), and Franklin (1991)

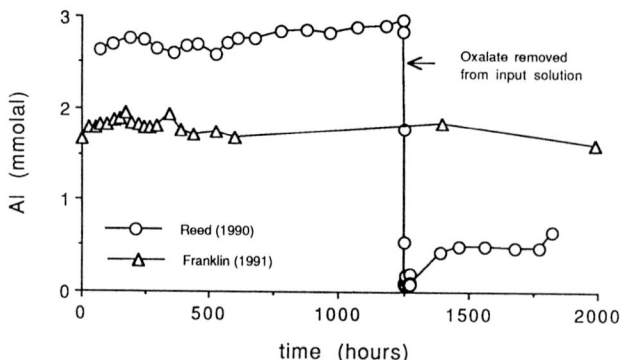

Fig. 2. Aluminum concentrations in solution as a function of time for portions of flow-through experiments of Reed (1990) and Franklin (1991). Solution used by Reed (1990) contained 4100 ppm acetate + 1000 ppm oxalate (pH$_f$ ≈ 4.5) until 1250 h; at that time, pore fluid was rapidly replaced with the acetate buffer (no oxalate). Solution used by Franklin (1991) contained 4200 ppm acetate + 500 ppm oxalate with pH$_f$ ≈ 4.6

1991) described later. Under these conditions, steady-state concentrations of aluminum averaged 80 ppm in pH-buffered oxalic acid (≈1000 ppm oxalate). When the oxalic acid was replaced with acetate buffer solution at ≈1250 h (Fig. 2), aluminum dropped to a steady-state value of ≈0.5 mmolal (14 ppm). These concentrations are higher by factors of 800 and 130, respectively, than calculated values (Stoessell and Pittman 1990) for kaolinite solubility at the same pH in the absence of organic acids. Silica in the pore fluid also varied directly with oxalate and inversely with flow rate. Oxalate concentrations in the reacted solution remained high throughout the oxalic-acid portion of the experiment. Acetate was also constant and near the initial value throughout the experiment. In addition, the quench pH of the reacted solution (pH$_f$) was essentially constant (pH$_f$ ≈ 4.5), only slightly higher (0.3 pH units) than the initial pH at 25 °C. Formate, a possible product of degradation of oxalate, was not detected in the reacted solutions; this indicates that decarboxylation of oxalate was not significant during the experiment, consistent with the recent observations of Crossey (1991). Also, maximum fluoride concentrations were ≈5 ppm, indicating that aluminum-fluoride complexes cannot account for the high concentrations of aluminum in solution.

Reed (1990) also found that molar values of Al/Si, Al/Na, and Al/K in solution were higher than stoichiometric values in feldspar. Al-favored ratios may be produced by either incongruent dissolution of feldspars producing an Al-poor "leached" layer, as was observed by Shotyk (1990), or by congruent dissolution followed by precipitation of Al-poor secondary minerals containing Si, Na, and K. Because no such phases were observed, Reed (1990) concluded that the elevated Al/cation molar ratios were due to preferential formation of Al-oxalate complexes.

Another experiment was conducted (Reed 1990; Reed and Hajash 1992) using only unbuffered 0.01 M oxalic acid (no solid) to monitor potential reaction with the experimental system and/or thermal degradation of the oxalate. In this experiment, which ran for ≈ 30 days at 100 °C and 345 bar, oxalate in the reacted solution remained near initial values and was apparently not removed by reaction with the system, precipitation, or decarboxylation. The pH_f was also essentially constant at 2.1. Reaction with the experimental system was limited to minor leaching of Fe (<7 ppm) and Ni (<10 ppm). Ti was below detection limits (<2 ppm).

Reed (1990) and Reed and Hajash (1990, 1992) concluded that geologically reasonable concentrations of both monofunctional and difunctional carboxylic acids can enhance substantially the solubility of aluminum at $pH \approx 4.5$–4.7, although oxalate has a much greater effect than acetate. Observed elevated metal concentrations were attributed to formation of organic anion complexes and/or increased reaction kinetics, which could allow rapid dissolution of aluminosilicates and creation of secondary porosity through reaction with carboxylic acid anions. They also suggested that the acetate system may buffer the pH in formation waters.

Similar flow-through experiments (Franklin et al. 1990, 1991; Franklin 1991) were conducted at 100 °C and 347 bar to investigate the solubility and dissolution kinetics of albite in pH-buffered acetic and oxalic acids. These experiments used a monomineralic starting material (Amelia albite) to simplify interpretation of the results. The primary pore fluids used for the solubility experiments were a 4200 ppm acetate buffer solution ($pH_i = 4.7$) and a 4200 ppm acetate-500 ppm oxalate solution ($pH_i = 4.4$). Other pore fluids used in kinetics studies include the following: 4200 ppm acetate buffer solutions at pH_i values of 2.5, 3.5, 4.7, and 5.5; 4200 ppm acetate-500 ppm oxalate solutions at pH_i values of 2.2, 3.2, 4.4, and 5.4; 4200 ppm acetate solution with 50, 100, 500, and 1000 ppm oxalate at pH_i values from 4.3 to 4.7. Dissolution kinetics were also studied using distilled-deionized water through which 5% CO_2 gas was bubbled until the pH stabilized. The pH_i of the CO_2 solution was about 4.5. Pore-fluid chemistry was monitored through time to obtain equilibrium and steady-state concentrations as well as dissolution rate constants.

Franklin (1991) found that both acetate and oxalate significantly increased albite solubility at temperatures, pressures, and pH values (4.6–4.7) that may exist under diagenetic conditions. For example, aluminum reached a steady-state concentration of 13 ppm in the acetate buffer solution, essentially identical to that observed by Reed (1990) in acetic acid at the same pH (4.7) and total acetate concentration. Dissolved aluminum was even higher in the acetate-oxalate solution ($pH_f = 4.6$) reaching a steady-state concentration of 48 ppm (Fig. 2), whereas aluminum concentrations in distilled water and the CO_2-water solution were <1 ppm. These high concentrations of aluminum in the acetate-oxalate solution did not vary substantially with flow rate and were maintained for pore-fluid residence times of up to 3 months, indicating that the elevated aluminum values were not short-term experimental artifacts. However, the high aluminum concentrations could be due to slow precipitation kinetics. To test this hypothesis, closed-system experiments were conducted to determine if the presence of kaolinite, which would provide nucleation sites for precipitation, would affect pore-fluid chemistry. The acetate-oxalate solution was reacted with albite and a mixture of albite plus kaolinite (10% kaolinite by weight) at 100 °C, 345 bar for 21 days in welded gold capsules at W/R = 3. Solutions from experiments with and without kaolinite had essentially the same concentration of aluminum (19–24 ppm); this indicates that the high aluminum concentrations from both flow-through and closed-system experiments were not caused by a lack of clay nucleation sites. Pore fluids from the flow-through experiments contained more aluminum (48 ppm) than those from the closed-system experiments; this was probably because the A/V for the flow-through experiments (0.1 m^2/ml) was an order of magnitude higher than that for the closed-system experiments. Consequently, at low flow rates and high A/V, pore fluids in the flow-through system increased in aluminum more than those in the closed systems over the same time period and, in fact, may have reached equilibrium.

Franklin (1991) also observed that both acetate and oxalate increased albite dissolution rates. The kinetics of albite dissolution in acetate solutions as a function of pH were investigated by using acetate buffer solutions having pH_i values of 2.5, 3.5, 4.7, and 5.5. Values for pH_f were 3.4, 3.8, 4.7, and 5.3, respectively. As expected, the difference between initial and final pH increased with increased departure from the acetate pK at 100 °C of ≈4.9. This reflects the decreased buffer capacity of solutions farther away from the pK. Concentration data vs. time (Fig. 3) show that, at essentially constant total acetate, apparent dissolution rates for Na, Al, and Si increased with decreasing pH_i from 5.5 to 2.5.

Dissolution rates for albite in acetate-oxalate solutions (Fig. 4) were significantly higher than those in acetate solutions and distilled water. The effect of pH on dissolution rate in the acetate-oxalate system was investigated using 4200 ppm acetate-500 ppm oxalate solutions with pH_i values of 2.2, 3.2, 4.4, and 5.4. The pH_f values were 2.8, 3.3, 4.6, and 5.1, respectively. Initial pH values were slightly lower than those for the acetate buffer alone because of addition of the oxalic acid. Apparent dissolution rates for Na, Al, and Si increased significantly as pH decreased from 5.4 to 2.2 (Fig. 4). H$^+$ activity appears to be an important factor in albite dissolution kinetics in acetate-oxalate solutions; however, the rates do not show as large a pH dependence as that observed in the acetate experiments (Fig. 3).

Franklin (1991) also investigated the effect of oxalate concentration on dissolution rate using 4200 ppm acetate solutions with 50, 100, 500, and 1000 ppm oxalate at pH values of 4.3–4.7. Apparent dissolution rates for Na, Al, and Si increased significantly as initial total oxalate increased from 50 to 1000 ppm (Fig. 5). However, initial dissolution rates in solutions with

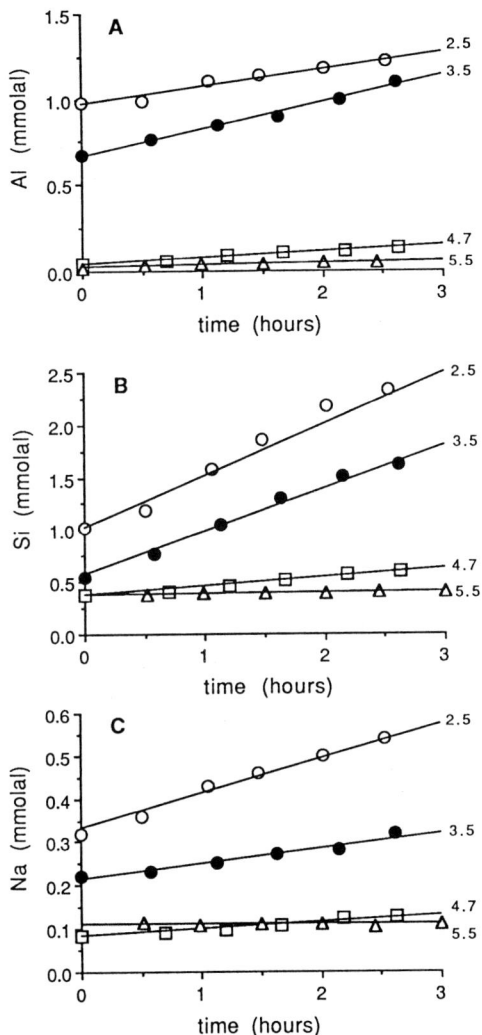

Fig. 3. Solute concentration versus time for albite-acetate experiments as a function of pH. *Numbers on the right* indicate the initial pH of each solution. Apparent dissolution rate is equal to the slope of each line. Data from Franklin (1991)

500 and 1000 ppm oxalate were essentially identical (within analytical uncertainty). These data indicate that, under the experimental conditions, the rate of change in dissolution kinetics with respect to oxalate concentration is greatest at low total oxalate and decreases as total oxalate exceeds values of ≈500 ppm. In contrast, long-term, steady-state aluminum concentrations described earlier (Fig. 2) increase directly with oxalate reaching ≈80 ppm Al in solutions containing 1000 ppm total oxalate.

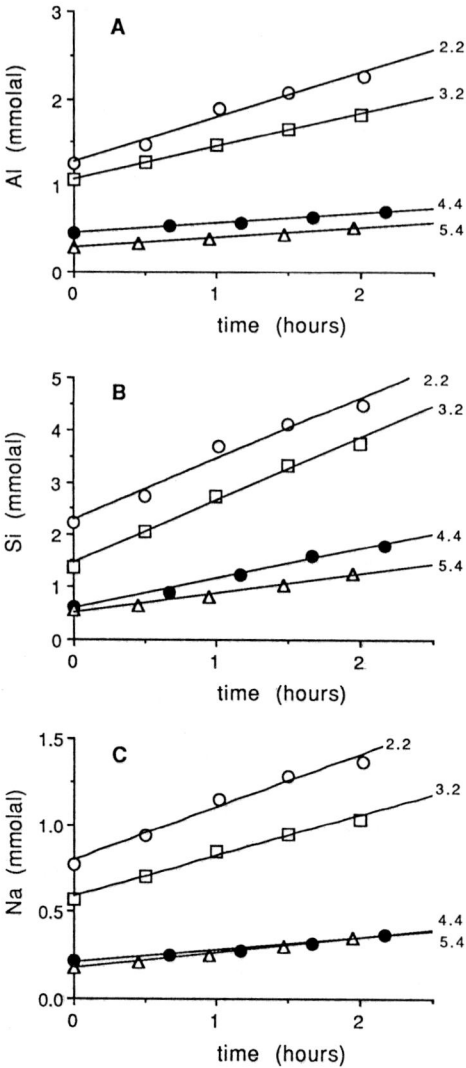

Fig. 4. Solute concentration versus time for albite-acetate-oxalate experiments as a function of pH. *Numbers on the right* indicate the initial pH of each solution. Apparent dissolution rate is equal to the slope of each line. Data from Franklin (1991)

Partial rate laws obtained for albite dissolution in acetate-oxalate solutions (Franklin 1991) indicate that dissolution rate is less dependent on pH relative to oxalate concentration. This suggests that oxalate adsorption and Al-oxalate complex formation are dominant processes controlling dissolution. Dissolution rate constants for acetate and acetate-oxalate solutions are as much as an order of magnitude higher than those for distilled water (pH_f

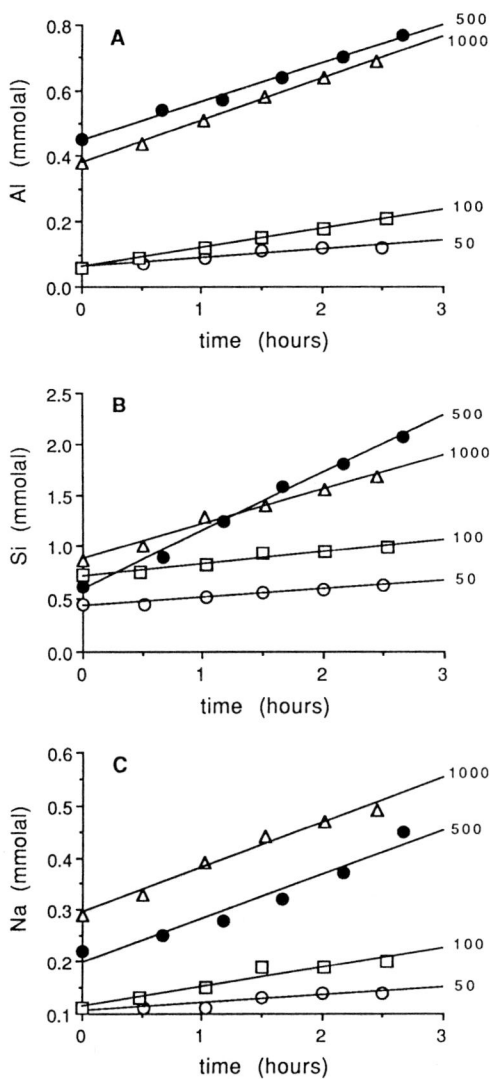

Fig. 5. Solute concentration versus time for albite-acetate-oxalate experiments as a function of initial oxalate concentration. *Numbers on the right* indicate the initial oxalate concentration of each solution. Apparent dissolution rate is equal to the slope of each line. Data from Franklin (1991)

$\approx$ 7) and the 5% CO_2-water solution ($pH_f \approx 5.9$). Apparent dissolution rates for albite in four representative solutions are shown in Fig. 6.

Kinetics data were collected before and after steady-state portions of the experiments to determine the effect of long-term dissolution on reaction rate. Comparison of slopes indicate that apparent reaction rates exhibit no systematic variation between studies conducted before and after the steady-

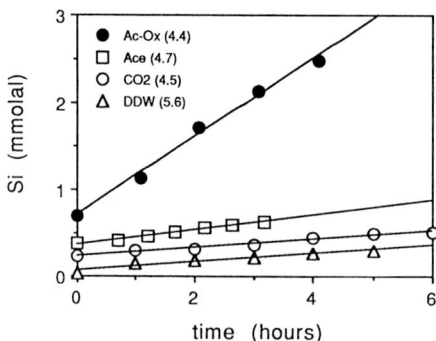

Fig. 6. Silica concentrations versus time showing apparent dissolution rates for albite in acetate-oxalate, acetate, 5% CO_2-water, and distilled-deionized water (*DDW*). Details of the fluid chemistry are given in text. The initial pH of each solution is given in *parentheses*. Data from Franklin (1991)

state experiments. This suggests that there are no major variations in dissolution mechanism during the experiments. In addition, there appeared to be no change in dissolution rate as saturation state increased to ≈ 0.5. Constant dissolution rate with increasing saturation suggests that the reaction is surface-controlled rather than solution-controlled. Extensive dissolution features observed by scanning electron microscopy on reacted albite surfaces were concentrated along structural dislocations such as twin planes, cleavage planes, and fluid inclusions. This also suggests surface-controlled reaction kinetics. No authigenic aluminosilicate minerals were observed.

Franklin (1991) also found that molar values of Al/Si and Al/Na ratios exceeded stoichiometric values in albite early in albite-acetate-oxalate experiments but gradually converged on stoichiometric levels. These data are very similar to the transient elevated Al/Si ratios observed by Mast and Drever (1987) and Bevan and Savage (1989) when oxalate was first introduced into the experimental systems. The data of Mast and Drever (1987) and Bevan and Savage (1989) do not show evidence for steady-state incongruent dissolution. The long-term behavior of the cation ratios from Franklin (1991) also do not support extensive continuous preferential leaching of Al. Early stages of dissolution in organic acids may be incongruent due to preferential complexing of Al (Reed 1990; Franklin 1991) followed by extended congruent dissolution.

3   Summary and Evaluation of Experimental Results

Experiments reacting carboxylic acids with feldspar and clay minerals result in solutions of highly variable aluminum content. The aluminum concentration in carboxylic acid solutions appears to be a complex function of tem-

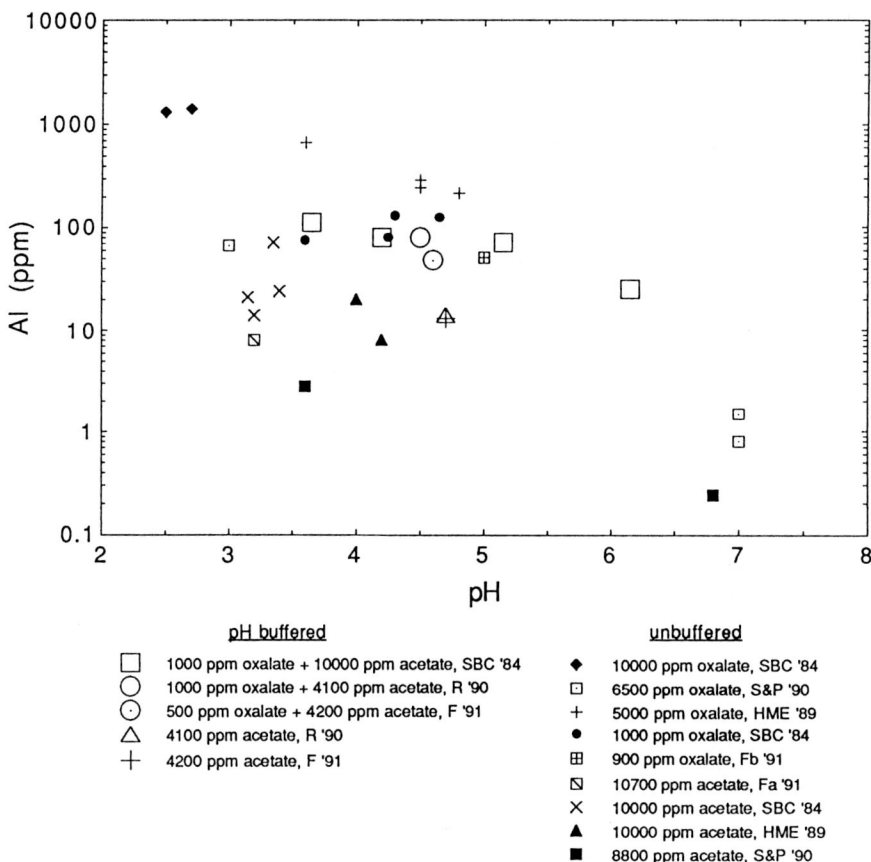

Fig. 7. Aluminum concentrations versus pH from carboxylic acid dissolution experiments conducted at elevated temperatures (80–100 °C). Final quench pH values were used when available; calculated pH values were used for the results from Fein (1991a,b). *SBC'84* = Surdam et al. (1984); *HME'89* = Hajash et al. (1989); *R'90* = Reed (1990); *S&P'90* = Stoessell and Pittman (1990); *F'91* = Franklin (1991); *Fa'91* = Fein (1991a); *Fb'91* = Fein (1991b)

perature, pH, as well as type and concentration of organic acid anion. However, most of the experimental data support the model of enhanced dissolution of feldspar and aluminum mobility proposed by Surdam et al. (1984).

Figure 7 summarizes the aluminum data from many of the experiments conducted at elevated temperatures (80–100 °C) described above. Although there is considerable scatter reflecting different initial solids, fluids, and experimental systems and procedures, some general trends are clear.

First, oxalate solutions are more effective at dissolving aluminum than acetate solutions and, in general, the aluminum content of the reacted solutions varies inversely with pH and directly with the concentration of the organic acid anion. The data from three experiments conducted by Stoessell and Pittman (1990) suggest that oxalate and acetate will not maintain high concentrations of aluminum in solution at pH $\geq$ 7.

Second, much of the recent experimental work (Hajash et al. 1989; Reed 1990; Reed and Hajash 1990, 1992; Fein 1991a,b; Franklin 1991; Franklin et al. 1991) is consistent with earlier work (Surdam et al. 1984; Surdam and MacGowan 1987; MacGowan and Surdam 1988); all of these studies found that aluminum concentrations in organic acid solutions were enhanced compared to calculated or measured values in the absence of organics, especially at pH $\approx$ 4–6. For example, the pH-buffered experiment (1000 ppm oxalate) of Surdam et al. (1984) contained 80 ppm aluminum at pH_f = 4.2; similar pH-buffered, flow-through experiments contained 80 ppm aluminum with 1000 ppm oxalate at pH_f $\approx$ 4.5 (Reed 1990) and 48 ppm aluminum with 500 ppm oxalate at pH_f $\approx$ 4.6 (Franklin 1991; Franklin et al. 1991). In contrast, the oxalic acid experiment of Stoessell and Pittman (1990) had only 67 ppm aluminum, despite the fact that the pH was lower (pH_i = 1.5, pH_f = 3) and the initial oxalate content was higher (6600 ppm) than in the other studies mentioned above. This relatively low aluminum concentration (compared to other work) is probably due to reaction between the oxalic acid and the titanium reaction vessel. Reported titanium concentrations (290–460 ppm) in reacted solutions far exceed aluminum values, indicating that a Ti-oxalate complex may have formed. In fact, they reported that a "titanium oxide" precipitate coated the feldspar and the walls of the pressure vessel; this precipitate may have been Ti-oxalate, which could account for the observed >50% decrease in oxalate in solution. Similar problems were encountered by Reed (1990) in her first flow-through experiment that reacted unbuffered 0.01 M oxalic acid (1000 ppm oxalate) with granitic sand. During slow flow (0.13 ml/h), oxalate was completely removed from solution and aluminum concentrations were <1 ppm; however, during rapid flow (24 ml/h), oxalate remained near its initial value and aluminum reached 62 ppm. Examination of the solid run products revealed abundant humboltine ($FeC_2O_4 \cdot 2H_2O$) covering small iron particles generated during the grinding of the granite. Apparently, the oxalic acid reacted rapidly with the iron and precipitated Fe-oxalate leaving little oxalate to complex with aluminum. In subsequent experiments (Reed 1990), using sand from which iron particles had been removed by magnetic separation, oxalate remained near its initial concentration and aluminum reached $\approx$ 80 ppm. These data (Reed 1990; Stoessell and Pittman 1990) demonstrate clearly the importance of careful sample preparation and the need for nonreactive materials in the construction of experimental systems used with carboxylic acids, especially with difunctional types at low pH and elevated temperatures.

High aluminum concentrations in carboxylic acid/silicate dissolution experiments have been interpreted in terms of formation of organic anion-aluminum complexes (Surdam et al. 1984; Surdam and MacGowan 1987;

MacGowan and Surdam 1988; Reed 1990; Reed and Hajash 1990, 1992; Fein 1991a,b; Franklin 1991; Franklin et al. 1991, especially in the pH range of 3.5 to 5.5; however, Stoessell and Pittman (1990) proposed three alternate interpretations, especially for the data of Surdam et al. (1984). They suggested that the high aluminum values could be due to (1) not filtering the fluids, (2) formation of complexes between fluoride (derived from the Teflon-lined reaction vessels) and aluminum, or (3) "kinetic inhibitions on the rapid precipitation of gibbsite and kaolinite." In light of the recent experimental work, these interpretations seem unlikely. For example, Reed (1990) and Franklin (1991) found high aluminum concentrations in reacted solutions that were passed through $0.2\,\mu m$ syringe filters and contained <5 ppm fluoride. Hajash et al. (1989) also reported high aluminum in filtered ($0.2\,\mu m$) solutions reacted with granitic sand in gold capsules (no fluoride from Teflon). Also, the recent work of Fein (1991b) on gibbsite solubility in oxalic acid indicates that the observed aluminum concentrations are not short-term metastable concentrations or experimental artifacts caused by a lack of nucleation sites.

Although difunctional acids apparently form stronger metal complexes than monofunctional types, the acetate-acetic acid system may mobilize substantially more aluminum under diagenetic conditions than predicted by some investigators (e.g., Stoessell and Pittman 1990; Fein 1991a). Data from Reed (1990), Reed and Hajash (1992), and Franklin (1991) show that acetate can significantly increase the amount of aluminum in solution. Steady-state values of 13–15 ppm aluminum obtained by dissolving feldspar in pH-buffered acetate solutions ($pH_f \approx 4.7$, acetate ≈ 4200 ppm) clearly indicate that aluminum-acetate complexes are more efficient at maintaining aluminum in solution than previously recognized. The acetate-acetic acid system may also be important as a pH buffer (e.g., Willey et al. 1975). Reed (1990) and Franklin (1991) have shown experimentally that acetate salt-acetic acid solutions reacting with feldspar are capable of effectively buffering fluid pH over a time scale of months, even at very high A/V and low W/R. These data suggest that monofunctional carboxylic acids, especially in high concentrations, should not be ignored in models of secondary-porosity development in sandstones, especially if reservoir fluids have in situ pH values in the range 4.0–5.5 as suggested by Kharaka et al. (1986).

Oxalate and acetate significantly increase feldspar dissolution rates at elevated temperatures (Bevan and Savage 1989; Franklin 1991; Franklin et al. 1991). Oxalate has a greater effect on the dissolution rate than acetate and measured rates increase with increasing oxalate at essentially constant pH. At essentially constant total oxalate, dissolution kinetics increase with decreasing pH, similar to the recent observations at low temperatures (Amrhein and Suarez 1988). At pH ≈ 4.4–4.7, oxalate and acetate significantly increase the albite dissolution rate over that observed for distilled water or CO_2-charged water (Fig. 6).

Geologically reasonable concentrations of both mono- and difunctional carboxylic acids can enhance substantially the solubility and dissolution kinetics of feldspar under moderately acidic conditions (pH ≈ 4–5), although oxalate has a much greater effect than acetate. The acetate system may buffer the pH in formation waters and allow rapid dissolution of aluminosilicates and creation of secondary porosity through reaction with carboxylic acid anions. However, the effectiveness of the dissolution and buffering processes will be a function of a number of parameters including type and availability of organic acids, pH, temperature, reaction kinetics, and flow rate. In addition, the presence of other phases such as carbonates or mafics and the activity of CO_2 and other components in the fluid may affect substantially the mobility of aluminum during diagenesis.

Much additional experimental work is required to constrain and quantify the effects of the parameters mentioned above. For example, mineral solubilities (silicates and carbonates) are needed in more complex fluids (including brines) containing organic acids at elevated temperatures and pressures with buffered pH. Under similar conditions, mineral dissolution kinetics should be studied at various temperatures (to determine activation energies) and saturation states. Also, data are needed on the stoichiometry and thermal stability of aluminum complexes with various organic acid anions, including the kinetics of thermal degradation of organic acids in the presence of minerals, especially at high A/V.

References

Amrhein C, Suarez DL (1988) The use of a surface complexation model to describe the kinetics of ligand-promoted dissolution of anorthite. Geochim Cosmochim Acta 52: 2785–2793

Barth T, Borgund AE, Riis M (1990) Organic acids in reservoir waters – relationship with inorganic ion composition and interactions with oil and rock. Org Geochem 16: 489–496

Bennett PC (1991) Quartz dissolution in organic-rich aqueous systems. Geochim Cosmochim Acta 55: 1781–1797

Bennett P, Melcer ME, Siegel DE, Hassett JP (1988) The dissolution of quartz in dilute aqueous solutions of organic acids at 25 °C. Geochim Cosmochim Acta 52: 1521–1530

Bevan J, Savage D (1989) The effect of organic acids on the dissolution of K-feldspar under conditions relevant to burial diagenesis. Mineral Mag 53: 415–425

Carothers WW, Kharaka YK (1978) Aliphatic acid anions in oilfield waters – implications for the origin of natural gas. Am Assoc Pet Geol Bull 62: 2241–2453

Chin P-K F, Mills GL (1991) Kinetics and mechanisms of kaolinite dissolution: effects of organic ligands. Chem Geol 90: 307–317

Chou L, Wollast R (1984) Study of the weathering of albite at room temperature and pressure with a fluidized bed reactor. Geochim Cosmochim Acta 48: 2205–2217

Crossey LJ (1985) The origin and role of water soluble organic compounds in clastic diagenetic systems. PhD Thesis, University of Wyoming, Laramie, 115 pp

Crossey LJ (1991) Thermal degradation of aqueous oxalate species. Geochim Cosmochim Acta 55: 1515–1527

Crossey LJ, Surdam RC, Lahann RW (1986) Application of organic/inorganic diagenesis to porosity prediction. In: Gautier D (ed) Roles of organic matter in sediment diagenesis, Soc Econ Paleontol Mineral Spec Publ 38, pp 147–155

Dibble WE Jr, Potter JM (1982) Effect of fluid flow on geochemical processes. SPE10994, Society of Petroleum Engineers, Richardson, Texas. 57th Annu Meet, New Orleans, 8 pp

Drummond SE, Palmer DA (1986) Thermal decarboxylation of acetate. Part II. Boundary conditions for the role of acetate in the primary migration of natural gas and the transportation of metals in hydrothermal systems. Geochim Cosmochim Acta 50: 825–833

Fein JB (1991a) Experimental study of aluminum-, calcium-, and magnesium-acetate complexing at 80 °C. Geochim Cosmochim Acta 55: 955–964

Fein JB (1991b) Experimental study of aluminum-oxalate complexing at 80 °C: implication for aluminum mobility in sedimentary basin fluids. Geology 19: 1037–1040

Fisher JB (1987) Distribution and occurrence of aliphatic acid anions in deep subsurface waters. Geochim Cosmochim Acta 51: 2459–2468

Fisher JB, Boles JR (1990) Water-rock interaction in Tertiary sandstones, San Joaquin Basin, California, USA: diagenetic controls on water composition. Chem Geol 82: 83–101

Franklin SP (1991) The role of carboxylic acids in feldspar and quartz dissolution and secondary porosity development: an experimental study under diagenetic conditions. PhD Thesis, Texas A&M University, College Station, 88 pp

Franklin SP, Hajash A Jr, Tieh TT (1990) Experimental dissolution of albite and quartz sands in a buffered oxalate-acetate solution at 100 °C and 347 bars in a flow-through system. Geol Soc Am Abstr Programs 22: 314

Franklin SP, Hajash A Jr, Tieh TT (1991) The role of organic acids in feldspar and quartz dissolution at 100 °C/347 bars: an experimental study. Am Assoc Pet Geol Bull 73: 576

Franks SG, Forester RW (1984) Relationships among secondary porosity, pore-fluid chemistry and carbon dioxide, Texas Gulf Coast. In: McDonald DA, Surdam RC (eds) Clastic diagenesis. Am Assoc Pet Geol Mem 37, pp 63–72

Grandstaff DE (1986) The dissolution rate of forsteritic olivine from Hawaiian beach sand. In: Colman SM, Dethier DP (eds) Rates of chemical weathering of rocks and minerals. Academic Press, New York, pp 41–59

Hajash A (1986) Marine diagenesis of feldspathic sand: an experimental investigation in a flow-through system. Geol Soc Am Abstr Programs 18: 624

Hajash A, Bloom MA (1991) Marine diagenesis of feldspathic sand: a flow-through experimental study at 200 °C, 1 kbar. Chem Geol 89: 359–377

Hajash A, Mahoney AJ, Elias BP (1989) Role of carboxylic acids in the dissolution of silicate sands: an experimental investigation at 100 °C and 345 bars. Geol Soc Am Abstr Programs 21(6): 49

Hanor JS, Workman AL (1986) Distribution of dissolved volatile fatty acids in some Louisiana oil field brines. Appl Geochem 1: 37–46

Heald MT, Larese RE (1973) The significance of solution of feldspar in porosity development. J Sediment Pet 43: 458–460

Huang WH, Kiang WC (1972) Laboratory dissolution of plagioclase feldspars in water and organic acids at room temperature. Am Mineral 57: 1849–1859

Kharaka YK, Law LM, Carothers WW, Goerlitz DF (1986) Role of organic species dissolved in formation waters from sedimentary basins in mineral diagenesis. In: Gautier DL (ed) Roles of organic matter in sediment diagenesis. Soc Econ Paleotol Mineral Spec Publ 38, pp 111–122

Lind CJ, Hem JD (1975) Effects of organic solutes on chemical reaction of aluminum. US Geol Surv Water-Supply Pap 1827-G, pp 1–83

Loucks RG, Dodge MM, Galloway WE (1984) Regional controls on diagenesis and reservoir quality in Lower Tertiary sandstones along the Texas Gulf Coast. In:

McDonald DA, Surdam RC (eds) Clastic diagenesis. Am Assoc Pet Geol Mem 37, pp 15–46

Lundegard PD, Senftle JT (1987) Hydrous pyrolysis: a tool for the study of organic acid synthesis. Appl Geochem 2: 605–612

MacGowan DB, Surdam RC (1988) Difunctional carboxylic acid anions in oilfield waters. Org Geochem 12: 245–259

MacGowan DB, Surdam RC (1990) Carboxylic acid anions in formation waters, San Joaquin Basin and Louisiana Gulf Coast, USA. Appl Geochem 5: 687–701

Manley EP, Evans LJ (1986) Dissolution of feldspars by low-molecular-weight aliphatic and aromatic acids. Soil Sci 141: 106–112

Mast MA, Drever JI (1987) The effect of oxalate on the dissolution rates of oligoclase and tremolite. Geochim Cosmochim Acta 51: 2559–2568

McMahon DA (1989) Secondary porosity in sandstones and diagenesis of adjacent shales, Oligocene, South Texas. Trans Gulf Coast Assoc Geol Soc 32: 94

Means JL, Hubbard N (1987) Short-chain aliphatic acid anions in deep subsurface brines: a review of their origin, occurrence, properties, and importance and new data on their distribution and geochemical implications in the Palo Duro Basin, Texas. Org Geochem 11: 177–191

Oxburgh R, Drever JI (1990) Effect of composition, pH and oxalate ion on the dissolution rate of plagioclase feldspars. Geol Soc Am Abstr Programs 22: 291

Palmer DA, Drummond SE (1986) Thermal decarboxylation of acetate, Part I. The kinetics and mechanism of reaction in aqueous solution. Geochim Cosmochim Acta 50: pp 813–823

Potter JM, Pohl DC, Rimstidt JD (1987) Fluid-flow systems for kinetic and solubility studies. In: Ulmer GC, Barnes HL (eds) Hydrothermal experimental techniques. Wiley-Interscience, New York, pp 240–260

Reed CL (1990) The role of oxalic acid on the dissolution of granitic sand: an experimental investigation in a hydrothermal flow-through system. MS Thesis, Texas A&M University, College Station, 59 pp

Reed CL, Hajash A (1990) Effect of flow rate on dissolution of granitic sand in oxalic acid: Flow-through experiments at 100 °C, 345 bars. Geol Soc Am Abstr Programs 22: A291

Reed CL, Hajash A (1992) Dissolution of granitic sand by pH-buffered carboxylic acids: a flow-through experimental study at 100 °C, 345 bars. Am Assoc Pet Geol Bull 76: 1402–1416

Rouxhet PG, Robin PL, Nicaise G (1980) Characterization of kerogens and their evolution by infrared adsorption spectroscopy. In: Durand B (ed) Kerogen – insoluble organic matter from sedimentary rocks. Technip, Paris, pp 163–190

Schmidt V, McDonald DA (1979) The role of secondary porosity in the course of sandstone diagenesis. In: Scholle PA, Schluger RR (eds) Aspects of diagenesis. Soc Econ Paleontol Mineral Spec Publ 26, pp 175–208

Seyfried WE Jr, Gordon PC, Dickson FW (1979) A new reaction cell for hydrothermal solution equipment. Am Mineral 64: 646–649

Seyfried WE Jr, Janecky DR, Berndt ME (1987) Rocking autoclaves for hydrothermal experiments. II. The flexible reaction-cell system. In: Ulmer GC, Barnes HL (eds) Hydrothermal experimental techniques, Wiley-Interscience, New York, pp 216–239

Shotyk W (1990) Incongruent and congruent dissolution of plagioclase feldspar: direct evidence from SIMS analyses of bytownite and anorthite. Geol Soc Am Abstr Programs 22: A291

Siebert RM, Moncure GK, Lahann RW (1984) A theory of framework grain dissolution in sandstones. In: McDonald DA, Surdam RC (eds) Clastic diagenesis. Am Assoc Pet Geol Mem 37, pp 163–175

Stillings LL, Perry CA, Voigt DE, Brantley SL. (1990) The effect of oxalic acid on feldspar dissolution rates. Geol Soc Am Abstr Programs 22: 291

Stoessell RK, Pittman ED (1990) Secondary porosity revisited: the chemistry of feldspar dissolution by carboxylic acids and anions. Am Assoc Pet Geol Bull 74: 1795–1805

Stumm W, Furrer G, Wieland E, Sinder B (1985) The effects of complex-forming ligands on the dissolution of oxides and aluminosilicates. In: Drever JI (ed) The chemistry of weathering. Reidel, Dordrecht, pp 55–74

Surdam RC, Crossey LJ (1985) Organic-inorganic reactions during progressive burial: key to porosity/permeability enhancement and/or preservation. Philos Trans R Soc Lond A 315: 135–156

Surdam RC, MacGowan DB (1987) Oilfield waters and sandstone diagenesis. Appl Geochem 2: 613–619

Surdam RC, Boese SW, Crossey LJ (1984) The chemistry of secondary porosity. In: McDonald DA, Surdam RC (eds) Clastic diagenesis. Am Assoc Pet Geol Mem 37, pp 317–345

Surdam RC, Crossey LJ, Hagen ES, Heasler HP (1989) Organic-inorganic interactions and sandstone diagenesis. Am Assoc Pet Geol Bull 73: 1–23

Willey LM, Kharaka YK, Rapp JB, Barnes I (1975) Short chain aliphatic acid anions in oil field waters and their contribution to the measured alkalinity. Geochim Cosmochim Acta 39: 1707–1711

Chapter 9 Experimental Studies of Organic Acid Decomposition

Julie L.S. Bell[1] and Donald A. Palmer[1]

Summary

This chapter presents a brief review of the decarboxylation and oxidation modes of spontaneous decomposition of carboxylic acids with the main emphasis on the reactions of acetic acid. The kinetics of thermal decarboxylation of short-chain carboxylic acids are characterized by means of an isokinetic relationship into two general mechanistic classes: aliphatic monocarboxylic acids, which react via a heterogeneously catalyzed mechanism, and dicarboxylic acids, which undergo homogeneous C—C bond cleavage. In this chapter, emphasis is placed on the former class of reactions. A detailed critical discussion of the experimental techniques used to derive the rate constants is given together with suggestions for future experiments that in the authors' view would serve to test the preliminary mechanistic assignments. Attention will also be focused on the need and methods used to characterize the solid surface component of heterogeneous catalysis. Specific mention is made of experiments involving oxidatrion reactions that may help to rationalize the correlations proposed in the literature on the basis of field data, as well as clarifying the nature and reversibility of these processes.

1 Introduction

The role played by short-chain, mono- and dicarboxylic acids in the evolving chemistry of sedimentary basins – processes such as the mass transport of metals, the dissolution/precipitation of carbonates and aluminosilicates, and the migration of methane – can only be assessed given knowledge of their stability under prevailing hydrothermal conditions, specifically temperatures up to 200 °C. Substantial concentrations of water-soluble carboxylic acids are

---

[1] Chemistry Division, Oak Ridge National Laboratory, P.O. Box 2008, Building 4500S, Oak Ridge, Tennessee 37831-6110, USA

found, for the most part, in association with organic matter and are therefore thought to have been produced from kerogen and/or petroleum during maturation (Zinger and Kravchik 1972; Surdam et al. 1984; Surdam and Crossey 1985; Kawamura and Kaplan 1987; Lundegard and Senftle 1987; Surdam and MacGowan 1987). The five carboxylic acids most frequently observed in basin brines are aliphatic mono- or dicarboxylic acids. The order of their decreasing abundance is: acetic, propionic butyric, malonic, and succinic (Zinger and Kravchik 1972; Carothers and Kharaka 1978; Surdam et al. 1984; Means and Hubbard 1985, 1987; Hanor and Workman 1986; Barth 1987; Fisher 1987; Kharaka et al. 1987; MacGowan and Surdam 1988; Fisher and Boles 1990). Conflicting data exist for oxalic acid in basin brines. Discrepancies arise when ion chromatography is used in the analysis; namely, the choice of the conventional ICE column does not allow oxalate to be identified unambiguously. On the other hand, the known strong complexation of oxalate by such metal ions as Al^{3+}, Mg^{2+}, Ca^{2+}, etc. at high temperatures may cause precipitation of oxalate compounds near the well head as the brine cools. Therefore, although it is acknowledged that oxalic acid may be present at significant levels in basin brines, it is not possible to rank it in terms of abundance with the other acids. Once cleaved from their organic precursors, the acids presumably migrate away from their point of origin in the aqueous phase; and it is this mobility that is an essential factor in the processes mentioned above. Thus, the need arises to establish the stability of these acids under conditions of temperature, pH, f_{O2}/f_{H2}, and mineralogy encountered during migration. Certainly carboxylic acids persist in some sedimentary environments as evidenced by analyses of oil-field brines (Kharaka et al. 1987). Possible reasons for the disappearance of these acids with increasing distance from their source (Zinger and Kravchik 1972; Matusevich and Shvets 1973) are: thermal decomposition and/or chemical transformations; adsorption onto mineral surfaces, especially smectitic clays (Bingham et al. 1965; Al-Owais et al. 1986; Kawamura et al. 1986; Purnell et al. 1987); and consumption by bacteria.

Concentration versus depth profiles for water-soluble carboxylic acids found in formation waters show a rapid rise from about 80 to 100 °C, followed by a continuous decrease in concentration through 200 °C (Carothers and Kharaka 1978; Bell 1991; Crossey 1991). Degradation by bacteria is probably the dominant process by which carboxylic acids are destroyed in shallow sediments where temperatures are generally less than 80 °C (Carothers and Kharaka 1978). Bacteria are believed to be largely inactive at higher temperatures and it is the combined effects of decreasing bacterial action and increasing production rate that are thought to account for the rapid rise in carboxylic acid concentration near 100 °C. The emphasis of this chapter is on the kinetics and mechanisms for thermal decomposition of the n-C_2 to n-C_4 carboxylic acids in (largely) sedimentary environments at depths sufficient for temperatures to exceed 100 °C. There is debate as to the survival of bacteria at higher temperatures (Crossey 1991), although until they are identified and found to be commonplace at higher temperatures, bacterial degradation of carboxylic acids will not be considered significant at temperatures over 100 °C.

Two types of reactions, decarboxylation and oxidation, exist that account for observed decreases in carboxylic acid concentrations in natural waters at temperatures above 100 °C. The process by which a carboxyl group is cleaved from the root molecule is defined as decarboxylation. In the acidic form, the simplest of such reactions is written:

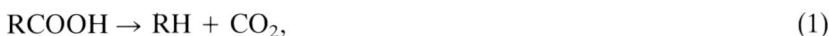

$$RCOOH \rightarrow RH + CO_2, \tag{1}$$

while in the basic form the equivalent reaction is:

$$RCOO^- + H_2O \rightarrow RH + HCO_3^- \tag{2}$$

These reactions have been reported to occur in the gas and solution phases in laboratory experiments, albeit at conditions not encountered in basin environments. The kinetics and products of decarboxylation of aliphatic, monocarboxylic acids have been shown to depend on the nature of the solid surface present (Palmer and Drummond 1986; Bell 1991; Bell et al. 1993). In fact, decarboxylation of these acids is used as a probe for the catalytic properties of metals and metal oxides (Rajadurai 1987; Kim and Barteau 1988).

Oxidation is analogous to decarboxylation [e.g., Eq. (1)], in the sense that both reactions involve breakage of the C—C bond in the activated complex, resulting in liberation of the carboxyl group as CO_2. For example, complete oxidation of acetic acid may be written:

$$CH_3COOH + 2O_2 \rightarrow 2CO_2 + 2H_2O, \tag{3}$$

whereas decarboxylation of acetic acid is represented by:

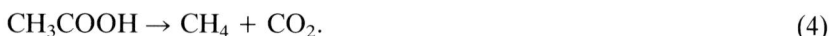

$$CH_3COOH \rightarrow CH_4 + CO_2. \tag{4}$$

In fact, oxidative decarboxylation of aliphatic carboxylic acids induced by strong oxidizing agents such as Pb(IV), Ag(II), Mn(III), Co(III), Ce(IV), etc. has been studied in H_2O and carboxylic acid solvents and was shown to yield CO_2 plus alkanes and/or other products resulting from the partial oxidation of the root molecule (Mosher and Kehr 1953; Kochi 1965a,b; Sheldon and Kochi 1968; Anderson and Kochi 1970a,b; Tinker 1970; Mittal et al. 1973; Adinarayana et al. 1975, 1976; Mehrotra 1981; Nazar and Wells 1985; Davies 1989). Complete oxidation of aliphatic carboxylic acids [e.g., Eq. (3)] has been studied as a method for removal of organic pollutants from wastewater streams (Baldi et al. 1974; Levec and Smith 1976; Levec et al. 1976; Imamura et al. 1982a; Pintar and Levec 1992). Partial oxidation of the root molecule frequently occurs during heterogeneously catalyzed decomposition of aliphatic carboxylic acids, yielding ketones, saturated and unsaturated aldehydes, alkenes, etc., depending on the complexity of the original molecule and the availability and type of oxidant (for formic, Mars et al. 1963; Trillo et al. 1972; McCarty et al. 1973; Falconer and Madix 1974; Ai 1977; Benziger and Madix 1979; Benziger et al. 1979; Sexton 1979; Barteau et al. 1980; Bowker and Madix 1981a; Sexton and Madix 1981; Avery et al. 1982; Hayden et al. 1983; Miles et al. 1983; Benziger and Schoofs 1984; Madix 1984; Clavilier and Sun 1986; Shustorovich and Bell 1989; Goddard et al. 1992: for acetic, Hurd and Martin 1929; Mitchell and Reid 1931; Bamford and Dewar 1949; Child and Hay 1963; Rubinshtein et al. 1964; Blake and Jackson 1968, 1969; Madix et al. 1976; Kuriacose and Jewur 1977; Gonzalez et al. 1978; Kraeutler and Bard 1978; Parrott et al. 1978; Ying and Madix 1979; Benziger and Madix 1980; Bos et al. 1980; Barteau et al. 1981; Bowker and Madix 1981b; Imamura et al. 1982b; Bowker et al. 1983; Schoofs and Benziger 1984; Servotte et al. 1985; Kim and Barteau 1988: for propionic, Kim and Barteau 1988: for butyric – Demorest et al. 1951; Millet et al. 1989; Virely et al. 1992). The distinction between decarboxylation, as given in Eqs. (1) and (2), and oxidation can sometimes be difficult to discern. For example, oxidative decarboxylation induced by one electron transfer to a strong oxidant (e.g., one of those mentioned above) also yields CO_2 and the hydrogenated root molecule.

It is important to remember that carboxylic acids in natural hydrothermal settings are metastable with respect to their constituents, such as C, CO, CH_4, or CO_2. The obvious reason for the persistence of these acids in basin brines is the existence of high activation energy barriers that inhibit spontaneous reaction. Indeed, the kinetic model formulated by Drummond and Palmer (1986) established that a thermal window could exist in which acetate could survive for a significant period in developing sedimentary basins and yet undergo subsequent decarboxylation to yield methane within the active lifetime of the basin. Pathways leading to other metastable species may have lower energy barriers and hence be favored kinetically over the thermodynamically predicted products. It is impossible to know, a priori, the relevant reactions affecting carboxylic acids under natural hydrothermal conditions; however, by applying Occam's Razor it is reasonable to first consider simple decarboxylation and oxidation reactions known to occur under experimental hydrothermal conditions.

In this chapter, the water-soluble carboxylic acids are divided into two homologous series, because aliphatic mono- and dicarboxylic acids behave quite differently with respect to thermal decomposition. Specifically, dicarboxylic acids readily undergo thermally induced homogeneous decarboxylation; whereas, in all experiments conducted to date, unimolecular decarboxylation of monocarboxylic acids effectively would not occur under relevant hydrothermal conditions in the absence of a heterogeneous catalyst. Evidence for this duality of mechanism will be presented in the forthcoming discussion, but will be restricted primarily to acetic, oxalic, and malonic acids for which extensive experimental information is available. Moreover, a critical discussion will be presented of the current state of experimental kinetic data and suggestions as to the direction of possible future experimental studies will be given.

In the following discussion, it is assumed that the reader has a basic understanding of chemical kinetics. For broad, basic treatments of the subject refer to Gardiner (1969), Moore and Pearson (1981), and Atkins (1982). Advanced reviews pertinent to the discussion of isokinetic relationships (also referred to as compensation laws) and transition state theory are to be found in Leffler (1955), Leffler and Grunwald (1963), Bunnett (1986), Wilkins (1991), Laidler and King (1983), and Truhlar et al. (1983). The general field of heterogeneous catalysis has been reviewed extensively with excellent articles by Bond (1987) and Boudart and Djéga-Mariadassou (1984). Examples of the numerous treatises dealing with catalytic surfaces are Hubbard (1989) and Minachev and Shpiro (1990).

2   Background Information

A significant amount of kinetic data exists for the decarboxylation and oxidation of carboxylic acids. However, a relatively small fraction of these results deals with n-C_2 to n-C_4 aliphatic mono- and dicarboxylic acids under conditions pertinent to geological interests. For example, the early studies of the decarboxylation kinetics of acetic acid utilized flow-though silica tubes in which the anhydrous gas was exposed to very high temperatures for only seconds (Bamford and Dewar 1949; Blake and Jackson 1968, 1969). Nevertheless, it is useful to consider all of these results because it reveals trends common for structural classes of carboxylic acids. In this background discussion, a brief introduction to the subject of isokinetic relationships is given, as well as an overview of the decarboxylation and oxidation of carboxylic acids in which isokinetic relationships are used to establish trends and gross variations in reaction mechanisms between structural classes of acids.

2.1   Isokinetic Relationships

In transition state theory (a review of which is given by Kreevoy and Truhlar 1986), the energy of activation determines the rate constant at a given temperature, $T(K)$:

$$\Delta G^{\ddagger} = \Delta H^{\ddagger} + T\Delta S^{\ddagger}. \tag{5}$$

The enthalpy, $\Delta H^{\ddagger}$, and entropy, $\Delta S^{\ddagger}$, of activation can be calculated from the temperature dependence of the first-order rate constant using the relationship derived by Eyring and Polanyi. The rate constant is obtained from the expression:

$$k = (k_P T/h)K^{\ddagger}, \tag{6}$$

where k_P and h are the Boltzman and Planck constants, and $K^{\ddagger}$ is the quasi-equilibrium constant involving the transition state and reactants. The energy of activation is derived from $K^{\ddagger}$ in the pseudo-thermodynamic relationship:

$$\Delta G^{\ddagger} = -RT \ln K^{\ddagger}, \tag{7}$$

where R is the universal gas constant. The combination of Eqs. (5), (6), and (7) yields the "Eyring and Polanyi" relationship:

$$\ln k = \ln(k_P T/h) + (\Delta S^{\ddagger}/R) - (H^{\ddagger}/RT). \tag{8}$$

Linear relationships between the enthalpy and entropy of activation have been verified for various series of organic reactions involving moderate changes in reactant structure or solvent type. The interpretation of this empirical observation (Leffler 1955; Leffler and Grunwald 1963; Galwey 1977; Galwey and Brown 1979; Conner 1982; Bunnett 1986), which is well supported by the preponderance of rate data, is not without serious statistical shortcomings (Brown 1962; Exner 1964a,b; Petersen 1964; Leffler 1965, 1966; Exner 1970, 1972; Banks et al. 1972; Wold 1972; Exner and Beránek 1973; Wold and Exner 1973; Krug 1980; Bunnett 1986). Nevertheless, the empirical conclusion is that a linear relationship between $\Delta H^{\ddagger}$ and $\Delta S^{\ddagger}$ may exist only if changes within the reaction series do not result in changes in mechanism, or in the nature of the transition state. The activation parameters in a linear isokinetic relationship are related as follows:

$$\Delta H^{\ddagger} = \Delta H_0^{\ddagger} + \beta \Delta S^{\ddagger}, \tag{9}$$

where $\Delta H_0^{\ddagger}$ simply represents a pseudo-activation enthalpy when $\Delta S^{\ddagger} = 0$; β is termed the isokinetic temperature because the reactions within the series proceed at the same rate when $T = \beta$, and $\Delta G^{\ddagger}$ is constant and equal to $\Delta H_0^{\ddagger}$.

2.2 Decarboxylation

Decarboxylation reactions may be induced (depending on the acid) in a variety of ways: thermally, bacteriologically, photochemically, or even by electrolysis as in the anodic reaction of the Kolbe synthesis. Thermally induced decarboxylation of many carboxylic acids in solution proceeds by a bimolecular mechanism involving addition of a nucleophilic solvent molecule to an electrophilic carbon atom on the root molecule – preferably at a carbon adjacent to (α) or one removed from (β) the carboxyl carbon (Fraenkel et al. 1954; Clark 1958, 1969). An electrophile-nucleophile pair is formed in the transition state, which subsequently undergoes heterolytic fission (i.e., decomposition of a molecule into two ions of opposite charge) to yield CO_2, a proton, and a carbanion; the latter two species are reactive intermediates, which then combine rapidly (Brown 1951). The solvent molecule departs unaffected and in this sense the solvent may be considered a catalyst. The rate for the decarboxylation of malonic acid in quinoline/ dioxane mixtures was shown to depend on the concentration of quinoline (a nucleophilic solvent) and provides a good example of this type of decomposition reaction (Fraenkel et al. 1954; Clark 1958).

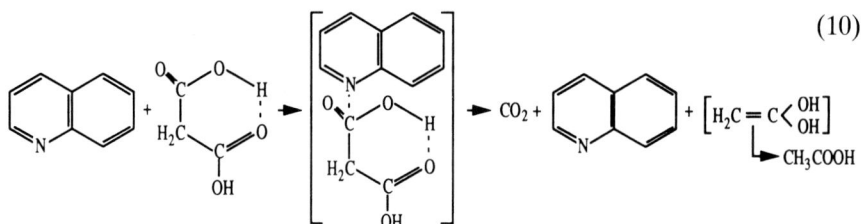

(10)

Several observations seem to be consistent with the kinetics of decarboxylation reactions (Gould 1959; Clark 1969; Richardson and O'Neal 1972). Firstly, decarboxylation occurs more readily if the carboxyl molecule incorporates a strong electron-withdrawing group, such as $-C=O$, $-NO_2$, $-C=N$, $-COOH$, $-Cl$, etc. The presence of an electron-attracting group induces a partial positive charge on the carboxyl carbon and thereby provides a site for nucleophilic attack by a solvent molecule (Clark 1969). This inductive effect also leads to an increase in the acidity of the acid (Dewar and Krull 1990) and to a weakening of the $R-CO_2H$ bond – the breaking of which is at the heart of the decarboxylation process, as it is believed to be the rate-determining step in reaching the transition state. Secondly, carboxylic acids with a doubly bonded oxygen or nitrogen bound to an α- or β-carbon (i.e., α- or β-keto, carboxyl, or imino groups) readily undergo decarboxylation because the root molecule, R, normally departs with the electron pair freed by the breaking of the $R-CO_2H$ bond. Functional groups such as these in the structure of R, particularly ones in the α- or β-position, stabilize the resulting carbanion, R^- (Clark 1969). In addition, the activated state for molecules of this type is thought to involve intramolecular hydrogen bonding of the carboxyl hydrogen to the oxygen or nitrogen of these groups, thereby facilitating the transfer of the carboxyl proton. This reasoning is directly applicable for the α- and β-carboxyl acids found in basin brines, namely, oxalic and malonic acid, respectively. Finally, the rates of decarboxylation for the acidic and anionic forms of carboxylic acids are different, with the rate for the acid usually exceeding that for the anion. The overall rate of decarboxylation of the mixture is given simply by:

$$\Sigma Rate = k_{acid}[RCO_2H] + k_{anion}[RCO_2^-], \tag{11}$$

where k_{acid} and k_{anion} are the end-member, first-order rate constants. Carboxylic acids can be divided into two classes (Clark 1969), based on whether decarboxylation is faster for the acidic or anionic form. The former class generally has doubly bonded oxygen or nitrogen groups in the α-, β-, or γ-positions, whereas members of the latter class have an α-aromatic group or electron-attracting groups (other than keto, carboxyl, or imino) in the α- or β-position (Clark 1969). Decarboxylation of acids in both classes proceeds

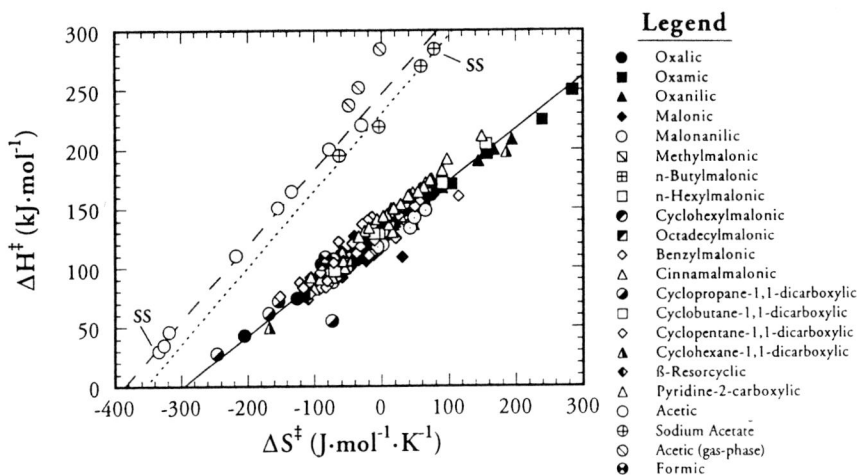

Legend

- ● Oxalic
- ■ Oxamic
- ▲ Oxanilic
- ◆ Malonic
- ○ Malonanilic
- ◲ Methylmalonic
- ⊞ n-Butylmalonic
- □ n-Hexylmalonic
- ◑ Cyclohexylmalonic
- ◪ Octadecylmalonic
- ◇ Benzylmalonic
- △ Cinnamalmalonic
- ◕ Cyclopropane-1,1-dicarboxylic
- ◰ Cyclobutane-1,1-dicarboxylic
- ◇ Cyclopentane-1,1-dicarboxylic
- ▲ Cyclohexane-1,1-dicarboxylic
- ◆ ß-Resorcyclic
- △ Pyridine-2-carboxylic
- ○ Acetic
- ⊕ Sodium Acetate
- ⊘ Acetic (gas-phase)
- ❻ Formic

Fig. 1. Isokinetic plot for the decarboxylation of the carboxylic acids listed in Table 1, where acetic acid, acetate, and the combined remaining data are represented by the *dashed, dotted,* and *solid lines,* respectively. *SS* refers to reactions over stainless steel surfaces

by similar mechanism (i.e., nucleophilic bimolecular addition followed by heterolytic fission). Differences in decarboxylation rate for the anionic versus the acidic forms arise from their relative abilities to achieve the structure of the activated complex (Clark 1969).

Enthalpies and entropies of activation for the decarboxylation of a wide variety of carboxylic acids are shown in the isokinetic plot in Fig. 1. The data given in Table 1 were taken mainly from a compilation of experimental kinetic results (Richardson and O'Neal 1972) for the thermally induced decarboxylation of carboxylic acids in different solvents. Despite the varying nature of the acids and solvents used, the activation parameters with the exception of acetic show similar linear trends, indicating that the mechanism for these reactions is likely to involve common rate-determining steps. Indeed, the linearity of the isokinetic plot is undeniable to the extent that outlying data points may be considered either as suspect experimental results, or as a reflection of a variation in mechanism, due to steric hindrance effects, oxidation, etc. In such a relation it is obviously important that the individual rate constants used to derive the activation parameters be measured over a wide range of temperature, preferably far from the isokinetic temperature. Moreover, the activation parameter values, particularly $\Delta S^{\ddagger}$, are difficult to determine accurately; therefore, in order to establish the isokinetic relationship unambiguously, the values of $\Delta H^{\ddagger}$ and $\Delta S^{\ddagger}$ should vary sufficiently to compensate for the experimental uncertainties. It is clear from Fig. 1 that $\Delta H^{\ddagger}$ ranges in value by $200 \, kJ \cdot mol^{-1}$, whereas the typical experimental uncertainty is on the order of several $kJ \cdot mol^{-1}$. It is important to note, with the exception of acetic acid, that all of the carboxylic acids plotted in Fig. 1 either have electron-withdrawing

Table 1. Activation parameters for the decarboxylation of some carboxylic acids

Solvent/conditions	$\Delta H^{\ddagger a}$	$\Delta S^{\ddagger b}$	Source[c]	Solvent/conditions	$\Delta H^{\ddagger a}$	$\Delta S^{\ddagger b}$	Source[c]
Oxalic				*Oxanilic*			
Glycerol	115	−20.50	1	Melt	168	89.5	1
Glycerol	110	−32.6	1	Aniline	208	194	1
Ethylene glycol	73.6	−126	1	0-Toluidine	200	167	1
propylene glycol	43.1	−205	1	0-Ethylaniline	190	144	1
Quinolene	163	65.9	1	N,N-dimethylaniline	157	64.0	1
6-Methylquinolene	135	5.10	1	Quinoline	162	66.9	1
8-Methylquinolene	158	57.3	1	8-Methylquinoline	149	41.8	1
1,4-Butanediol	107	−66.1	1	n-Hexyl ether	168	89.5	1
Dimethylsulfoxide	170	86.6	1	n-Amyl ether	118	−27.6	1
1,3-Butanediol	123	−20.5	1	n-Butyl ether	105	−58.2	1
Triethylphosphate	121	−24.3	1	bis-(2-Chloroethyl)ether	89.5	−93.7	1
2,3-Butanediol	94.6	−84.5	1	Dibenzyl ether	154	59.4	1
Aniline	169	67.8	2	Anisole	136	46.4	1
Methyl-aniline	149	34.7	2	Phenetole	136	16.7	1
Dimethyl-aniline	136	5.23	2	b-Chlorophenetole	131	2.93	1
Acetophenone	141	20.0	3	Benzoic acid	139	19.7	1
o-Cresol	134	3.77	4	Hexanoic acid	154	55.6	1
p-Cresol	131	−7.95	4	o-Cresol	172	100	1
m-Cresol	113	−46.0	4	n-Decyl alcohol	151	53.6	1
Dioxane	122	−23.0	5				
Dioxane	120	−28.3	6	*Malonic*			
Vapor	128	−12.1	7	Melt	150	49.8	1
H_2SO_4 (0.6% water)	122	41.5	6	Aniline	113	−18.8	1
H_2SO_4 (3% water)	119	8.14	6	N-methylaniline	111	22.3	1
Water/I = 0.2 pH 5	144	−32.0	8	N,N-dimethylaniline	110	27.4	1
Water/I = 0.2 pH 4	218	98.5	8	N,N-diethylaniline	106	34.5	1
Water/formic I = 0.2	161	−21.9	8	Pyridine	109	30.1	1
pH 5				2-Picoline	87.4	−72.8	1
Water/I = 0.3 pH 5	196	82.9	8	3-Picoline	91.6	−59.4	1
Water/formic I = 0.3	210	111	8	4-Picoline	105	−23.4	1
pH 5				Quinoline	112	−9.92	1
Water/I = 0.2 pH 6	123	−94.2	8	6-Methylquinoline	110	−12.6	1
Water/I = 0.3 pH 6	119	−104	8	8-Methylquinoline	102	−43.9	1
Water/I = 0.2 pH 7	218	98.5	8	N,N-diethylcyclohexyl-	108	−23.9	1
Water/formic I = 0.2	161	−21.9	8	amine			1
pH 7				n-Butyl alcohol	114	−18.4	1
Water/I = 0.3 pH 7	96.4	−158	8	n-Amyl alcohol	109	−31.8	1
Water/formic I = 0.3	167	−7.28	8	Isoamyl alcohol	113	−189	1
pH 7				n-Hexyl alcohol	109	−31.8	1
Water	165	63.1	6	2-Ethyl-1-hexanol	104	−43.5	1
Water	154	40.2	9	Diisobutylcarbinol	104	−44.8	1
				Cyclohexanol	96.2	−62.8	1
Dideuterio-oxalic				Benzyl alcohol	124	4.18	1
Dioxane	134	5.34	5	Glycerol	103	−51.0	1
Vapor	141	17.2	9	Phenol	114	−37.2	1
D_2O	150	32.0	5	m-Cresol	135	13.4	1
				p-Cresol	125	−10.0	1
Oxamic				0-Cresol	101	−69.0	1
Aniline	250	285	1	Thiophenol	144	28.9	1
0-Toluidine	225	239	1	Propionic acid	141	25.5	1
Quinoline	197	157	1	b-Mercaptopropionic	127	−41.4	1
8-Methylquinoline	151	51.0	1	n-Butyric acid	135	10.5	1
Dimethyl sulfoxide	158	62.3	1	n-Valeric acid	135	10.0	1
Triethyl sulfoxide	171	103	1	Isovaleric acid	136	15.1	1
Hexanoic acid	128	−5.23	1	Pivalic acid	162	76.6	1
Octanoic acid	128	−5.23	1	Hexanoic acid	136	13.4	1

Table 1. *Continued*

Solvent/conditions	$\Delta H^{\ddagger a}$	$\Delta S^{\ddagger b}$	Source[c]	Solvent/conditions	$\Delta H^{\ddagger a}$	$\Delta S^{\ddagger b}$	Source[c]
Heptanoic acid	124	−14.2	1	*Cyclohexylmalonic*			
Octanoic acid	146	37.2	1	Hexanoic acid	133	8.12	1
Decanoic acid	111	−46.0	1	Octanoic acid	141	23.5	1
d,1-2-Methylpentanoic	111	−46.4	1				
Benzoic acid	127	−7.53	1	*Octadecylmalonic*			
Anisole	126	−15.4	1	Melt	128	−5.44	1
β-Chlorophenetole	116	−33.0	1				
Benzaldehyde	117	−29.3	1	*Benzylmalonic*			
Nitrobenzene	118	−30.1	1	Melt	123	−10.9	1
0-Nitrotoluene	98.3	−74.9	1	Aniline	82.8	−90.4	1
2-Nitro-m-xylene	126	−13.1	1	N-ethylaniline	91.6	−66.1	1
Dimethyl sulfoxide	93.3	−62.8	1	N-sec-butylaniline	111	−15.1	1
Triethyl phosphtate	109	−25.9	1	N,N-dimethylaniline	161	115	1
Water pH 0.42	124	−4.94	10	0-Toluidine	125	20.9	1
Water pH 1.18	123	−8.04	10	Quinoline	83.3	−83.3	1
Water pH 1.98	121	−14.7	10	8-Methylquinoline	110	−19.2	1
Water pH 2.09	122	−12.0	10	n-Butyric acid	96.2	−79.1	1
Water pH 2.11	123	−9.12	10	Decanoic acid	113	−37.7	1
Water pH 2.18	123	−9.29	10	m-Cresol	92.0	−105	1
Water pH 2.27	118	−21.1	10	p-Cresol	113	−59.4	1
Water pH 2.40	120	−18.9	10				
Water pH 2.51	124	−7.85	10	*Cinnamalmalonic*			
Water pH 2.72	124	−9.01	10	Aniline	99.6	−55.2	1
Water pH 3.28	122	−19.8	10	o-Chloroaniline	82.0	−100	1
Water pH 3.68	124	−23.4	10	o-Toluidine	91.6	−73.2	1
Water pH 4.02	123	−18.7	10	N,N-dimethylaniline	131	15.9	1
Water pH 4.89	122	−24.3	10	Quinoline	98.3	−67.8	1
from k_{anion} calculated	110	−61.3	10	8-Methylquinoline	90.4	−91.2	1
				Phenol	110	−66.9	1
Malonanilic				m-Cresol	92.0	−105	1
Aniline	115	−6.28	1	p-Cresol	113	−59.4	1
N-ethylaniline	133	41.8	1				
Quinoline	87.9	−73.2	1	*Cyclopropane-1,1-dicarboxylic*			
8-Methylquinoline	119	1.68	1	H_2SO_4	54.9	−74.0	1
Acetaniline	88.3	−80.3	1	H_3PO_4	107	−72.6	1
o-Cresol	149	64.9	1	Melt	61.3	−168	1
m-Cresol	139	39.3	1	Collidine	27.8	−246	1
p-Cresol	142	48.1	1	Diethylene glycol	71.6	−153	1
1,4-Butanediol	99.2	−53.1	1				
2,3-Butanediol	118	−6.28	1	*Cyclobutane-1,1-dicarboxylic*			
1,3-Butanediol	136	43.5	1	H_2SO_4	172	89.6	1
				H_3PO_4	167	65.3	1
Methylmalonic				Melt	61.3	−168	1
Melt	146	40.2	1	Collidine	27.8	−246	1
				Diethylene glycol	129	−11.5	1
n-Butylmalonic							
Melt	135	12.1	1	*Cyclopentane-1,1-dicarboxylic*			
Hexanoic acid	139	21.3	1	H_2SO_4	88.5	−81.3	1
Octanoic acid	137	16.0	1	H_3PO_4	78.1	−113	1
Phenol	151	54.4	1	Melt	156	55.1	1
m-Cresol	124	−9.62	1	Collidine	73.1	−109	1
p-Cresol	100	−66.1	1	Diethylene glycol	77.6	−107	1
n-Hexylmalonic				*Cyclohexane-1,1-dicarboxylic*			
Melt	135	11.7	1	H_2SO_4	198	186	1
o-Cresol	108	−47.7	1	H_3PO_4	106	−57.3	1
Octanoic acid	129	−4.18	1	Melt	144	28.0	1

Table 1. *Continued*

Solvent/conditions	$\Delta H^{\ddagger a}$	$\Delta S^{\ddagger b}$	Source[c]	Solvent/conditions	$\Delta H^{\ddagger a}$	$\Delta S^{\ddagger b}$	Source[c]
Collidine	49.7	−167	1	p-Bromoanisole	137	10.5	1
Diethylene glycol	106	−45.1	1	Phenetole	150	14.2	1
				Nitrobenzene	144	2.09	1
b-Resorcyclic				Nitrobenzene	143	4.18	1
Hexanoic acid	143	−14.8	1	o-Nitrotoluene	162	40.4	1
Heptanoic acid	141	−20.9	1	p-Nitrotoluene	147	2.6	1
Octanoic acid	137	−28.9	1	Octanoic acid	144	2.09	1
Decanoic acid	122	−64.4	1	2,4-Dimethylbenzophenol	161	38.9	1
Resorcinol	121	−43.9	1	Diphenyl ether	162	40.8	1
o-Cresol	75.7	−151	1	Triethyl phosphate	169	62.3	1
p-Cresol	87.4	−121	1	p-Cymene	164	53.6	1
2,3-Butanediol	104	−71.5	1	p-Dibromobenzene	145	7.95	1
1,3-Butanediol	82.8	−116	1				
1,4-Butanediol	152	49.0	1	*Formic*			
Ethylene glycol	118	−31.8	1	Formic acid	103	−89.5	1
Quinoline	144	24.9	1	99% formic-1% H_2O	109	−83.7	1
8-Methylquinoline	95.8	−91.2	1				
				Acetic			
Pyridine-2-carboxylic				Water/TiO$_2$	165	−133	11
Melt	167	55.3	1	Water/TiO$_2$	151	−154	12
Aniline	131	−25.6	1	Water/quartz	200	−77.6	11
N-ethylaniline	159	39.7	1	Water/calcium-	45.9	−319	11
N,N-diethylaniline	154	28.0	1	montmorillonite	45.9	−319	
5-Chloro-2-methylaniline	164	45.6	1	Water/pyrite	221	−30.0	11
p-Nitroaniline	151	17.6	1	Water/gold	110	−218	11
Tributylaniline	183	89.5	1	Water/stainless steel	29.8	334	13
Quinoline	141	−6.99	1	Water/magnetite	34.8	−326	11
Phenol	211	149	1	Vapor	252	−34.3	14
p-Cresol	150	18.8	1	Vapor	237	−49.2	15
p-Nitrophenol	192	97.1	1	Vapor	284	−2.51	16
Isopropylphenol	172	64.9	1				
t-Butyl-o-thiocresol	175	74.1	1	*Sodium Acetate*			
1,3-Butanediol	175	71.1	1	Water/calcite	147	−3.95	11
1,4-Butanediol	160	42.7	1	Water/gold	270	58.0	12
p-Dimethyloxybenzene	134	−19.2	1	Water/TiO$_2$	195	−62.9	12
p-Dimethyloxybenzene	126	−34.3	1	Water/stainless steel	281	78.3	12
β-Chlorophenetole	138	8.37	1				

[a] $(kJ \cdot mol^{-1})$.
[b] $(J \cdot mol^{-1} \cdot K^{-1})$.
[c] Original source references may be found in reference 1, Richardson and O'Neal (1972), unless otherwise indicated by specific references.
1 = (Richardson and O'Neal 1972); 2 = (Clark 1957); 3 = (Adams et al. 1978); 4 = (Clark 1963); 5 = (Lutgert and Schroer 1940); 6 = (Dinglinger and Schroer 1937, 1938); 7 = (Lapidus et al. 1964); 8 = (Crossey 1991); 9 = (Lapidus et al. 1966a,b); 10 = (Hall 1949); 11 = (Bell et al. 1993); 12 = (Palmer and Drummond 1986); 13 = (Kharaka et al. 1983); 14 = (Bamford and Dewar 1949); 15 = (Blake and Jackson 1968); 16 = (Blake and Jackson 1969).

groups on the root molecule; or α- or β-keto, carboxyl, or imino groups; or an aromatic group in the α-position. Therefore, these acids probably decarboxylate via a common mechanism such as the C—C bond cleavage mechanism described above.

The isokinetic results for acetic acid in Fig. 1 are obviously different from the relationship for the remaining acids. The greater stability of aliphatic monocarboxylic acids may be due to: (1) their inability to form a chelate structure via intramolecular hydrogen bonding as alluded to above; (2) the lack of electron-attracting substituents, which would render the carboxyl carbon susceptible to nucleophilic attack; and (3) their inability to form the stable intermediate required for heterolytic fission. It has been shown (Kharaka et al. 1983; Palmer and Drummond 1986; Bell 1991; Bell et al. 1993) that acetic acid does not readily undergo thermally induced decarboxylation unless a suitable catalytic surface is present. For example, the projected half-life for 1 M acetic acid solution at 100 °C in the presence of a weak catalyst, such as that provided by titanium oxide, is ca. 6 billion years, whereas in a stainless steel vessel the projected half-life is ca. 12 years. Thus, the data in Fig. 1 indicate that acetic acid, and probably other aliphatic monocarboxylic acids, undergo decarboxylation via a heterogeneous catalytic mechanism different from the general homogeneous mechanism postulated for the other acids presented in this figure.

2.3 Oxidation

Oxidation of the n-C_2 to n-C_4 mono- and dicarboxylic acids has been observed under laboratory conditions in the presence of a variety of oxidants (see references in Sect. 1). The mechanisms and products of these reactions vary depending on the nature of the oxidant and the conditions under which the reaction takes place. Many of these "oxidation reactions" result in decarboxylation of the acid by radical chain reactions to yield CO_2 plus the hydrogenated root molecule rather than involving complete oxidation of the acid molecule to CO_2. For example, 3500 Å radiation induces a redox reaction in which a carboxyl functional group is oxidized to CO_2, while $Ce^{IV}(O_2CCH_3)_4$ is reduced to $Ce^{III}(O_2CCH_3)_3$ (Sheldon and Kochi 1968).

Oxidation of carboxylic acids has not been studied in aqueous solution in the presence of geologically relevant oxidants. In the field, water having intimate contact with organic matter is not likely to contain strongly oxidizing species. However, acids migrating in the aqueous phase may eventually encounter oxidizing mineral surfaces or dissolved oxidizing species such as sulfate. For example, the reduction of transition metals, such as Fe^{3+} and Mn^{3+} in oxide or hydroxide mineral surfaces, has been shown to occur via a coupled redox reaction in which organic reductants, such as catechol, phenol, salicylate, and benzoate (Stone 1986), supply the electron to the metal. These results raise the question whether carboxylates coordinated to metal complexes in solution undergo enhanced rates of reaction (e.g., Buckingham and Clark 1982) and decarboxylation.

3 Stability of Aliphatic Monocarboxylic Acids

As mentioned previously, acetic acid is the most abundant naturally occurring carboxylic acid and is an ideal model for the class of short-chain, aliphatic, monocarboxylic acids. The reactivity of the n-C_2 to n-C_4 aliphatic, monocarboxylic acids is dominated by the carboxyl group and it has been shown that the addition of carbons to the alkyl group has little effect on the acids' chemical behavior. For example, the pK_a for the dissociation of the acids under standard conditions changes very little between acetic acid (pK_a = 4.75) and n-butyric acid (pK_a = 4.81). Therefore, it seems reasonable to expect that the hydrothermal stability of the n-C_3 and n-C_4 monocarboxylic acids should be similar to that of acetic acid in terms of both the nature and kinetics of relevant reactions. The kinetics of thermally induced decarboxylation of aliphatic monocarboxylic acids, especially formic and acetic acids, have been studied extensively in the gas phase (see references in Sect. 1), but only recently in the aqueous phase (Kharaka et al. 1983; Palmer and Drummond 1986; Bell 1991; Bell et al. 1993). Apart from one experiment in which the rate of butyric acid decarboxylation at 359 °C was found to be similar to that for acetic acid (Palmer and Drummond 1986), the decomposition kinetics of the n-C_3 and longer chain monocarboxylic acids have not been investigated in aqueous solution. In this section, particular attention is given to hydrothermal experiments involving acetic acid. Discussion of the decomposition of the other acids will be limited to a review of gas phase kinetic studies and analogy to aqueous acetic acid decomposition.

3.1 Acetic Acid

The stability of acetic acid in basin brines at temperatures above which bacteria are active depends on the rates of the various reactions in which the acid participates. Under geologically relevant hydrothermal conditions, carboxylic acids are likely to undergo either decarboxylation or oxidation. Thermodynamic calculations of the decarboxylation reaction in Eq. (4) at 100 °C (Shock 1988, 1989) indicate that, given methane and carbon dioxide fugacities achievable in natural waters, virtually no acetic acid/acetate should persist at equilibrium. It will be shown here that the persistence of acetic acid in basin brines for geologically meaningful time periods is in keeping with experimentally observed rates for the decarboxylation process.

3.1.1 Decarboxylation

The decarboxylation of acetic acid was first studied as a side reaction in the high temperature (400 to 600 °C) dehydration reaction to acetone or ketene (Nef 1901; Hurd and Martin 1929; Mitchell and Reid 1931). When heated, acetic acid decomposes according to several competing pathways, one of which is decarboxylation:

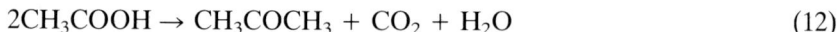

$$2CH_3COOH \rightarrow CH_3COCH_3 + CO_2 + H_2O \qquad (12)$$

$$CH_3COOH \rightarrow H_2O + CH_2{=}C{=}O \qquad (13)$$

$$CH_3COOH \rightarrow CH_4 + CO_2. \tag{14}$$

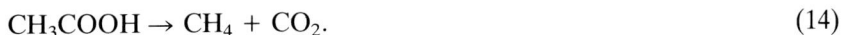

In these early studies, the decarboxylation reaction was found to compete with the other reactions with increasing success as temperature decreased from 600 to 400 °C. Subsequent gas-phase studies at lower temperature (Bamford and Dewar 1949; Demorest et al. 1951; Blake and Jackson 1968, 1969; Kuriacose and Swaminathan 1969; Swaminathan and Kuriacose 1970; Kuriacose and Jewur 1977) showed that the temperature dependence of the rate constant for decarboxylation was such that the projected half-life at 100 °C was on the order of 100 billion years. Note that, with the exception of experiments involving acetic acid in water vapor (Blake and Jackson 1969), these experiments were anhydrous and at temperatures far in excess of those relevant to the sedimentary environment.

Hydrothermal kinetic experiments in stainless steel autoclaves (Kharaka et al. 1983) demonstrated that acetic acid can readily decarboxylate at temperatures commonly found in sedimentary basins. The activation energy for decarboxylation derived from their experiments, $33.9 \, kJ \cdot mol^{-1}$ ($8.1 \, kcal \cdot mol^{-1}$), is extremely low for a process involving the breaking of a carbon-carbon bond (Palmer and Drummond 1986), and is certainly quite different from that derived from the gas-phase experiments of Blake and Jackson (i.e., $240 \, kJ \cdot mol^{-1}$). Palmer and Drummond concluded that the reaction was catalyzed by the vessel surface and, therefore, conducted similar experiments in titanium and gold vessels. The resulting activation energies of $146 \, kJ \cdot mol^{-1}$ (Palmer and Drummond 1986) and $270 \, kJ \cdot mol^{-1}$ (Bell 1991; Bell et al. 1993), respectively, indicate that the reaction rate is indeed dependent on the nature of the solid surfaces present.

Catalysis by Mineral Surfaces. The fact that decarboxylation of acetic acid is catalyzed heterogeneously has important implications for the stability of acetic acid in basin brines that are in intimate contact with sediments. The catalytic abilities of quartz, amorphous silica, calcite, calcium montmorillonite, iron-bearing montmorillonite, pyrite, magnetite, and hematite with respect to the decarboxylation of aqueous acetic acid and sodium acetate have been studied (Bell 1991; Bell et al. 1993). These mineral surfaces are among the most abundant present in sediments and are therefore the surfaces that brines are likely to encounter during migration. The results from experiments involving iron-bearing montmorillonite and hematite were tentatively interpreted as being representative of oxidation rather than decarboxylation and will be discussed in that context. An extremely wide variation in catalytic ability was observed. Quartz, calcite, and pyrite were found to have no effect (above the rate imposed by the reactor vessel walls) on the decarboxylation of acetic acid or acetate (projected half-lives at 100 °C are far in excess of 5 billion years), whereas calcium montmorillonite and magnetite were found to have a profound effect (projected half-lives at 100 °C are 400 and 11 000 years, respectively).

Two reaction series were investigated by Bell et al. (1993): aqueous acetic acid/surface and aqueous sodium acetate/surface. Activation enthalpies and entropies for decarboxylation of acetic acid and sodium acetate are shown in the isokinetic plot in Fig. 1. The dashed and dotted lines represent least-squares fits to the data for acetic acid and acetate, respectively. The statistical validity of isokinetic relationships has been challenged (Banks et al. 1972; Bunnett 1986), particularly when the range of experimental temperatures, and hence the variation in the rate constants, is limited. This limitation is typical of most kinetic studies where a relatively narrow window exists over which reactions can be monitored in the laboratory using conventional experimental techniques (e.g., ca. 30 °C variation in temperature and about $10^3 \, s^{-1}$ in rate constant). However, the long history of studies of acetic acid decomposition has involved flow techniques at temperatures in the range 460 to 900 °C (Bamford and Dewar 1949; Blake and Jackson 1969) and more recently batch methods (Kharaka et al. 1983; Drummond and Palmer 1986; Bell 1991), which embraced the range 200 to 420 °C. Consequently, rate constants vary from $3.7 \, s^{-1}$ to $5.8 \times 10^{-9} \, s^{-1}$, thereby adding significantly to the statistical validity of the linear relationship. Several inferences regarding acetic acid/acetate decarboxylation may be drawn from Fig. 1. The similarity of these two fits suggests that reactions within the two series proceed by a similar mechanism or at least have a similar rate-limiting step. The sequence of reactions with respect to surface type is reversed from the acetic acid series to the sodium acetate series. Finally, the very different isokinetic relationship for acetic acid, as compared to the other acids in Fig. 1, is consistent with the observation that acetic acid (and the other aliphatic monocarboxylic acids) does not have the prerequisite structure necessary to undergo decarboxylation via the homogeneous mechanism applicable to the other acids. Note that for heterogeneously catalyzed decarboxylation of acetic acid, the observed activation parameters, $\Delta H^{\ddagger}_{obs}$ and $\Delta S^{\ddagger}_{obs}$, actually represent combined contributions from the processes of adsorption and formation of the activated complex:

$$\Delta H^{\ddagger}_{obs} = \Delta H_{ads} + \Delta H^{\ddagger}_{act} \tag{15}$$

$$\Delta S^{\ddagger}_{obs} = \Delta S_{ads} + \Delta S^{\ddagger}_{act} \tag{16}$$

and as such correspond to the difference between the energy of the initial or ground state, where the solvated reactant molecule has not yet adsorbed on the catalyst surface, and that of the adsorbed molecule in the activated state.

The Effect of pH on Reaction Rate. Hydrothermal experiments in gold bag liners, as well as stainless steel and titanium vessels, have shown that the rate of decarboxylation of the acidic form of acetic acid is faster than the rate for the anion form (Kharaka et al. 1983; Palmer and Drummond 1986; Bell 1991; Bell et al. 1993). In addition, decarboxylation was found to be first order with respect to either acetic acid or acetate (Bamford and Dewar

1949; Blake and Jackson 1968, 1969; Kharaka et al. 1983; Palmer and Drummond 1986; Bell 1991; Bell et al. submitted). The fact that the acidic and anionic forms of acetic acid have different decarboxylation rates is consistent with the established behavior of other carboxylic acids (Gould 1959; Clark 1969; Richardson and O'Neal 1972). However, in experiments conducted with starting solutions of intermediate pH (6–8), in which both the acid and anion forms were present, a dramatic increase in the overall rate was observed. This observation is of great importance in modeling the stability of acetic acid in sedimentary basin fluids where the pH falls between 4 and 8. A mechanism was proposed (Bell et al. 1993) that describes adequately the pH dependence in the observed first-order rate constant; however, this mechanism requires further verification.

3.1.2 Oxidation

The simplest reaction for the oxidation of acetic acid is represented by Eq. (3) in which the acid was completely oxidized to CO_2. In a geological setting where oxygen fugacity is buffered at very low values by mineral assemblages, it is doubtful that complete oxidation is accomplished. Partial oxidation has been observed in solution with strong oxidants, such as Ce^{4+}, Pb^{4+}, and Mn^{3+} (see references in Sect. 1). Partial oxidation of carboxylic acids on solid surfaces is known to occur, particularly when the supply of oxygen is limited. For example, partial oxidation of isobutyric acid has been reported by Virely et al. (1992). In these experiments, isobutyric acid was converted predominantly to methacrylacrylic acid with lesser amounts of acetone and propene at 385 to 410 °C over Fe–P–O catalysts. The extent of catalysis was shown to depend on the presence of Fe^{3+} within the catalyst. In this process reduction to Fe^{2+} occurs so that, in the absence of added O_2 to rejuvenate the catalyst, the reaction proceeds until the Fe^{3+} sites are expended.

The ability of naturally occurring oxidants such as SO_4^{2-}, Fe^{3+}, and Mn^{3+} associated with mineral surfaces to affect the oxidation of acetic acid has yet to be established. Oxidation of acetate by an electron acceptor such as Fe(III) has been suggested by Surdam et al. (1984):

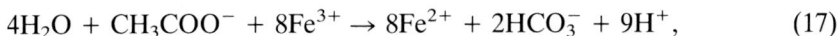

$$4H_2O + CH_3COO^- + 8Fe^{3+} \rightarrow 8Fe^{2+} + 2HCO_3^- + 9H^+, \tag{17}$$

and although the reaction is written as a homogeneous solution phase process, it was suggested that the reaction may actually involve ferric iron present in mineral surfaces. Reduction of iron(III) during acetic acid decomposition was observed by Bell et al. (1993) in hydrothermal experiments in which significant, and in some cases almost complete, conversion of hematite to magnetite occurred. The acetic acid decomposition rate was found to be very rapid initially and then decreased to a rate commensurate with the rate in the presence of magnetite. It was speculated that the rapid initial decomposition rate was related to the reduction of the Fe(III) at the hematite surface via a redox reaction such as:

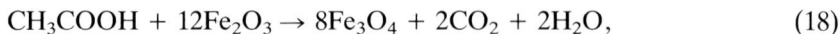

$$CH_3COOH + 12Fe_2O_3 \rightarrow 8Fe_3O_4 + 2CO_2 + 2H_2O, \tag{18}$$

or by partial oxidation of acetic acid to ketene or acetone. Rapid acetic acid decomposition rates were also observed in the presence of synthetic magnetite, synthetic pyrite, and an iron-bearing montmorillonite and were tentatively attributed to similar redox reactions involving iron(III) present either in the structure of the latter or as defects in the structure of the synthetic minerals. An immiscible organic phase was observed and semi-quantitative analyses of gas samples collected at the termination of these experiments showed that, although CO_2 predominated, CH_4 was also present. The presence of CH_4 was attributed to increasing decarboxylation of acetic acid (catalyzed by magnetite in experiments involving hematitie and by the montmorillonite surface in experiments involving iron-bearing montmorillonite) as the redox capacity of the mineral surfaces was exceeded. The immiscible organic phase was attributed tentatively to polymerization/condensation of ketene or acetone intermediates. Detailed quantitative information regarding the nature of the products, both gaseous and solid, is required in order to establish whether, in fact, such redox reactions take place.

The equilibrium constants for oxidation [Eq. (3)] and for decarboxylation [Eq. (4)] have been calculatd by Shock (1988, 1989) for conditions of 100°C and 300 bar and were then used to generate the activity diagrams given in Figs. 2 and 3. This information provides a framework for assessing the relative importance of oxidation and decarboxylation reactions in determining the stability of acetic acid in basin brines. These data show that,

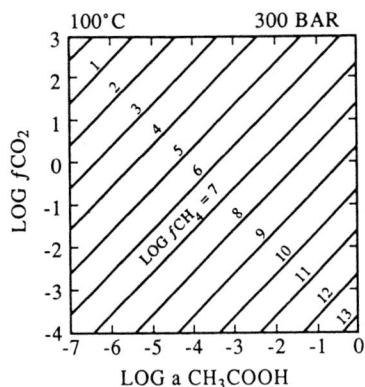

Fig. 2. Activity diagram for the decarboxylation of acetic acid, as given in Eq. (4), at 100°C and 300 bar (Shock 1988), showing $\log(f_{CO_2})$ versus $\log(a_{CH_3COOH})$. The *solid lines* are contours for $\log(f_{CH_4})$. (With permission by E.L. Shock and The Geological Society of America)

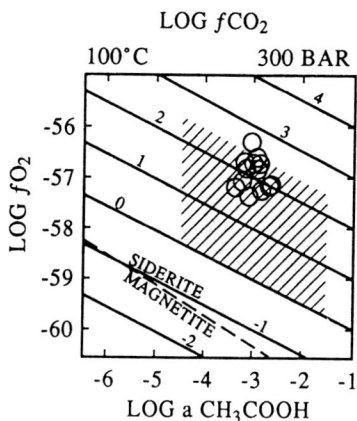

Fig. 3. Activity diagram taken from Shock (1989) for the oxidation of acetic acid to CO_2 and H_2O, as given in Eq. (3), at $100\,°C$ and 300 bar, showing $\log(f_{O_2})$ versus $\log(a_{CH_3COOH})$. The *solid lines* are contours for $\log(f_{CO_2})$. The *dashed line* separates the stability fields of magnetite and siderite. The *ruled area* corresponds to possible reservoir conditions. *Open symbols* represent natural samples for which hypothetical oxygen fugacities were calculated by using values for a_{CH_3COOH} and $a_{CH_3CH_2COOH}$ obtained from analyses of brine samples from the San Joaquin and Gulf Coast Basins (Carothers and Kharaka 1978) in Eq. (24). (With permission by E.L. Shock and The Geological Society of America)

given a sufficient source of oxygen and in the absence of kinetic barriers, oxidation of acetic acid should be favored over decarboxylation (i.e., $K_3 \gg K_4$).

The values for the fugacity of methane (f_{CH_4}) or carbon dioxide (f_{CO_2}) required in order for the observed activities of acetic acid (a_{CH_3COOH}) to be in equilibrium according to Eq. (4) are unobtainable (see Fig. 2) in that they exceed expected total lithostatic pressures (Shock 1988, 1989). Conversely, values of a_{CH_3COOH} predicted for equilibrium with obtainable f_{CH_4} and f_{CO_2} values correspond to concentrations of acetic acid on the order of $10^{-7}\,\mathrm{mol \cdot kg^{-1}}$, which is below the detection limit of conventional ion chromatography. Therefore, acetic acid observed in basin brines has not achieved equilibrium with respect to decarboxylation. This fact provided the driving force for research into the kinetics of the decarboxylation reaction (Kharaka et al. 1983; Palmer and Drummond 1986; Bell 1991; Bell et al. 1993).

On the other hand, geologically reasonable values of f_{O_2} and f_{CO_2} are predicted for equilibrium to be established with respect to oxidation of acetic acid [Eq. (3); Fig. 3]. Therefore, it is possible that acetic acid in basin brines exists in a state of equilibrium with respect to oxidation to CO_2 and H_2O. The analogous oxidation reaction for propionic acid is simply:

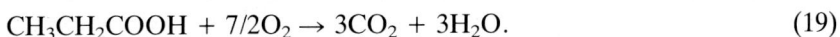

$$CH_3CH_2COOH + 7/2O_2 \rightarrow 3CO_2 + 3H_2O. \tag{19}$$

The fugacity of carbon dioxide in equilibrium with a solution of acetic acid and propionic acid can be determined from the following expressions (Shock 1988, 1989):

$$6\log f_{CO_2} = 3\log a_{CH_3COOH} + 6\log f_{O_2} - 6\log a_{H_2O} - 3\log K_3. \tag{20}$$

$$6\log f_{CO_2} = 2\log a_{CH_3CH_2COOH} + 7\log f_{O_2} - 6\log a_{H_2O} - 2\log K_{19}. \tag{21}$$

These expressions may be equated in the following form:

$$\log f_{O_2} = -2\log a_{CH_3CH_2COOH} + 2\log K_{19}$$
$$+ 3\log a_{CH_3COOH} - 3\log K_3. \tag{22}$$

Oxygen fugacities predicted by Shock (1988, 1989) using acetic and propionic acid data from Carothers and Kharaka (1978) for brines within 5 of 100 °C are shown plotted on an activity diagram for the oxidation of acetic acid in Fig. 4. These yield an approximate straight line with a slope of 3/2 on a plot of $\log a_{CH_3COOH}$ versus $\log a_{CH_3CH_2COOH}$ (Fig. 4) as would be expected for equilibrium with respect to oxidation according to:

$$3CH_3COOH \rightarrow 2CH_3CH_2COOH + O_2 \tag{23}$$

$$\log a_{CH_3CH_2COOH} = 3/2\log a_{CH_3COOH} + 1/2(\log K - \log f_{O_2}), \tag{24}$$

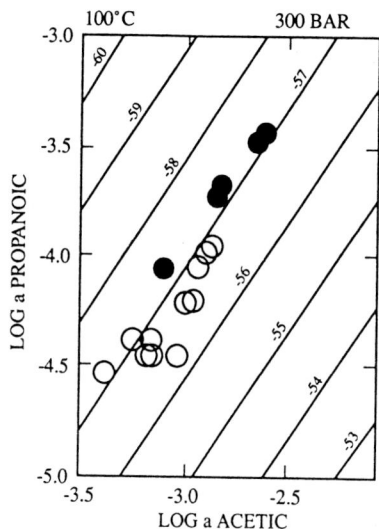

Fig. 4. Activity diagram for the equilibrium between acetic and propionic acids, as given in Eq. (23), at 100 °C and 300 bar (Shock 1989), showing $\log(a_{CH_3CH_2COOH})$ versus $\log(a_{CH_3COOH})$. The *solid lines* are contours of $\log(f_{O_2})$. Both *open* and *solid symbols* represent activities calculated from reported concentrations of acetic and propionic acids in natural brine samples from the San Joaquin and Gulf Coast Basins (Carothers and Kharaka 1978). *Solid symbols* indicate values for which a correlation with iodide is shown in Fig. 7. (With permission by E.L. Shock and The Geological Society of America)

which is Eq. (22) rearranged when $\log K_{19} - 3/2\log K_3 = \log K_{23}$. It appears that acetic and propionic acids in these particular brines from the San Joaquin and Gulf Coast Basins may well be in equilibrium with respect to the oxidation equilibria in Eqs. (3) and (19) (Shock 1988, 1989).

The fact that acetic acid appears to be at equilibrium with respect to the oxidation reaction given in Eq. (3) is consistent with experimental results. Fast reaction rates have been observed for the decomposition of acetic acid in the presence of hematite, magnetite, and iron-bearing montmorillonite and were tentatively attributed to oxidation (Bell et al. 1993). If the oxidation process is indeed rapid and sufficient oxygen is present, detection of acetic and propionic acids in basin brines would be impossible because values of a_{CH_3COOH} and $a_{CH_3CH_2COOH}$ at equilibrium with carbon dioxide would be below the detection limits of ion chromatography.

It has been suggested (Shock 1988, 1989) that, under the hydrothermal conditions prevalent in sedimentary basins, oxidation rather than decarboxylation controls the concentration of monocarboxylic acids. In this scenario, if the net reaction in Eq. (23) was significantly faster than other reactions buffering the oxygen fugacity, this equilibrium could control the oxygen fugacity of the system. The following discussion is hinged on finding viable reaction pathways by which equilibrium [Eq. (23)] could occur rapidly under the constraints imposed by our present understanding of the mechanism for decarboxylation. First, take the case where Eqs. (3) and (19) provide the pathway for equilibration of acetic and propionic acids as given in Eq. (23). Both the forward and reverse reactions of (3) and (19) must then proceed substantially faster than decarboxylation reactions, which will otherwise control the eventual destruction of the acids. Taking acetic acid as an example, k_{-3} [from Eq. (3)] must be greater than k_4 [from Eq. (4)]. The situation is illustrated in the energy profile diagram in Fig. 5. The Gibbs energy of the reactants and products for both the decarboxylation and oxidation of acetic acid are shown on the ordinate. The progress along the hypothetical reaction coordinate (abscissa) of the decarboxylation and oxidation reactions are shown by the solid and dashed lines, respectively. Although the activation energy for the decarboxylation reaction in the forward direction may be larger than that for the forward oxidation reaction, it is certainly much smaller than that of the reverse oxidation reaction, and decarboxylation is therefore expected to control the ultimate destruction of acetic acid.

The above discussion considered the simple oxidation of monocarboxylic acids in which the acids compete for oxygen and proceed through to the products CO_2 and H_2O as shown in Eqs. (3) and (19). In this case, Eqs. (23) and (24) are merely mathematic constructs for calculating the oxygen fugacity when the reactions given in Eqs. (3) and (19) reach equilibrium. Formation of significant amounts of acetic acid by redox equilibria with propionic acid via CO_2 and H_2O, and vice versa, is unlikely if equilibrium for Eqs. (3) and (19) lies far toward products and if the rates of the back

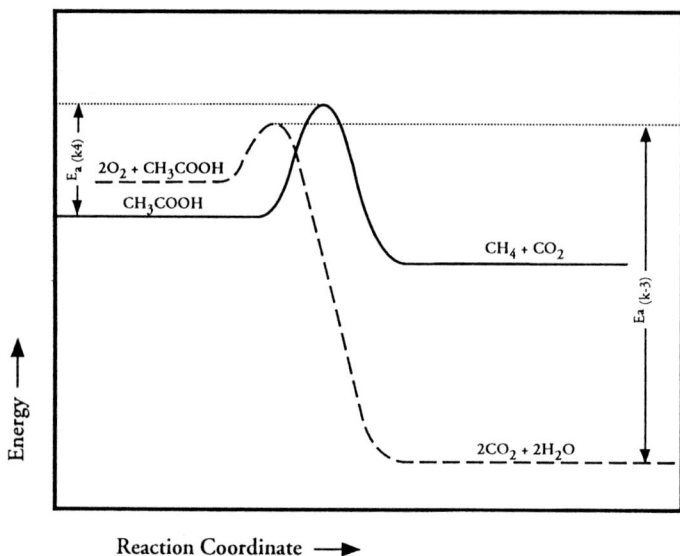

Fig. 5. Schematic reaction profile diagram for the decarboxylation (*solid line*) of acetic acid [Eq. (4)] and for the oxidation (*dashed line*) of acetic acid [Eq. (3)]. The activation energy for the reverse oxidation reaction is E_a (k_{-3}) and for the forward decarboxylation reaction it is E_a (k_4)

reactions, k_{-3} and k_{-19}, are too slow. Significant rates for the back reactions would imply that monocarboxylic acids should be expected in any geologic settings where high pressures of CO_2 and H_2O exist under conditions of very low oxygen fugacity – a relatively common set of conditions under which carboxylic acids have not yet been observed in the clear absence of an organic precursor. Perhaps more significantly, this mechanistic pathway would lead to an equally rapid isotopic equilibrium of the carbon atoms within the acid molecules and possibly to rapid equilibration with adsorbed methane and oxygen. Rapid isotopic equilibration of carbon isotopes between CH_4 and CO_2 is not observed at such low temperatures in natural systems.

An alternative oxidation mechanism (Shock 1988, 1989) would be for the reaction in Eq. (23) to proceed directly without involving CO_2 or H_2O intermediates as required in the coupled oxidation reactions discussed above. The trimolecular interaction of acetic acid molecules is statistically unlikely; instead, the reaction in Eq. (23) would have to occur as a series of reaction steps taking place on a surface or as a chain reaction in solution. Reaction involving cleavage of the carboxyl carbon-alkyl carbon bond in aliphatic monocarboxylic acids is likely to require a catalyst for the same structural reasons as discussed in the context of decarboxylation. Although it is possible that propionate absorbed on a surface could more readily lose an alkyl group to form acetate, it is unlikely that the reverse reaction could occur more rapidly than decarboxylation for the following reason. This reaction requires the breaking of both the carbon-carbon bond and the carbon-oxygen bonds in at least one acetic acid molecule. This is the same carbon-carbon bond that is broken during decarboxylation; an event that is believed to be the rate-limiting step for this process.

If such a "metastable" equilibrium between acetic and propionic acids exists in basin brines, plots of log(acetate, mol $\cdot$ kg^{-1}) versus log(propionate, mol $\cdot$ kg^{-1}) for data from all basins would be expected to display slopes of 3/2. Acetate and propionate concentrations in brines near 100 °C, reported by Carothers and Kharaka (1978) and used by Shock and similar results from other authors, are shown in Fig. 6. Taken as a whole, the results indicate a linear trend. However, it is apparent that for each basin, the data sets rarely display a slope of 3/2. Explanations for the weak overall linear trend, other than the metastable oxidation reaction between acetate and propionate, abound. For example, the overall linear correlation may be an artifact of short-chain carboxylic acid formation, which in turn is a manifestation of the relative abundance of suitable precursor molecules within the parent kerogen. In addition, mixing of brines with meteoric water, or with brines from other hydrologic units, would also result in linear trends on log(acetate)/log(propionate) correlation plots. Conversely, the linear trend may originate from an uneven distribution of kerogen/oil in the host sediment. Specifically, if a cubic meter of sediment contains kerogen with

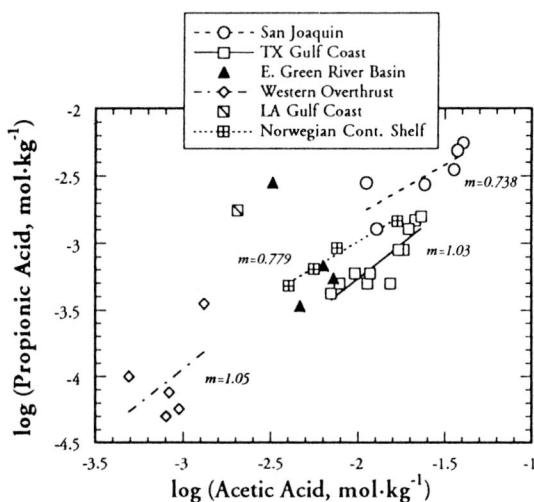

Fig. 6. Correlation plot of log(propionic acid, mol·kg^{-1}) versus log(acetic acid, mol·kg^{-1}) for natural brines within 5 degrees of 100 °C. The data are differentiated by basin: O, San Joaquin (Carothers and Kharaka 1978; MacGowan and Surdam 1988); □, Texas Gulf Coast (Carothers and Kharaka 1978); ▲, eastern Green River (Fisher 1987); ◇, Western Overthrust (Fisher 1987); □, Louisiana Gulf Coast (MacGowan and Surdam 1988); □, Norwegian continental shelf (Barth 1987). Slopes (m) of the linear regressions are given in *italics*

the potential to produce x moles of acetic acid and y moles of propionic acid, and if the mass of kerogen is distributed unevenly with the sediment, brine samples extruded from the sediment will have different concentrations of both acids, but the ratio will be x/y, yielding a linear correlation plot such as shown in Figs. 4 and 6. Furthermore, when plotted together, worldwide acetic/propionic acid ratios are liable to show a weak linear trend of positive slope because any kerogen that forms acetic acid is likely to form propionic acid, and visa versa.

Information with regard to fluid mixing may be obtained from correlation plots for three species, which do not interact significantly with the petrologic environment (Hanor 1987); in this case acetate, propionate, and iodide. Iodide is thought to be produced by the breakdown of iodine-rich marine organic matter (Carothers and Kharaka 1978) and, like bromide, it does not interact significantly with minerals in rocks with which the brines have contact (Hanor 1987). Careful review of the data for brines from the San Joaquin Basin (Carothers and Kharaka 1978) illustrates the point that the hydrologic source of brine samples must be taken under consideration in order to make sense of linear correlations between various dissolved species. Correlation plots between data for acetate, propionate, and iodide from the Temblor production zones in the San Joaquin Basin, California (Carothers and Kharaka 1978) show a distinct correlation indicative of fluid mixing or a heterogeneous kerogen distribution as discussed above (Fig. 7; Table 2). These data are identified by solid symbols in Fig. 4. Unfortunately, very little published data exist for brines that have been analyzed for iodide and carboxylates and even less data are published in which multiple samples were taken from one production zone at temperatures above those where biological action is prevalent.

Polymerization/Condensation Reactions. Immiscible oils were observed on opening autoclaves at the end of acetic acid decomposition experiments that

Table 2. Dilution factors and data from Carothers and Kharaka (1978) for acetate, propionate, and iodide in brines from the Temblor production zone at Coalinga and Kettleman North Dome, California

ID	Temperature (°C)	Acetate (mol·kg^{-1})	Dilution factor	Propionate (mol·kg^{-1})	Dilution factor	Iodide (mol·kg^{-1})	Dilution factor
Coalinga							
Av. 15 wells	54	9.99×10^{-6}	1.81×10^{-4}	2.70×10^{-4}	0.0435	3.94×10^{-5}	0.065
Kettleman NorthDome							
31-18Q	81	5.52×10^{-2}	1.00	6.21×10^{-3}	1.00	6.07×10^{-4}	1.000
33-IP	98	4.04×10^{-2}	0.732	5.52×10^{-3}	0.889	4.49×10^{-4}	0.740
74-13P	95	2.41×10^{-2}	0.437	2.71×10^{-3}	0.436	4.10×10^{-4}	0.675
32-32J	101	3.74×10^{-2}	0.678	4.83×10^{-3}	0.778	3.86×10^{-4}	0.636
61-33J	99	1.28×10^{-2}	0.232	1.27×10^{-3}	0.205	1.65×10^{-4}	0.271
25-33J	107	1.91×10^{-2}	0.346	2.07×10^{-3}	0.333		

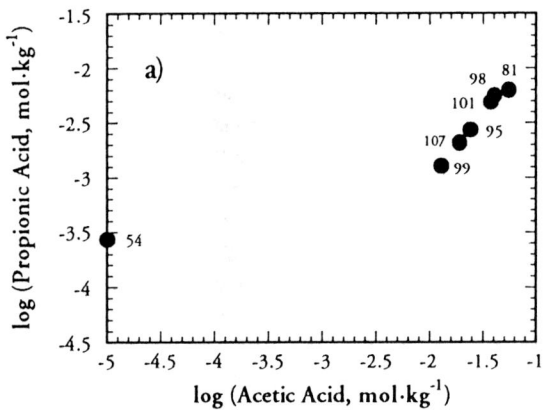

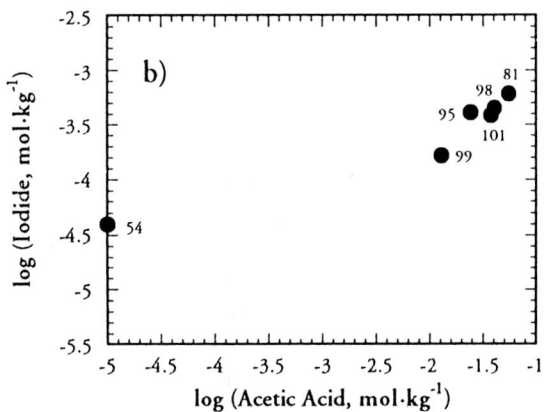

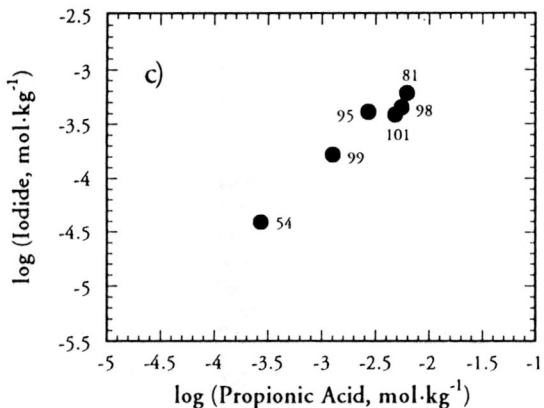

Fig. 7. Correlation plots for brine analysis data for **a** propionic and acetic acids, **b** iodide and acetic acid, and **c** iodide and propionic acid from the Temblor production zones, Coalinga and Kettleman North Dome, California. Data are taken from Carothers and Kharaka (1978). Individual points are labeled with the measured or estimated bore-hole temperatures in °C. Note the similarity in pattern between the plots (with the exception of the 54 °C point for which preferential consumption of acetic acid by bacteria may have been a factor) and compare with the dilution factors given in Table 2

contained materials such as silica, hematite, and magnetite (Kharaka et al. 1983; Palmer and Drummond 1986; Bell 1991; Bell et al. 1993). Obviously, polymerization/condensation-type reactions had occurred, which were able to compete with decarboxylation. It is interesting to note that these organic products must be thermally stable at temperatures in excess of 330 °C in order to have survived the experimental process. The anhydrous ketonization of acetic acid has been found to be a surface-catalyzed reaction (surfaces studied include hematite, magnetite, titanium oxide, manganese oxide, chromium oxide, silica, zeolites, and carbon black) involving a ketene or acetone precursor and forms, depending on catalyst and conditions, a variety of unsaturated alkene and carbonyl products (Hurd and Martin 1929; Kuriacose and Swaminathan 1969; Swaminathan and Kuriacose 1970; Kuriacose and Jewur 1977; Gonzalez et al. 1978; Barteau et al. 1981; Servotte et al. 1985). Similar reactions may take place when experiments are carried out in the presence of water. This is clearly a very significant reaction sequence that may have application to oil formation and implications for Archean prebiotic budgets of organic compounds if it occurs at a significant rate at temperatures below 200 °C.

3.2 Other Aliphatic Monocarboxylic Acids

Propionic and butyric acids have structures similar to acetic acid and are expected to display much the same chemical and kinetic behavior. Virtually no experimental studies of their decarboxylation kinetics in aqueous solutions have been reported, although it is expected that these reactions are also catalyzed heterogeneously. The rate constant for decarboxylation of a $1.1 \, \text{mol} \cdot \text{kg}^{-1}$ solution of n-butyric acid is $4.20 \times 10^{-8} \, \text{s}^{-1}$ (titanium oxide surface at 359 °C), whereas the comparable value for acetic acid is $3.89 \times 10^{-8} \, \text{s}^{-1}$ (Palmer and Drummond 1986). However, there are no corresponding data on the activation parameters for these acids so that the expected similarity of the linear isokinetic plots for these and acetic acid has not yet been tested.

The stability of formic acid has been studied extensively in the gas phase in dynamic experiments, which indicate that it would also be far less stable than acetic acid in aqueous solution (Mars et al. 1963; McCarty et al. 1973; Falconer and Madix 1974; Ai 1977; Benziger and Madix 1979; Barteau et al. 1980; Bowker and Madix 1981a; Sexton and Madix 1981; Miles et al. 1983). Formic acid is either absent, or occurs at very low concentrations in analyses of most basin brines (Zinger and Kravchik 1972; Surdam et al. 1984; Barth 1987; MacGowan and Surdam 1988; Fisher and Boles 1990). Some brines emanating from submarine hydrothermal vents in the Guaymas basin contain formate (formed from decomposition of organic matter such as algae in the vicinity of the vents) at somewhat higher concentrations (Martens et al. 1988; Martens 1990). Formic acid has been produced at concentrations nearing those of acetic acid during hydrous pyrolysis experiments. Presumably, formic acid produced under the above circumstances is sampled shortly after its formation, compared to those samples taken from more mature petroleum brines. In keeping with the lower thermal stability of formic acid compared to acetic acid, formic acid produced in the latter brines may have succumbed to thermal decomposition.

Decomposition of several higher molecular weight acids have been studied. The thermal decomposition of behenic acid (n-C22) was studied (Jurg and Eisma 1964; Eisma and Jurg 1969) as a model substance for the decomposition of fatty acids to form n-alkanes during petroleum production or maturation. The results from the hydrous and anhydrous pyrolysis of behenic acid over a clay catalyst established that decarboxylation yielding an n-C22 alkane was the predominant reaction. However, small amounts of longer carbon-chain species were detected, indicating that side reactions also occurred.

The thermal degradation of gallic, vanillic, phthalic, elegiac, and tannic acids has been shown to be rapid at temperatures ranging from 90 to 250°C, suggesting that these acids are short-lived in deep formation waters (Boles et al. 1988).

4 Stability of Dicarboxylic Acids

The dicarboxylic acids found in basin brines (i.e., oxalic, malonic, and succinic) are expected to be less stable under hydrothermal conditions than monocarboxylic acids of comparable chain lengths. The stability of these acids has been discussed previously to the extent that structural factors make α-, β-, and γ-carboxyl acids susceptible to homogeneous decarboxylation. The mechanisms for decarboxylation of β-carboxyl acids and their derivatives in solvents of varying polarity have been especially well studied and the results are believed to be generally applicable to α- and γ-carboxyl acids as well (Clark 1969). For this reason, the following detailed discussion of the mechanism for homogeneous decarboxylation of dicarboxylic acids is based primarily on malonic acid. Finally, oxidation of dicarboxylic acids may be predicted, although the process has not been well studied.

4.1 Decarboxylation

The decarboxylation of dicarboxylic acids yields carbon dioxide and a monocarboxylic acid of one carbon shorter chain length:

$$HOOCCOOH \rightarrow CO_2 + HCOOH \tag{25}$$

$$HOOCCH_2COOH \rightarrow CO_2 + CH_3COOH \tag{26}$$

$$HCOOCH_2CH_2COOH \rightarrow CO_2 + CH_3CH_2COOH. \tag{27}$$

It is interesting to note that dicarboxylic acids were observed initially during hydrous pyrolysis of kerogen (MacGowan and Surdam 1988). However, as heating continued the formation of monocarboxylic acids mirrored decreases in dicarboxylic acid concentrations – suggesting that a portion of the monocarboxylic acids found in basin brines may be the product of decarboxylation of less stable dicarboxylic acid precursors. Of the three dicarboxylic acids found in basin brines, the decarboxylation of malonic acid has been studied most extensively in hydrous systems and in various other solvent systems.

The mechanism for the homogeneous decarboxylation of malonic acid and its derivatives is thought to depend, at least in part, on the ability of the molecule to form an intramolecular hydrogen bond between the hydroxyl hydrogen of one carboxyl group and an oxygen on the other carboxyl group (Westheimer and Jones 1941; King 1947; Fraenkel et al. 1954; Clark 1958, 1969; Richardson and O'Neal 1972). The resulting six-member cyclic structure facilitates the transfer of the proton upon liberation of CO_2. The rapid decarboxylation of malonic acid derivatives (e.g., dimethylacetoacetic and dimethylmalonic acids), which cannot form an enolic structure $(-C{=}C{-}OH)$, is indicative of the formation of a keto structure $(-C{-}C{=}O)$ in the activated state (Pedersen 1932; Westheimer and Jones 1941; Clark 1969). Rates for the decarboxylation of malonic acid in quinoline/dioxane mixtures were shown to depend on the concentration of the nucleophile, quinoline (Fraenkel et al. 1954; Clark 1958), implying that the activated state results from nucleophilic attack by a solvent molecule [see Eq. (10)]. Similar structures for the activated states of oxalic and succinic acids are possible; i.e., five- and seven-member cyclic structures, respectively, and it is likely that decarboxylation of these acids follows similar mechanisms to that of malonic acid.

Ionization of dicarboxylic acids has been shown to have a profound effect on the decarboxylation rate. The disodium salt of malonic acid $(NaOOCCH_2COONa)$ was found to be relatively stable with respect to decarboxylation (Fairclough 1938) up to temperatures of 125 °C, whereas the monosodium salt decomposed by a first-order reaction. Rates for the decarboxylation of both malonic and oxalic acids were slower in polar solvents in which a high degree of ionization was expected (Richardson and O'Neal 1972). Similarly, the rate of decomposition of dibromomalonic acid was found to be proportional to the concentration of the undissociated acid molecule in solution (Muus 1935).

An extensive study of the rate of malonic acid decarboxylation in aqueous solutions of pH 0.5 to 5 (Hall 1949) yielded a rate expression involving rate constants for both the undissociated species $(HOOCCH_2COOH)$ and the monovalent anion $(HOOCCH_2COO^-)$:

$$\text{Rate} = k_{obs}[a_s] = k_{acid}[a\text{-}x] + k_{anion}[x], \tag{28}$$

where k_{obs} is the observed rate constant at a given stoichiometric concentration of malonic acid, a_s; k_{acid} and k_{anion} are the rate constants for decarboxylation of the acid and monovalent anion species, respectively; [a-x] is the concentration of undissociated malonic acid; and [x] is the concentration of the monovalent anion species. The rate of decarboxylation of the undissociated acid was found to be faster than the rate for the monoprotonated malonate ion (i.e., $k_{acid} > k_{anion}$). The activation parameters for the decarboxylation of malonic acid were found not to change significantly with changes in pH. Extrapolation of the observed rate constants determined by Hall at 80 and 90–100 °C yields an estimated half-life of 3 days for the decarboxylation of $0.01\ mol \cdot kg^{-1}$ malonic acid ($pK_1 = 3.02$, $pK_2 = 6.14$ at 100 °C; Kettler et al. 1992) in a solution of pH 5. Clearly, malonic acid cannot be expected to survive geologically meaningful lengths of time in basin brines. An explanation of the persistence of dicarboxylic acids in basin brines is given Sect. 5.4).

A similar study was carried out by Crossey (1991) on oxalic acid over the pH range 3.7 to 7. The observed rate of oxalic acid decarboxylation was found to slow with increasing pH throughout the range studied. Extrapolation of the rate data determined by Crossey (1991) at 160, 180, and 200–100 °C yields an estimated half-life of 81 years for the decarboxylation of a $0.01\ mol \cdot kg^{-1}$ solution of oxalic acid ($pK_1 = 1.58$, $pK_2 = 4.79$ at 100 °C, Kettler et al. 1991) at pH 5. Like malonic acid, oxalic acid cannot be expected to persist over geologically meaningful times in basin brines.

Enthalpies and entropies of activation for the decarboxylation of oxalic, malonic, and acetic acids are listed in Table 1 and are shown separately on the isokinetic plots in Fig. 8. Linear trends are observed for: (1) aqueous acetic acid and sodium acetate in the presence of various catalysts; (2) aqueous oxalic acid at several pH values; (3) oxalic acid in different solvents; and (4) malonic acid in different solvents and in aqueous solutions having a different pH. Note that the isokinetic trend for the decarboxylation of malonic acid in aqueous solutions at various pH is identical to that for the reaction in nonaqueous solvents, i.e., there is one isokinetic trend for malonic acid. Moreover, the effect of pH on the activation parameters for the decarboxylation of malonic acid in aqueous solution is minimal. On the other hand, the activation data for the decarboxylation of oxalic acid in aqueous solutions determined by Crossey (1991) do not follow the same isokinetic trend as do the corresponding data for this reaction in other solvents. By contrast, activation data calculated from the rate constants determined by Dinglinger and Schroer (1937) for oxalic acid in water (pH 0.5) fall on the isokinetic trend set by the decarboxylation of oxalic acid in nonaqueous solvents, as well as the rate data determined by Lapidus et al. (1964) in the vapor phase. The cause of the disparity between the isokinetic relationships determined by Crossey (1991) and the remainder of the oxalic acid results requires further investigation. The reaction could have been surface-catalyzed, but this is doubtful because some of the oxalic acid

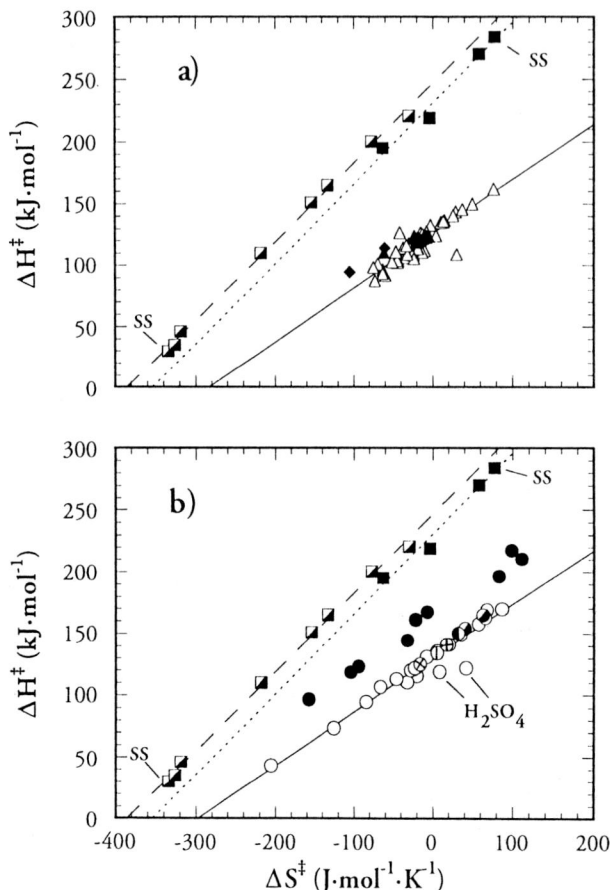

Fig. 8. Isokinetic plots for the decarboxylation of **a** malonic and **b** oxalic acids. Activation data for acetic acid are plotted for spatial reference. *Filled* and *partially filled symbols* represent aqueous systems. Linear least-squares fits to acetic acid, acetate, and the combined remaining data are given by the *dashed, dotted,* and *solid lines,* respectively. Note that with the exception of Crossey's (1991) results, the data for malonic and oxalic acids would superimpose. The data are given in Table 1. The acids and sources are as follows: ▲, malonic (Richardson and O'Neal 1972); △, malonic (Hall 1949); ○, oxalic (Richardson and O'Neal 1972); ●, oxalic (Crossey 1991); ◗, oxalic (Dinglinger and Schroer 1937, 1938; Lutgert and Schroer 1940); ⊗, oxalic vapor (Lutgert and Schroer 1940); ⬙, dideuterio-oxalic (Lutgert and Schroer 1940); ⊕, dideuterio-oxalic vapor (Lapidus et al. 1966a,b); ◖, dideuterio-oxalic in D_2O (Lutgert and Schroer 1940); ◪, acetic (Bell et al. 1993); ■, sodium acetate (Bell et al. 1993); and ♦, gallic (Boles et al. 1988). A more detailed version of this plot appears in Bell et al. (1993)

decarboxylation experiments using other solvents were also performed in Pyrex vessels; however, this possibility should be pursued further. In the absence of reliable rate data at geologically relevant pH, the rate constants reported by Dinglinger and Schroer (1937) at pH 0.5 may be considered as maximum values because the rate is expected to decrease with increasing pH. The half-life calculated from these rate constants at 100 °C is 64 days.

5   Future Experiments

Although many hydrothermal experiments have been undertaken in an effort to delimit the probable rates of acetic acid and oxalic acid decarboxylation in natural settings, the results have merely been sufficient to define the nature of various factors which affect the decarboxylation of the n-C_2 to n-C_4 mono- and dicarboxylic acids. The subjects of the kinetics of oxidation and polymerization/condensation of these species remain virtually untouched. It will not be possible to speculate with certainty about the stability of these acids in basin brines until kinetic information is acquired allowing comparison of decarboxylation, oxidation, and condensation of the acids with the relevant mineral transformation, i.e., the rate at which mineral oxygen buffers can be expected to adjust f_{O_2}/f_{H_2}. The following is a summary of some of the experiments that are necessary to provide such information. Hindsight applied to the previous experimental kinetic studies also provides suggestions for improved approaches to experimental and analytical procedures.

5.1   Analysis of Decomposition Products

From a mechanistic standpoint, and to shed light on many of the controversial issues surrounding this field, there is an overwhelming need to determine qualitatively and quantitatively the products of decarboxylation as a function of time, surface, and temperature. It is necessary to carry out kinetic experiments in a fashion such that both reactants and products can be sampled representatively. This can be accomplished either by sampling a system having a single aqueous phase, such as in a gold bag apparatus, or by quenching a series of capsules that typically contain both liquid and vapor phases. The former method is more indicative of natural, pressurized hydrothermal systems and provides samples unaffected by partitioning to the gas phase, but is of limited use in cases where water-insoluble "oils" are formed.

The principal products of decarboxylation are carbon dioxide and the root alkyl group (or corresponding monocarboxylic acid for decarboxylation of dicarboxylic acids) formed by the simple cleavage of the C−C bond; however, other (usually) minor products have been noted and need to be identified. Typical minor gaseous products are carbon monoxide, alkenes, alkanes, and hydrogen. In the past (Palmer and Drummond 1986; Bell et al. 1993), bulk gas samples taken at the end of an experiment were analyzed by mass spectrometry because this provided a convenient first approach. However, this approach only provided information on the volatile gaseous species and ignored any isoluble organic phases or water-soluble species. In general, mixtures of the complex nonvolatile products can only be identified tentatively by mass spectrometry from their daughter, or fragment, components. Recent advances in ion chromatography provide a convenient method of identifying and quantifying the amounts of carboxylates formed, as well as some other water-soluble organics. Extraction of soluble organics from the aqueous phase and sampling of the water-insoluble organics would permit identification of longer chain molecules by HPLC or GC/MS. These analyses need to be conducted periodically throughout the duration of the experiment.

Ideally, the intermediates and products need to be identified if the reaction mechanisms are to be determined unambiguously. In order to establish a complete picture of the sequence of product formation, the observed or suspected intermediate species need to be studied independently to establish their breakdown pattern and thermal stability. For example, the decomposition sequence for aqueous morpholine, which is used as a pH buffer in the water/steam cycle of many power plants, was studied in this manner and found to involve two independent, complex reaction schemes (one of which yields acetic acid). Each reactive intermediate was identified and its decomposition kinetics studied independently to verify that its decomposition rate was in keeping with the projected overall mechanism for morpholine decom-

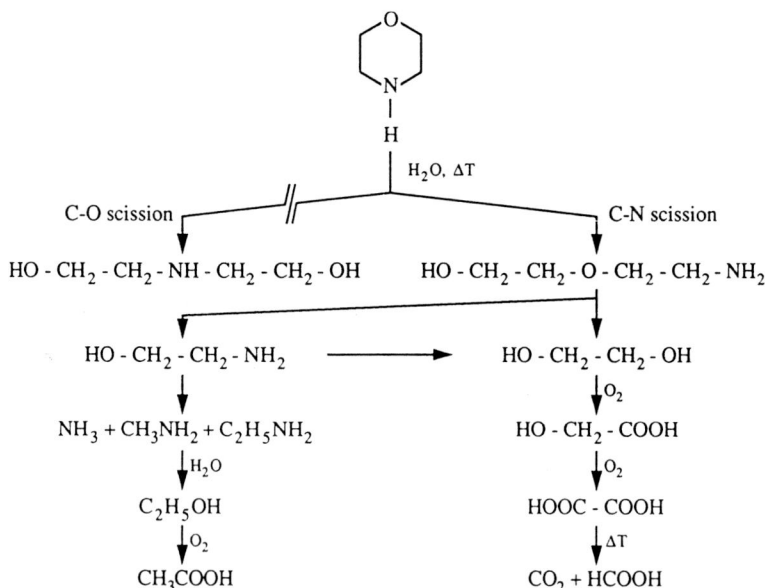

Fig. 9. Reaction scheme for the thermal decomposition of morpholine. (Gilbert et al 1990)

position and that its decomposition products comply with those given in the morpholine scheme (Fig. 9).

As a final point, in view of the existence of competing thermal "decarboxylation" reactions and oxidation pathways, it is essential to purge the solutions and vessel with an inert gas prior to placing the vessel in the furnace. The choice of reaction vessel is also critical and "blank" experiments need to be carried out under each experimental condition to allow for this contribution to the overall process.

5.2   Effect of Surface on Reaction Rate

The field of heterogeneous catalysis is not only vast, but is also evolving rapidly as industry strives to find new, energy-efficient pathways for chemical production, e.g., great emphasis has been focused on the water-gas shift reaction. The following discussion is couched in empirical terms without the pretense of a thorough review of this field.

One important empirical observation is the existence of so-called volcano curves (Balandin 1969), which describe the variation in the efficiency of a catalyst as the nature of the surface is changed. It is assumed that in the majority of cases resulting in heterogeneous catalysis, adsorption is chemical rather than physical and that it is the strength of the adsorbate-surface bond (i.e., chemisorption) that determines, or at least strongly affects, reaction rate. Typically, as the adsorption of the reactant molecule onto the surface becomes stronger, the rate of reaction increases to a maximum beyond which the adsorption is so strong that the reactant is no longer readily released and the reaction rate decreases. Conversely, if chemisorption is weak, the reactant may not be sufficiently perturbed to exhibit significant catalysis. Thus, dependence of the reaction rate on chemisorption is not a simple linear function, but may have a volcano-shaped form. This observation provides one explanation for the existence of the completely different effects of stainless steel surfaces on the decarboxylation rates of acetic acid and acetate (see Table 1). Stainless steel presents the most effective catalytic surface studied to date for the decarboxylation of acetic acid and yet it has a minimal effect on the rate of acetate decarboxylation (Table 1; Fig. 8). Acetate may be so strongly adsorbed on such a surface as to reduce the catalytic effect.

The large variations in rate constant observed for decarboxylation of acetic acid and acetate, the very low activation energies observed by Kharaka et al. (1983) in their pioneering work on this reaction in stainless steel vessels, the observation of first- and zero-order kinetics with respect to acetic acid concentration depending on surface type (Palmer and Drummond 1986), and the reversal in the trend in the values for ΔH^+ for acetic acid and acetate over a number of different surfaces attest to the surface playing a vital role in controlling the stability of acetic acid (and probably other aliphatic monocarboxylic acids) in aqueous and anhydrous systems. In view of the obvious existence of a dominant heterogeneously catalyzed pathway, it is perhaps puzzling that, with the exception of the reaction of acetic acid over pyrite (Bell 1991), no surface area dependence has been observed; although similar observations are not uncommon in other heterogeneous catalytic systems. Even in the case of pyrite, the effect was found to be substantially less than first-order in surface area. The answer has to lie in one of two mechanisms. The first possibility is that the number of surface-active sites in the systems studied to date far exceed the number occupied by adsorbed substrate so that, unless the surface area is reduced to minimal levels, no effect is observable. Conversely, zero-order kinetics with respect to adsorbate concentration are expected if the surface area is reduced to sufficiently low values. In practical situations, reducing the surface area of a desired catalyst to a minimum usually means that the catalytic effect of the vessel walls overshadows the desired result. The second possibility is that the adsorption process and rearrangement of the substrate on the surface site are slow, whereas reaction at that site is fast so that, as long as a minimum number of such sites exist, no surface area effect will be observed. Based on limited experiments involving acetic acid solutions in contact with a commercial magnetite, where zero-order kinetics were observed (Palmer and Drummond 1986), it appears that the former mechanism is operative. It was postulated (Bell et al. 1993) that a limited number of either impurities or surface defects acted as active sites in this particular magnetite sample and, because these were all occupied by acetic acid molecules, no dependence on the acetic acid concentration was observed.

More experiments of this nature are required; however, future experiments should be performed with careful characterization of the surface before and after exposure to the carboxylic acid in order to identify the nature of the catalytic site. This information would be invaluable in understanding the detailed mechanism of adsorption, particularly the role of defects as sites of specific activity. In natural systems, where the concentrations of carboxylic acids are lower and the surface area much larger than in these laboratory studies, the conventional first-order reaction rate can be anticipated to operate. There are many methods of varying sophistication for determining the surface area of materials (Satterfield 1980; Boudart and Djéga-Mariadassou 1984; Hubbard 1989; Minachev and Shpiro 1990). The question always arises whether a particular method accesses the entire surface (pore size restrictions, etc.) and whether the active sites are uniformly distributed over a given surface. There is also the problem of recrystallization of the catalytic surface during the experiment, not only changing the surface area, but perhaps occurring preferentially at dislocations, vacancies, etc. that are the catalytically active sites. Thus, physical measurements of surface area need not necessarily correlate with surface activity.

Another aspect of surface area effects deals with the poisoning of catalytically active sites by competing adsorbates. The species most pertinent to oil-field brines would be chloride ion. Indeed, there is evidence that chloride, bromide, and iodide inhibit the adsorption of aromatics on platinum surfaces (Hubbard 1989). If species such as chloride ion compete with carboxylic acids for reactive surface sites, some degree of surface pacification appears likely, because chloride can be present in large excess over organic acids in natural brines. The adsorption of ions is pH-dependent, and this pH profile will vary from one type of surface to another depending on the zero-point charge (zpc) for that material. Moreover, nothing is known about the behavior of the zpc at such extreme temperatures. Thus, in any detailed mechanistic study, attention should be paid to such surface properties as defects, impurities, and the adsorbed cation/anion profile as a function of pH and temperature. Note that the effect of changes in the ionic strength of the solution on the rate constant for decarboxylation by addition of salt will be trivial, particularly as little charge development occurs in forming the transition state, i.e., the activity coefficient changes are generally small compared to the accuracy of the rate measurement.

Presently, it is not known whether dissolution of a surface layer serves to create or remove active catalytic sites. In earlier work (Palmer and Drummond 1986), acetic acid solutions were exposed to a well-crystallized (yellow) titanium oxide surface and significantly slower kinetics were observed; although with continuous use the surface returned to the typical "blue" oxide sheen, and the kinetics of decarboxylation reverted to the faster rates observed originally. Acetate forms strong complexes with transition metal ions, particularly at elevated temperatures (Palmer and Drummond 1988; Giordano and Drummond 1991). The stability of these complexes is directly dependent on acetate concentration, pH, and the inverse of ionic strength.

5.3 pH Effects

In previous acetic acid decarboxylation experiments, the pH was controlled by the acetic acid/acetate buffer so that, with the preferential consumption of acetic acid observed in these experiments, it is obvious that the pH varied considerably during the course of decarboxylation (Bell 1991). In oxalic acid decarboxylation studies (Crossey 1991), the less reactive acetic acid/acetate buffer was used to control pH; although even in this case, some significant increases in pH (0.5 units) occurred during a given experiment. It would be of interest to conduct a number of experiments of this type in a "nonreactive buffer mixture", such as CO_2/CHO_3^-, to gauge the pH effect independent of the reactive substrate. In addition, in light of the pH dependence of k_{obs} for malonic acid shown in Eq. (28) (Hall 1949), it would be constructive to carry out a similar study of monocarboxylic acids using a wider range of pH values than investigated previously (Palmer and Drummond 1986).

5.4 Effect of Carboxylate/Metal Complexing

Metal ions that enter into complexation equilibria with carboxylic acids may affect the decarboxylation rate of the coordinated acid. The rate may be retarded or enhanced, depending on the acid and metal involved. For example, the decarboxylation of nitroacetate is retarded by complexation (Pedersen 1927, 1949; Finkbeiner and Stiles 1963), whereas the decarboxylation of both oxalacetic and dimethyloxalacetic acids is promoted by complexation with metal ions (Steinberger and Westheimer 1951; Gelles 1956; Kosicki and Kipovac 1964; Tsai 1967). This may be an especially important aspect for the stability of dicarboxylic acids, which have marginal thermal stabilities and form complexes that are generally orders of magnitude more stable than their monodentate counterparts. The fact that the unprotonated forms of carboxylic acids generally react significantly slower than the protonated forms, suggests that complexation of anionic form may render them relatively inert. Therefore, studies of the decarboxylation of complexed oxalate in homogeneous solutions [e.g., aluminum oxalate, $Al(C_2O_4)^+$ has a formation constant at $25\,^{\circ}C$ of ca. $10^6\,kg \cdot mol^{-1}$, so that virtually all the oxalate in solution would be in the complexed form] would be of interest.

Ion chromatographic analyses of carboxylic acids in brine samples would not be affected by complexation if the complex is destroyed either during the original sampling process (e.g., destabilization on cooling), or by interaction with the eluent (e.g., change in pH) during analysis. On the other hand, strong complexes of the type formed between highly charged metal ions and polycarboxylates, may readily survive these events and result in smaller, or negligible, peaks.

5.5 Isotope Studies

Kharaka et al. (1983) performed one experiment at $300\,^{\circ}C$ in a stainless steel vessel where the carbon isotopic composition of the products, CH_4 and CO_2, of acetic acid decarboxylation was monitored with time. These results confirm that the lighter $^{12}C-^{12}C$ bonds are broken preferentially. It would be of interest to conduct similar experiments using enriched ^{13}C isotopes to verify whether the methane and carbon dioxide are actually produced in isotopic equilibrium with each other. The reaction mechanisms presented above would suggest that this is not the case. Indeed, conducting such experiments over other catalytic surfaces where more complex products are formed would provide a valuable mechanistic tool for determining the extent to which the alkyl groups are stripped of their hydrogen atoms in the adsorbed activated state. Finally, the proposed rapid oxidation equilibrium between acetic and propionic acids [Eq. (23)] must provide a path for isotopic scrambling of the carbon atoms involved within these acids. This type of experiment may help to address the validity of the proposed rapid redox couple, which is otherwise difficult to confirm by conventional thermal experiments within the restricted lifetime of laboratory experiments.

6 Conclusions

The existence of the n-C_2 to n-C_4 mono- and dicarboxylic acids in hydrothermal sedimentary environments depends upon the rates of their production and the rates of decomposition and/or oxidation. These two classes of acids exhibit very different rates and mechanisms of decarboxylation. Decarboxylation of acetic acid, and probably of other aliphatic monocarboxylic acids, proceeds by a heterogeneous catalytic mechanism apparently very different from the homogeneous mechanism for decarboxylation of the dicarboxylic acids. Due to the limited amount of experimental information regarding the kinetics of oxidation or condensation for both classes of acids, no definitive mechanistic trends can be postulated for this process. Nevertheless, it is possible to place constraints on the kinetics and mechanism for the oxidation reaction(s) if this process were assumed to control the ultimate decomposition of acetic acid. Results from studies of mono- and dicarboxylic acid decarboxylation are summarized below.

1. In the absence of an effective catalytic surface, acetic acid may be expected to survive indefinitely (>5 billion years). Potential catalysts studied to date include: stainless steel, titanium oxide, gold, Pyrex, fused quartz, quartz, pyrite, calcium montmorillonite, an iron-bearing montmorillonite, calcite, hematite, and magnetite. The extremely wide range of catalytic abilities is manifested by the activation energies for the decarboxylation of acetic acid, varying from the most active surface, stainless steel, $34 \, \text{kJ} \cdot \text{mol}^{-1}$, to the least active surface, titanium oxide, $170 \, \text{kJ} \cdot \text{mol}^{-1}$. Only stainless steel, magnetite, Pyrex, and calcium montmorillonite demonstrated appreciable catalytic activity. The decomposition rate in the presence of hematite and iron-bearing montmorillonite was rapid, but the kinetics were complicated by competing reactions (possibly oxidation) and by alteration of the mineral surfaces.

2. The kinetics of decarboxylation of dicarboxylic acids appear to be relatively unaffected by the nature of the containment vessel, suggesting that it proceeds via a homogeneous reaction involving interaction with a nucleophilic solvent molecule. Consistent with this assignment of a common mechanism is the existence of similar isokinetic relationships for this class of acid.

3. Reaction rates for the decarboxylation of dicarboxylic acids are extremely fast in comparison with even the fastest catalyzed rate for acetic acid decarboxylation. It is unlikely that uncomplexed acids (at least in the case of oxalic and malonic acids for which kinetic data exist) are able to survive geologically meaningful lengths of time. These results conflict with the fact

that oxalic and malonic acids have been observed in basin brines. Unless the acids are formed continuously, during, or after sampling (perhaps from hydrolysis of esters in petroleum), they must have some degree of thermal stability. It is suggested that the acids may have been stabilized by complexation with metal ions.

4. The kinetics of monocarboxylic acid decarboxylation are first-order with respect to total carboxylate concentration. The rate constants for the acid form are greater than for the anionic species, whereas the rate in solutions of intermediate pH is faster than either of the end-member rates.

5. Decarboxylation of dicarboxylic acids is shown to be first-order with respect to total carboxylate and, like monocarboxylic acids, it is found that the rate for the acid form is faster than for the anionic form. However, unlike monocarboxylic acids, the overall rate in solutions of median pH is a linear function of the end-member rates.

6. The kinetics for the decarboxylation of the monocarboxylic acids can be used to place constraints on the kinetics of the oxidation of monocarboxylic acids and their interrelated equilibria if the oxidation process is to control the ultimate fate of the acids. Only in the event that the equilibrium given in Eq. (23) does not involve CO_2 and water intermediates, but rather proceeds by some surface-catalyzed process in which the forward and back reaction(s) are both faster than the decarboxylation reaction, will decarboxylation not control the stability of monocarboxylic acids in basin brines.

7. The experimental observations of polycondensation reactions involving aqueous acetic acid open an important area for future research. Thermally induced reactions of this type may involve similar catalytic pathways to the oxidation process in (6).

8. Directions for future research initiatives are suggested that could provide the information regarding the kinetics and detailed mechanisms of both oxidative and decarboxylative decomposition of the mono- and dicarboxylic acids. Clearly, the stability of the acids is affected by interaction with the inorganic species, both aqueous ions and solid mineral surfaces, present in the sedimentary environment.

Acknowledgments. Financial support for this work was provided by the Office of Basic Energy Sciences of the US Department of Energy under contract DE-AC05-84OR21400 with Martin Marietta Energy Systems, Inc. The original literature survey was carried out with financial support of the National Science Foundation under grant number EAR-8903750. The authors are grateful to D.J. Wesolowski for many insightful comments.

References

Adams LJ, Franzus B, Huang TT-S (1978) On the decarboxylation of oxalic acid solutions. Int J Chem Kinet 10: 669–675

Adinarayana M, Sethuram B, Navaneeth Rao T (1975) Kinetics of Ag$^+$ catalysed oxidative decarboxylation of some organic acids by Ce^{4+} in H$_2$SO$_4$ medium. Curr Sci 44:
581–583

Adinarayana M, Saiprakash PK, Sethuram B, Navaneeth Rao T (1976) Kinetics of Ag$^+$
catalysed oxidative decarboxylation of acetic acid by Ce^{4+} in H$_2$SO$_4$ medium. J Indian
Chem Soc LIII: 255–257

Ai M (1977) Activities for the decomposition of formic acid and the acid-base properties
of metal oxide catalysts. J Catal 50: 291–300

Al-Owais AA, Ballantine JA, Purnell JH, Thomas JM (1986) Thermogravimetric study of
the intercalation of acetic acid and of water by Al^{3+}-exchanged montmorillonite. J Mol
Catal 35: 201–212

Anderson JM, Kochi JK (1970a) Manganese(III) complexes in oxidative decarboxylation
of acids. J Am Chem Soc 92: 2450–2460

Anderson JM, Kochi JK (1970b) Silver(I)-catalyzed oxidative decomposition of acids by
peroxydisulfate. The role of silver(II). J Am Chem Soc 92: 1651–1659

Atkins PW (1982) Physical chemistry. Freeman, San Fransisco, 1095 pp

Avery NR, Toby BH, Anton AB, Weinberg WH (1982) Decomposition of formic acid on
Ru(001): an EELS search for a formic anhydride intermediate. Surface Sci 122:
L574–L578

Balandin AA (1969) Modern state of the multiplet theory of heterogeneous catalysis. Adv
Catal Related Subjects 19: 1–210

Baldi G, Goto S, Chow C-K, Smith JM (1974) Catalytic oxidation of formic acid in water.
Intraparticle diffusion in liquid-filled pores: Ind Eng Chem Process Design Dev 13:
447–452

Bamford CH, Dewar JS (1949) The thermal decomposition of acetic acid. J Chem Soc
(B): 2877–2882

Banks BEC, Damjanovic V, Vernon CA (1972) The so-called thermodynamic compensation law and thermal death. Nature 240: 147–148

Barteau MA, Bowker M, Madix RJ (1980) Acid-base reactions on solid surfaces: the
reactions of HCOOH, H$_2$CO, and HCOOCH$_3$ with oxygen on Ag(110). Surface Sci
94: 303–322

Barteau MA, Bowker M, Madix RJ (1981) Formation and decomposition of acetate
intermediates on the Ag(110) surface. J Catal 67: 118–128

Barth T (1987) Quantitative determination of volatile carboxylic acids in formation waters
by isotachophoresis. Anal Chem 59: 2232–2237

Bell JLS (1991) Acetate decomposition in hydrothermal solutions. PhD Thesis, The
Pennsylvania State University, University Park, 228 pp

Bell JLS, Palmer DA, Barnes HL, Drummond SE (1993) Thermal decomposition of
acetate. Part III. Catalysis by mineral surfaces. Geochim Cosmochim Acta (in press)

Benziger JB, Madix RJ (1979) The decomposition of formic acid on Ni(100). Surface Sci
79: 394–412

Benziger JB, Madix RJ (1980) Reactions and reaction intermediates on iron surfaces. II.
Hydrocarbons and carboxylic acids. J Catal 65: 49–58

Benziger JB, Schoofs GR (1984) Influence of absorbate interactions on heterogeneous reaction kinetics. Formic acid decomposition on nickel. J Phys Chem 88: 4439–
4444

Benziger JB, Ko EI, Madix RJ (1979) The decomposition of formic acid on W(100) and
W(100)-(5x1) C surfaces. J Catal 58: 149–153

Bingham FT, Sims JR, Page HL (1965) Retention of acetate by montmorillonite. Soil Soc
Proc, pp 670–672

Blake PG, Jackson GE (1968) The thermal decomposition of acetic acid. J Chem Soc (B):
1153–1155

Blake PG, Jackson GE (1969) High- and low- temperature mechanisms in the thermal
decomposition of acetic acid. J Chem Soc (B): 94–96

Boles JS, Crerar A, Grissom G, Key TC (1988) Aqueous thermal degradation of gallic acid. Geochim Cosmochim Acta 52: 341–344

Bond GC (1987) Heterogeneous catalysis. Principles and applications, 2nd edn. Oxford Chemistry Series, Oxford, 176 pp

Bos U, Herzog W, Leupold E-I (1980) Die Reaktionen von Acetaldehyd, Ethanol und Essigsäure an einem Rhodium/Kieselsäure-Katalysator. Ber Bunsen-Ges Phys Chem 84: 182–186

Boudart M, Djéga-Mariadassou G (1984) Kinetics of heterogeneous catalytic reactions. Princeton University Press, Princeton, 222 pp

Bowker M, Madix RJ (1981a) XPS, VPS, and thermal desorption studies of the reactions of formaldehyde and formic acid with the Cu(110) surface. Surface Sci 102:542–565

Bowker M, Madix RJ (1981b) The adsorption and oxidation of acetic acid and acetaldehyde on Cu(110). Applications Surface Sci 8:299–317

Bowker M, Houghton H, Waugh KC (1983) The interaction of acetaldehyde and acetic acid with the ZnO surface. J Catal 79: 431–444

Brown MA (1951) The mechanism of thermal decarboxylation. Q Rev Chem Soc Lond 5: 131–146

Brown RA (1962) The linear enthalpy-entropy effect. J Org Chem 27: 3015–3026

Buckingham DA, Clark C (1982) Metal-hydroxide promoted hydrolysis of activated esters. Hydrolysis of 2,4-dinitrophenylacetate and 4-nitrophenylacetate. Aus J Chem 35: 431–436

Bunnett JF (1986) From kinetic data to reaction mechanism. In: Bernasconi CF (ed) Investigation of rates and mechanisms of reactions. Part 1. Techniques of chemistry, vol VI. Wiley, New York, pp 251–372

Carothers WW, Kharaka YK (1978) Aliphatic acid anions in oil-field waters – implications for origin of natural gas. Am Assoc Pet Geol Bull 62: 2441–2453

Child WC Jr, Hay AJ (1963) The thermodynamics of the thermal decomposition of acetic acid in the vapor phase. J Am Chem Soc 86: 182–187

Clark LW (1957) The kinetics of the decomposition of oxalic acid in non-aqueous solutions. J Phys Chem 61: 699–701

Clark LW (1958) The effect of quinoline and its derivatives on malonic acid. J Phys Chem 62: 500–502

Clark LW (1963) The decarboxylation of oxalic acid in cresols and glycols. J Phys Chem 67: 1355–1358

Clark LW (1969) The decarboxylation reaction. In: Patai S (ed) The chemistry of carboxylic acids and esters. The chemistry of functional groups series. Wiley, New York, pp 589–622

Clavilier J, Sun SG (1986) Electrochemical study of the chemisorbed species formed from formic acid dissociation at platinum single crystal electrodes. J Electroanal Chem 199: 471–480

Conner WC (1982) A general explanation for the compensation effect: the relationship between ΔS^+ and activation energy. J Catal 78: 238–246

Crossey LJ (1991) Thermal degradation of aqueous oxalate species. Geochim Cósmochim Acta 55: 1515–1527

Davies G (1989) Correlation of activation parameters and the case for substitution controlled redcution of $CoOH_{aq}^{2+}$ and $Co(NH_3)_2OH_{aq}^{2+}$. Implications for electrocatalysis by aquocobalt(III) and other strongly oxidizing metal species. Inorg Chim Acta 60: 83–86

Demorest M, Mooberry D, Danforth JD (1951) Decomposition of ketones and fatty acids by silica-alumina composites. Ind Eng Chem 43: 2560–2572

Dewar MJ, Krull KL (1990) Acidity of carboxylic acids: due to delocalization or induction? J Chem Soc Chem Commun: 333–334

Dinglinger A, Schroer E (1937) The kinetics of the thermal decomposition of oxalic acid in solution. Z Phys Chem A179: 401–426

Dinglinger A, Schroer E (1938) Supplement to the investigation of the thermal decomposition of oxalic acid in solution: decomposition in aqueous solutions at 99.4°. Z Phys Chem A181: 375–378

Drummond SE, Palmer DA (1986) Thermal decarboxylation of acetic acid. Part II. Boundary conditions for the role of acetate in the primary migration of natural gas and the transportation of metals in hydrothermal systems. Geochim Cosmochim Acta 50: 825–833

Eisma E, Jurg JW (1969) Fundamental aspects of the generation of petroleum. In: Egington G, Murphy MTJ (eds) Organic chemistry. Methods and results. Springer, Berlin Heidelberg New York, pp 676–698

Exner O (1964a) Concerning the isokinetic relationship. Nature 201: 488–490

Exner O (1964b) On the enthalpy-entropy relationship. Collect Czech Chem Commun 29: 1094–1113

Exner O (1970) Determination of the isokinetic temperature. Nature 227: 366–367

Exner O (1972) Statistics of the enthalpy-entropy relationship. I. The special case. Collect Czech Chem Commun 37: 1425–1444

Exner O, Beránek V (1973) Statistics of the enthalpy-entropy relationship. II. The general case. Collect Czech Chem Commun 38: 781–798

Fairclough R A (1938) Kinetics of decarboxylation of certain organic acids. J Chem Soc: 1186–1190

Falconer JL, Madix RJ (1974) The kinetics and mechanism of the autocatalytic decomposition of HCOOH on clean Ni(110). Surface Sci 46: 473–504

Finkbeiner HL, Stiles M (1963) Chelation as a driving force in organic reactions. α-Nitro acids by control of the carboxylation-decarboxylation equilibrium. J Am Chem Soc 85: 616–622

Fisher JB (1987) Distribution and occurrence of aliphatic acid anions in deep subsurface water. Geochim Cosmochim Acta 51: 2459–2468

Fisher JB, Boles JR (1990) Water-rock interaction in Tertiary sandstones, San Joaquin Basin, California, USA: diagenetic controls on water composition. Chem Geol 82: 83–101

Fraenkel G, Belford RL, Yankwich PE (1954) Decarboxylation of malonic acid in quinoline and related media. J Am Chem Soc 76: 15–18

Galwey AK (1977) Compensation effect in heterogeneous catalysis. Adv Catal 26: 247–322

Galwey AK, Brown ME (1979) Compensation parameters in heterogeneous catalysis. J Catal 60: 335–338

Gardiner WC Jr (1969) Rates and mechanisms of chemical reactions. Benjamin/Cummings, Menlo Park, 284 pp

Gelles E (1956) Kinetics of the decarboxylation of oxalacetic acid. J Chem Soc: 4736–4739

Gilbert R, Lamarre C, Dunbar Y, MacNeil CK, Eatock JW (1990) Identification and distribution of the morpholine breakdown products in different steam-condensate cycles of CANDU-PHW nuclear power plants. In: Riddle JM, Passell T (eds) Workshop on use of amines in conditoning steam water circuits. Electric Power Research Institute, Tampa, FL, pp 19-1–19-25

Giordano TH, Drummond SE (1991) The potentiometric determination of stability constants for zinc acetate complexes in aqueous solutions to 295°C. Geochim Cosmochim Acta 55: 2401–2415

Goddard JD, Yamaguchi Y, Schaefer HF III (1992) The decarboxylation and dehydration reactions of monomeric formic acid. J Chem Phys 96: 1158–1166

Gonzalez F, Munuera G, Prieto JA (1978) Mechanism of ketonization of acetic acid on anatase TiO_2 surfaces. J Chem Soc Faraday Trans 74: 1517–1529

Gould ES (1959) Mechanism and structure in organic chemistry. Holt, New York, pp 346–353

Hall GA (1949) The kinetics of the decomposition of malonic acid in aqueous solution. J Am Chem Soc 71: 2691–2693

Hanor JS (1987) Origin and migration of subsurface, sedimentary brines. Soc Econ Paleontol Mineral Short Course Notes 21, 247 pp

Hanor JS, Workman AL (1986) Dissolved fatty acids in Louisiana oil-field brines. Appl Geochem 1: 37–46

Hayden BE, Prince K, Woodruff DP, Bradshaw AM (1983) An IRAS study of formic acid and surface formate adsorbed on Cu(110). Surface Sci 133: 589–604

Hubbard AT (1989) Structure of the solid-liquid interface. In: Compton R G (ed) Comprehensive chemical kinetics, reactions at the liquid-solid interface, vol 41. Elsevier, New York, 286 pp

Hurd CD, Martin KE (1929) Ketene from acetic acid. J Am Chem Soc 51: 3614–3617

Imamura S-I, Hirano A, Kawabata N (1982a) Wet oxidation of acetic acid catalyzed by Co-Bi complex oxides. Ind Eng Chem Product Res Dev 21: 570–575

Imamura S, Matsushige H, Kawaabata N, Inui T, Takegami Y (1982b) Oxidation of acetic acid on Co-Bi composite oxide catalysts. J Catal 78: 217–224

Jurg JW, Eisma E (1964) Petroleum hydrocarbons: generation from fatty acid. Science 144: 1451–1452

Kawamura K, Kaplan IR (1987) Dicarboxylic acids generated by thermal alteration of kerogen and humic acids. Geochim Cosmochim Acta 51: 3201–3207

Kawamura K, Tannenbaum E, Huizinga BJ, Kaplan IR (1986) Volatile organic acids generated from kerogen during laboratory heating. Geochem J 20: 3201–3207

Kettler RM, Palmer DA, Wesolowski DJ (1991) Dissociation quotients of oxalic acid in aqueous sodium chloride media to 175 °C. J Solution Chem 20: 905–927

Kettler RM, Wesolowski DJ, Palmer DA (1992) Dissociation quotients of malonic acid in aqueous sodium chloride media to 100 °C. J Solution Chem 21: 883–900

Kharaka YK, Carothers WW, Rosenbauer RJ (1983) Thermal decarboxylation of acetic acid: implications for origin of natural gas. Geochim Cosmochim Acta 47: 397–402

Kharaka YK, Maest AS, Carothers WW, Law LM, Lamothe PJ, Fries TL (1987) Geochemistry of metal-rich brines from central Mississippi salt dome basin, USA. Appl Geochem 2: 543–561

Kim KS, Barteau MA (1988) Pathways for carboxylic acid decompositions on TiO_2. Langmuir 4: 945–953

King JA (1947) A new synthesis of dl-serine. J Am Chem Soc 69: 2738–2741

Kochi JK (1965a) The mechanism of oxidative decarboxylation with lead(IV) acetate. J Am Chem Soc 87: 1811–1812

Kochi JK (1965b) Formation of alkyl halides from acids by decarboxylation with lead(IV) acetate and halide salts. J Org Chem 22A: 3265–3271

Kosicki GW, Kipovac SN (1964) The pH and pD dependence of the spontaneous and magnesium-ion-catalyzed decarboxylation of oxalacetic acid. Can J Chem 42: 403–415

Kraeutler B, Bard AJ (1978) Heterogeneous photocatalytic decomposition of saturated carboxylic acids on TiO_2 powder. Decarboxylative route to alkanes. J Am Chem Soc 100: 5985–5992

Kreevoy MM, Truhlar DG (1986) Transition state theory. In: Bernasconi CF (ed) Investigation of rates and mechanisms of reactions, Part 1. Techniques of chemistry, vol VI. Wiley, New York, pp 13–96

Krug RR (1980) Detection of the compensation effect (Θ rule). Ind Eng Chem Fundamentals 19: 50–59

Kuriacose JC, Jewur SS (1977) Studies on the surface interaction of acetic acid on iron oxide. J Catal 50: 330–341

Kuriacose JC, Swaminathan R (1969) Studies on the ketonization of acetic acid on chromia. I. The adsorbate-catalyst interaction. J Catal 14: 348–354

Laidler KJ, King MC (1983) The development of transition state theory. J Phys Chem 87: 2657–2664

Lapidus G, Barton D, Yankwich PE (1964) Kinetics and stoichiometry of the gas-phase decomposition of oxalic acid. J Phys Chem 68: 1863–1865

Lapidus G, Barton D, Yankwich PE (1966a) Reversing hydrogen isotope effect on the rate of the gas phase decomposition of oxalic acid. J Phys Chem 70: 407–411

Lapidus G, Barton D, Yankovich PE (1966b) Reversing the intramolecular kinetic carbon isotope effect in the gas phase decomposition of oxalic acid. J Phys Chem 70: 3155–3159

Leffler JE (1955) The enthalpy-entropy relationship and its implications for organic chemistry. J Org Chem 20: 1202–1231

Leffler JE (1965) Concerning the isokinetic relationship. Nature 205: 1101–1102

Leffler JE (1966) The interpretation of enthalpy and entropy data. J Org Chem 31: 533–537

Leffler JE, Grunwald E (1963) Rates and equilibria of organic reactions. Wiley, New York, 458 pp

Levec J, Smith JM (1976) Oxidation of acetic acid solutions in a trickle-bed reactor. AIChE J 22: 159–168

Levec J, Herskowitz M, Smith JM (1976) An active catalyst for the oxidation of acetic acid solutions. AIChE J 22: 919–920

Lundegard PD, Senftle JT (1987) Hydrous pyrolysis; a tool for the study of organic acid synthesis. Appl Geochem 2: 605–612

Lutgert I, Schroer E (1940) The kinetics of the thermal decomposition oxalic acid in solution. Z Phys Chem A187: 133–148

MacGowan DB, Surdam RC (1988) Difunctional carboxylic acid anions in oilfield waters. Org Geochem 12: 245–259

Madix RJ (1984) Reaction kinetics and mechanism: model studies on metal single crystals. Catal Rev Sci Eng 26: 281–297

Madix RJ, Falconer JL, Suszko AM (1976) The autocatalytic decomposition of acetic acid on Ni(110). Surface Sci 54: 6–20

Mars P, Scholten JJF, Zwietering P (1963) The catalytic decomposition of formic acid. Adv Catal 14: 35–113

Martens CS (1990) Generation of short-chain organic acid anions in hydrothermally altered sediments of the Guaymas Basin, Gulf of California. Appl Geochem 5: 71–76

Martens CS, Albert DB, Chanton JP, Pauly GG, Canuel EA (1988) Organic acids and light hydrocarbons in hydrothermally altered Guaymas Basin sediments. 1988 Annu Meet Geological Society of America, Abstr 21422, vol 20, p A296

Matusevich VM, Shvets VM (1973) Significance of organic acids of subsurface waters for oil-gas exploration in West Siberia. Geol Nefti i Gaza 10: 459–464

McCarty J, Falconer J, Madix RJ (1973) Decomposition of formic acid on Ni(110). I. Flash decomposition from the clean surface and flash desorption of reaction products. J Catal 30: 235–249

Means JL, Hubbard NJ (1985) The organic chemistry of deep ground waters from the Palo-Duro Basin, Texas: implications for radionuclide complexation, ground-water origin, and petroleum exploration. Tech Rep Batelle, BMI/ONWI-578 Distribution Category UC-70, 75 pp

Means JL, Hubbard N (1987) Short-chain aliphatic acid anions in deep subsurface brines: a review of their origin, occurrence, properties, and importance and new data on their distribution and geochemical implications in the Palo Duro Basin, Texas. Org Geochem 11: 177–191

Mehrotra RN (1981) Kinetics and mechanisms of redox reactions in aqueous solution, Part 7. Decarboxylation of aliphatic acids by aquasilver(II) ions. J Chem Soc Dalton Trans: 897–901

Miles SL, Bernasek SL, Gland JL (1983) The effects of substrate oxidation on the adsorption and decomposition of HCOOH on Mo(100). Surface Sci 127: 271–282

Millet M, Virely M, Forissier M, Bussiere P, Vedrine JC (1989) Mössbauer spectroscopic study of iron phosphate catalysts used in selective oxidation. Hyperfine Interact 46: 619–628

Minachev Kh M, Shpiro ES (1990) Catalytic surface: physical methods of studying. CRC Press, Boston, 375 pp

Mitchell JA, Reid EE (1931) The decomposition of acetic acid in the presence of silica gel. J Am Chem Soc 53: 338–342

Mittal L, Mittal JP, Hayon E (1973) Photo-induced decarboxylation of aliphatic acids and esters in solution. Dependence upon state of protonation of the carboxyl group. J Phys Chem 77: 1482–1487

Moore JW, Pearson RG (1981) Kinetics and mechanism. Wiley, New York, 455 pp

Mosher WA, Kehr CL (1953) The decomposition of organic acids in the presence of lead tetraacetate. J Am Chem Soc 75: 3172–3176

Muus J (1935) Carbon dioxide cleavage from dibromomalonic acid. J Phys Chem 39: 343–353

Nazar AFM, Wells CF (1985) Kinetics of the oxidation of substrate ligands by transition-metal cations. J Chem Soc Faraday Trans I, 81: 801–812

Nef JU (1901) Dissoziationsvorgänge bei den einatomigen Alkoholen, Aethern und Salzen. Justus Liebigs Ann Chem 318: 220–226

Palmer DA, Drummond SE (1986) Thermal decarboxylation of acetate. Part I. The kinetics and mechanism of reaction in aqueous solution. Geochim Cosmochim Acta 50: 813–823

Palmer DA, Drummond SE (1988) Potentiometric determination of the molal formation constants of ferrous acetate complexes in aqueous solutions to high temperatures. J Phys Chem 92: 6795–6800

Parrott SL, Rogers JW Jr, White JM (1978) The decomposition of ethanol, propanol and acetic acid chemisorbed on magnesium oxide. Application Surface Sci 1: 443–454

Pedersen KJ (1927) The velocity of the decomposition of nitroacetic acid in aqueous solution. Trans Faraday Soc 23: 316–326

Pedersen KJ (1932) The decomposition of α-nitrocarboxylic acids with some remarks on the decomposition of β-ketocarboxylic acids. J Phys Chem 38: 559–571

Pedersen KJ (1949) The effect of metal ions on the rate of decomposition of nitroacetic acid. Acta Chem Scand 3: 676–696

Petersen RC (1964) The linear relationship between enthalpy and entropy of activation. J Org Chem 29: 3133–3135

Pintar A, Levec J (1992) Catalytic oxidation of organics in aqueous solutions. I. Kinetics of phenol oxidation. J Catal 135: 345–357

Purnell JH, Al-Owais A, Ballantine JA (1987) Thermogravimetric analysis study of the adsorption of ethanoic acid vapor by ion-exchanged montmorillonite. In: Schulz LG, van Olphen H, Mumpton FA (eds) Proc Int Clay Conf, Denver, 1985, pp 335–339

Rajadurai S (1987) Synthesis, structural characterization, and catalytic study of $ZnCrFeO_4$ spinel. Mater Chem Phys 15: 459–466

Richardson WH, O'Neal HE (1972) The unimolecular decomposition of oxygenated organic compounds (other than aldehydes and ketones). In: Bamford CH, Tipper CFH (eds) Comprehensive chemical kinetics, vol 5. Decomposition and isomerization of organic compounds. Elsevier, Amsterdam, pp 381–555

Rubinshtein AM, Yakerson VI, Lafer LI (1964) Catalytic ketonization of acetic acid over a mixed $CaCO_3$-Li_2CO_3 catalyst. Kinet Catal 5: 319–323

Satterfield CN (1980) Heterogeneous catalysis in practice. McGraw-Hill, New York, 416 pp

Schoofs R, Benziger JB (1984) Decomposition of acetic acid monomer, acetic acid dimer, and acetic anhydride on Ni(111). Surface Sci 143: 359–368

Servotte Y, Jacobs J, Jacobs PA (1985) Selectivity in the conversion of acetic acid over the MFI-type zeolites. Acta Phys Chem 31: 609–618

Sexton BA (1979) Observation of formate species on a copper(100) surface by high resolution electron energy loss spectroscopy. Surface Sci 88: 319–330

Sexton BA, Madix RJ (1981) A vibrational study of formic acid interaction with clean and oxygen-covered silver(110) surfaces. Surface Sci 105: 177–195

Sheldon RA, Kochi JK (1968) Photochemical and thermal reduction of cerium(IV) carboxylates. Formation and oxidation of alkyl radicals. J Am Chem Soc 90: 6687–6698

Shock EL (1988) Organic acid metastability in sedimentary basins. Geology 16: 886–890

Shock EL (1989) Corrections to "Organic acid metastability in sedimentary basins". Geology 17: 572–573

Shustorovich E, Bell AT (1989) An analysis of formic acid decomposition on metal surfaces by the bond-order-conservation-order-potential approach. Surface Sci 222: 371–382

Steinberger R, Westheimer FH (1951) Metal ion-catalyzed decarboxylation: a model for an enzyme system. J Am Chem Soc 73: 429–435

Stone AT (1986) Adsorption of organic reductants and subsequent electron transfer on metal oxide surfaces. In: Davis J A, Hayes K F (eds) Geochemical processes at mineral surfaces. American Chemical Society, Washington DC, pp 446–461

Surdam RC, Crossey LJ (1985) Mechanisms of organic/inorganic interactions in sandstone/shale sequences. Soc Econ Paleontol Mineral Short Course Notes 17: 177–232

Surdam RC, MacGowan DB (1987) Oilfield waters and sandstone diagenesis. Appl Geochem 2: 613–619

Surdam RC, Boese SW, Crossey LJ (1984) The chemistry of secondary porosity. In: McDonald DA, Surdam RC (eds) Clastic diagenesis. Am Assoc Pet Geol Mem 37, pp 127–149

Swaminathan R, Kuriacose JC (1970) Studies on the ketonization of acetic acid on chromia. II. The surface reaction. J Catal 16: 357–362

Tinker HB (1970) The decarboxylation of carboxylic acids during the autoxidation of cyclohexane. J Catal 19: 237–244

Trillo JM, Munuera G, Criado JM (1972) Catalytic decomposition of formic acid on metal oxides. Catal Rev 7: 51–86

Truhlar DG, Hase WL, Hynes JJ (1983) Current status of transition-state theory. J Phys Chem 87: 2664–2682

Tsai CS (1967) Spontaneous decarboxylation of oxalacetic acid. Can J Chem 45: 873–880

Virely C, Forissier M, Millet JM, Vedrine JC, Huchette D (1992) Kinetic study of isobutyric acid oxydehydrogenation on various Fe-P-O catalysts: proposal for the reaction mechanism. J Mol Catal 71: 119–213

Westheimer FH, Jones WA (1941) The effect of solvent on some reaction rates. J Am Chem Soc 63: 3283–3286

Wilkins R (1991) Kinetics of reactions of transition metal complexes. VCH, Weinheim, pp 87–89

Wold S (1972) The abnormal behaviour of enthalpy-entropy plots. Chem Scr 2: 145–147

Wold S, Exner O (1973) Statistics of the enthalpy-entropy relationship. Chem Scr 3: 5–11

Ying DHS, Madix RJ (1979) Thermal desorption study of the acetic acid decomposition on clean Ni/Cu(110) alloy surfaces. J Catal 60: 441–451

Zinger AS, Kravchik TE (1972) The simpler organic acids in ground water of the Lower Volga Region (genesis and possible use in prospecting for oil). Dokl Akad Nauk SSSR 202: 693–696

Chapter 10   Application of Thermodynamic Calculations to Geochemical Processes Involving Organic Acids

Everett L. Shock[1]

Summary

This chapter summarizes some of the insights gained about the behavior of organic acids in geochemical processes through thermodynamic calculations. Such calculations are possible because of the combination of numerous experimental studies on aqueous organic acids and theoretical equations of state, which allow accurate extrapolations of the measurements as well as predictions of thermodynamic properties of aqueous organic acids for which data have not been measured. Estimates of thermodynamic data for aqueous organic acids allow quantitative tests of several hypotheses concerning the role of organic acids in geochemical processes. For example, thermodynamic calculations described in this chapter indicate that decarboxylation of organic acids is unlikely to proceed to any significant extent under sedimentary basin conditions. At the partial pressures of CO_2 and CH_4 associated with oil-field brines, equilibrium constants for the decarboxylation of acetic acid require the acid concentrations to be *many orders of magnitude* lower than reported values.

Although the concentrations of acetic acid in oil-field brines cannot be in equilibrium with *both* CO_2 and CH_4, they may be in redox equilibrium with CO_2 as demonstrated by additional calculations described in this chapter. This means that there is an enormous kinetic barrier blocking reactions between acetic acid and CH_4 under sedimentary basin conditions. Therefore, acetic acid is preserved in a metastable state in oil-field brines, and appears to be in metastable equilibrium with CO_2.

In addition, thermodynamic evaluation of acetic acid and propanoic concentrations in many brines indicates that these acids are in homogeneous metastable equilibrium. As a consequence, the ratio of the concentrations of acetic and propanoic acids in basinal brines can be used as a tracer of the oxidation state of sedimentary basins. Additional thermodynamic calculations described in this chapter allow tests of the plausibility of the hypothesis

---

[1] Department of Earth and Planetary Sciences, Washington University, St. Louis, Missouri 63130, USA

that the complex mixture of liquid hydrocarbons found in petroleum may buffer the oxidation state recorded by the acid ratios. It is found that this is a plausible argument not only for sedimentary basins but for hydrous pyrolysis experiments as well.

Thermodynamic calculations can only demonstrate whether compounds are in equilibrium with one another (stable or metastable) and reveal nothing about the reaction mechanisms through which such equilibrium states are reached and maintained. In the case of metastable equilibrium among petroleum hydrocarbons, organic acids, and CO_2 in sedimentary basins, thermophilic microorganisms may catalyze otherwise sluggish reactions so that the geologically observable metastable state is reached. If so, then many of the reactions shown to be in metastable equilibrium may represent overall metabolic processes, and the application of thermodynamic calculations enters the area of geochemical bioenergetics. Some preliminary examples for dehydrogenation, hydrogenation, sulfate reduction, and methanogenesis reactions involving organic acids are discussed in this context at the end of this chapter. It appears that chemical reactions, which supply energy to microorganisms at low temperatures, provide *considerably more* energy at the elevated temperatures encountered in sedimentary basins.

1 Introduction

Organic acids have been invoked to account for the differences between measured alkalinities for oil-field brines and those calculated based on carbonate equilibria (Willey et al. 1975), as a source and transport mechanism for natural gas through decarboxylation reactions (Carothers and Kharaka 1978), in the transport of metals through metal-organic complexes (Giordano and Barnes 1981), in the development of secondary porosity (Surdam et al. 1984), as pH buffers that could potentially control mineral-fluid reactions (Lundegard and Land 1989), and as tracers of metastable equilibrium states among organic compounds in natural systems (Shock 1988, 1989, 1990a). These acids have also been used as indicators of oil migration (Jaffe et al. 1988a,b) and in petroleum prospecting techniques (Zinger and Kravchik 1970). The relatively high concentrations of organic acids in some natural fluids (up to ~10 000 ppm in oil-field brines; MacGowan and Surdam 1990) suggest that these compounds may be major aqueous species in biogeochemical cycles of carbon, hydrogen, oxygen, and nitrogen between detrital organic matter, petroleum, minerals, natural aqueous solutions, and the atmosphere. The consumption and production of aqueous organic acids by microorganisms in soil, groundwater, and petroleum reservoir environments indicate that these compounds are involved in subsurface metabolic processes. Understanding the wide variety of processes that may involve organic acids, including the energetics of biologically mediated reactions, requires thermodynamic and kinetic properties of reactions in which organic acids take part. The goals of this chapter are to summarize the available thermodynamic data for aqueous organic acids found in sedimentary basin brines, highlight the consistencies in these data, which permit estimation of values not yet collected in experiments, and show how these data can be applied to develop quantitative tests of several of the geochemical processes thought to involve organic acids.

2   Calculation of Standard State Thermodynamic Properties of Aqueous Organic Acids

Examples of the wide variety of reactions involving organic acids, which could occur in geochemical processes, include dissociation,

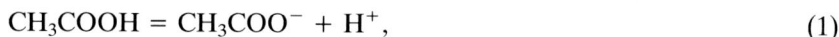

$$CH_3COOH = CH_3COO^- + H^+, \tag{1}$$

complexation,

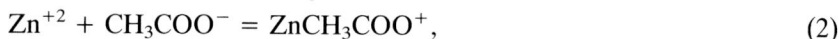

$$Zn^{+2} + CH_3COO^- = ZnCH_3COO^+, \tag{2}$$

oxidation (or dehydrogenation in metabolic reactions),

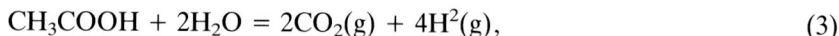

$$CH_3COOH + 2H_2O = 2CO_2(g) + 4H^2(g), \tag{3}$$

reduction (or hydrogenation in metabolic reactions),

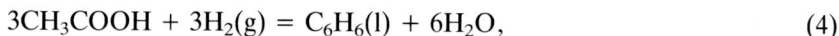

$$3CH_3COOH + 3H_2(g) = C_6H_6(l) + 6H_2O, \tag{4}$$

and decarboxylation,

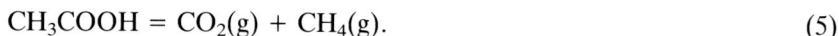

$$CH_3COOH = CO_2(g) + CH_4(g). \tag{5}$$

Reactions (1–5) are written for acetic acid, but analogous reactions could be written for other organic acids. Having data for each organic acid species insures the greatest flexibility for evaluating the thermodynamic properties of these reactions either individually or simultaneously.

One way of reaching this goal is to employ an equation of state for aqueous species, which allows theoretical explanations for experimentally collected data. The revised Helgeson-Kirkham-Flowers (HKF) equation of state is based on electrostatic theory of aqueous solutions and has been shown to accurately represent a wide variety of standard state data for aqueous ions and electrolytes (Helgeson et al. 1981; Shock and Helgeson 1988; Tanger and Helgeson 1988; Sassani and Shock 1992; Shock et al. 1992), inorganic acids and dissolved gases (Shock et al. 1989), inorganic metal-ligand complexes (Sassani and Shock 1990, 1993; Sverjensky et al. submitted), aqueous hydrocarbons (Shock and Helgeson 1990; Shock submitted), aqueous aldehydes (Schulte and Shock 1993), aqueous ketones, alcohols, amines and amino acids (Shock and Helgeson 1990), aqueous peptides (Shock 1992a), aqueous carboxylic and hydroxyacids (Shock and Helgeson 1990; Shock 1993a), aqueous chloroacetic acids (Helgeson 1992), and aqueous metal-acetate complexes (Shock and Koretsky 1993). By taking account of the considerable systematic behavior of aqueous species, as well as the wealth of experimental data at high temperatures and pressures, the revised HKF equation of state allows accurate extrapolation of standard state data beyond the ranges of experimental measurements, and predictions for many aqueous species for which limited or no experimental data are available. Calculations are currently possible from 0 to 1000°C and pressures from 1 to 5000 bars. Predictions are facilitated by correlations among equation-of-state parameters obtained by regression of standard state thermodynamic data. As additional experimental data become available they can be regressed with the revised HKF equations to yield equation-of-state parameters, which help refine the accuracy of extrapolations and correlation algorithms. Examples of data regression and correlations are given in this section, together with examples of the predictive power of the revised HKF equations.

Volume and heat capacity data at various temperatures for aqueous solutions of the Na-salts of formate, acetate, propanoate, heptanoate, decanoate, benzoate, glycolate, lactate, 2-hydroxybutanoate, 2-hydroxypentanoate, 2-hydroxyhexanoate, oxalate, and succinate, as well as similar measurements for solutions of formic, acetic, propanoic, glycolic, lactic, 2-hydroxybutanoic, 2-hydroxypentanoic, and 2-hydroxyhexanoic acids were regressed with the revised HKF equations of state by Shock (1993a). Comparison of regression results with experimental data for some of these aqueous species can be made with the plots in Fig. 1. Note that the calculated values are consistent with the approach to ∞ or $-\infty$ in the theoretical limit at the critical point of H_2O, depending on the type of aqueous species. An example of the correlations among equation-of-state parameters, which are possible as a result of these regression calculations, is shown in Fig. 2, where values of the σ equation-of-state parameter are plotted against values of $\Delta\overline{V}_n^\circ$, the structural contribution to the standard partial molal volume at 25°C and 1 bar. The correlation shown in Fig. 2 is given by (Shock 1993a)

$$\sigma = 1.07\,\Delta\overline{V}_n^\circ + 3.0, \tag{6}$$

which can be used for inorganic and organic aqueous ions and nonelectrolytes.

The predictive capabilities of the revised HKF equations are demonstrated in Fig. 3 where calculated values of log K for dissociation reactions can be compared with experimental data. Results shown in Fig. 3 are typical of comparisons between calculations and experimental data for several acids for which high-temperature/pressure data have been collected (Shock 1993a). However, the group of acids for which adequate experimental data are available is remarkably limited, especially when compared to the myriad

Na-benzoate

Na-decanoate

Acetic Acid

Acetic Acid

Propanoic Acid

Propanoic Acid

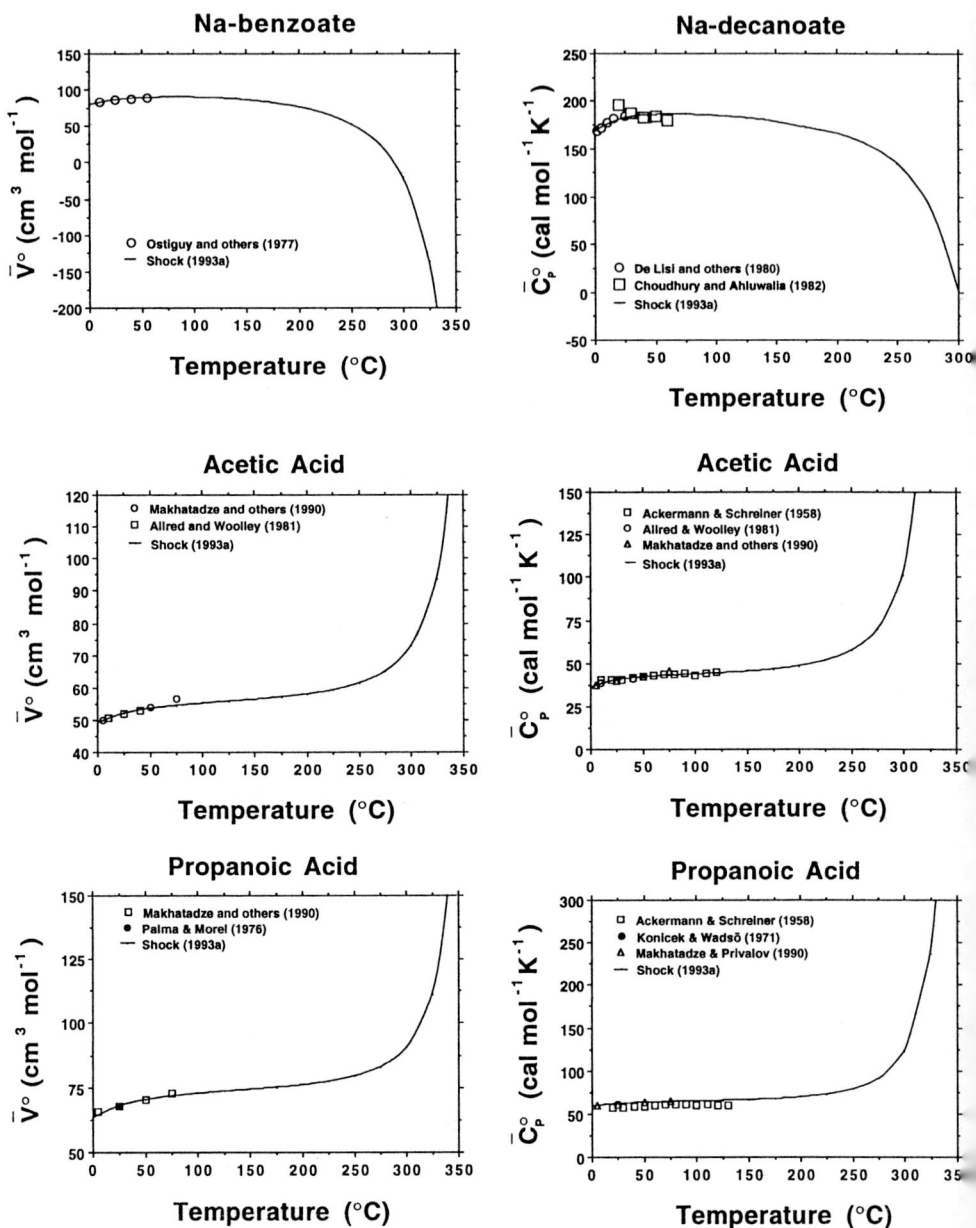

Fig. 1. Standard partial molal volumes and isobaric heat capacities of aqueous organic acids and aqueous Na-carboxylate electrolytes as functions of temperature at the vapor-liquid saturation pressure of H_2O (P_{sat}). The *symbols* represent experimental values from the references indicated in the figure, but the curve corresponds to values calculated with the revised HKF equations of state using data and parameters from Shock (1993a)

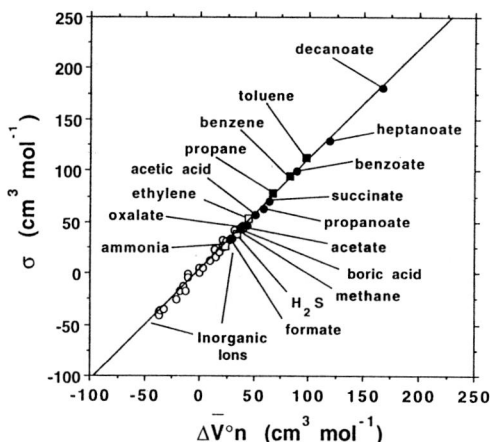

Fig. 2. Correlation of the σ equation-of-state parameter with the structural contribution to the standard partial molal volume of aqueous species at 25 °C and 1 bar ($\Delta \overline{V}_n^{\circ}$). The *symbols* represent values obtained from regression of experimental data for aqueous electrolytes and neutral species at elevated temperatures taken from Shock and Helgeson (1988, 1990), Shock et al. (1989), and Shock (1993a). The correlation is given by Eq. (6)

organic acids, which could take part in geochemical processes. Predictions of thermodynamic data with the revised HKF equations help to fill this void. Organic acids for which thermodynamic data and revised HKF parameters are available are listed in Table 1. Thermodynamic data for these aqueous organic acids can be calculated at temperatures to 1000 °C and pressures up to 5 kb. These calculations are facilitated by the SUPCRT92 computer code (Johnson et al. 1992).

One reason why close agreement is possible between experiments and calculations, such as that shown in Figs. 1 and 3, is that the revised HKF equations incorporate the systematic behavior of aqueous organic acids and other organic solutes. Examples of the systematic behavior of standard state volumes and heat capacities of aqueous organic acids at 25 °C and 1 bar are shown in Fig. 4. In these plots, values of $\overline{V}^{\circ}$ and $\overline{C}_p^{\circ}$ taken from the critical compilation of Shock (1993a) are plotted against the number of carbons in the alkyl chain ($\overline{n}$) for straight-chain homologous series. Note that the 1-carbon members of these homologous series deviate from the correlation lines. A conceptual explanation for this behavior can be found in the difference in the effect of inserting a CH_2 group into a compound between the functional group and a hydrogen atom, as in

$$HCOOH + CH_2 \rightarrow CH_3COOH,$$

formic   acetic   (7)

and inserting a CH_2 group into the alkyl chain, as in

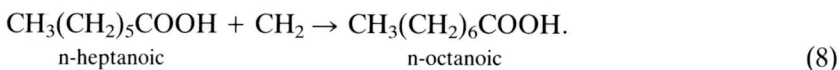

$$CH_3(CH_2)_5COOH + CH_2 \rightarrow CH_3(CH_2)_6COOH.$$

n-heptanoic   n-octanoic   (8)

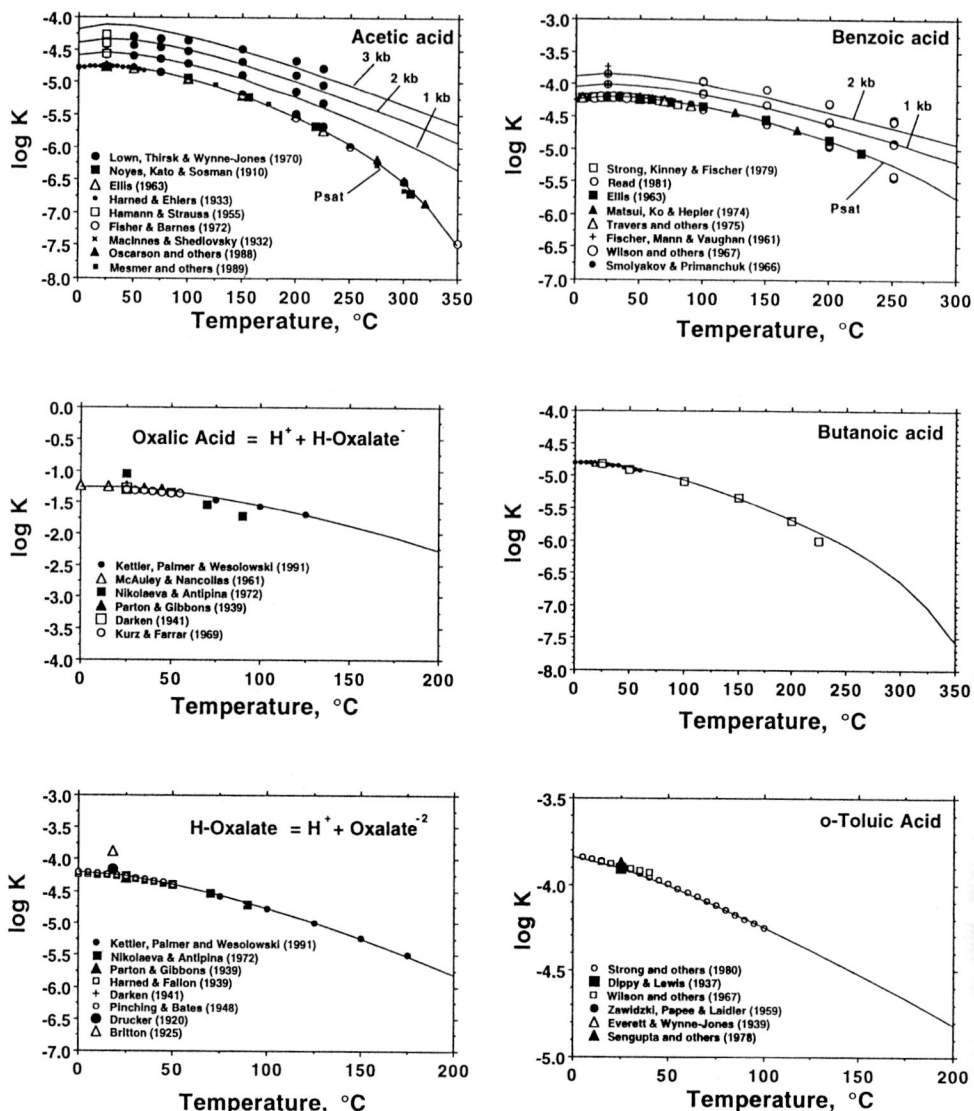

Fig. 3. Plots of log K for acid dissociation reactions as functions of temperature at P_{sat}, and for acetic and benzoic acids at elevated pressure labeled in kilobars. *Symbols* represent experimental data, but curves are calculated with the revised HKF equations of state using data and parameters from Shock (1993a)

The consequences of these differences are discussed by Shock and Helgeson (1990) and Shock (1993a).

Tables of calculated equilibrium constants for dissociation reactions of the monocarboxylic, dicarboxylic, hydroxy, and aromatic acids listed in Table 1

Table 1. Aqueous organic acids for which thermodynamic properties can be calculated at high pressures and temperatures with the revised HKF equations of state (Shock and Helgeson 1990; Shock 1993a). At present, similar calculations are possible for anions of all except and amino acids

Monocarboxylic acids	Dicarboxylic acids	Hydroxyacids	Amino acids
Formic	Oxalic	Glycolic	Glycine
Acetic	Malonic	Lactic	Alanine
Propanoic	Succinic	2-Hydroxybutanoic	α-Aminobutyric
n-Butanoic	Glutaric	2-Hydroxypentanoic	Valine
n-Pentanoic	Adipic	2-Hydroxyhexanoic	Leucine
n-Hexanoic	Pimelic	2-Hydroxyheptanoic	Isoleucine
n-Heptanoic	Suberic	2-Hydroxyoctanoic	Serine
n-Octanoic	Azelaic	2-Hydroxydetanoic	Threonine
n-Nonanoic	Sebacic		Aspartic
n-Decanoic			Glutamic
n-Undecanoic	*Aromatic acids*		Asparagine
n-Dodecanoic	Benzoic		Glutamine
	o-Toluic		Phenylalanine
	p-Toluic		Tryptophan
	m-Toluic		Tyrosine
			Methionine

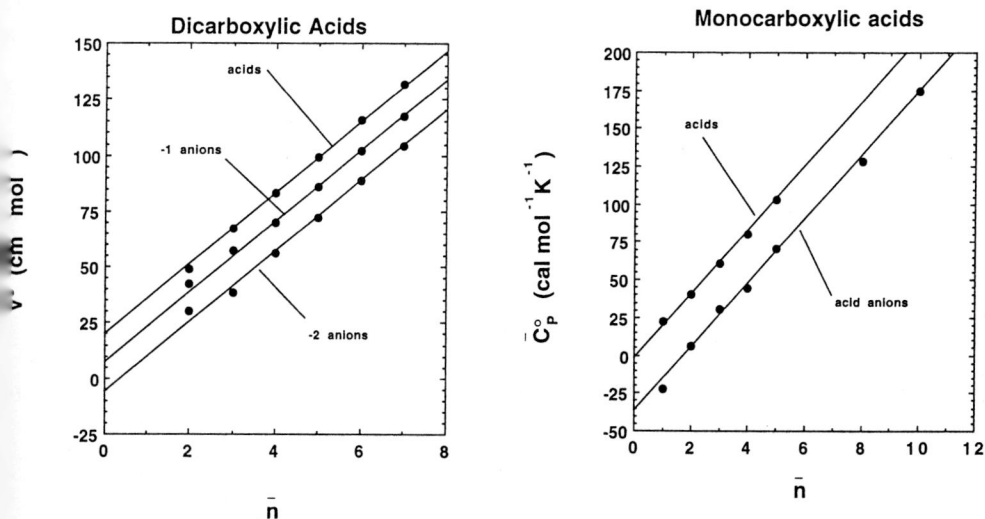

Fig. 4. Standard partial molal volumes and isobaric heat capacities of aqueous organic acid species at 25 °C and 1 bar from the critical compilation of Shock (1993a) plotted against the number of moles of carbon in 1 mol of the species ($\bar{n}$). Equations for the correlation curves are given by Shock (1993a)

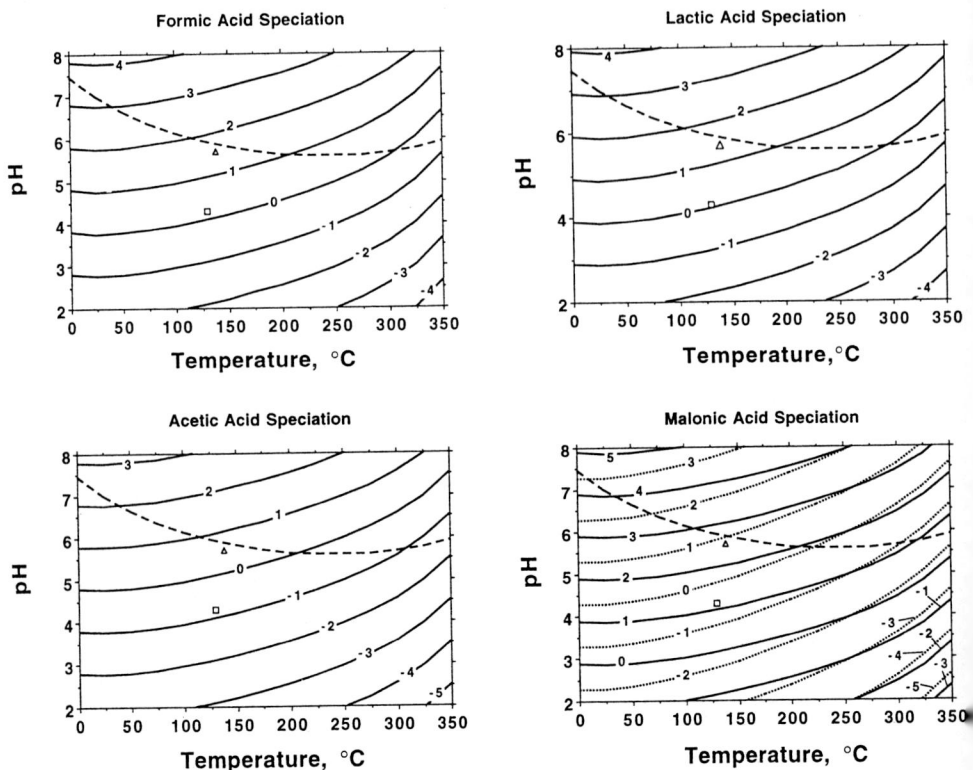

Fig. 5. Organic acid speciation plots showing log (a anion/a neutral acid) as functions of pH and temperature. In the case of malonic acid the *dotted curves* correspond to log (a malonate^{-2}/a malonic acid). The *symbols* show the location of two oil-field brines discussed by Sverjensky (1984) and in the text

are given by Shock (1993a). These dissociation constants were used to construct the speciation plots shown in Fig. 5. These plots show contours of log (a anion$^-$/a acid) as functions of pH and temperature at the corresponding vapor-liquid saturation pressures for H_2O (P_{sat}) where a represents activity. The dotted contours in the malonic acid plot indicate values of log (a malonate^{-2}/a malonic acid). Also shown in these plots are dashed curves corresponding to neutral pH at P_{sat}. It can be seen that in each case the acids become more associated [lower values of log (a anion$^-$/a acid)] with increasing temperature at neutral pH.

Calculations of the type used to construct the plots in Fig. 5 allow a rapid assessment of the speciation of aqueous organic acids under the conditions of interest to a particular geochemical process. For example, the symbols in the plots shown in Fig. 5 correspond to pH values calculated by Sverjensky (1984) for two brine compositions reported by Kharaka et al. (1979). In one case, a brine from Pleasant Bayou, Texas, he concluded that the pH is less

Table 2. Calculated speciation of organic acids in two oil-field brines reported by Kharaka et al. (1980) and interpreted by Sverjensky (1984)

	Pleasant Bayou, Texas	Rayleigh Field, Mississippi
Temperature, °C	138	130
pH	5.7	4.3
Acetic:acetate	14.4:85.6	79.6:20.4
Propanoic:propanoate	15.0:85.0	80.6:19.4
Butanoic:butanoate	23.4:76.6	87.1:12.9
Oxalic:H-oxalate:oxalate	~0: 4.8:95.2	0.1:51.7:48.2
Malonic:H-malonate:malonate	0.2:44.7:55.1	6.4:88.7:4.9

than 5.7 at 138 °C by assuming calcite saturation. In the other case he calculated pH = 4.3 (±0.3) for a brine from Rayleigh Field, central Mississippi, by assuming equilibration between the brine and the muscovite-kaolinite-quartz assemblage at 130 °C. Comparison of the two temperature-pH pairs plotted in Fig. 5 shows that the degree of association can vary enormously for different brine compositions and reservoir conditions. Calculated speciation for several organic acids in these brines are listed in Table 2. The variation of organic acid speciation with pH is likely to control the extent to which organic acids can enhance mineral solubility during the generation of secondary porosity (Crossey et al. 1986; Edman and Surdam 1986; Giles and Marshall 1986; Lundegard and Land 1986; Giles and de Boer 1989, 1990; Surdam et al. 1989; Lundegard and Kharaka 1990), as well as the effectiveness of organic acids as complexing agents for metals in hydrothermal fluids (Giordano and Barnes 1981; Giordano 1985; Shock and Sverjensky 1989; Harrison and Thyne 1992; Shock and Koretsky 1993). In the calculations described above, equilibration with respect to pH in the homogeneous aqueous solution was assumed. This assumption is warranted because the rates of acid dissolution reactions are not likely to be rate-limiting for geochemical processes as they are much more rapid than mineral dissociation and precipitation reactions. On the other hand, other reactions involving organic acids in the homogeneous phase appear to be kinetically inhibited and could be rate-limiting in other geochemical processes as discussed below.

3 Thermodynamic Analysis of Geochemical Processes Involving Organic Acids

In the examples selected below, thermodynamic calculations are applied to decarboxylation reactions, redox reactions between organic acids, and redox reactions among organic acids, CO_2, and hydrocarbons. In each case, considerable insight can be gained into the ways in which organic acids respond to changes in temperature, pressure, oxidation state, pH, fluid composition, and partical pressures of gases during geochemical processes. In addition, stable equilibrium calculations provide a framework for quantifying kinetic barriers, which must exist if stable equilibrium is avoided in natural systems. If evidence for barriers to stable equilibrium can be found, it is possible to characterize metastable equilibrium states using subsets of the thermodynamic data and implicit kinetic constraints.

The versatility of the revised HKF equations, together with data and parameters for individual species, allows calculations for many types of reactions. As examples, predicted log K values at P_{sat} for oxidation, reduction, and decarboxylation reactions of acetic acid and propanoic acid corresponding to reactions (3), (5),

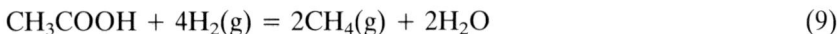

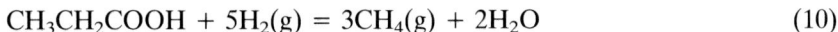

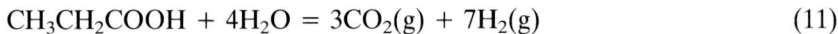

$$CH_3COOH + 4H_2(g) = 2CH_4(g) + 2H_2O \tag{9}$$

$$CH_3CH_2COOH + 5H_2(g) = 3CH_4(g) + 2H_2O \tag{10}$$

$$CH_3CH_2COOH + 4H_2O = 3CO_2(g) + 7H_2(g) \tag{11}$$

and

$$CH_3CH_2COOH = CO_2(g) + C_2H_6(g) \tag{12}$$

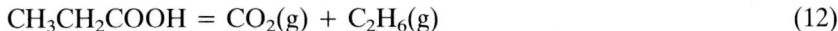

are plotted in Fig. 6 and listed in Table 3. These data are used below in discussions of decarboxylation and redox reactions in geochemical processes. They can also be used to study the effects of organic acids on mineral solubility and dissolution rates, which many investigators have studied in natural systems and in the laboratory (Schalscha et al. 1967; Huang and Keller 1971, 1972a,b,c; Huang and Kiang 1972; Antweiler and Drever 1983; Manley and Evans 1986; Mast and Drever 1987; Bennett et al. 1988; Hennet et al. 1988; Bevan and Savage 1989; Fein 1991; Wogelius and Walther 1991; among others).

3.1 Decarboxylation

Decarboxylation of organic acids has been a popular idea among sedimentary petrologists and aqueous geochemists since it was proposed by Carothers and Kharaka (1978) to explain a trend of decreasing concentration of carboxylic acids with increasing temperature from 80 to 200°C, which they observed in samples from California and Texas. Evidence for such a universal trend has all but evaporated as a result of numerous analyses of organic acids in oil-field brines, which were inspired largely by Carothers and Kharaka's landmark paper. The combined results from these studies are depicted in Fig. 7 where it can be seen that any trend of concentration with temperature is obscure. Nevertheless, several investigators have drawn upper limit curves on plots of this type and advocate high-temperature

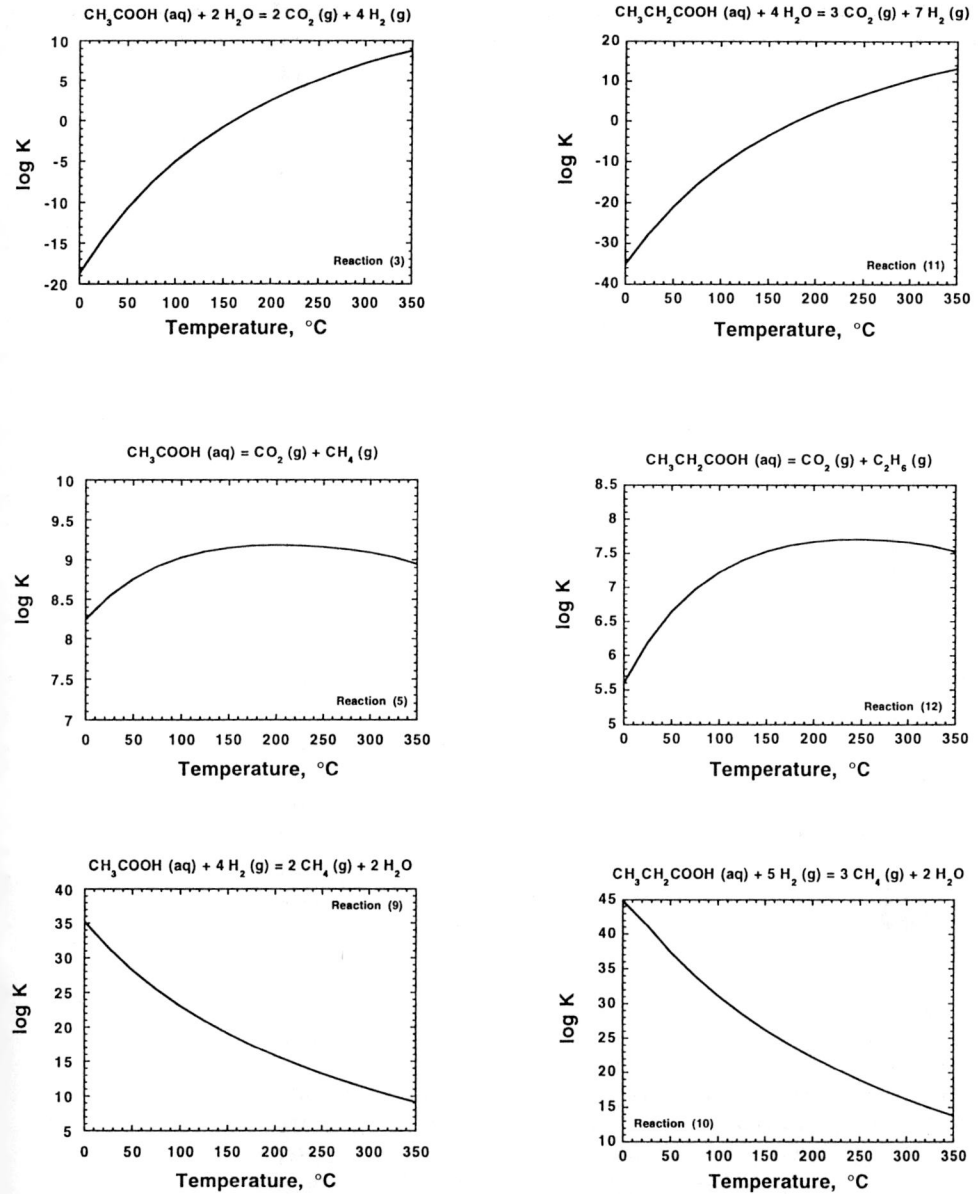

Fig. 6. Plots of log K for reactions (3), (5), (9), (10), (11), and (12) as functions of temperature at P_{sat}. Values of log K for these reactions are also listed in Table 3

Table 3. Log K values for reactions (3), (5), and (9–12) calculated with data and equation of state parameters from Shock et al. (1989), Shock and Helgeson (1990), and Shock (1993a)

P_{SAT}

Temp. °C	0.0	25	50	75	100	125	150	175	200	225	250	275	300	325	350
Reaction															
3	-18.74	-14.38	-10.73	-7.65	-5.02	-2.74	-0.76	0.98	2.51	3.87	5.09	6.17	7.15	8.02	8.78
5	8.24	8.54	8.76	8.92	9.03	9.10	9.15	9.18	9.18	9.18	9.16	9.13	9.09	9.03	8.94
9	35.23	31.46	28.25	25.48	23.07	20.95	19.06	17.37	15.86	14.48	13.23	12.09	11.03	10.04	9.11
10	45.85	41.32	37.45	34.09	31.14	28.54	26.21	24.12	22.23	20.51	18.94	17.50	16.16	14.92	13.74
11	-35.10	-27.43	-21.02	-15.61	-10.99	-7.00	-3.52	-0.48	2.21	4.59	6.72	8.63	10.34	11.88	13.25
12	5.60	6.19	6.64	6.98	7.22	7.40	7.53	7.62	7.67	7.70	7.71	7.69	7.66	7.61	7.53

500 bar

Temp. °C	0	25	50	75	100	125	150	175	200	225	250	275	300	325	350	375	400
Reaction																	
3	-17.93	-13.61	-10.01	-6.97	-4.38	-2.13	-0.18	1.52	3.03	4.37	5.56	6.62	7.58	8.43	9.19	9.87	10.47
5	8.71	8.99	9.19	9.32	9.41	9.46	9.49	9.50	9.49	9.47	9.43	9.39	9.34	9.27	9.20	9.12	9.01
9	35.36	31.60	28.39	25.61	23.19	21.06	19.16	17.47	15.94	14.56	13.3	12.15	11.10	10.12	9.21	8.36	7.55
10	46.10	41.59	37.71	34.34	31.38	28.75	26.41	24.31	22.4	20.67	19.09	17.63	16.29	15.05	13.89	12.80	11.77
11	-33.83	-26.22	-19.89	-14.54	-9.98	-6.04	-2.61	0.39	3.04	5.38	7.47	9.34	11.01	12.52	13.87	15.07	16.14
12	6.19	6.77	7.19	7.50	7.71	7.87	7.97	8.03	8.06	8.07	8.05	8.02	7.97	7.91	7.83	7.74	7.62

1000 bar

Temp. °C	50	100	150	200	250	300	350	400	450	500	550	600
Reaction												
3	−9.30	−3.74	0.39	3.56	6.05	8.05	9.67	11.00	12.07	12.95	13.64	14.19
5	9.61	9.78	9.83	9.79	9.72	9.61	9.47	9.32	9.15	8.95	8.71	8.46
9	28.53	23.31	19.27	16.03	13.38	11.16	9.27	7.65	6.22	4.94	3.79	2.74
10	37.95	31.59	26.60	22.57	19.23	16.42	14.01	11.92	10.07	8.42	6.92	5.56
11	−18.80	−9.00	−1.73	3.85	8.24	11.76	14.61	16.94	18.86	20.43	21.70	22.74
12	7.72	8.18	8.39	8.45	8.41	8.31	8.16	7.98	7.77	7.53	7.25	6.96

2000 bar

Temp. °C	50	100	150	200	250	300	350	400	450	500	550	600	650	700	750
Reaction															
3	−7.90	−2.51	1.50	4.58	7.00	8.95	10.53	11.84	12.93	13.84	14.62	15.27	15.83	16.31	16.72
5	10.46	10.53	10.49	10.39	10.26	10.12	9.96	9.79	9.62	9.45	9.28	9.11	8.94	8.77	8.16
9	28.81	23.56	19.48	16.21	13.53	11.29	9.38	7.74	6.31	5.06	3.94	2.94	2.05	1.23	0.49
10	38.38	31.97	26.93	22.86	19.49	16.64	14.21	12.10	10.25	8.61	7.15	5.83	4.64	3.56	2.57
11	−16.69	−7.13	−0.03	5.41	9.69	13.13	15.93	18.25	20.17	21.79	23.16	24.33	25.32	26.18	26.92
12	8.71	9.06	9.18	9.16	9.06	8.19	8.74	8.54	8.33	8.11	7.89	7.67	7.44	7.22	7.01

Organic Acids in Oilfield Brines

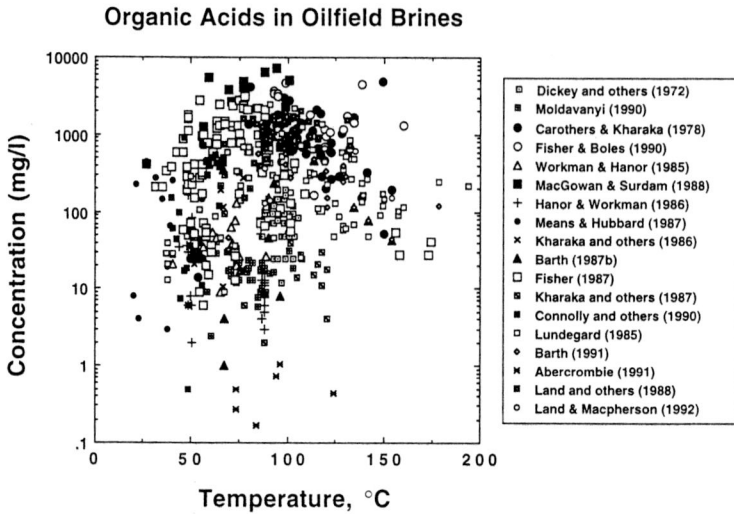

Fig. 7. Total concentrations of organic acids in oil-field brines at the temperatures reported for each sample in the references indicated

decarboxylation (Crossey et al. 1984; Surdam et al. 1984; Surdam and Crossey 1985; Surdam and MacGowan 1987; MacGowan and Surdam 1988). A curve of this type usually has a maximum near 90 °C. This implies that

$$\frac{d \log m_T}{dT} = 0,\tag{13}$$

where m_T stands for the total concentration of organic acids, at the temperature which corresponds to the greatest range of concentration values as shown in Fig. 7. It should therefore be apparent that a curve of this type is in direct conflict with the compositions of natural brines.

Comparison of thermodynamic calculations for organic acid decarboxylation reactions with observations from nature and results of laboratory experiments provides an illustration of the application of stable equilibrium calculations to identify the magnitude of kinetic barriers. The law of mass action expression for the acetic acid decarboxylation reaction [Eq. (5)] is given by

$$K_5 = \frac{fCO_2 fCH_4}{aCH_3COOH},\tag{14}$$

where K_5 refers to the equilibrium constant for reaction (5), f represents fugacity, and a stands for activity. The logarithmic form of Eq. (14) can be rearranged to yield

$$\log fCO_2 = -\log fCH_4 + (\log K_5 + \log aCH_3COOH),\tag{15}$$

which corresponds to an equation of a line with a slope of -1 and an intercept equal to the parenthetical term. The values of $\log K_5$ from Table 3 allow construction of fugacity diagrams contoured for acetic acid activity such as that shown in Fig. 8. The dashed lines on this plot indicate contours of $\log a CH_3COOH$, and the solid curve outlines the region of the diagram where the sum of the fugacities of CO_2 and CH_4 are geologically attainable for a total pressure of 500 bar. As shown in Fig. 6, $\log K$ for reaction (5) is only slightly temperature-dependent, so plots at other temperatures are similar to that shown for 150 °C in Fig. 8.

It can be seen in Fig. 8 that the high concentrations of acetic acid in oil-field brines (activities typically from 10^{-2} to 10^{-4}, see below) are preserved metastably with respect to the decarboxylation reaction. In other words, if stable equilibrium was reached in sedimentary basin brines, concentrations of acetic acid would be several orders of magnitude lower than they are. Similar results can be obtained for propanoic acid with the data in Table 3. This disequilibrium with respect to the decarboxylation reaction indicates that large kinetic barriers to decarboxylation of organic acids exist under sedimentary basin conditions.

Several investigators have studied the kinetics of reaction (5) and other organic acid decarboxylation reactions as functions of temperature and fluid composition, as well as in the presence of minerals, which may catalyze the reaction (Kharaka et al. 1983; Drummond and Palmer 1986; Palmer and Drummond 1986; Schleusener et al. 1987, 1988; Crossey 1991). Early

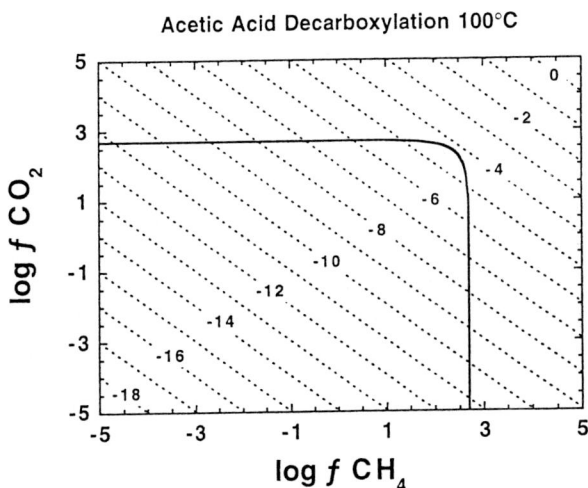

Fig. 8. Contours of log activity of acetic acid (*dotted lines*) at stable equilibrium with the decarboxylation reaction (5) in terms of the fugacities of CO_2 and CH_4 at 100 °C. The *solid curve* indicates the upper limits of the sum of the two gas fugacities, which is attainable at 500 bar. Note that activities of acetic greater than $\sim 10^{-4}$ are metastable with respect to the decarboxylation reaction

reports of rapid decarboxylation rates were accounted for by the catalyzed destruction of acetic acid in stainless steel bombs, and the resulting un-catalyzed rates are consistent with the metastable preservation of acetic acid for geologic time spans under sedimentary basin conditions.

3.2   Oxidation/Reduction Reactions

The apparent inhibition of the decarboxylation reaction of acetic acid under sedimentary basin conditions does not require disequilibrium with both CO_2 and CH_4. This can be inferred from Fig. 8, where it can be seen that the high concentrations of acetic acid in basinal brines could be in equilibrium with either CO_2 or CH_4 if either reaction (3) or (9) were reversible under sedimentary basin conditions. Exploring this possibility requires an assessment of the fugacities of H_2 which are likely to prevail in sedimentary basins.

Mineral assemblages, which can buffer fH_2 in fluid-driven processes, are familiar to geochemists and petrologists as oxygen fugacity (fO_2) buffer assemblages. For example, the magnetite-hematite buffer reaction,

$$2Fe_3O_4 + 1/2O_2(g) = 3Fe_2O_3,$$
$$\text{magnetite} \qquad\qquad\qquad \text{hematite} \tag{16}$$

can be combined with the water disproportionation reaction,

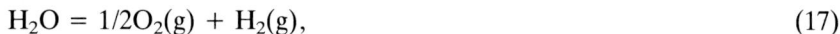

$$H_2O = 1/2O_2(g) + H_2(g), \tag{17}$$

to yield

$$2Fe_3O_4 + H_2O = 3Fe_2O_3 + H_2(g). \tag{18}$$

The law of mass action expression for reaction (18) simplifies to

$$\log K_{18} = \log fH_2 \tag{19}$$

for pure mineral phases, and unit activity of H_2O, which is a justifiable assumption up to NaCl concentrations of about 3 molal at sedimentary basin temperatures and pressures (Helgeson 1969, 1985; Sverjensky 1984). Analogous expressions for other mineral buffer assemblages are given by

$$3/2FeS + H_2O = 3/4FeS_2 + 1/4Fe_3O_4 + H_2(g) \tag{20}$$
$$\text{pyrrhotite} \qquad\qquad \text{pyrite} \quad\ \text{magnetite}$$

and

$$3/2FeSiO_4 + H_2O = Fe_3O_4 + 3/2SiO_2 + H_2(g). \tag{21}$$
$$\text{fayalite} \qquad\qquad \text{magnetite} \quad \text{quartz}$$

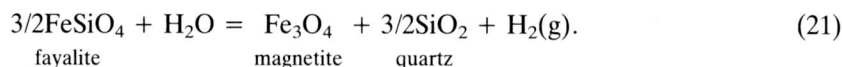

These reactions, together with overall reactions among the hydrocarbon components of petroleum that may characterize differences in the relative abundances of alkanes and aromatics such as

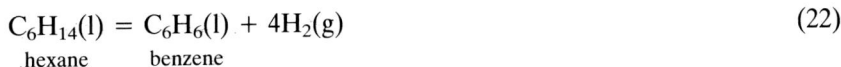

$$C_6H_{14}(l) = C_6H_6(l) + 4H_2(g) \tag{22}$$
$$\text{hexane} \qquad \text{benzene}$$

and

$$8C_5H_{12} = 5C_8H_{10} + 5H_2(g), \tag{23}$$
$$\text{pentane} \qquad \textit{m}\text{-xylene}$$

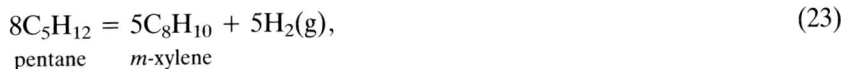

among many others, provide a framework for characterizing fugacities of H_2 which may be attained in fluid-driven geochemical processes.

Law of mass action expression for reactions (3) and (9) are given by

$$\log K_3 = 4\log fH_2 + 2\log fCO_2 - 2\log aH_2O - \log aCH_3COOH \tag{24}$$

and

$$\log K_9 = 2\log fCH_4 + 2\log aH_2O - 4\log fH_2 - \log aCH_3COOH. \tag{25}$$

At unit activity of H_2O, these expressions can be rearranged to yield

$$\log fH_2 = -1/2\log fCO_2 + 1/4(\log K_3 + \log aCH_3COOH), \tag{26}$$

and

$$\log fH_2 = 1/2\log fCH_4 - 1/4(\log K_9 + \log aCH_3COOH), \tag{27}$$

respectively. These expressions correspond to equations of lines on plots of $\log fH_2$ vs. $\log fCO_2$ and $\log fH_2$ vs. $\log fCH_4$. Plots of these types are shown in Fig. 9 for 100, 150, and 200 °C and were constructed with the appropriate equilibrium constants from Table 3. These plots are contoured for $\log aCH_3COOH$ (dotted lines), and values of $\log fH_2$ set by the fayalite-magnetite-quartz (FMQ), pyrhotite-pyrite-magnetite (PPM), and magnetite-hematite (MH) assemblages are indicated by dashed lines. The solid, vertical line indicates the values of $\log fCO_2$ or $\log fCH_4$ corresponding to a total pressure of 500 bar. The plots shown in Fig. 9 can be used to construct plausibility arguments to determine whether acetic acid may be in meta-stable equilibrium with either CO_2 or CH_4, because it is evident from the discussion of decarboxylation that it cannot be in stable equilibrium with both.

If we assert that oxidation states in sedimentary basin brines are likely to be intermediate between those set by MH and PPM, then it is possible to determine whether CO_2 or CH_4 may be in metastable equilibrium with acetic acid. Using an activity of 10^{-3} for CH_3COOH (see below), the 100 °C plots show that attainable fugacities of CO_2 are less than the total pressure of 500 bar at all fH_2 values above about $10^{-3.2}$. In contrast, the same activity of CH_3COOH can only be in equilibrium with high fugacities of CH_4 between values of fH_2 set by MH and about $10^{-3.4}$ which is a very narrow range. At higher temperatues, the range of fH_2 consistent with equilibration with CH_4 expands slightly, but that for equilibration with CO_2 remains large. Therefore, in the context of this plausibility argument, it appears that

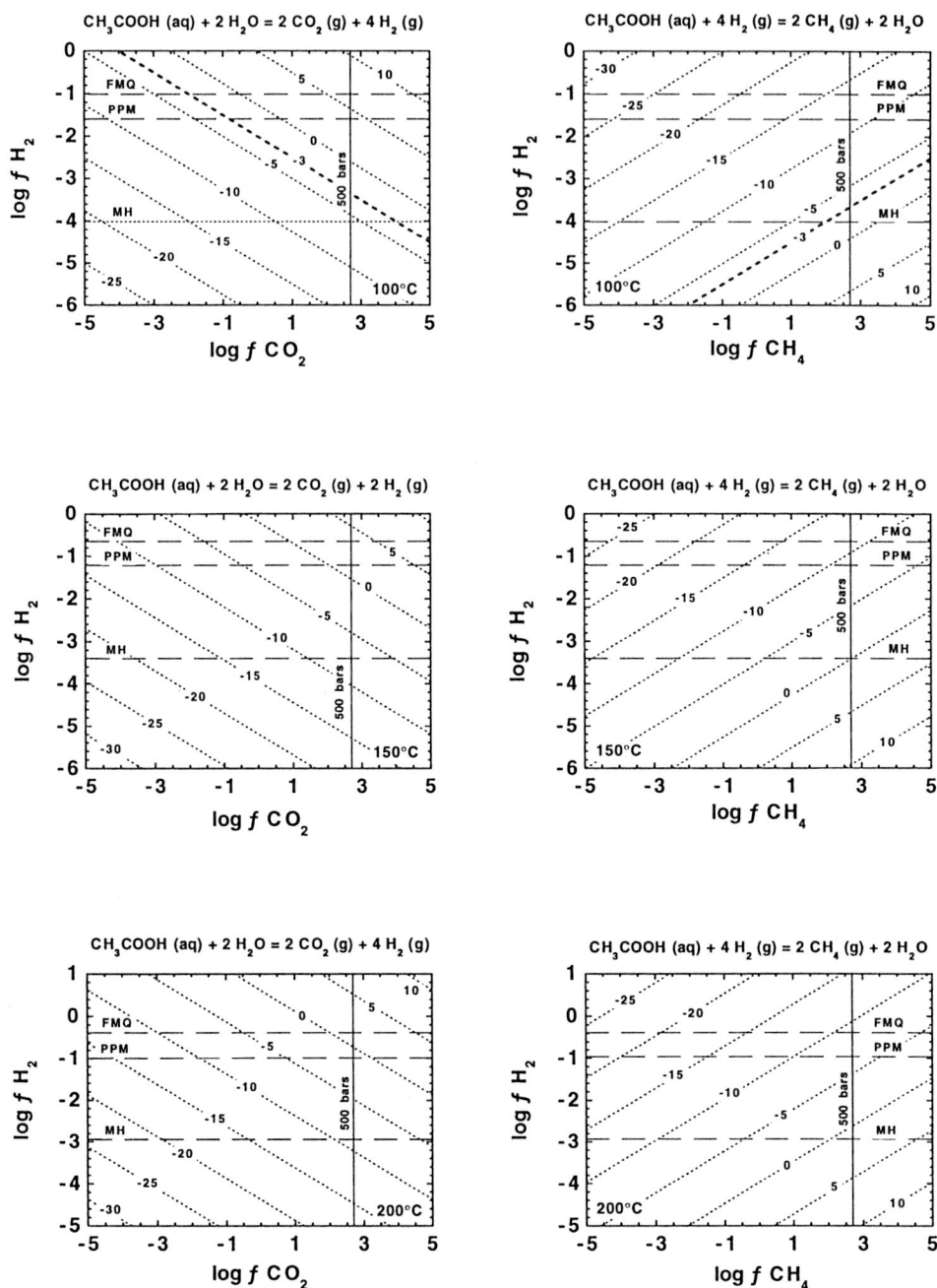

Fig. 9. Contours of log activity of acetic acid (*dotted lines*) in terms of log fH_2, and log fCO_2 or log fCH_4 at 100, 150, and 200 °C. *Bold contours* in 100 °C plots refer to log activity of −3 (see text). *Dashed lines* indicate values of log fH_2 set by various mineral assemblages, and *vertical solid lines* indicate the value of log fCO_2 or log fCH_4 corresponding to 500 bar, the total pressure considered in these calculations

the potential for metastable equilibrium between acetic acid and CO_2 is large under geologically realistic conditions, but that the potential for metastable equilibrium involving acetic acid and CH_4 is severely limited. As shown below, the values of fH_2 obtained from the abundances of acetic and propanoic acids in natural brines are consistent with metastable equilibrium involving CO_2, but not CH_4.

3.3 Organic Acids as Redox Tracers

The argument outlined above shows that metastable equilibrium between CO_2 and acetic acid is plausible at fH_2 values buffered by mineral assemblages, but it is desirable to place more rigorous constraints on an argument of this type. The assumption that oxidation states are set by a mineral buffer assemblage is difficult to test in the natural system, because it is all too often that rocks are not available from the reservoirs from which brines have been collected. However, it is possible that the organic acids in the brines can be used as tracers of the prevailing oxidation state under the reservoir conditions (Shock 1988, 1989, 1990a). If we consider homogeneous metastable equilibrium between acetic and propanoic acids, we can test the possibility that the ratios of the acids provide such a tracer.

Homogeneous metastable equilibrium between acetic and propanoic acids implies reversibility of the overall reaction

$$3CH_3COOH + 2H_2(g) = 2CH_3CH_2COOH + 2H_2O. \tag{28}$$

The law of mass action expression for this reaction can be rearranged to yield

$$\log a\,CH_3CH_2COOH = 3/2\log a\,CH_3COOH + 1/2(\log K_{28} + 2\log fH_2). \tag{29}$$

If we adopt the likely assumption that conversions between molality and activity will be nearly the same for the two acids, we can test the hypothesis of homogeneous metastable equilibrium by constructing isothermal, logarithmic plots of organic acid molalities from analyses of natural solutions, which should exhibit slopes of 3/2. Examples of such plots are shown in Fig. 10. The data reported by Martens (1990) are for 80 °C pore fluids from hydrothermally altered sediments of the Guaymas Basin, Gulf of California. The samples from the Kettleman Dome reported by Carothers and Kharaka (1978) range from 95 to 101 °C. Both of these sets of data for natural solutions are consistent with the 3/2 slope required by Eq. (29). Other examples are given by Shock (1988, 1989).

The law of mass action expression for Eq. (28) can also be rearranged to yield

$$\log fH_2 = \log a\,CH_3CH_2COOH - 3/2\log a\,CH_3COOH - 1/2\log K_{28}. \tag{30}$$

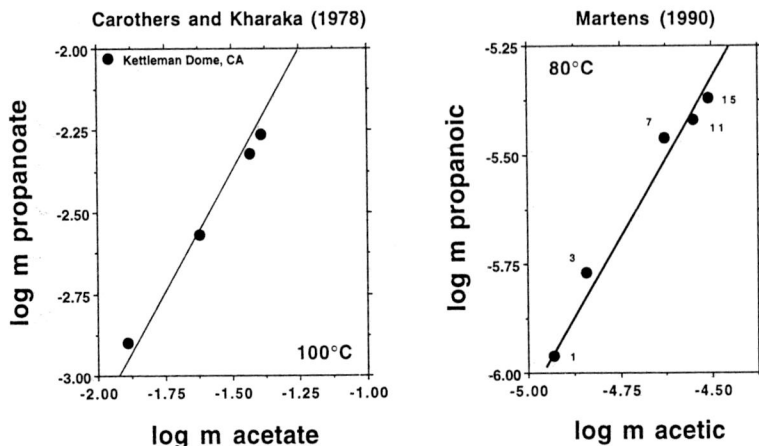

Fig. 10. Isothermal logarithmic plots of organic acid concentrations for natural samples. The curves are drawn with slopes of 3/2, which is indicative of metastable equilibrium with respect to reaction (28). *Numbers* on plot from data of Martens (1990) indicate depths (in cm) below the sediment/seawater interface

Values of $\log K_{28}$ can be calculated from the values of $\log K_3$ and $\log K_{11}$ from Table 3. These values can then be combined with activities of acetic and propanoic acids calculated from measured concentrations to evaluate $\log fH_2$. In this sense, the reported concentrations can be used as tracers of the oxidation state as revealed by the calculated fH_2 values. Methods for calculating activities from concentration measurements are described in the Appendix. Application of this methodology to the samples reported by Carothers and Kharaka (1978), for which both acetic and propanoic acids were measured, yields the values of $\log fH_2$ plotted in Fig. 11. Each symbol in Fig. 11 corresponds to an individual sample and is plotted against the temperature for that sample as reported by Carothers and Kharaka (1978). Also shown in Fig. 11 are curves corresponding to the values of $\log fH_2$ set by the PPM and MH assemblages [reactions (20) and (21)]. It can be seen that the calculated values of $\log fH_2$ for these oil-field brines are intermediate between those set by the PPM and MH assemblages, but they exhibit a trend, which is closely parallel to the mineral-buffered values.

Values of fH_2 revealed by the relative concentrations of acetic and propanoic acid can be used to test explicitly whether the organic acids can be in metastable equilibrium with CO_2, CH_4, hydrocarbons, or other organic compounds. At present, little is known about the concentrations of hydrocarbons and other organic compounds in sedimentary basin brines. Little is also known about the mixing of hydrocarbons in petroleum and how to evaluate activities of individual components in such complex mixtures. Nevertheless, thermodynamic calculations allow the construction of a variety of plausibility arguments regarding the compositional extent of metastable

Carothers & Kharaka (1978)

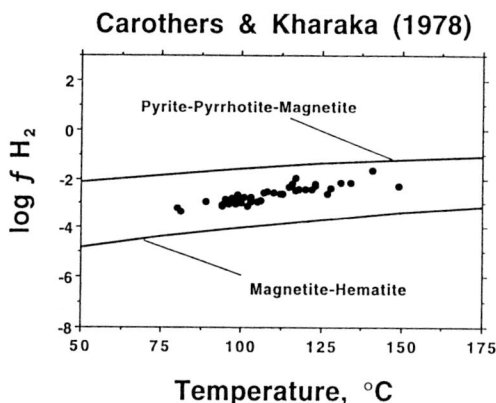

Fig. 11. Values of $\log f H_2$ calculated for samples reported by Carothers and Kharaka (1978) at the temperatures reported for each sample assuming that the relative concentrations of acetic and propanoic acid can be used as a redox tracer. Also shown are the positions of two mineral buffers

equilibrium states in sedimentary basins and other hydrothermal systems. Some of these arguments are developed below and others are presented by Shock (1988, 1989, 1990a,b, 1992b), Shock and Schulte (1990), Shock and McKinnon (1993), and Helgeson et al. (1991).

Values of $\log f H_2$, evaluated as described above and shown in Fig. 11, can be combined with the calculated activities of CH_3COOH and the values of $\log K_3$ from Table 3 to evaluate $\log f CO_2$ for any sample for which acetic and propanoic acids are reported. The samples from around 100 °C (Fig. 11) yield the values of $\log f H_2$ and $\log a CH_3COOH$ plotted in Fig. 12, which contains two contour plots like those shown in Fig. 9 on which calculated values of $\log f CO_2$ and $\log f CH_4$ are plotted. It can be seen that the values of $\log f CO_2$ correspond to partial pressures of CO_2 that are attainable, but that the corresponding partial pressures of CH_4 greatly exceed the total pressure. Analogous values of $\log f CO_2$ and $\log f CH_4$, evaluated for all of the samples reported by Carothers and Kharaka (1978), for which both acetic and propanoic acids are reported, are represented at the measured temperatures by the symbols in Fig. 13. Comparison to the 500-bar boundary indicates that equilibrium values of $\log f CH_4$ are unattainable at all temperatures. In contrast, with the exception of a single sample, calculated values of $\log f CO_2$ range from 0.7 to 2.5 and could be attained in sedimentary basins.

3.4   Controls on $f H_2$ in Sedimentary Basin Brines

As mentioned above, the trend of the $\log f H_2$ values with temperature (Fig. 11) is nearly parallel to those set by mineral buffer assemblages. Analysis of

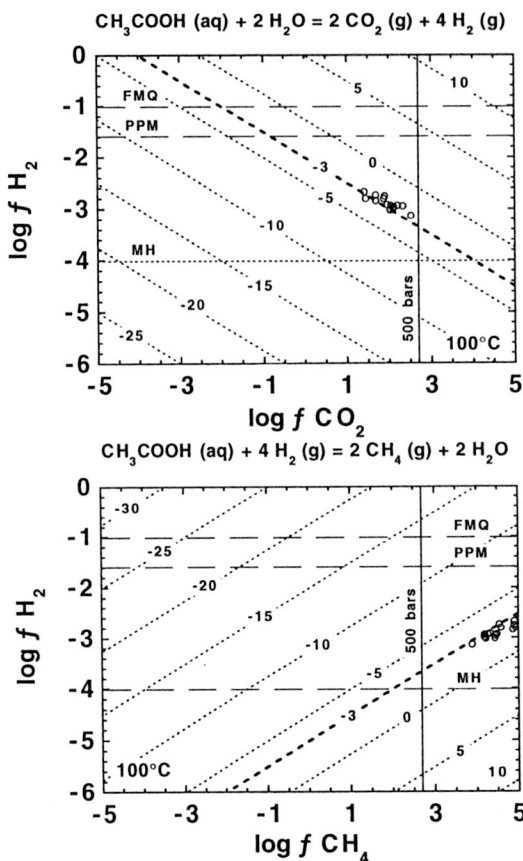

$$CH_3COOH\ (aq) + 2\ H_2O = 2\ CO_2\ (g) + 4\ H_2\ (g)$$

$$CH_3COOH\ (aq) + 4\ H_2\ (g) = 2\ CH_4\ (g) + 2\ H_2O$$

Fig. 12. Plots like those shown in Fig. 9 at 100 °C including samples reported by Carothers and Kharaka (1978) at 100 ± 5 °C. Locations of points are calculated with the assumption of metastable equilibrium with respect to Eq. (28) and the reported analyses for acetic and propanoic acids. The resulting values of $\log f CO_2$ are less than the total pressure for these calculations (500 bar) and indicate that the organic acids may be in metastable equilibrium with CO_2. By the same arguments, the samples are not in metastable equilibrium with CH_4

Fig. 12 indicates that metastable equilibrium with respect to reactions (3) and (28) does not appear to operate at the mineral-buffered values of $\log f H_2$ at 100 °C. It is possible that something else in the natural system might set the $f H_2$ values that the acids are tracing. One likely candidate is the mixture of hydrocarbons in petroleum. The plausibility of this hypothesis can be tested with additional calculations as described below.

To test the hypothesis that the hydrocarbons in petroleum control the $f H_2$, which the organic acids record, it would be desirable to have quantitative analyses of the petroleum, which coexists with each brine sample. Data of this type are not generally available. Nevertheless, we can construct

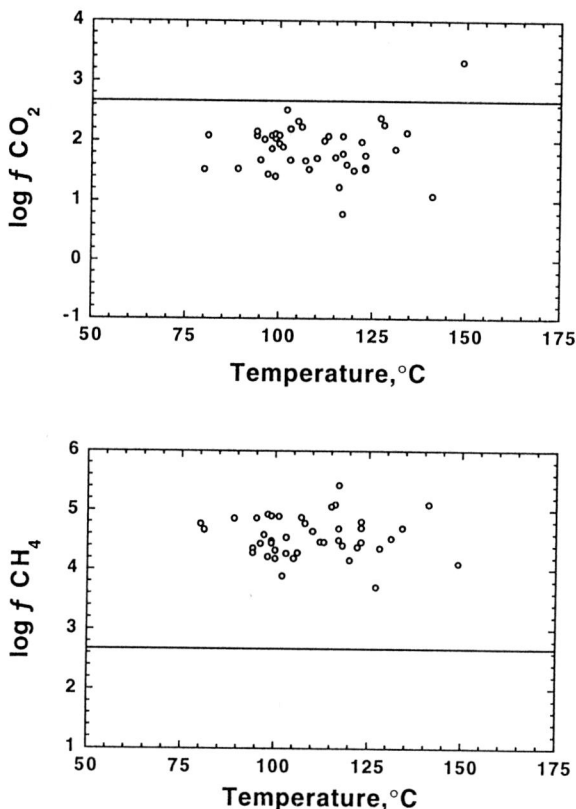

Fig. 13. Values of log fCO_2 and log fCH_4 calculated for each sample reported by Carothers and Kharaka (1978) for which both acetic and propanoic acids are given at the temperature reported for each sample. Fugacities are calculated with the equilibrium constants for reactions (3) and (9), using the values of fH_2 obtained from the acid concentrations and the assumption of metastable equilibrium with respect to reaction (28)

a plausibility argument to identify those hydrocarbons that may be in metastable redox equilibrium with the organic acids. As shown above, CH_4 cannot be part of this metastable assemblage. We can test various other hydrocarbons in a similar manner.

Examples of the many reactions that could be written involving oxidation/reduction reactions between organic acids and hydrocarbons are listed in Table 4. As in the tests of metastable equilibrium with respect to CO_2 and CH_4 described above, law of mass action expressions for the reactions in Table 4 can be rearranged to yield equations of lines in $\log fH_2$ vs. $\log aCH_3COOH$ space. An example is given by

$$\log fH_2 = -3/7\log aCH_3COOH + 1/7(\log K + \log aC_6H_{14}), \qquad (31)$$

294 E.L. Shock

Table 4. Reactions considered in testing redox equilibria between organic acids and hydrocarbons

$CH_4 + H_2O = 1/2CH_3COOH(aq) + 2H_2(g)$ methane	$C_2H_6 + 2H_2O = CH_3COOH(aq) + 3H_2(g)$ ethane
$C_3H_8 + 3H_2O = 3/2CH_3COOH(aq) + 4H_2(g)$ propane	$C_4H_{10} + 4H_2O = 2CH_3COOH(aq) + 5H_2(g)$ butane
$C_5H_{12} + 5H_2O = 5/2CH_3COOH(aq) + 6H_2(g)$ pentane	$C_6H_{14} + 6H_2O = 3CH_3COOH(aq) + 7H_2(g)$ hexane
$C_7H_{16} + 7H_2O = 7/2CH_3COOH(aq) + 8H_2(g)$ heptane	$C_8H_{18} + 8H_2O = 4CH_3COOH(aq) + 9H_2(g)$ octane
$C_9H_{20} + 9H_2O = 9/2CH_3COOH(aq) + 10H_2(g)$ nonane	$C_{10}H_{22} + 10H_2O = 5CH_3COOH(aq) + 11H_2(g)$ 2-methylnonane
$C_{15}H_{32} + 15H_2O = 15/2CH_3COOH(aq) + 16H_2(g)$ pentadecane	$C_{16}H_{34} + 16H_2O = 8CH_3COOH(aq) + 17H_2(g)$ hexadecane
$C_{20}H_{42} + 20H_2O = 10CH_3COOH(aq) + 21H_2(g)$ icosane	$C_{32}H_{66} + 32H_2O = 16CH_3COOH(aq) + 33H_2(g)$ dotriacontane
$C_6H_6 + 6H_2O = 3CH_3COOH(aq) + 3H_2(g)$ benzene	$C_7H_8 + 7H_2O = 7/2CH_3COOH(aq) + 4H_2(g)$ toluene
$C_8H_{10} + 8H_2O = 4CH_3COOH(aq) + 5H_2(g)$ 1,2,4,5-tetramethylbenzene	$C_{10}H_{14} + 10H_2O = 5CH_3COOH(aq) + 7H_2(g)$ ethylbenzene
$C_6H_{12} + 6H_2O = 3CH_3COOH(aq) + 6H_2(g)$ cyclohexane	$C_7H_{14} + 7H_2O = 7/2CH_3COOH(aq) + 7H_2(g)$ methylcyclohexane
$C_8H_{16} = 8H_2O + 4CH_3COOH(aq) + 8H_2(g)$ ethylcyclohexane	$C_5H_{10} + 5H_2O = 5/2CH_3COOH(aq) + 5H_2(g)$ cyclopentane
$C_{16}H_{10} + 16H_2O = 8CH_3COOH(aq) + 5H_2(g)$ pyrene	$C_{10}H_8 + 10H_2O = 5CH_3COOH(aq) + 4H_2(g)$ naphthalene
$C_{14}H_{10} + 14H_2O = 7CH_3COOH(aq) + 5H_2(g)$ anthracene	$C_{11}H_{10} + 11H_2O = 11/2CH_3COOH(aq) + 5H_2(g)$ 1-methylnaphthalene

which will have a slope of $-3/7$ and an intercept that depends on $\log K$ for the reaction and the activity of hexane (see Fig. 14). Calculating values of $\log K$ for reactions of this type requires data for aqueous organic acids (see above), and standard state data for hydrocarbons. The standard state for hydrocarbons adopted in this study corresponds to the pure compound at any temperature and pressure. There are numerous studies that report standard state volumes, isobaric heat capacities, and entropies, as well as standard state Gibbs free energies and enthalpies of formation for hydrocarbons. Thermodynamic data for hydrocarbons used in this study are taken from compilations presented by (Stull et al. 1969; Domalski 1972; Domalski et al. 1984; Domalski and Hearing 1990).

The plots shown in Fig. 14 are constructed with equilibrium constants at 100 °C for reactions from Table 4. In each case, contours of $\log$ (a hydrocarbon) at 2, 0, and -2 are shown. Also shown in these plots are values of $\log f H_2$ and $\log a$ acetic acid from oil-field brine samples within a few degrees of 100 °C calculated from analyses reported by Carothers and Kharaka (1978), Kharaka et al. (1986), Fisher (1987), Barth (1987b),

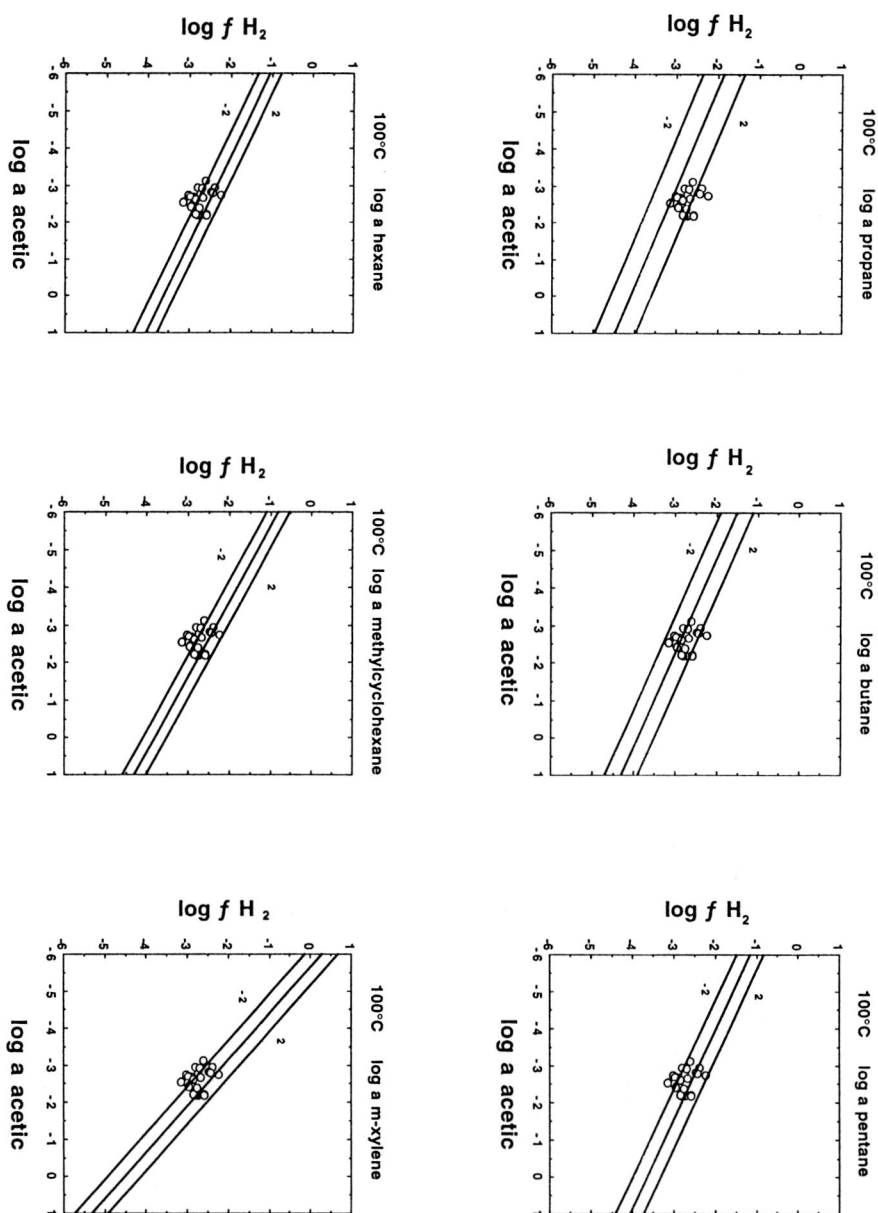

Fig. 14. Plots of log fH$_2$ vs. log a acetic acid showing samples (*symbols*) at 100 ± 5 °C calculated from acetic and propanoic concentrations from several analytical studies (see text). Also shown are contours of the log activity of several hydrocarbons at 2, 0, and −2

MacGowan and Surdam (1988), and Fisher and Boles (1990). In the case of propane and butane, the natural samples are consistent with fugacities (1 to 1000 bar) that the probably higher than the partial pressures of these gases in oil reservoirs. Like CH_4, propane and butane appear to be excluded from the metastable equilibrium assemblage of organic compounds. It is not so easy to exclude the other hydrocarbons shown in Fig. 14, as many natural samples plot near activities around 10^{-2} for the hydrocarbons in the coexisting petroleum. Although it is difficult to test this hypothesis directly without a model for the mixing of hydrocarbons in petroleum, the results in Fig. 14 suggest that many hydrocarbons in petroleum may be at concentrations consistent with redox equilibria involving the coexisting oil-field brine.

The slopes of the contour lines in Fig. 14 vary with the stoichiometry of the hydrocarbons. This is emphasized in the plot shown in Fig. 15, where the straight lines correspond to activities of 10^{-2} for several alkanes, cycloalkanes, and aromatic hydrocarbons found at or below the 1% level in petroleum. An intersection between any two lines yields values of $\log fH_2$ and $\log a\, CH_3COOH$ consistent with the two hydrocarbons having activities of 10^{-2} in petroleum. Note that the lines tend to intersect in the region of the diagram where the symbols representing calculated fH_2 values traced by organic acid ratios for the natural samples fall. The correspondence of lines and symbols in Fig. 15 suggests that the mixture of hydrocarbons in petroleum influences, or controls, the ratios of the organic acids in the

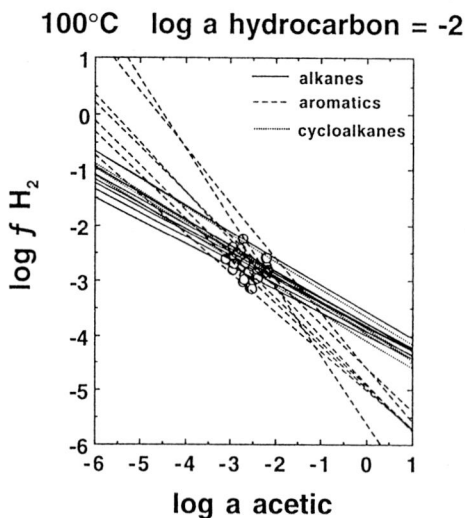

Fig. 15. Plot of $\log fH_2$ vs. $\log a$ acetic acid showing the same samples as Fig. 14, as well as location of activity contours at 10^{-2} for several alkanes, cycloalkanes, and aromatic hydrocarbons (see Table 4)

coexisting brine through a network of oxidation/reduction reactions. This hypothesis can be tested to a limited extent by comparison with the well-studied Ponca City #6 crude, which was designated as a representative crude petroleum by the American Petroleum Institute in 1928 (Rossini et al. 1953; Rossini and Mair 1959).

To illustrate this point, and to test the hypothesis that hydrocarbon mixtures can control the fH_2 revealed by organic acid ratios, we can use the composition of the Ponca City #6 crude and two pairs of hydrocarbons whose contours intersect in Fig. 15. For the sake of this illustration we can use benzene and hexane, and pentane and m-xylene. If we make the assumption that hydrocarbons mix ideally in petroleum, then we can estimate activities of hydrocarbons in the Ponca City #6 crude from their volume percentages. These estimates give $\log a$ benzene $= -1.78$, $\log a$ hexane $= -0.86$, $\log a$ pentane $= -0.81$ and $\log a$ m-xylene $= -1.38$. These activities, together with values of $\log K_{22}$ and $\log K_{23}$, allow evaluation of the corresponding $\log fH_2$ values and comparison to the values of fH_2 calculated from the organic acid concentrations described above. This type of comparison has to be done in the context of a plausibility argument where we substitute the well-studied Ponca City #6 crude for the real, but uncharacterized, crudes, which coexist with each oil-field brine sample. Results of this approach are plotted in Fig. 16 together with values of $\log fH_2$ evaluated from Carothers and Kharaka's (1978) samples. It can be seen that there is close agreement between the $\log fH_2$ estimates from the composition of the Ponca City #6 crude and those evaluated from the organic acids. This analysis suggests that the mixture of hydrocarbons in petroleum may indeed set the fH_2 values recorded by organic acids in high-temperature oil-field

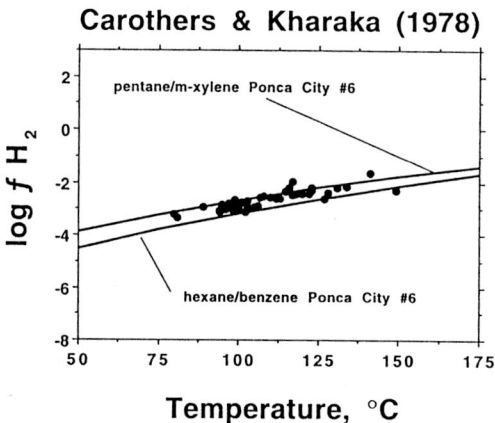

Fig. 16. Plot like that shown in Fig. 11 with curves showing values of $\log fH_2$ consistent with ratios of hydrocarbons in the Ponca City #6 crude

brines. Supporting evidence for such reactions comes from hydrous pyrolysis experiments.

3.5 Oxidation States in Hydrous Pyrolysis Experiments

Several investigators have conducted hydrous pyrolysis experiments on kerogen and source rocks and report yields of organic acids (Kawamura et al. 1986; Barth et al. 1987, 1989; Eglinton et al. 1987; Lundegard and Senftle 1987; Thornton and Seyfried 1987; Seewald et al. 1990). These results can be converted into molalities by accounting for the mass of reactants and volume of H_2O used in the experiments (see Appendix). The resulting molalities from several studies are depicted in the plot shown in Fig. 17. Temperatures used in these studies range from 250 to 350°C. The two dashed lines with slopes of 3/2 are consitent with the stoichiometry of reaction (28) and reflect, schematically, the change in equilibrium constant for reaction (28) with temperature. The close agreement with a 3/2 slope distribution suggests that metastable equilibrium states are achieved in these experiments, and that these states are similar to those attained naturally in sedimentary basins. Although simulation of the natural system is the goal of hydrous pyrolysis experiments, this is the first time that the results have

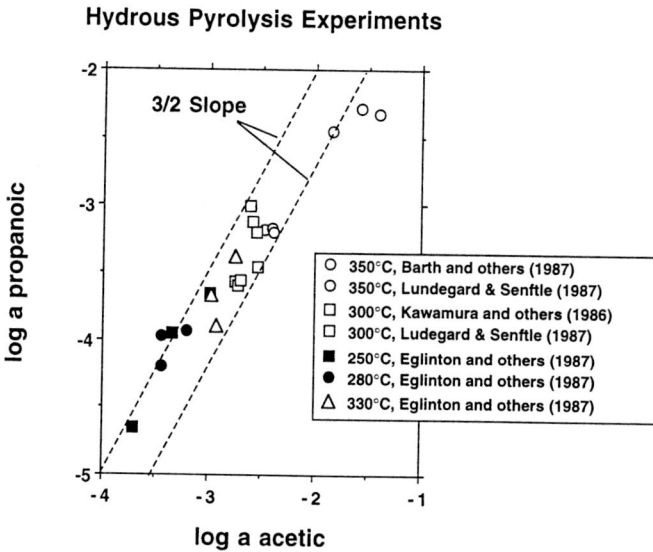

Fig. 17. Logarithmic plots of organic acid concentrations for hydrous pyrolysis experiments using kerogen (Eglinton et al. 1987) and source rocks (Kawamura et al. 1986; Barth et al. 1987; Lundegard and Senftle 1987). The lines drawn with 3/2 slopes are consistent with homogeneous metastable equilibrium between the two aqueous acids in these experiments (reaction 28) (see text)

been placed into the same thermodynamic frame of reference as the natural samples.

Although the pH at the temperatures of the experiments is never buffered and seldom considered in the experimental design, speciation calculations like those used to construct Fig. 5 indicate that organic acids are likely to be highly associated at the temperatures used in hydrous pyrolysis experiments. Therefore, it can be assumed that the molalities calculated from the yields reported for the experiments (see Appendix) can be equated with activities without taking account of acid dissociation (unlike the situation described above for sedimentary basin brines). These activities, together with values of $\log K$ for reaction (28) at the experimental temperatures and pressures, allow estimates of $\log f H_2$ appropriate for each experiment.

Results of these calculations for data reported by Lundegard and Senftle (1987) are shown in Table 5 for 72-h experiments at five temperatures, and isothermal, 270 °C, experiments run from 72 to 360 h. Note that the calculated values of $\log f H_2$ decrease with longer duration in the experiments conducted at 270 °C, reaching negative values after ~240 h. These more negative values of $\log f H_2$ are consistent with those calculated from the ratios of hydrocarbons in the Ponca City #6 crude at the same temperature and pressure as shown in Fig. 18, which allows comparison of values of $\log f H_2$ calculated from Carothers and Kharaka's (1987) analyses together with those appropriate for several hydrous pyrolysis studies. It can also be seen in Fig. 18 that experiments of the same duration (72 h, uppermost symbols from 250 to 350 °C) approach the crude-oil curves more closely with

Table 5. Calculated fugacities of H_2 consistent with concentrations of actetate and propanoate from hydrous pyrolysis experiments reported by Lundegard and Senftle (1987)

72-h Experiments	
T (°C)	Calculated $\log f H_2$
250	0.770
270	0.670
300	0.420
325	0.460
350	0.600

270 °C Experiments	
Hours	Calculated $\log f H_2$
72	0.670
120	0.030
240	−0.450
360	−0.380

Sedimentary Basins and Hydrous Pyrolysis

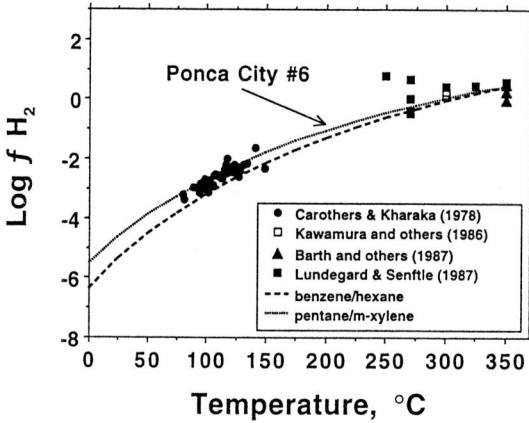

Fig. 18. Plot of log fH$_2$ against temperature shown in the trajectories of the two Ponca City #6 hydrocarbon curves at P$_{sat}$ together with results for natural samples below 150 °C and hydrous pyrolysis experiments at temperatures >250 °C

increasing temperature. This analysis suggests that long durations may allow metastable states to develop between hydrocarbons and organic acids in hydrous pyrolysis experiments. The extent to which metastable states appear to prevail in natural environments at temperatures ≥80 °C indicates that longer-duration hydrous pyrolysis experiments may simulate processes occurring in the natural setting. This type of quantitative analysis should be more rigorous than comparison of gas chromatograms and other semi-quantitative methods often used in support of hydrous pyrolysis experiments. Close agreement between both types of data and the trajectory of the Ponca City #6 crude suggests that oxidation states in both settings are controlled by the mixture of hydrocarbons in petroleum. This hypothesis should be tested by detailed analyses of brines, petroleums, gases, and mineral assemblages in sedimentary basins and in carefully constrained laboratory studies.

3.6 Oxidation States in Low-Temperature Brines

There are numerous studies of organic acids in sedimentary basin solutions, which report both acetic and propanoic acid concentrations. As in the case of the samples collected by Carothers and Kharaka (1978), these are generally not accompanied by analyses of the coexisting petroleum, natural gas, or mineral assemblages. In many cases, little is known about the inorganic composition of the same solutions. Nevertheless, by applying the same assumptions and calculation methods used above for the oil-field

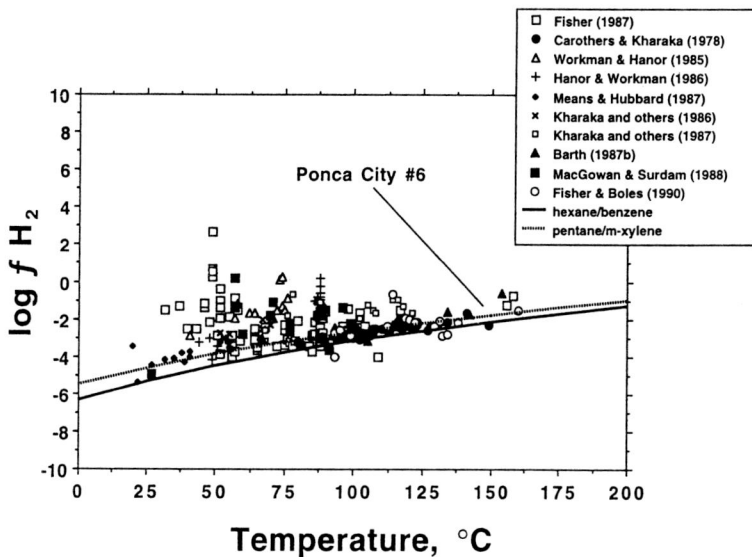

Fig. 19. Plot of log fH$_2$ against temperature showing the values of fH$_2$ calculated for each of the samples reported in the references given for which both acetic and propanoic acids were measured (see text and appendix). Also shown are the values of log fH$_2$ consistent with ratios of hydrocarbons in the Ponca City #6 crude

brines reported by Carothers and Kharaka (1978), estimates of log$f$H$_2$ can be obtained from each of these analyses. The results of these calculations are shown in Fig. 19 for 232 samples for which acetic and propanoic acid concentrations are reported. Note that with increasing temperature the values of log$f$H$_2$ calculated from organic acid analyses tend to converge on those estimated from pairs of hydrocarbon concentrations in the Ponca City #6 crude. This may indicate that metastable states between hydrocarbons and organic acids are more easily or rapidly attained at higher temperatures. On the other hand, at low temperature there is considerably more scatter in the distribution of calculated fH$_2$ values. Uncertainties in the calculation of pH for these solutions (see Appendix), as well as analytical uncertainties affecting low concentrations of acids, may contribute to this scatter. Note, however, that few samples plot at fH$_2$ values that are more oxidized than those indicated by the crude oil curves. Either the kinetics of reactions among organic compounds in these low-temperature environments are so slow that metastable states do not form (at temperatures $\leq$80 °C), or acetic acid is actively depleted in these solutions relative to propanoic acid. Micro-organisms may flourish under these low-temperature conditions, and many are known to metabolize carboxylic acids (Warford et al. 1979; Shaw et al. 1984; Gelwicks et al. 1989; Pronk et al. 1991; Blair and Carter 1992; among many others). Perhaps these organisms have a preference for acetic acid or acetate in these environments.

4   Organic Acids in Subsurface Metabolic Processes

Evidence summarized above indicates that using organic acids to evaluate H_2 fugacities in natural solutions, or those generated in hydrous pyrolysis experiments, is most reliable if the isothermal distribution of the acid concentrations is consistent with reaction (28) and the type of plots shown in Figs. 10 and 17. The thermodynamic analyses described above reveal nothing about the specific mechanisms through which the observed metastable states are achieved. One possibility is that bacteria enable these metastable states through their combined metabolic processes. On the other hand, bacteria may also be responsible for departures from the metastable equilibria suggested above in the discussion of the low-temperature distribution of points in Fig. 19. This may be possible if sufficient energy can be provided by other metabolic reactions which allow the system to be perturbed away from the local energy minimum characteristic of a metastable state.

Table 6. Extreme thermophiles, their preferred temperature ranges, metabolic energy sources, and locations

Sulfolobus (sulfur oxidizers)	85–90 °C Hot springs, Yellowstone	Brock et al. (1972)
Pyrodictium (anaerobic sulfur reducers)	82–110 °C Hydrothermal vents, Volcano, Italy	Stetter et al. (1983)
Hyperthermus (fermenting sulfur reducers)	95–112 °C Hydrothermal vent, Azores	Zillig et al. (1990)
Methanococcus (methanogens)	50–86 °C Hydrothermal vents, 21 °N East Pacific Rise	Jones et al. (1993)
Methanothermus (methanogens)	57–97 °C Continental solfatara fields, Iceland	Lauerer et al. (1986)
Methanopyrus (methanogens)	85–110 °C Hydrothermal vents, Guaymas Basin	Huber et al. (1989)
Thermococcus (organotrophic sulfur reducers)	70–95 °C Submarine hydrothermal systems	Zillig et al. (1983)
Pyrococcus (organotrophic sulfur reducers)	70–103 °C Submarine hydrothermal systems	Fiala and Stetter (1986)
Staphylothermus (heterotrophic sulfur reducers)	85–92 °C Volcano, Italy; East Pacific Rise	Fiala et al. (1986)

Organic acids are central to several well-studied metabolic processes involved in methanogenesis and sulfur metabolism in surface environments and during sediment diagenesis (see Thauer et al. 1977, for a review). Similar metabolic processes may operate at higher temperatures. Considerable evidence concerning archaebacteria, which thrive at temperatures $\geqslant 100\,°C$, is now available from continental and submarine hot springs (see Table 6), and organisms may be present at this and higher temperatures throughout sedimentary basins. Thermodynamic data for metabolic reactions involving organic acids are widely used at low temperatures in the study of bioenergetics (Thauer et al. 1977; Thauer 1990; Goldberg et al. 1991; Tewari and Goldberg 1991; Alberty 1992; among many others). It is unlikely that data at 25 °C and 1 bar are applicable to deep subsurface environments because the thermodynamic properties of such reactions change dramatically with temperature and pressure, as shown by the examples described in this section.

4.1 Dehydrogenation

Numerous dehydrogenation reactions involving organic acids are employed by bacteria at 25 °C (Thauer et al. 1977). Examples include:

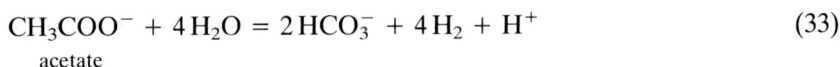

$$HCOO^- + H_2O = HCO_3^- + H_2 \tag{32}$$

formate

$$CH_3COO^- + 4\,H_2O = 2\,HCO_3^- + 4\,H_2 + H^+ \tag{33}$$

acetate

and

$$C_2H_5COO^- + 7H_2O = 3HCO_3^- + 7H_2 + 2H^+. \tag{34}$$

propanoate

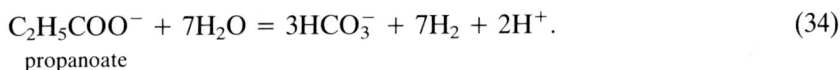

Values of $\Delta\bar{G}_r^°$ at elevated temperatures and pressures can be calculated for these and many other reactions from data and parameters given by Shock and Helgeson (1988, 1990) and Shock (1993a,b). Standard partial molal Gibbs free energies of reaction are typically recalculated in biochemical applications according to (Alberty 1992)

$$\Delta\bar{G}_r^{°\prime} = \Delta\bar{G}_r^° + \nu_{H^+,r}\Delta\bar{G}'_{H^+}, \tag{35}$$

where

$$\Delta\bar{G}'_{H^+} = 2.303\,RT\log a_{H^+}, \tag{36}$$

$\nu_{H^+,r}$ corresponds to the stoichiometric number of moles of H^+ in the reaction, which is positive when H^+ is a product and negative when H^+ is a reactant, and the activity of H^+ used is that of neutrality at the temperature and pressure of interest. These calculations are usually done at 25 °C with

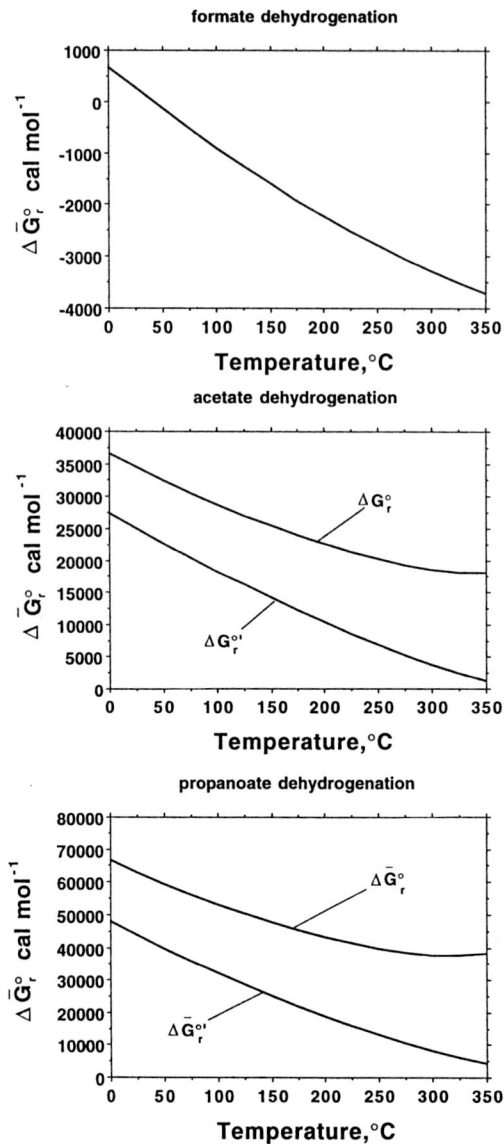

Fig. 20. Plots of standard Gibbs free energies for dehydrogenation reactions of formate [reaction (32)], acetate [reaction (33)], and propanoate [reaction (34)] as functions of temperature at P_{sat}. Values of $\Delta\bar{G}_r^{o\prime}$ used in biochemical reaction networks are also shown (see text)

pH 7, and $\Delta\bar{G}'_{H^+} = -9551 \, \text{cal} \, \text{mol}^{-1}$. The pH of neutrality changes with temperature and pressure, and it follows that $\Delta\bar{G}'_{H^+}$ can be calculated from equilibrium constants for the reaction

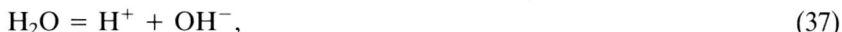

$$H_2O = H^+ + OH^-, \tag{37}$$

and combined with values of $\Delta\bar{G}°_r$ through Eq. (35) to obtain $\Delta\bar{G}°'_r$. Plots of $\Delta\bar{G}°_r$, for reaction (32), and $\Delta\bar{G}°_r$ and $\Delta\bar{G}°'_r$ for reactions (33) and (34), are shown in Fig. 20. It can be seen that in each case $\Delta\bar{G}°_r$ decreases with increasing temperature [note that $\Delta\bar{G}°_r = \Delta\bar{G}°'_r$ for reaction (32)], indicating that *less* energy is required to drive these dehydrogenation reactions at elevated temperatures than at 25 °C. Values of $\Delta\bar{G}°_r$ for reaction (32) are negative at temperatures above ~40 °C, which means that energy could be gained from this dehydrogenation reaction. By the same argument, acetate and propanoate dehydrogenation reactions *would not* be energy sources at any temperature encountered in sedimentary basin brines. These calculations may help to explain the observation that formate and formic acid are usually found at much lower concentrations in oil-field brines than acetate, pro- panoate and, in many instances, other organic acids.

4.2 Hydrogenation

Dehydrogenation reactions are coupled through enzymes to hydrogenation and other reactions in order to drive metabolic processes. Several hydro- genation reactions have negative values of $\Delta\bar{G}°'_r$ at low temperature (referred to as an exergonic reaction), and are employed as energy sources in me- tabolic processes. One example of a hydrogenation reaction is given by

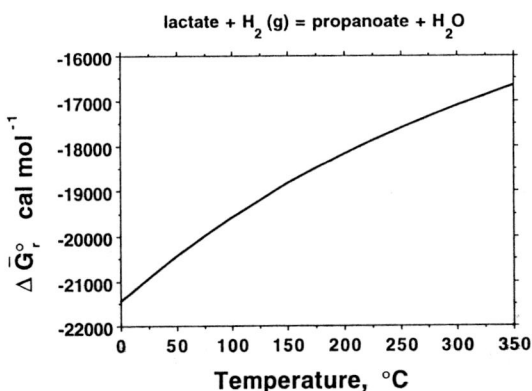

Fig. 21. The standard partial molal Gibbs free energy for reaction (38) as a function of temperature at P_{sat}. This hydrogenation reaction can be coupled with the dehydrogenation reactions depicted in Fig. 20 to drive metabolic processes

$$C_2H_4OHCOO^- + H_2(g) = C_2H_5COO^- + H_2O, \tag{38}$$

$$\qquad\text{lactate} \qquad\qquad\qquad\qquad \text{propanoate}$$

and values of $\Delta G_r^{o'}$ are plotted for this reaction in Fig. 21. Because H^+ does not appear in this reaction, $\Delta \bar{G}_r^{o'} = \Delta \bar{G}_r^{o}$. Note that $\Delta \bar{G}_r^{o'}$ becomes less negative with increasing temperature, but that the increase in $\Delta G_r^{o'}$ is relatively small (about $3000\,\text{cal mol}^{-1}$ from 25 to 300 °C). Therefore, reaction (38) remains exergonic over a wide range of temperatures that might be encountered in the subsurface. If reaction (38) were coupled with dehydrogenation reaction (32), the overall process would also remain exergonic at elevated temperatures, indicating that energy would be gained by an organism, or group of organisms, which consume lactate and formate and produce propanoate and bicarbonate.

4.3 Sulfate Reduction

Lactate is also involved in overall sulfate reduction reactions such as these employed by the *Desulfovibrio* bacteria. One such reaction is given by (Decker et al. 1970)

$$2C_2H_4OHCOO^- + SO_4^{-2} = 2CH_3COO^- + 2HCO_3^- + H_2S(g). \tag{39}$$

This reaction does not involve H^+, so $\Delta \bar{G}_r^{o'} = \Delta \bar{G}_r^{o}$ as in the case of reaction (38). Values of $\Delta \bar{G}_r^{o'}$ for this reaction are plotted against temperature in Fig. 22 where it can be seen that the reaction becomes increasingly exergonic with increasing temperature. As a result, sulfate reduction appears to be an excellent source of metabolic energy throughout sedimentary basin con-

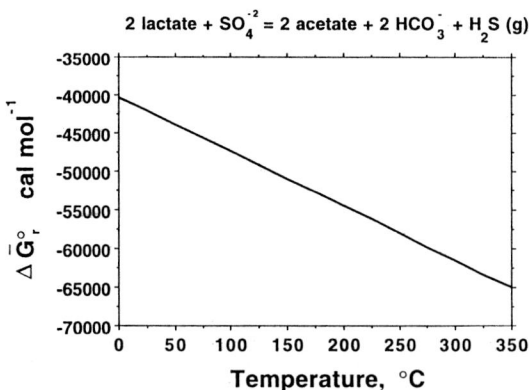

Fig. 22. The standard partial molal Gibbs free energy of reaction (39) as a function of temperature at P_{sat}. This overall sulfate reduction reaction is used at low temperatures by *Desulfovibrio* organisms, and it can be seen that more energy is available from this reaction at elevated temperatures

ditions. Comparison with Fig. 20 shows that reaction (39) becomes increasingly exergonic with increasing temperature as the dehydrogenation reactions are decreasingly endergonic, indicating that coupled metabolic reactions may provide more energy for organisms living at elevated temperatures than the same reactions can at 25 °C and 1 bar. Calculations of this type may help to elucidate the geochemical controls on microbial sulfate reduction which can apparently occur at temperatures up to at least 110 °C (Jørgensen et al. 1992).

4.4 Methanogenesis

As in the case of sulfate reduction, methanogenesis can produce more energy at elevated temperatures than at 25 °C and 1 bar. Consider the overall acetate reduction reaction

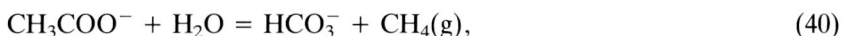

$$CH_3COO^- + H_2O = HCO_3^- + CH_4(g), \tag{40}$$

for which values of $\Delta \bar{G}_r^{o\prime}$ (equal to $\Delta \bar{G}_r^o$) are plotted in Fig. 23. From the discussion of the decarboxylation of CH_3COOH above, together with the $\log K$ values for reaction (5) shown in Table 3, it would be expected that reaction (40) becomes increasingly exergonic with increasing temperature. However, it might not be expected that methanogensis by way of reaction (40) would produce nearly 50% more energy at 100 than at 25 °C, and that methanogenesis is likely to be an energetically profitable metabolic reaction throughout sedimentary basin conditions. In addition, consideration of the calculations shown in Figs. 22 and 23 for reactions (39) and (40) suggests that consortia of sulfate reducing and methanogenic microorganisms could

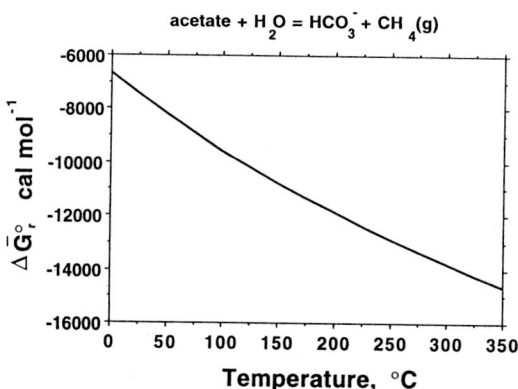

Fig. 23. The standard partial molal Gibbs free energy of reaction (40) as a function of temperature at P_{sat}. Note that the available energy from this overall methanogenesis reaction increases with increasing temperature

thrive at elevated temperatures. Their combined effects could convert organic compounds to methane and carbonate given a steady supply of sulfate. Perhaps it is not surprising that many of the extremely thermophilic microorganisms listed in Table 6 are either methanogens or sulfur metabolizers.

5 Concluding Remarks

The examples described above reveal some of the many applications of thermodynamic calculations to understanding geochemical processes involving organic acids. Many more applications in the area of metabolic processes are required before a full appreciation of the role of metabolism in geochemical processes can be obtained. Nevertheless, it appears that the thermodynamic properties of organic acids and the reactions in which they are involved do not pose barriers to life at high temperatures (see also Amend and Helgeson 1991; Shock 1992a). It also appears that at temperatures $\geqslant 100\,°C$ organic acids, petroleum, and CO_2 are converging toward metastable equilibrium states in sedimentary basin fluids. Identifying a link between the organic compounds and CO_2 allows reactions involving hydrocarbons and silicate minerals to be explicitly linked through pH. At the same time, organic redox reactions can be explicitly tied to redox reactions involving Fe, S, Cu, U, and other elements. The direction of progress of these coupled reactions will depend on the relative abundance of petroleum, aqueous fluids, and mineral phases in each environment. Therefore, it may be possible to isolate examples of natural systems where the petroleum composition can drive mineral dissolution and precipitation, or cases where a mineral assemblage, especially one capable of truly buffering the oxidation state without being exhausted, can effectively alter the relative abundances of organic acids in the coexisting aqueous phase, as well as the composition of the coexisting petroleum. These latter possibilities have implications for the overall cycle of carbon and nitrogen between the mantle, crust, oceans, and atmosphere, as well as the origin of life as discussed by Shock (1990a, 1992b). In any event, the results described above provide evidence that the theoretical framework is now in place to identify crucial experiments, and help to guide the sampling and analysis of natural solutions.

Acknowledgments. The research leading to this chapter has benefitted from many useful and enthusiastic discussions with Harold Helgeson who is responsible for getting me involved with the thermodynamic properties of organic compounds and their behavior in geochemical processes. Discussions with many other investigators have influenced the way I have approached research on organic acids including Dimitri Sverjensky, J.K. Böhlke, Laura Crossey, Dick Kettler, Barbara Ransom, Eric Oelkers, Bill Murphy, Peter

Lichtner, Jim Johnson, Ken Jackson, Nick Rose, Wendy Harrison, Leigh Price, John Kerridge, Sherwood Chang, Lynn Walter, Tom Giordano, Jeff Hanor, Lynton Land, Mike Lewan, Paul Lundegard, Bert Fisher, Julie Bell, Dave Wesolowski, Ed Drummond, Mike Engel, Steve Macko, Jan Amend, Jeremy Fein, Geoff Thyne, Dave Sassani, Mitch Schulte, Tom McCollom, Ron Surdam, and Yousef Kharaka, many of whom may disagree with some of the arguments presented in this chapter. Technical assistance from Patty DuBois, Carla Koretsky, and Michelle Morgan is greatly appreciated, as are critical comments of an earlier version of this chapter by Mike Lewan and Ed Pittman. This research was funded by NSF grants EAR-8803822 and EAR-9018468, as well as the donors of the Petroleum Research Fund of the American Chemical Society through grants 20266-G2 and 23870-AC8.

References

Abercrombie HJ (1991) Reservoir processes in steam-assisted recovery of bitumen, Leming pilot, Cold Lake, Alberta, Canada; compositions, mixing and sources of co-produced water. Appl Geochem 6: 495–508

Ackermann T, Schreiner F (1958) Molwärmen und Entropien einiger Fettsäuren und ihrer Anionen in wässriger Lösung. Z Elektrochem 62: 1143–1151

Alberty RA (1992) Equilibrium calculations on systems of biochemical reactions at specified pH and pMg. Biophys Chem 42: 117–131

Allred GC, Woolley EM (1981) Heat capacities of aqueous acetic acid, sodium acetate, ammonia, and ammonium chloride at 283.15, 298.15 and 313.15 K: ΔC_p° for ionization of acetic acid and for dissociation of ammonium ion. J Chem Thermodynamics 13: 155–164

Amend J, Helgeson HC (1991) Calculation of the relative stabilities at elevated temperatures and pressures of aqueous nucleosides, nucleotides, and other biochemical molecules required for bacterial metabolism in diagenetic processes. Geol Soc Am Abstr Programs 23: A212

Antweiler RC, Drever JI (1983) The weathering of a late Tertiary volcanic ash: importance of organic solutes. Geochim Cosmochim Acta 47: 623–629

Barth T (1987a) Quantitiative determination of volatile carboxylic acid in formation waters by isotachophoresis. Anal Chem 59: 2232–2237

Barth T (1987b) Multivariate analysis of aqueous organic acid concentrations and geological properties of North Sea reservoirs. Chemometrics Intelligent Lab Syst 2: 155–160

Barth T (1991) Organic acids and inorganic ions in waters from petroleum reservoirs, Norwegian continental shelf: a multivariate statistical analysis and comparison with American reservoir formation water. Appl Geochem 6: 1–15

Barth T, Borgund AE, Hopland AL, Graue A (1987) Volatile organic acids produced during kerogen maturation – amounts, composition and role in migration of oil. Adv Org Geochem 13: 461–465

Barth T, Borgund AE, Hopland AL (1989) Generation of organic compounds by hydrous pyrolysis of Kimmeridge oil shale-bulk results and activation energy calculations. Org Geochem 14: 69–76

Bennett PC, Melcer ME, Siegel DI, Hassett JP (1988) The dissolution of quartz in dilute aqueous solutions of organic acids at 25 °C. Geochim Cosmochim Acta 52: 1521–1530

Bevan J, Savage D (1989) The effect of organic acids on the dissolution of K-feldspar under conditions relevant to burial diagenesis. Mineral Mag 53: 415–425

Blair NE, Carter WD Jr (1992) The carbon isotope biogeochemistry of acetate from a methanogenic marine sediment. Geochim Cosmochim Acta 56: 1247–1258

Britton HTS (1925) Hydrogen and oxygen electrode titrations of some dibasic acids and of dextrose. J Chem Soc 127: 1896–1917

Brock TD, Brock KM, Belly RT, Weiss RL (1972) *Sulfolobus*: a new genus of sulfur-oxidizing bacteria living at low pH and high temperature. Arch Mikrobiol 84: 54–68

Carothers WW, Kharaka YK (1978) Aliphatic acid anions in oil-field waters-implications for origin of natural gas. Am Assoc Pet Geol Bull 62: 2441–2453

Choudhury NR, Ahluwalia JC (1982) Temperature dependence of heat capacities of sodium decanoate, sodium dodecanoate, and sodium dodecyl sulphate, in water. J Chem Thermodynamics 14: 281–289

Connolly CA, Walter LM, Baadsgaard H, Longstaffe FJ (1990) Origin and evolution of formation waters, Alberta Basin, Western Canada Sedimentary Basin. Appl Geochem 5: 375–395

Crossey LJ (1991) Thermal degradation of aqueous oxalate species. Geochim Cosmochim Acta 55: 1515–1527

Crossey LJ, Frost BR, Surdam RC (1984) Secondary porosity in laumontite-bearing sandstones. In: McDonald DA, Surdam RC (eds) Clastic diagenesis. Am Assoc Pet Geol Mem 37, pp 225–237

Crossey LJ, Surdam RC, Lahann RW (1986) Application of organic/inorganic diagenesis to porosity prediction. In: Gautier D (ed) Roles of organic matter in sediment diagenesis. Soc Econ Paleontol Mineral Spec Publ 38, pp 147–156

Darken LS (1941) The ionization constants of oxalic acid at 25° from conductance measurements. J Am Chem Soc 63: 1007–1011

Decker K, Jungermann K, Thauer RK (1970) Energy production in anaerobic organisms. Angew Chem Int Ed 9: 138–158

De Lisi R, Perron G, Desnoyers JE (1980) Volumetric and thermochemical properties of ionic surfactants: sodium decanoate and octylamine hydrobromide in water. Can J Chem 58: 959–969

Dickey PA, Collins AG, Fajardo I (1972) Chemical composition of deep formation waters in southwestern Louisiana. Am Assoc Pet Geol Bull 56: 1530–1533

Dippy JFJ, Lewis RH (1937) Studies of the ortho-effect. Part II. The dissociation constants of some o-substituted acids. J Chem Soc 1937: 1426–1429

Domalski ES (1972) Selected values of heats of combustion and heats of formation of organic compounds containing the elements C H N O P, and S. J Phys Chem Ref Data 1: 221–277

Domalski ES, Evans WH, Hearing ED (1984) Heat capacities and entropies of organic compounds in the condensed phase. J Phys Chem Ref Data 13(Suppl 1): 286 pp

Domalski ES, Hearing ED (1990) Heat capacities and entropies of organic compounds in the condensed phase, vol II. J Phys Chem Ref Data 19: 881–1047

Drucker C (1920) Weitere Untersuchungen über die Dissoziation ternärer Elektrolyte. Z Phys Chem 96: 381–427

Drummond SE, Palmer DA (1986) Thermal decarboxylation of acetate. Part II. Boundary conditions for the role of acetate in the primary migration of natural gas and the transportation of metals in hydrothermal systems. Geochim Cosmochim Acta 50: 825–833

Edman JD, Surdam RC (1986) Organic-inorganic interactions as a mechanism for porosity enhancement in the Upper Cretaceous Ericson sandstone, Green River Basin, Wyoming. In: Gautier D (ed) Roles of organic matter in sediment diagenesis. Soc Econ Paleontol Mineral Spec Publ 38: 85–109

Eglinton TI, Curtis CD, Rowland SJ (1987) Generation of water-soluble organic acids from kerogen during hydrous pyrolysis: implications for porosity development. Mineral Mag 51: 495–503

Ellis AJ (1963) The ionizaton of acetic, propionic, n-butyric and benzoic acid in water, from conductance measurements up to 225°. J Chem Soc 1963: 2299–2310

Everett DH, Wynne-Jones WFK (1939) The thermodynamics of acid-base equilibria. Trans Faraday Soc 35: 1380–1401

Fein JB (1991) Experimental study of aluminum-, calcium-, and magnesium-acetate complexing at 80 °C. Geochim Cosmochim Acta 55: 955–964

Fiala G, Stetter KO (1986) *Pyrococcus furiosus* sp. nov. represents a novel genus of a heterotrophic archaebacteia growing optimally at 100 °C. Arch Microbiol 145: 56–61

Fiala G, Stetter KO, Jannasch HW, Langworthy TA, Madon J (1986) *Staphylothermus marinus* sp. nov. represents a novel genus of extremely thermophilic submarine heterotrophic archaebacteria growing up to 98 °C. Syst Appl Microbiol 8: 106–113

Fischer A, Mann BR, Vaughan J (1961) Influence of pressure on the Hammett reaction constant: dissociation of benzoic acids and phenylacetic acids. J Chem Soc 1961: 1093–1097

Fisher JB (1987) Distribution and occurrence of aliphatic acid anions in deep subsurface waters. Geochim Cosmochim Acta 51: 2459–2468

Fisher JR, Barnes HL (1972) The ion-product constant of water to 350°. J Phys Chem 76: 90–99

Fisher JB, Boles JR (1990) Water-rock interaction in Tertiary sandstones, San Joaquin Basin, California, USA: diagenetic controls on water composition. Chem Geol 82: 83–101

Gelwicks JT, Risatti JB, Hayes JM (1989) Carbon isotope effects associated with auto-trophic acetogenesis. Org Geochem 14: 441–446

Giles MR, deBoer RB (1989) Secondary porosity: creation of enhanced porosities in the subsurface from the dissolution of carbonate cements as a result of cooling formation waters. Mar Pet Geol 6: 261–269

Giles MR, deBoer RB (1990) Origin and significance of redistributional secondary porosity. Mar Pet Geol 7: 378–397

Giles MR, Marshall JD (1986) Constraints on the development of secondary porosity in the subsurface: re-evaluation of processes. Mar Pet Geol 3: 243–255

Giordano TH (1985) A preliminary evaluation of organic ligands and metal-organic complexing in Mississippi Valley-type ore solutions. Econ Geol 80: 96–106

Giordano TH, Barnes HL (1981) Lead transport in Mississippi Valley-type ore solutions. Econ Geol 76: 2200–2211

Goldberg RN, Bella D, Tewari YB, McLaughlin MA (1991) Thermodynamics of hydrolysis of oligosaccharides. Biophys Chem 40: 69–76

Hamann SD, Strauss W (1955) The chemical effects of pressure. Part 3. Ionization constants at pressures up to 1200 atm. Trans Faraday Soc 51: 1684–1690

Hanor JS, Workman AL (1986) Distribution of dissolved volatile fatty acids in some Louisiana oil field brines. Appl Geochem 1: 37–46

Harned HS, Ehlers RW (1933) The dissociation constant of acetic acid from 0 to 60° centigrade. J Am Chem Soc 55: 652–656

Harned HS, Fallon LD (1939) The second ionization constant of oxalic acid from 0 to 50 degrees. J Am Chem Soc 61: 3111–3113

Harned HS, Sutherland RO (1934) The ionization constant of n-butyric acid from 0 to 60°. J Am Chem Soc 56: 2039–2041

Harrison WJ, Thyne GD (1992) Predictions of diagenetic reactions in the presence of organic acids. Geochim Cosmochim Acta 56: 565–586

Helgeson HC (1969) Thermodynamics of hydrothermal systems at elevated temperatures and pressures. Am J Sci 267: 729–804

Helgeson HC (1985) Some thermodynamic aspects of geochemistry. Pure Appl Chem 57: 31–44

Helgeson HC (1992) Calculation of the thermodynamic properties and relative stabilities of aqueous acetic and chloroacetic acids, acetate and chloracetates, and acetyl and chloroacetyl chlorides at high and low temperatures and pressures. Appl Geochem 7: 291–308

Helgeson HC, Kirkham DH, Flowers GC (1981) Theoretical prediction of the thermodynamic behavior of aqueous electrolytes at high pressures and temperatures. IV. Calculation of activity coefficients, osmotic coefficients, and apparent molal and standard and relative partial molal properties to 600°C and 5 KB. Am J Sci 281: 1249–1516

Helgeson HC, Knox A, Shock EL (1991) Petroleum, oil field brines and authigenic mineral assemblages: are they in metastable equilibrium in hydrocarbon reservoirs? 15th Int European Assoc Meet of Organic Geochemists, Manchester, Program and Abstracts, p 39

Hennet R, Crerar DA, Schwartz J (1988) Organic complexes in hydrothermal systems. Econ Geol 83: 742–764

Huang WH, Keller WD (1971) Dissolution of clay minerals in dilute organic acids at room temperature. Am Mineral 56: 1082–1095

Huang WH, Keller WD (1972a) Kinetics and mechanisms of dissolution of Fithian illite in two complexing organic acids. In: Serratosa JM (ed) Proc Int Clay Conf in Madrid, Spain. Tipografia Artistica, Madrid, pp 321–331

Huang WH, Keller WD (1972b) Organic acids as agents of chemical weathering of silicate minerals. Nature 239: 149–151

Huang WH, Keller WD (1972c) Geochemical mechanics for the dissolution, transport, and deposition of aluminum in the zone of weathering. Clays Clay Minerals 20: 69–74

Huang WH, Kiang WC (1972) Laboratory dissolution of plagioclase feldspars in water and organic acids at room temperature. Am Mineral 57: 1849–1859

Huber R, Kurr M, Jannasch HW, Stetter KO (1989) A novel group of abyssal methanogenic archaebacteria (*Methanopyrus*) growing at 110°C. Nature 342: 833–834

Jaffe R, Albrecht P, Oudin J-L (1988a) Carboxylic acids as indicators of oil migration. I. Occurrence and geochemical significance of C-22 diastereoisomers of the $(17\beta H, 21\beta H)$ C^{30} hopanoic acid in geological samples. Adv Org Geochem 13: 483–488

Jaffe R, Albrecht P, Oudin JL (1988b) Carboxylic acids as indicators of oil migration. II. Case of the Mahakam Delta, Indonesia. Geochim Cosmochim Acta 52: 2599–2607

Johnson JW, Oelkers EH, Helgeson HC (1992) SUPCRT92: a software package for calculating the standard molal thermodynamic properties of minerals, gases, aqueous species, and reactions from 1 to 5000 bars and 0° to 1000°C. Comput Geosci 18: 899–947

Jones WJ, Leigh JA, Mayer F, Woese CR, Wolfe RS (1983) *Methanococcus jannaschii* sp. nov., an extremely thermophilic methanogen from a submarine hydothermal vent. Arch Microbiol 136: 254–261

Jørgensen BB, Isaksen MF, Holger WJ (1992) Bacterial sulfate reduction above 100°C in deep-sea hydrothermal vent sediments. Science 258: 2756–2758

Kawamura K, Tannenbaum E, Huizinga, BJ, Kaplan IR (1986) Volatile organic acids generated from kerogen during laboratory heating. Geochem J 20: 51–59

Kettler RM, Palmer DA, Wesolowski DJ (1991) Dissociation quotients of oxalic acid in aqueous sodium chloride media to 175°C. J Solution Chem 20: 905–927

Kharaka YK, Lico MS, Wright VA, Carothers WW (1979) Geochemistry of formation waters from Pleasant Bayou No. 2 well and adjacent areas in coastal Texas. In: Dorfman MH, Fisher WL (eds) 4th Proc United States Gulf Coast Geopressured-Geothermal Energy Conf, Austin, Texas, pp 168–193

Kharaka YK, Carothers WW, Rosenbauer RJ (1983) Thermal decarboxylation of acetic acid: implications for origin of natural gas. Geochim Cosmochim Acta 47: 397–402

Kharaka YK, Law LM, Carothers WW, Goerlitz DF (1986) Role of organic species dissolved in formation waters from sedimentary basins in mineral diagenesis. In: Gautier D (ed) Roles of organic matter in sediment diagenesis. Soc Econ Paleontol Mineral Spec Publ 38, pp 111–122

Kharaka YK, Maest AS, Carothers WW, Law LM, Lamothe PJ, Fries TL (1987) Geo-chemistry of metal-rich brines from central Mississippi Salt Dome Basin, USA. Appl Geochem 2: 543–561

Konicek J, Wadsö I (1971) Thermochemical properties of some carboxlic acids, amines and n-substituted amides in aqueous solution. Acta Chem Scand 25: 1541–1551

Kurz JL, Farrar JM (1969) The entropies of dissociation of some moderately strong acids. J Am Chem Soc 91: 6057–6062

Land LS, MacPherson GL (1992) Origin of saline formation waters, Cenozoic section, Gulf of Mexico sedimentary basin. Am Assoc Pet Geol Bull 76: 1344–1362

Land LS, MacPherson GL, Mack LE (1988) The geochemistry of saline formation waters, Miocene, offshore Louisiana. Trans Gulf Coast Assoc Geol Soc 38: 503–511

Larsson E, Adell B (1931) Die elektrolytische Dissoziation von Säuren in Salzlösungen II. Die Dissoziationskonstanten einiger Fettsäuren und die Aktivitätsverhältnisse ihrer Ionen in Natriumchlorid- und Kaliumchloridlösungen. Z Phys Chem A157: 381–396

Lauerer G, Kristjansson JK, Langworthy TA, König H, Stetter KO (1986) Methano-thermus sociabilis sp. nov., a second species within the Methanothermaceae growing at 97°C. Syst Appl Microbiol 8: 100–105

Lown DA, Thirsk HR, Lord Wynne-Jones (1970) Temperature and pressure dependence of the volume of ionization of acetic acid in water from 25 to 225°C and 1 to 3000 bars. Trans Faraday Soc 66: 51–73

Lundegard PD (1985) Carbon dioxide and organic acids: origin and role in burial diagenesis (Texas Gulf Coast Tertiary). PhD Thesis, University of Texas, Austin, 144 pp

Lundegard PD, Kharaka YK (1990) Geochemistry of organic acids in subsurface waters: field data, experimental data and models. In: Melchior DC, Bassett RL (eds) Chemical modeling of aqueous systems II. Am Chem Soc Symp Ser 416, Washington DC, pp 169–189

Lundegard PD, Land LS (1986) Carbon dioxide and organic acids: their role in porosity enhancement and cementation, Paleogene of the Texas Gulf Coast. In: Gautier D (ed) Role of organic matter in mineral diagenisis. Soc Econ Paleontol Mineral Spec Publ 38, pp 129–146

Lundegard PD, Land LS (1989) Carbonate equilibria and pH-buffering by organic acids – response to changes in pCO$_2$. Chem Geol 74: 277–287

Lundegard PD, Senftle JT (1987) Hydrous pyrolysis: a tool for the study of organic acid synthesis. Appl Geochem 2: 605–612

MacGowan DB, Surdam RC (1988) Difunctional carboxylic acid anions in oilfield waters. Org Geochem 12: 245–259

MacGowan DB, Surdam RC (1990) Importance of organic-inorganic reactions to modeling water-rock interactions during progressive clastic diagenesis. In: Melchior DC, Bassett RL (eds) Chemical modeling of aqueous systems II. Am Chem Soc Symp Ser 416, Washington DC, pp 494–507

MacInnes DA, Shedlovsky T (1932) The determination of the ionization constant of acetic acid, at 25 degrees, from conductance measurements. J Am Chem Soc 54: 1429–1438

Makhatadze GI, Privalov PL (1990) Heat capacity of proteins. I. Partial molar heat capacity of individual amino acid residues in aqueous solution: hydration effect. J Mol Biol 213: 375–384

Makhatadze GI, Medvedkin VN, Privalov PL (1990) Partial molar volumes of poly-peptides and their constituent groups in aqueous solution over a broad temperature range. Biopolymer Chem 30: 1001–1010

Manley EP, Evans LJ (1986) Dissolution of feldspars by low-molecular-weight aliphatic and aromatic acids. Soil Sci 141: 106–112

Martens CS (1990) Generation of short chain organic acid anions in hydrothermally altered sediments of the Guaymas Basin, Gulf of California. Appl Geochem 5: 71–76

Mast MA, Drever JI (1987) The effect of oxalate on the dissolution rates of oligoclase and tremolite. Geochim Cosmochim Acta 51: 2559–2568

Matsui T, Ko HC, Hepler LG (1974) Thermodynamics of ionization of benzoic acid and substituted benzoic acids in relation to the Hammett equation. Can J Chem 52: 2906–2911

McAuley A, Nancollas GH (1961) Thermodynamics of ion association. Part VII. Some transition-metal oxalates. J Chem Soc 1961: 2215–2221

Means JL, Hubbard N (1987) Short-chain aliphatic acid anions in deep subsurface brines: a review of their origin, occurrence, properties, and importance and new data on their distribution and geochemical implications in the Palo Duro Basin, Texas. Org Geochem 11: 177–191

Mesmer RE, Patterson CS, Busey RH, Holmes HF (1989) Ionization of acetic acid in NaCl(aq) media: a potentiometric study to 573 K and 130 bar. J Phys Chem 93: 7483–7490

Moldovanyi EP (1990) Evolution of basinal brines: elemental and isotopic evolution of formation waters and diagenetic minerals during burial of carbonate sediments, Upper Jurassic Smackover Formation, southwest Arkansas, US Gulf Coast. PhD Thesis, Washington University, St. Louis, 247 pp

Nikolaeva NM, Antipina VA (1972) The dissociation constants of oxalic acid in water at temperatures from 25 to 90 °C. Izv Sib Otd Akad Nauk SSSR Ser Khimicheskikh Nauk 6: 13–17 (in Russian)

Noyes AA, Kato Y, Sosman RB (1910) The hydrolysis of ammonium acetate and the ionization of water at high temperatures. J Am Chem Soc 32: 159–178

Oscarson JL, Gillespie SE, Christensen JJ, Izatt RM, Brown PR (1988) Thermodynamic quantities for the interaction of H^+ and Na^+ with $C_2H_3O_2^-$ and Cl^- in aqueous solution from 275 to 320 °C. J Solution Chem 17: 865–885

Ostiguy C, Ahluwalia JC, Perron G, Desnoyers JE (1977) Heat capacities, volumes, and expansibilities of sodium phenyl carboxylates in water. Can J Chem 55: 3368–3370

Palma M, Morel J-P (1976) Viscosite des solutions aqueuses d'acides carboxyliques aliphatiques et des carboxylates de potassium a 25 °C. J Chim Phys 73: 643–649

Palmer DA, Drummond SE (1986) Thermal decarboxylation of acetate. Part I. The kinetics and mechanism of reaction in aqueous solution. Geochim Cosmochim Acta 50: 813–823

Parton HN, Gibbons RC (1939) The thermodynamic dissociation constants of oxalic acid. Trans Faraday Soc 35: 542–545

Pinching GD, Bates RG (1948) Second dissociation constant of oxalic acid from 0 to 50 °C, and the pH of certain oxalate buffer solutions. J Res Natl Bur Standards 40: 405–416

Pronk JT, Liem K, Bos P, Kuenen JG (1991) Energy transduction by anaerobic ferric iron respiration in *Thibacillus* ferrooxidans. Appl Environ Microbiol 57: 2063–2068

Read AJ (1981) Ionization constants of benzoic acid from 25 to 250 °C and to 2000 bar. J Solution Chem 10: 437–450

Rossini FD, Mair BJ (1959) The work of the API research project 6 on the composition of petroleum. 5th World Petroleum Congr, Proc New York, 1959, vol 5, pp 223–245

Rossini RD, Mair BJ, Streiff AJ (1953) Hydrocarbons from petroleum. Am Chem Monogr Ser 121. Reinhold, New York, 556 pp

Sassani DC, Shock EL (1990) Speciation and solubility of palladium in aqueous magmatic-hydrothermal solutions. Geology 18: 925–928

Sassani DC, Shock EL (1992) Estimation of standard partial molal entropies of aqueous ions at 25 °C and 1 bar. Geochim Cosmochim Acta 56: 3895–3908

Sassani DC, Shock EL (1993) Solubility and transport of platinum-group elements in supercritical aqueous fluids: thermodynamic properties of Ru, Rh, Pd, and Pt solids, aqueous ions, and aqueous complexes to 5 kbar and 1000 °C. Geochim Cosmochim Acta (in press)

Schalscha EB, Appelt H, Schatz A (1967) Chelation as a weathering mechanism. I. Effect of complexing agents on the solubilization of iron from minerals and granodiorite. Geochim Cosmochim Acta 31: 587–596

Schleusener JL, Drummond SE, Palmer DA, Barnes HL (1987) Effects of common minerals on acetate decarboxylation kinetics. Geol Soc Am Abstr Programs 19: 832–833

Schleusener JL, Barnes HL, Drummond SE, Palmer DA (1988) Activation parameters and low temperature half-lives for the decarboxylation of acetate in sedimentary basin fluids. Geol Soc Am Abstr Programs 20: 150

Schulte MD, Shock EL (1993) Aldehydes in hydrothermal solution: standard partial molal thermodynamic properties and relative stabilities at high temperatures and pressures. Geochim Cosmochim Acta (in press)

Seewald JS, Seyfried WE Jr, Thornton EC (1990) Organic-rich sediment alteration: an experimental and theoretical study at elevated temperatures and pressures. Appl Geochem 5: 193–209

Sengupta M, Pal K, Chakravarti A, Mahapatra P (1978) Dissociation constants of toluic acids in aqueous solution at different temperatures and study of related thermodynamic parameters. J Chem Eng Data 2: 103–107

Shaw DG, Alperin MJ, Reeburgh WS, McIntosh DJ (1984) Biogeochemistry of acetate an anoxic sediment of Skan Bay, Alaska. Geochim Cosmochim Acta 48: 1819–1825

Shock EL (1988) Organic acid metastability in sedimentary basins. Geology 16: 886–890

Shock EL (1989) Corrections to "Organic acid metastability in sedimentary basins". Geology 17: 572–573

Shock EL (1990a) Geochemical constraints on the origin of organic compounds in hydrothermal systems. Origins Life Evol Biosphere 20: 331–367

Shock EL (1990b) Do amino acids equilibrate in hydrothermal fluids? Geochim Cosmochim Acta 4: 1185–1189

Shock EL (1992a) Stability of peptides in high temperature aqueous solutions. Geochim Cosmochim Acta 56: 3481–3491

Shock EL (1992b) Chemical environments of submarine hydrothermal systems. Origins Life Evol Biosphere 22: 66–107

Shock EL (1993a) Organic acids in hydrothermal solutions: standard molal thermodynamic properties of carboxylic acids, and estimates of dissociation constants at high temperatures and pressures. Am J Sci (in press)

Shock EL (1993b) Hydrothermal dehydration of aqueous organic compounds. Geochim Cosmochim Acta (in press)

Shock EL, Helgeson HC (1988) Calculation of the thermodynamic and transport properties of aqueous species at high pressures and temperatures: correlation algorithms for ionic species and equation of state predictions to 5 kb and 1000 °C. Geochim Cosmochim Acta 52: 2009–2036

Shock EL, Helgeson HC (1990) Calculation of the thermodynamic and transport properties of aqueous species at high pressures and temperatures: standard partial molal properties of organic species. Geochim Cosmochim Acta 54: 915–945

Shock EL, Koretsky CM (1993) Metal-organic complexes in geochemical processes: calculation of standard partial molal thermodynamic properties of aqueous acetate complexes at high pressures and temperatures. Geochim Cosmochim Acta (in press)

Shock EL, McKinnon WB (1993) Hydrothermal processing of cometary volatiles – application to triton. Icarus (in press)

Shock EL, Schulte MD (1990) Summary and implications of reported amino acid concentrations in the Murchison meteorite. Geochim Cosmochim Acta 54: 3159–3173

Shock EL, Sverjensky DA (1989) Hydrothermal organometallic complexes of base metals. Geol Soc Am Abstr Programs 21: A8

Shock EL, Helgeson HC, Sverjensky DA (1989) Calculation of the thermodynamic and transport properties of aqueous species at high pressures and temperatures: standard partial molal properties of inorganic neutral species. Geochim Cosmochim Acta 53: 2157–2183

Shock EL, Oelkers EH, Johnson JW, Sverjensky D A, Helgeson H C (1992) Calculation of the thermodynamic properties of aqueous species at high pressures and temperatures: effective electrostatic radii, dissociation constants, and standard partial molal properties to 1000 °C and 5 kb. J Chem Soc Faraday Trans 88: 803–826

Smolyakov BS, Primanchuk MP (1966) Dissociation constants of benzoic acid at temperatures between 25° and 90°. Russian J Phys Chem 40: 331–333

Stetter KO, König H, Stackebrandt E (1983) *Pyrodictium* gen. nov., a new genus of submarine disc-shaped sulphur reducing archaebacteria growing optimally at 105 °C. Syst Appl Microbiol 4: 535–551

Strong LE, Kinney T, Fischer P (1979) Ionization of aqueous benzoic acid: conductance and thermodynamics. J Solution Chem 8: 329–345

Strong LE, Copeland TG, Darragh M, van Waes C (1980) Ionization of aqueous toluic acids: conductance and thermodynamics. J Solution Chem 9: 109–128

Stull DR, Westrum EF Jr, Sinke GC (1969) The chemical thermodynamics of organic compounds. Wiley, New York, 865 pp

Surdam RC, Crossey LJ (1985) Organic-inorganic reactions during progressive burial: key to porosity and permeability enhancement and preservation. Philos Trans R Soc Lond Ser A 315: 135–156

Surdam RC, MacGowan DB (1987) Oilfield waters and sandstone diagenesis. Appl Geochem 2: 613–619

Surdam RC, Boese SW, Crossey LJ (1984) The chemistry of secondary porosity. In: McDonald DA, Surdam RC (eds) Clastic diagenesis. Am Assoc Pet Geol Mem 37, pp 127–149

Surdam RC, Crossey LJ, Hagen ES, Heasler HP (1989) Organic-inorganic interactions and sandstone diagenesis. Am Assoc Pet Geol Bull 73: 1–23

Sverjensky DA (1984) Oil field brines as ore-forming solutions. Econ Geol 79: 23–37

Sverjensky DA, Hemley JJ, D'Angelo WM (1991) Thermodynamic assessment of hydrothermal alkali feldspar-mica-aluminosilicate equilibria. Geochim Cosmochim Acta 55: 989–1004

Tanger JC, Helgeson HC (1988) Calculation of the thermodynamic and transport properties of aqueous species at high pressures and temperatures: revised equations of state for the standard partial molal properties of ions and electrolytes. Am J Sci 288: 19–98

Tewari YB, Goldberg RN (1991) Thermodynamics of hydrolysis of disacchrides: lactulose, α-D-melibiose, palatinose, D-trehalose, D-turanose and 3-o-β-D-galactopyranosyl-D-arabinose. Biophys Chem 40: 59–67

Thauer RK (1990) Energy metabolism of methanogenic bacteria. Biochim Biophys Acta 1018: 256–259

Thauer RK, Jungermann K, Decker K (1977) Energy conservation in chemotrophic anaerobic bacteria. Bacteriol Rev 1977: 100–180

Thornton EC, Seyfried WEJr (1987) Reactivity of organic-rich sediment in seawater at 350 °C, 500 bars: experimental and theoretical constraints and implications for the Guaymas Basin hydrothermal system. Geochim Cosmochim Acta 51: 1997–2010

Travers JG, McCurdy KG, Dolman D, Hepler LG (1975) Glass-electrode measurements over a wide range of temperatures: the ionization constants (5–90 °C) and thermodynamics of ionization of aqueous benzoic acid. J Solution Chem 4: 267–274

Warford AL, Kosiur DR, Doose PR (1979) Methane production in Santa Barbara Basin sediments. Geomicrobiol J 1: 117–137

Willey LM, Kharaka YK, Presser TS, Rapp JB, Barnes I (1975) Short chain aliphatic acid anions in oil field waters and their contribution to the measured alkalinity. Geochim Cosmochim Acta 39: 1707–1711

Wilson JM, Gore NE, Sawbridge JE, Cardenas-Cruz F (1967) Acid-base equilibria of substituted benzoic acids. Part I. J Chem Soc (B) 1967: 852–859

Wogelius RA, Walther JV (1991) Olivine dissolution at 25 °C: effects of pH, CO_2, and organic acids. Geochim Cosmochim Acta 55: 943–954

Workman AL, Hanor JS (1985) Evidence for large-scale vertical migration of dissolved fatty acids in Louisiana oil field brines: Iberia field, south-central Louisiana. Trans Gulf Coast Assoc Geol Soc 35: 293–300

Zawidzki TW, Papèe HM, Laidler KJ (1959) Thermodynamics of ionization processes in aqueous solution. Trans Faraday Soc 55: 1743–1745

Zillig WI, Holz I, Janekovic D, Schäfer W, Reiter W-D (1983) The archaebacterium *Thermococcus celer* represents a novel genus within the thermophilic branch of the archaebacteria. Syst Appl Microbiol 4: 88–94

Zillig W, Holz I, Janekovic D, Klenk H-P, Imsel E, Trent J, Wunderl S, Forjaz VH, Coutinho R, Ferreira T (1990) *Hyperthermus butylicus*, a hyperthermophilic sulfur-reducing archaebacterium that ferments peptides. J Bacteriol 172: 3959–3965

Zinger AS, Kravchik TE (1970) The simpler organic acids in ground water of the lower Volga region (genesis and possible use in prospecting for oil). Dokl Akad Nauk SSSR 202: 218–221

Appendix: Calculation of Activities of Organic Acids from Reported Concentrations

Organic acid concentrations are reported in a wide variety of units. In this study, all concentrations were converted to molalities. Reports of concentrations from hydrous pyrolysis experiments are typically given as yields relative to some initial concentration of organic matter or total organic carbon in a sample. In some instances, incomplete descriptions of the samples inhibit accurate calculation of molalities for the resulting aqueous phase. For example, Lundegard and Senftle state that they used 20 to 30 g of source rock in each experiment. Since they report their yields of organic acids in terms of grams of organic carbon per gram of carbon in the sample, it is not possible to evaluate the actual mass of organic acids produced without assuming the total mass of sample. In this study, 25 g of source rock was assumed for each experiment. Similarly, Kawamura et al. (1986) state that 0.3 to 0.5 g of kerogen was used in each experiment, and an assumption of 0.4 g was made in the present study. Eglinton et al. (1987) state that they used 250 mg of kerogen in some experiments, but also state that in experiments involving minerals, the ratio (by weight) of kerogen to mineral is 1 : 20 without giving the total mass of material used in the experiments. It is therefore impossible to convert the yields of organic acids from their kerogen and limonite experiments to useful concentration units. In contrast, Barth et al. (1987, 1989) report yields for 5-g samples, which allow accurate conversion to molalities.

Reports of concentration of organic acids in natural solutions are often given as sums of all forms of each acid owing to the analytical methods employed (Willey et al. 1975; Carothers and Kharaka 1978; Barth 1987a; Fisher 1987). Therefore, the first step in calculating activities of individual species under the natural conditions is to convert these sums into molalities of anionic and neutral forms of the aqueous acids. This is done by comparing the pH in the natural fluid with the pK_a values of each acid at the temperature and pressure of the environment sampled. A consistent set of pK_a values of many organic acids at elevated temperatures and pressures is given by Shock (submitted). On the other hand, calculations of pH require making some assumptions.

Major element fluid compositions, total salinities, pH values measured in the field, and descriptions of mineral assemblages coexisting with each brine are generally not reported in the literature sources on organic acids. In the present study, pH values were estimated by assuming equilibrium between a 1.0 molal NaCl solution and the albite-kaolinite-quartz assemblage for all samples, using mineral data consistent with Sverjensky et al. (1991) standard state data for aqueous ions from Shock and Helgeson (1988), and activity coefficients for Na^+ and Cl^- from Helgeson et al. (1981). This is an over-simplification, which is probably not strictly appropriate for many samples, especially those from carbonate reservoirs. Nevertheless, in the absence of the necessary analytical data for the natural samples, it provides a unified method for treating the 232 samples used in this study for which concentrations of both acetic and propanoic acids are reported. Activity coefficients for acids and acid anions were evaluated as described by Shock (1988, 1989). The resulting activities of CH_3COO^-, CH_3COOH, $CH_3CH_2COO^-$, and CH_3CH_2COOH were employed in the calculations described in this text. For example, calculations of fH_2 for each sample using Eq. (30) involve the calculated activities of CH_3COOH and CH_3CH_2COOH and not the reported concentrations.

Chapter 11   Metal Transport in Ore Fluids by Organic Ligand Complexation

Thomas H. Giordano[1]

Summary

The association of organic matter with ore minerals, gangue, and host rock in many low- to moderate-temperature ore deposits has been known for many years. As a reductant or oxidant, organic matter may function as an active ore-forming agent up to magmatic temperatures, however, most other roles for organic matter, including metal complexation, are limited to less than about 200 °C.

Metal-humate and metal-fulvate complexes probably contribute significantly to metal transport and speciation in interstitial waters of subaqueous sediments, in shallow sediments undergoing early diagenesis, and in the supergene environment. It is, therefore, likely that these complexes are involved in ore-forming processes responsible for syngenetic fixation of metals in young sediment as well as epigenetic transport and deposition of metals in shallow subsurface environments. Because concentrations of amino acids are typically an order of magnitude lower than concentrations of humic and fulvic acids in interstitial waters, metal-amino acid complexes are probably less important than humate or fulvate complexes in most low temperature ore-forming environments. Organosulfur ligands may contribute to the speciation of metals and sulfur in ore-forming processes responsible for syngenetic deposition of metal sulfides in low-temperature subaqueous environments. If they are sufficiently stable at elevated temperatures, thiols and other organosulfur ligands could conceivably contribute to both metal and sulfur transport in moderate-temperature (50 to 250 °C) hydrothermal ore fluids. The concentrations of carboxylate ions in surface and shallow subsurface interstitial waters are sufficiently low to preclude the importance of metal-carboxylate complexes as significant metal-transporting agents in low-temperature ore fluids responsible for supergene mineralization, syngenetic deposition in anoxic environments, and low-temperature epigenetic ores. However, the observed elevated concentrations of organic acids in oil-field brines strongly suggest that metal-carboxylate complexes

[1] Department of Geological Sciences, New Mexico State University, Las Cruces, New Mexico 88003, USA

may contribute significantly to metal transport in ore fluids with temperatures ranging from roughly 75 to 150 °C.

To evaluate the ore transport potential of a particular metal-organic complex, the theoretically estimated concentration of that species in the ore solution of interest must be determined by calculation. Concentrations of Ca, Mg, Na, Pb, Zn, Fe, and Al as acetate, oxalate, malonate, succinate, and catechol complexes were calculated for three reconstructed Mississippi Valley-type (MVT) ore solutions and a model composite ore fluid for red bed-related base metal (RBRBM) deposits. Based on these calculations, some important inferences can be made regarding metal transport by organic complexes in ore fluids of sedimentary origin. Significant amounts of lead and zinc cannot be mobilized as metal-organic complexes involving acetate or other carboxylate ligands in ore fluids containing greater than 10^{-5} molal reduced sulfur as hydrogen sulfide or bisulfide. However, in reduced ore fluids sufficient metal and reduced sulfur may be transported as complexes involving organosulfur ligands. On the other hand, significant quantities of dissolved lead and zinc can be transported by carboxylate complexes in oxidized ore fluids containing less than 10^{-9} molal reduced sulfur. It is hoped that these results will encourage investigators to evaluate the role of metal-organic complexing in the genesis of all those deposits with a clear genetic link to organic processes.

1 Introduction

The association of organic matter with ore minerals, gangue, and host rock in many low- to moderate-temperature ore deposits has been known for many years (Saxby 1976; Dean 1986). Specific roles of organic matter in ore-forming processes have been suggested in several early works (Krauskopf 1955; Barton 1967; Skinner 1967) and more recently outlined and reviewed by Saxby (1976), Macqueen (1980), Barton (1982), Giordano (1985), Leventhal (1986), Manning (1986), and Kribek (1989). Excellent discussions of specific biological processes related to ore genesis are given by Trudinger (1976, 1980), Westbroek and DeJong (1983), Trudinger et al. (1985), and Ehrlich (1986). Biological processes can affect ore formation in several ways by (1) generating organic matter; (2) accumulating ore constituents; (3) after death, making ore constituents available to the environment; (4) modifying physicochemical conditions in the environment; and (5) transforming elements from one state to another by metabolism. For the purpose of this chapter, I divide the potential roles of non-living organic matter in ore-forming processes into five categories: (1) aqueous complexation: (2) nonaqueous complexation; (3) substrate for microbial processes; (4) modification of the physicochemical environment; and (5) reducing or oxidizing agent.

Aqueous complexation involves the formation of water-soluble metal-organic complexes from metal cations and ligands. Such complexes can facilitate mobilization and transport of metals and other constituents (e.g., sulfur) in an ore fluid. If sufficient metal is transported in a metal-organic complex, the destabilization of the complex could cause ore deposition. Several types of nonaqueous complexation can affect the mobilization, transport, and deposition of ore metals. These include (1) homogeneous precipitation of insoluble metal-organic complexes from aqueous solutions; (2) chemisorption and physical sorption of metals by immobile organic matter; (3) chemisorption and physical sorption of metals by mobile colloidal and particulate organic matter or sorption of aqueous metal-organic complexes by suspended clays and oxides (deposition of metals takes place through flocculation and settling); and (4) partitioning of metals between an aqueous phase and a mobile or immobile oily phase in which metals would be present as oil-soluble metal-organic complexes. As a substrate, insoluble and dissolved organic matter is consumed as an energy source by heterotrophic microorganisms carrying out sulfate reduction, biomineralization, and other biogenic processes affecting metal transport and deposition. As a modifier of the environment, organic matter in a dissolved, gaseous, or condensed (liquid or solid) form can affect the pH, Eh, and other chemical parameters of ore-forming environments through participation in biogenic and abiogenic processes. If sufficient organic matter is present, reduced conditions can be maintained, thereby allowing dissimilatory anaerobic sulfate-reducing bacteria to generate hydrogen sulfide. Thermal degradation of sulfur-, oxygen-, and nitrogen-containing organic compounds can release H_2S, CO_2, NH_3, CH_4, and various organic ligands to a potential ore fluid. As a reducing agent, organic matter in the form of humate, bitumen, kerogen, or simple organic molecules (e.g., methane) can result in thermochemical (nonbiological) reduction of sulfate, thiosulfate, and polysulfide to hydrogen sulfide. Organic material can also reduce metals directly causing either deposition or mobilization.

As a reductant or oxidant, organic matter may function up to magmatic temperatures, however, most of the roles for organic matter outlined above are limited to less than about 200 °C (Barton 1982). In this chapter, I will touch upon some of these roles, but the primary focus will be on the first role listed, aqueous complexation of metals by organic ligands and the transport of metals by metal-organic complexes. First, I briefly review the occurrences of organic matter in ores and highlight those deposit types whose genesis is believed to have involved the active participation of organic matter. I then present a brief status report on current knowledge about organic ligands and metal-organic complexes of possible importance in ore-forming processes. Finally, I present several metal-transport models for specific ore-fluid types. The results of these models, although provisional, clearly establish the relative importance of specific organic ligands as complexing agents in ore fluids.

2 Organic Matter in Ores

The current interest in organic matter as an active chemical agent in ore-forming processes is based primarily on the presence of condensed organic compounds in a variety of sediment-hosted ore deposits formed at temperatures near or below 250 °C. Most of these deposits are hosted by cratonic sedimentary rocks and were formed by hydrochemical processes within sedimentary basins. In Table 1 are listed five major ore deposit types, which typically contain visible organic matter and that are unequivocally linked to sedimentary basins and related diagenetic processes. Organic matter is also a well-known constituent in Precambrian fluvial-deltaic placer deposits hosting gold-uranium mineralization (Zumberge et al. 1978; Mossman and Dyer 1985; Willingham et al. 1985) and in some deposits linked to hot spring activity. Specific deposits containing organic matter and genetically linked to continental hot spring activity include epithermal mercury deposits (Blumer 1975), epithermal gold-silver mineralization hosted by carbonaceous sediments within caldera structures at Creede, Colorado (Smith 1986) and Pueblo Viejo, Dominican Republic (Kesler et al. 1986; Kettler et al. 1990),

Table 1. Major types of hydrochemical ore deposits containing visible organic matter and hosted by cratonic basinal sedimentary rocks

Ore deposit type	Tectonic environment	Host rocks	Approx. deposition temperature (°C)	Reference[a]
Syngenetic chemical	Intracratonic basins	Near-shore marine: siliciclastic	25	1
Sandstone uranium	Intracratonic basins	Continental: siliciclastic	50	2
Red-bed copper	Intracratonic basins	Continental and marginal marine: siliciclastic	50–100	3
Mississippi Valley-type	Intracratonic basins	Shallow marine and shoreline: carbonate and siliciclastic	75–200	4
Sediment-hosted base metal	Incipient rift basins	Shallow marine: carbonate and siliciclastic	100–275	3

[a] 1, Vine and Tourtelot (1970); Tourtelot (1979); Coveney et al. (1987); Grauch and Huyck (1990); Ripley et al. (1990); Schultz (1991); 2, Nash et al. (1981); Turner-Peterson and Fishman (1986); Turner-Peterson et al. (1986); Maynard (1991a); Hansley and Spirakis (1991); 3, Gustafson and Williams (1981); Boyle et al. (1989); Heroux et al. (1989); Landais and Meyer (1989); Ho et al. (1990); Maynard (1991b); 4, Macqueen and Powell (1983); Gize (1984); Macqueen (1986); Sverjensky (1986); Price and Kyle (1986); Gize and Barnes (1987); Leventhal (1990); Henry et al. (1992).

and the auriferous carbonaceous ores of the Carlin trend in Nevada (Radtke and Scheiner 1970; Radtke et al. 1980; Hausen and Park 1986; Hulen 1991). Bituminous material has been found intimately associated with base-metal massive sulfide mineralization recently deposited in the Red Sea trough, Guaymas Basin, and other submarine hydrothermal vent sites (Saxby 1972; Simoneit and Lonsdale 1982; Welhan and Lupton 1987; Bazylinski et al. 1988; Kawka and Simoneit 1990; Simoneit 1990). To the author's knowledge, organic matter has not been found in Kuroko or other types of high-temperature volcanogenic massive sulfide deposits. One other common association should be noted. Organic matter in the form of coal often contains anomalous concentrations of metal, and in some cases coal deposits appear to be genetically related to deposits of iron, aluminum, and uranium. An excellent review of the association of coal and ores is given by Laznicka (1985).

Ores deposited at temperatures greater than about 300 °C or heated to these temperatures by later hydrothermal or metamorphic events are usually devoid of condensed organic carbon (Saxby 1976). Bituminous material, however, has been observed in some high-temperature (greater than about 250 °C) hydrothermal deposits (Germanov 1965; Zezin and Sokolova 1968; Germanov and Bannikova 1972; Sokolova et al. 1980; Gize 1986; Bannikova and Shirinskiy 1988). Although kerogen and bitumen in sedimentary rocks and ores are usually converted to graphite between 250 and 400 °C (Saxby 1976; Tissot and Welte 1984), the presence of organic matter in a mesophase state in some deposits suggests that graphitization may not occur below 350 ± 50 °C in some ore environments (Gize 1986). Helgeson (1991) presents thermodynamic evidence that suggests petroleum, containing hydrocarbons in metastable equilibrium with certain mineral assemblages, may persist to about 320 °C and that heavy crude oil and asphaltenes may persist to even higher temperatures.

Organic matter hosted by metallic deposits occurs in a variety of textural and compositional forms. Indigenous kerogen and bitumen are usually the dominant forms of carbonaceous material in mineralization of syngenetic origin, whereas introduced (epigenetic) organic material in the form of humate, bitumen, dissolved aqueous species, and gaseous hydrocarbons is more common in epigenetic deposits. Epigenetic humate and bitumen are typically deposited in pore spaces and fractures or as coatings on mineral surfaces. Introduced bitumen is also found as a component of inclusions in minerals, as are gaseous hydrocarbons and various other organic compounds dissolved in the aqueous phase of inclusions (Roedder 1976, 1984). Primary aqueous inclusions in ore or gangue minerals are generally thought to represent actual preserved samples of ore-forming fluids (Roedder 1979). Analysis of the aqueous phase in such inclusions for polar organic compounds potentially provides a source of information on the nature and concentration of possible organic ligands present in ore-forming solutions. Unfortunately, such determinations are difficult to obtain and only a few semiqualitative results have been reported in the literature. Germanov and Mel' Kanovitskaga (1975) analyzed fluid inclusions in polymetallic mineralization from northern Caucasus in the former Soviet Union. In crushed rock samples, they reported concentrations of C_1 to C_6 organic acids in the range 0.4 to 3.2 ppm, with acetic acid comprising 54 to 76% of the total. McLimans (1977) reports the detection of aqueous organic compounds in fluid inclusions from the Wisconsin zinc-lead district. Concentrations ranged from 0.05 to 0.001 molal for species with molecular weights of 15 to 80.

The observational evidence outlined above strongly suggests that organic matter is not a significant active agent in ore-forming processes operating near or above $250 \pm 50\,°C$. Further support of this suggestion is the availability of adequate inorganic models to account for transport and deposition of high-temperature ($>250 \pm 50\,°C$) hydrothermal ore deposits including skarn-, porphyry-, and Kuroko-type deposits and many kinds of vein and replacement ores (Barnes 1979). For these reasons, as well as thermal stability evidence discussed below, it is not likely that organic ligands significantly contribute to metal transport in ore fluids with temperatures greater than about $250\,°C$. On the other hand, there is usually sufficient and often compelling evidence suggesting that organic matter is an active agent in the genesis of those deposit types listed in Table 1, as well as other low-temperature ($<50\,°C$) and moderate-temperature ($50-250\,°C$) deposits. In many cases, inorganic mechanisms alone are insufficient to satisfactorily account for transport and deposition of these ores. In the following sections, I will focus on these low- and moderate-temperature deposits and particularly those ore-forming environments developed within or peripheral to cratonic sedimentary basins.

3 Metal-Organic Complexes

3.1 Coordination Compounds

Coordination compounds are substances in which a central atom is bonded to surrounding atoms or groups of atoms called ligands. Two types of coordination compounds involving central metal atoms and organic ligands are recognized: organometallic compounds and metal-organic complexes. Organometallic compounds are species in which a central metal atom is bonded covalently to at least one carbon atom of a surrounding organic molecule (e.g., tetraethyl lead and methyl mercury). Although most organometallic compounds are artificially produced and are not found in nature, some are known to be synthesized biogenically and a few may be produced abiogenically in natural systems (Stumm and Morgan 1981; Gill and Bruland

1990). To my knowledge, the only discussion of such compounds in relation to metal transport in ore fluids is that of Levitskiy et al. (1982), who only suggest the possibility of their importance. The lack of evidence for the presence of organometallic compounds in natural aqueous systems other than surface water and shallow groundwater strongly suggests that these metal-bearing coordination compounds are not involved in ore-forming processes, except perhaps in the fixation and accumulation of metals in organisms (Trudinger 1976).

Metal-organic complexes are structures in which a metal cation is attached to one or more organic ligands by direct bonding to electron-donor atoms other than carbon, most commonly oxygen, sulfur, and nitrogen (Langmuir 1979). If the ligand does not replace water molecules in the first hydration sphere of the metal cation, a weak outer-sphere complex (ion-pair) is formed. If a water molecule in the first hydration sphere is replaced by the organic ligand, a strong inner-sphere complex is formed. Strong complexes called chelate compounds are formed if two or more donor atoms from the same ligand bind the metal cation. Unlike organometallic compounds, metal-organic complexes are known to be present in a wide variety of geochemical environments, including soil solution, surface water, and groundwater (Aiken et al. 1985; Thurman 1985; Buffle 1988) and are presumed to be present in oil-field brines (Lundegard and Kharaka 1990; MacGowan and Surdam 1990a) and ore fluids (Giordano and Barnes 1981; Giordano 1985, 1990; Drummond and Palmer 1986; Hennet et al. 1988a). Idealized structures of selected ligand species are illustrated in Fig. 1 for many of the geochemically important organic ligands considered below.

To transport significant quantities of metal in ore solutions as a metal-organic complex, the participating organic ligand must meet the following criteria (Giordano 1985; Drummond and Palmer 1986): the ligand must (1) be present as a dissolved constituent of the ore fluid at a sufficient level of concentration; (2) have ionization constants that permit significant proportions of the anionic form at ore-fluid pH's; (3) have sufficient thermal stability to survive at least one pass of the ore fluid through the system; and (4) form metal-organic complexes of sufficient strength to favorably compete with inorganic and other organic ligands. This latter criterion is discussed below (Sect. 4), in terms of competing organic ligands and their affect on metal speciation and in terms of competing metal ions and their affect on the speciation of organic ligands. Several lines of evidence (Barnes and Czamanske 1967; Barnes 1979; Anderson 1983) indicate that concentrations of base and ferrous metals in ore-forming solutions must be at least 1 to 10 ppm (approximately 10^{-5} to 10^{-4} molal) to form a typical hydrothermal ore deposit. Minimum ore-fluid concentrations for rarer metals (e.g., Hg, Au, Ag) are probably several orders of magnitude lower. It follows that the total concentration of the ligand or ligands involved in significant ore-metal transport must also be near or above these same minimum concentrations.

LIGAND SPECIES

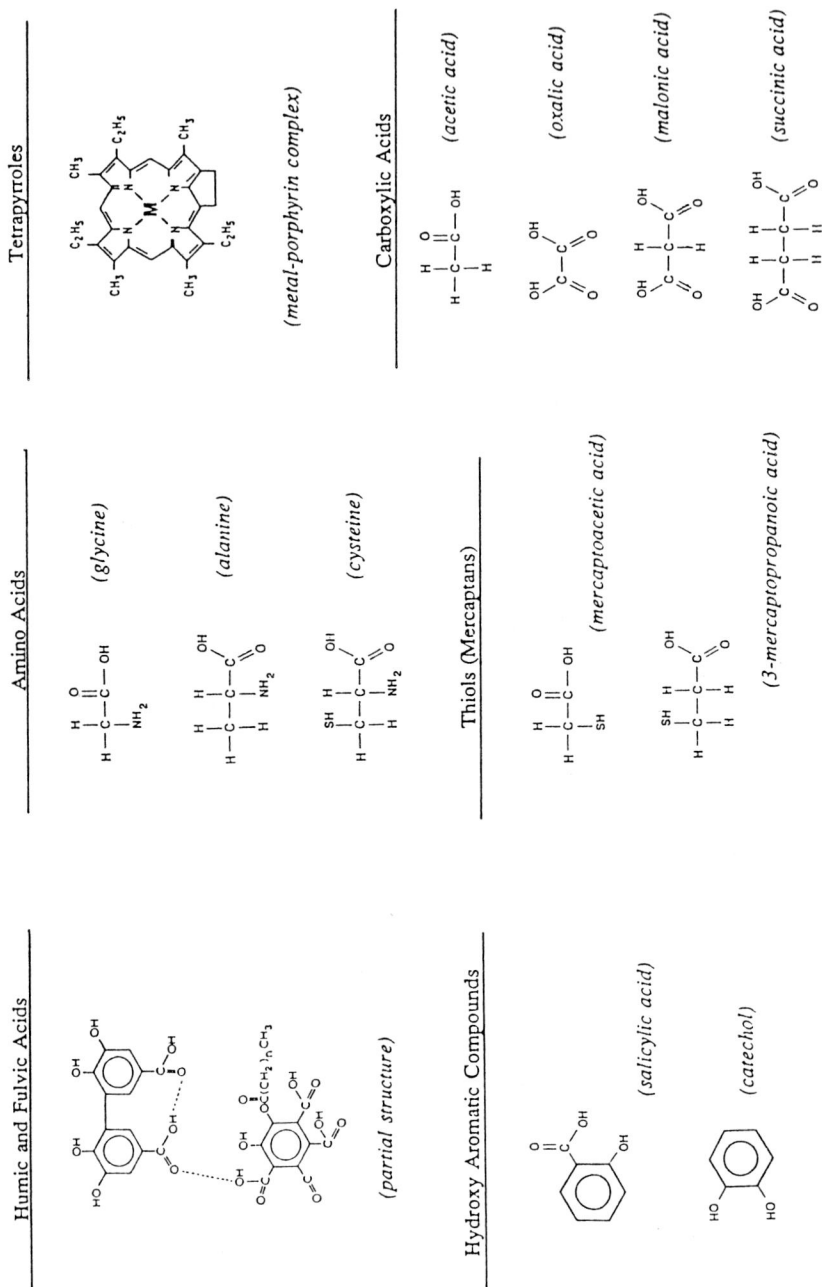

Tetrapyrroles

(metal-porphyrin complex)

Amino Acids

(glycine)

(alanine)

(cysteine)

Carboxylic Acids

(acetic acid)

(oxalic acid)

(malonic acid)

(succinic acid)

Thiols (Mercaptans)

(mercaptoacetic acid)

(3-mercaptopropanoic acid)

Humic and Fulvic Acids

(partial structure)

Hydroxy Aromatic Compounds

(salicylic acid)

(catechol)

Fig. 1. Idealized structures for selected organic ligand species found in sedimentary environments

3.2 Sources of Organic Ligands

Within cratonic sedimentary basins organic matter accumulates under subaqueous conditions along with carbonate sediment and siliciclastic sediment. While passing through successive stages of burial and maturation, organic matter can release to sediment pore water, which may develop into a potential ore fluid, large quantities of dissolved organic ligands (Surdam et al. 1984; Tissot and Welte 1984; Lundegard and Kharaka 1990). In Fig. 2, the progressive diagenetic changes in condensed organic matter (middle column) and the nature of dissolved organic material in interstitial pore water (right column) are shown as a function of depth interval in which transformations are initiated. If basinal pore waters represent potential ore fluids that through time evolve toward specific ore fluids, then it is likely that organic acids, phenols, thiols, and organic sulfides (Fig. 2, right column) are among the more important organic ligands in these evolving ore fluids. A rough correlation between sediment depth in Fig. 2 and deposition temperature in Table 1 provides a means of predicting which organic ligands should be expected in specific types of ore fluids. In Table 1, the deposition temperature for a specific deposit type is probably similar to the maximum temperature of water-rock interaction for the corresponding evolving ore fluid. These pore waters, which most likely have their source far from the site of deposition, scavenge organic ligands and metals from their source sediments as well as from rocks along the path of migration to the site of deposition. Thus, metal-rich ore solutions can be generated by the con-

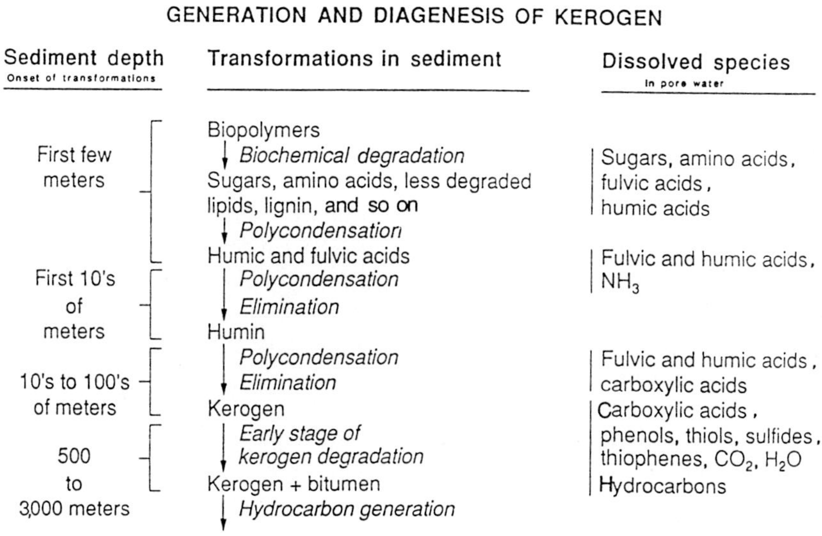

GENERATION AND DIAGENESIS OF KEROGEN

Sediment depth Onset of transformations	Transformations in sediment	Dissolved species In pore water
First few meters	Biopolymers ↓ *Biochemical degradation* Sugars, amino acids, less degraded lipids, lignin, **and so on** ↓ *Polycondensation* Humic and fulvic acids	Sugars, amino acids, fulvic acids, humic acids
First 10's of meters	*Polycondensation* ↓ *Elimination* Humin	Fulvic and humic acids, NH_3
10's to 100's of meters	*Polycondensation* ↓ *Elimination* Kerogen	Fulvic and humic acids, carboxylic acids
500 to 3,000 meters	*Early stage of* ↓ *kerogen degradation* Kerogen + bitumen ↓ *Hydrocarbon generation*	Carboxylic acids, phenols, thiols, sulfides, thiophenes, CO_2, H_2O Hydrocarbons

Fig. 2. Formation and diagenesis of organic matter in sediments as a function of depth

comitant formation of organic ligands and metal-organic complexes within a source sediment and subsequent complexation of metals (leaching) by organic ligands along a pore-fluid flow path.

In many hydrothermal systems the time-temperature relationship is approximated by the sinusoidal function (Drummond and Palmer 1986). In single-pass fluid flow systems (for example, sedimentary basins with compaction- and gravity-driven fluid migration) one-half of one cycle describes the time-temperature relationship, while in systems in which fluid migration is driven by convection (for example, near a cooling intrusive), many cycles are possible. Within a sedimentary basin, a single-pass potential ore fluid can migrate from or through the hotter parts of the basin. In both cases, the composition of dissolved organic matter in the ore fluid should be strongly influenced by reactions at the highest temperatures attained. For example, amino, humic, and fulvic acid ions (Fig. 2) should be the dominant organic ligands in low-tempetature (less than about 50 °C) ore fluids (Table 1) evolved at depths of less than about 1000 m. Carboxylic acid ions, phenols, and organosulfur species should be the dominant organic ligands produced, along with hydrocarbons and other products of kerogen degradation, in hydrothermal ore fluids (greater than about 50 °C) evolved in or passing through deeper parts of the basin.

3.2.1 Humic and Fulvic Acid Ligands

Humic and fulvic acids (Fig. 1) are the dominant dissolved organic constituents in surface waters, shallow subsurface waters derived by infiltration, and interstitial waters in young subaqueous sediments (Nissenbaum and Swaine 1976; Reuter and Perdue 1977; Aiken et al. 1985; Rashid 1985; Thurman 1985). These highly oxidized polymers occur in low-temperature (less than about 50 °C) fresh waters at concentrations up to about 10^{-4} molal (based on average molecular weight of about 1000) and in seawater typically to about 10^{-6} molal. Single molecules of these polyelectrolyte ligands have molecular weights typically between 500 and 5000 and contain a number of oxygen-bearing functional groups (especially, carboxyl and phenolic hydroxyl groups) capable of bonding with metal cations. Thus, humic and fulvic acids form strong metal-organic complexes (chelate compounds) with most multivalent metal cations (Reuter and Perdue 1977; Jackson et al. 1978; Schnitzer and Khan 1978; Stevenson 1983; Aiken et al. 1985). Metal-humate and metal-fulvate complexes probably contribute significantly to metal transport and speciation in interstitial waters of subaqueous sediments and in shallow sediments (less than about 1000 m) undergoing early diagenesis. It is, therefore, likely that these complexes are involved in ore-forming processes responsible for syngenetic fixation of metals in young sediment as well as epigenetic deposition of metals from low-temperature (less than about 50 °C) ore fluids to depths of about 1000 m. Aqueous humate and fulvate complexes of gold (Baker 1978; Varshal et al. 1984; Severson et al. 1986; Vlassopoulous et al. 1990; Coel et al. 1991), platinum (Wood 1990), palladium (Wood 1991), and the metals Cu, Zn, Al, Fe (Kribek et al. 1977) are thought to be important in supergene environments. Humate complexes are also thought to be important in the genesis of certain sandstone uranium deposits (Turner-Peterson et al. 1986) and may possibly transport metals in ore fluids responsible for red-bed copper and related deposits. Because humic and fulvic acids have low thermal stabilities under hydrothermal conditions (Schnitzer and Khan 1978; Aiken et al. 1985; Boles et al. 1988), they are not likely to be important complexing agents in ore fluids having temperatures much greater than 100 °C.

3.2.2 Amino Acid Ligands

Amino acids (Fig. 1) are a ubiquitous but minor dissolved organic constituent in surface waters, shallow groundwaters, and interstitial waters of sediments (Reuter and Perdue 1977; Jackson et al. 1978; Thurman 1985). Within the water column and during early diagenesis, amino acids are subject to intense microbial degradation and, therefore, their concentrations are normally low compared to the relatively inert humic and fulvic acids residing in the same environment. Typical maximum concentrations of total free amino acids in fresh water and marine water are approximately 10^{-5} and 10^{-6} molal, respectively, about an order of magnitude lower than concentrations of dissolved humic substances. Degens et al. (1964) and Rapp (1976) report concentrations of total amino acids of up to 7×10^{-6} mol/l in oil-field brines, with glycine, alanine, and serine typically the dominant species. These ligands, as well as other naturally occurring aqueous amino acids (e.g., aspartic acid, leucine, glutamic acid, cystine, and cysteine), form strong complexes with most multivalent metal cations except the alkaline earth cations (Martell and Smith 1974). Because ligand concentrations of amino acids are typically an order of magnitude lower than concentrations of humic and fulvic acids in interstital waters, metal-amino acid complexes are probably less important than humate or fulvate complexes in most low-temperature ore-forming environments. However, if concentrations of amino acids are sufficiently high in specific environments, amino acid complexation may play an important role in metal transport and deposition (Veitch and McLeroy 1972; Saxby 1976). Although theoretical calculations by Shock (1990) show that amino acids may survive metastably to elevated temperatures, the observational evidence, from the field (Degens et al. 1964; Rapp 1976; Haberstroh and Karl 1989) and from degradation experiments (Bernhardt et al. 1984; White 1984; Miller and Bada 1988; Bada et al. 1991), strongly suggests that amino acids are probably not present in natural waters above 100 °C at concentration levels high enough to significantly affect metal speciation. Thus, amino acid complexes are not likely to be important in ore fluids responsible for most types of epigenetic hydrothermal mineralization.

3.2.3 Organosulfur Ligands

Organosulfur compounds (Fig. 1) containing reduced sulfur are present as minor constituents in marine sediments (Tissot and Welte 1984; Kiene and Taylor 1989; Vairavamurthy and Mopper 1989), but concentrations of specific species dissolved in interstitial waters are not widely documented. Interestingly, the classes of organosulfur compounds found in young marine sdiments are the same as those found in petroleum and include thiols (mercaptans), sulfides, disulfides, and thiophene derivatives. To the author's knowledge, specific organosulfur compounds have not been identified in oil-field brines, but it is likely that such compounds partitioned into the interstitial aqueous phase of carbonaceous rocks during diagenesis (Tissot and Welte 1984). Kharaka et al. (1979) measured sulfide concentrations in oil-field brines from the Gulf coast by two methods and suggested the discrepancy in results may be due to organic sulfur species. Possibly important organosulfur ligands dissolved in interstitial waters in young sediments include sulfur-bearing amino acids (e.g., cystine and cysteine) and mercaptocarboxylic acids. Mercaptocarboxylate ligands and similar thiol compounds are found in crude oil (Tissot and Welte 1984) and are probably also released to deep, hot formation waters during the initial stages of kerogen degradation (Fig. 2). Mercaptocarboxylate ligands are known to form highly stable complexes with most ore-metal cations, but only weak complexes with cations of the alkaline earth metals. Although not strictly an organosulfur compound, the thiocarbonate ligand CO_2S^{2-} may be an important complexing agent of Pt, Pd, Zn, and Ni in moderate-temperature (50 to 250 °C) hydrothermal ore fluids (Hennet et al. 1988b).

An important advantage of organosulfur ligands over other organic complexing agents is the ability to transport both metals and reduced sulfur in the same ore fluid (Saxby 1976). This gives rise to an attractive version of the single ore-fluid hypothesis. Metals and sulfide are transported together in one solution as soluble metal-organic sulfide complexes (e.g., metal-mercaptocarboxylate complexes). Metals and sulfur could be transported in sufficient amounts to form economic deposits if concentrations of such complexes were greater than about 10^{-5} molal. Subsequent breakdown of these complexes at the site of deposition could yield the necessary metals and sulfide to form galena, sphalerite, and other minerals (Saxby 1976; Barnes 1983; Gize and Barnes 1989).

The evidence reviewed above suggests that cystine, cysteine, mercaptocarboxylic acids, and other organosulfur ligands may contribute to the speciation of metals and sulfur in ore-forming processes responsible for syngenetic deposition of metal sulfides in low-temperature subaqueous sediments. To the author's knowledge, the thermal stability of thiols, organic sulfides, and thiophene derivatives has not been determined under a wide range of hydrothermal conditions. Nevertheless, these compounds are found in petroleum, which is typically generated between 80 and 120 °C (Tissot and Welte 1984) and, therefore, it is possible that they have a similar thermal stability under hydrothermal conditions. If they are sufficiently stable at elevated temperatures, thiols and other organosulfur ligands could conceivably contribute to both metal and sulfur transport in moderate-temperature (50 to 250 °C) hydrothermal ore fluids.

3.2.4 Tetrapyrrole Ligands

Porphyrins and related tetrapyrrole ligands (Fig. 1) form metal-organic complexes in which a central metal cation is bonded to four nitrogen atoms of an aromatic tetrapyrrole structure (Lewan and Maynard 1982). Although such complexes are thermally stable and moderately inert to temperatures above 350 °C (Saxby 1976; Lewan 1984; Tissot and Welte 1984), they are strongly partitioned into condensed phases of organic matter relative to any coexisting aqueous phase. Thus, with increasing maturity of sediment-hosted organic matter, metal-tetrapyrrole complexes are concentrated in humic substances, kerogen, and finally bitumens, including expelled petroleum phases, which subsequent to generation in a source rock can migrate to sites of ore deposition. Tetrapyrrole ligands form strong complexes with nickel (Ni^{2+}) and vanadium (VO^{2+}) and it is now thought that such complexes are responsible for the fixation of nickel, vanadium, and perhaps other metals in young sediments during early diagenesis (Lewan and Maynard 1982). The amounts of individual metals complexed by tetrapyrrole ligands during diagenesis depend strongly on the composition of the condensed organic matter and minerals in the sediment, as well as the chemistry of the coexisting interstitial water, most importantly pH, Eh, and the concentrations of competing ligands and reduced inorganic sulfur (Lewan 1984; Kotova et al. 1987; Lipiner et al. 1988).

It has been suggested that petroleum, containing dissolved tetrapyrrole complexes of V, Ni, Co, Fe, Cu, Zn, and Pb, should be considered a potentially important ore-transporting agent (Manning 1986; Price and Kyle 1986). In a biphase system comprising a mobile petroleum phase and an aqueous ore fluid, metals may be released to the ore fluid along a flow path or at the site of deposition by thermal, oxidative, or biological degradation of metal-tetrapyrrole complexes or by favorable partitioning of metals from the petroleum phase (containing metals in tetrapyrrole complexes) to the aqueous phase (containing metals in organic and inorganic complexes) as a result of changes in ore fluid chemistry (Manning 1986; Hennet et al. 1988b). Gize and Barnes (1989) suggested that metalloporphyrin complexes may be the source of nickel and cobalt found in Mississippi Valley-type deposits. To properly evaluate the effects of metal-tetrapyrrole complexation on metal transport and deposition, a better understanding is required of (1) the stability and concentration of such complexes in humic substances, kerogen, and bitumen and (2) the nature of metal cation partitioning between tetrapyrrole complexes in these phases and complexes in a coexisting aqueous phase.

3.2.5   Carboxylic Acid Ligands

Individual carboxylic acids (Fig. 1) and their corresponding acid anions (carboxylate ions) are found in surface and shallow subsurface waters where they always form a minor component of the dissolved organic matter (Thurman 1985). In deeper formation waters, however, carboxylate species are the dominant dissolved organic constituents and in some deep subsurface waters carboxylate anions are present at sufficient concentrations to dominate the total alkalinity (Kharaka et al. 1986). Carboxylate ions form metal-organic complexes of moderate strength with most multivalent metal cations (Martell and Smith 1977) and if present in formation waters at concentrations near or above 10^{-4} molal, such ligands can be important complexing agents for metals in carbonaceous source rocks and basinal aquifers. In most oil-field brines, acetate is generally the dominant organic ligand and is followed in level of concentration by longer chained aliphatic carboxylate ions (Table 2). Although concentrations of acetate up to 10 000 ppm (0.17 molal) have been reported in oil-field brines (Surdam et al. 1984), high concentrations of acetate in formation waters are more typically on the order of 1000 ppm (0.017 molal). Some deep basinal waters contain high concentrations of difuctional carboxylate ions, especially malonate and oxalate, which are typically followed in level of concentration by longer chained homologues (Table 2). Concentrations of oxalate and malonate in formation waters are normally less that 100 ppm but maximum concentrations of 494 ppm (0.0055 molal) and 2540 ppm (0.025 molal) have been reported for oxalate and malonate, respectively, by MacGowan and Surdam (1988). Recently, the aromatic acid anions benzoate and salicylate have been reported at the millimolal level. In addition to those anions listed in Table 2, several other carboxylate species have been identified only qualitatively in oil-field brines (Kharaka et al. 1986; Fisher and Boles 1990).

The concentrations of oxalate, malonate, and other difunctional carboxylate ions are probably limited by the very low solubility of their respective calcium salts (Table 3) and their susceptibility to thermal degradation as discussed below. The monofunctional carboxylate ions listed in Table 2 have calcium salt solubilities that are orders of magnitude greater than those of the dicarboxylate ions (Table 3). Thus, maximum concentrations of acetate and other monofunctional carboxylate ligands are more likely controlled by their relative rates of production and thermal degradation

Table 2. Dissolved carboxylate ligands detected in sedimentary basin brines

Name of acid ion		Concentration maxima		Reservoir temp. (°C)	Reference[a]
common	IUPAC	ppm	Molality		
Aliphatic monocarboxylate anions					
Formate	Methanoate	174	$10^{-2.4}$	101	1
Acetate	Ethanoate	10 000	0.17	?	2
Propionate	Propanoate	4 400	0.06	88	3
Butyrate	Butanoate	682	$10^{-2.11}$	138	1
Valerate	Pentanoate	371	$10^{-2.44}$	114	1
	Hexanoate	107	$10^{-3.04}$	86	1
	Heptanoate	99	$10^{-3.12}$	106	1
	Octanoate	42	$10^{-3.54}$	86	1
Aliphatic dicarboxylate anions					
Oxalate	Ethanedioate	494	$10^{-2.26}$	88	3
Malonate	Propanedioate	2 540	0.025	101	3
Succinate	Butanedioate	63	$10^{-3.30}$	98	4
Glutarate	Pentanedioate	36	$10^{-3.52}$	98	4
	Hexanedioate	0.5	$10^{-5.5}$	77	4
	Heptanedioate	0.6	$10^{-5.4}$	77	4
	Octanedioate	5.0	$10^{-4.5}$	98	4
	Nonanedioate	6.0	$10^{-4.5}$	98	4
	Decanedioate	1.3	$10^{-5.2}$	98	4
Maleate	cis-Butenedioate	66	$10^{-3.25}$	72	1
Aromatic carboxylate anions					
	Benzoate	50	$10^{-3.37}$	?	5
Salicylate	o-Hydroxybenzoate	65	$10^{-3.33}$	96	1

[a] 1, MacGowan and Surdam (1990a,b); 2, Surdam et al. (1984); 3, MacGowan and Surdam (1988); 4, Kharaka et al. (1986); 5, Barth (1987).

Table 3. Solubilities of calcium salts of selected carboxylic acids

Ca salts of carboxylic acids		Temperature (°C)	Solubility[a] (molality)
Composition	Name		
$Ca(C_2H_3O_2)_2$	Ca acetate	100	1.5
$Ca(C_3H_5O_2)_2 \cdot H_2O$	Ca propionate	100	6.18
$Ca(C_5H_9O_2)_2$	Ca valerate	100	1.16
CaC_2O_4	Ca oxalate	96	0.0001
$CaC_3H_2O_4 \cdot 4H_2O$	Ca malonate	100	0.0228
$CaC_4H_6O_4 \cdot 3H_2O$	Ca succinate	80	0.0274
$Ca(C_7H_5O_2)_2 \cdot 3H_2O$	Ca benzoate	80	0.2776
$Ca(C_7H_5O_3)_2 \cdot 2H_2O$	Ca salicylate	25	0.1288

[a] Morrison and Boyd (1966); Weast (1977).

in the hydrothermal environment and not their calcium salt solubility. Carboxylate species are generally present at concentrations greater than 10 ppm in formation waters with termperatures ranging from 40 to 200 °C. However, the highest concentrations of total carboxylate ions correspond to temperatures in the range 80 to 120 °C with peak concentrations near 100 °C. This distribution of organic acid anions in formation waters is consistent with our understanding of kerogen maturation in sedimentary basins from both field and experimental hydrous pyrolysis studies (Surdam et al. 1984; Lundegard and Kharaka 1990). Carbon dioxide and low molecular weight organic acids should be the dominant oxygen-bearing products released to formation waters (potential ore fluids) during the early stages of kerogen degradation and bitumen generation (Fig. 2). Biological activity is primarily responsible for the generally low concentrations of dissolved carboxylic acids in formation waters below 80 °C, whereas thermal decarboxylation produces a trend of decreasing concentration with temperature at elevated temperatures. A major decrease in concentration of monofunctional carboxylic acid ions is observed beginning about 120 °C, whereas concentrations of difunctional acid ions begin dropping near 100 °C. In formation waters with temperatures above 200 °C, the total concentration of carboxylic acid anions probably never exceeds 100 ppm (Lundegard and Kharaka 1990).

Rates of decarboxylation under hydrothermal conditions have been determined experimentally for several organic acids of geochemical interest, including acetic acid (Kharaka et al. 1983; Palmer and Drummond 1986; Bell 1991), oxalic acid (Crossey 1991), malonic acid (Hall 1949), and gallic acid (Boles 1986; Boles et al. 1988). The results of these investigations demonstrate that decarboxylation rates are pH-dependent and extremely sensitive to temperature and the types of catalytic surfaces in the system. A comparison of the available data from thermal degradation studies of carboxylic acids and their anions indicates the following trend in thermal stability (Crossey 1991): acetate ≫ formate > oxalate > gallate = malonate. At temperatures near 100 °C, the half-life of acetate is tens of millions of years, whereas oxalate has a half-life on the order of thousands of years. Lundegard and Kharaka (1990) present the results of several field studies, which suggest that half-lives of total acetate in formation waters near 100 °C are between 25 and 60 million years, values which are consistent with the experimental results cited above. It is generally agreed that under most natural hydrothermal conditions carboxylate anions, although thermodynamically not stable, are likely to be in metastable equilibrium with respect to non-redox reactions. However, there is disagreement as to whether or not carboxylate anions under hydrothermal conditions are in metastable equilibrium with respect to specific redox reactions (Shock 1988, 1989; Capuano 1990; Bell 1991; Helgeson 1991). Shock (1988, 1989) proposes that the distribution and concentration of organic acids in oil-field brines are controlled by metastable equilibria within the aquifer and not individual rates of decarboxylation. Based on thermodynamic calculations, Shock suggests that CO_2 in sedimentary basins may be in equilibrium with organic acids, but that CH_4 is not; thus implying that the high concentrations of carboxylate ions found in oil-field brines are far from equilibrium with respect to decarboxylation reactions. He further suggests that individual carboxylate ions are probably in mutual homogeneous metastable equilibrium with respect to ambient oxygen fugacities.

The concentration of carboxylate ions in surface and shallow subsurface interstitial waters is sufficiently low to preclude the importance of metal-carboxylate complexes as significant metal-transporting agents in low-temperature ore fluids responsible for supergene mineralization, syngenetic deposition in anoxic environments, and low-temperature epigenetic ores such as sandstone uranium deposits. However, the observed elevated concentrations of organic acids in oil-field brines strongly suggest that metal-carboxylate complexes may contribute significantly to metal transport in ore fluids with temperatures ranging from roughly 75 to 150 °C. Although Lundegard and Kharaka (1990) report no detectable organic acids in water samples from two Salton Sea geothermal wells with temperatures greater than 250 °C, metal-acetate complexes may be important to 200 °C in ore-forming systems of short duration (Drummond and Palmer 1986). These studies along with the field and laboratory evidence presented above clearly indicate that carboxylic acid ligands are probably not present at sufficient concentration levels to be important complexing agents in hydrothermal ore fluids above 250 °C.

3.3 Thermodynamic Data

To evaluate the ore transport potential of a particular metal-organic complex, the theoretically estimated concentration of that species in the ore solution of interest must be determined by calculation. Various types of information are required to perform such a calculation including chemical and physical parameters of the ore fluid and its surrounding geochemical environment, composition and thermodynamic data of all possibly important nonaqueous phases, and stoichiometry and thermodynamic data for all possibly important aqueous species. This latter category will be discussed here, with specific attention given to organic acid ligands and their corresponding metal-organic complexes. To calculate metal-ligand speciation, stability constants or free energy data for protonated organic ligand species and specific metal-organic complexes are required. With respect to the acid-ion ligands discussed in the previous section, thermodynamic data bases are the least developed for humate and fulvate species. True thermodynamic equilibrium constants are difficult to obtain for reactions involving humate and fulvate species because of the nonuniform nature of naturally occurring humic substances and the difficulty in characterizing individual humate and fulvate ions. However, because of the importance of such compounds in the soil environment, conditional stability constants have been generated at temperatures near 25 °C for a large number of fulvate and humate complexes involving metal cations typically found in oxidized environments near the earth's surface. Several excellent reviews and compilations of metal-humate and metal-fulvate complexes are available (e.g., Schnitzer and Khan 1978; Fitch and Stevenson 1983; Perdue 1985; Perdue and Lytle 1986). Martell and Smith (1974, 1977, 1982) and Smith and Martell (1975, 1989) have developed an extensive compilation of stability constants and other thermodynamic data for a large number of nonhumate organic ligands including amino acids, carboxylic acids, and organosulfur species. Nearly all of the data in this compilation and in other recent compilations of stability constants (e.g., Perrin 1979) are for temperatures below 100 °C and mostly for temperatures near 25 °C. These data as well as those available for humate and fulvate complexes are sufficient to adequately model metal-ligand speciation in supergene environments and low-temperature ore fluids if all other pertinent data and information are available.

To accurately calculate speciation in hydrothermal ore fluids, thermodynamic data at the specified hydrothermal temperature are required. At present, experimentally determined high temperature stability constants are available for only a few species of interest; most notably, acetate and oxalate complexes of some rock-forming and ore-forming metals and for protonated species of several carboxylate ligands (Table 4). For most other organic acids and metal-organic complexes of interest, experimentally determined stability constants are available only at temperatures near 25 °C. Although experimental data may be lacking, high-temperature stability constants can be estimated using reference thermodynamic data at some lower temperature, usually 25 °C, and one of several extrapolation methods. In addition to the simple van't Hoff approach, other methods include those based on the volumetric properties of water (Marshall 1970; Mesmer et al. 1988; Anderson et al. 1991), those based on the isocoulombic principle (Lindsay 1980; Hennet et al. 1988a; Mesmer et al. 1988; Giordano and Drummond 1991), and the predictive method developed by Helgeson (1967, 1969) which uses the standard enthalpy and entropy of reaction (Harrison and Thyne 1992). If experimentally determined values for stability constants are completely lacking for a particular metal-organic complex of interest, it may be possible to estimate values using a variety of correlation methods described by Langmuir (1979). In the speciation models presented in the next section, activities for complexes were calculated using high-temperature data (Table 4) for acetic acid and acetate complexes of Zn^{2+}, Fe^{2+}, Pb^{2+}, Al^{3+}, Ca^{2+}, Mg^{2+}; high-temperature thermodynamic data (Table 4) for oxalic acid, malonic acid, and oxalate complexes of Pb^{2+} and Al^{3+}; low-temperature data (near 25 °C) for the Na-acetate complex, oxalate complexes (except for Pb^{2+}), malonate complexes, succinate complexes, and species involving the ligand catechol; and available thermodynamic data from the literature on chloride, carbonate, sulfate, sulfide, and hydroxide complexes.

Table 4. Experimental studies of stability constants for acetate, oxalate, and malonate species under hydrothermal conditions

Reaction	Temperature range (°C)	References[a] 1
Acetate ($Ac^- = CH_3COO^-$)		
$H^+ + Ac^- = HAc$	25–350	2, 3, 4, 5, 6
$Na^+ + Ac^- = NaAc$	275–320	5
$Ca^{2+} + Ac^- = CaAc^+$	80–350	7, 8
$Mg^{2+} + Ac^- = MgAc^+$	25–150	7, 9
$Al^{3+} + Ac^- = AlAc^{2+}$	25–125	7, 10
$Al^{3+} + 2Ac^- = Al(Ac)_2^+$	75–150	7, 10
$Fe^{2+} + Ac^- = FeAc^+$	50–300	11
$Fe^{2+} + 2Ac^- = Fe(Ac)_2$	50–300	11
$Fe^{2+} + 3Ac^- = Fe(Ac)_3^-$	50–300	11
$Pb^{2+} + Ac^- = PbAc^+$	25–85	12, 13
$Pb^{2+} + 2Ac^- = Pb(Ac)_2$	25–85	12, 13
$Zn^{2+} + Ac^- = ZnAc^+$	50–300	14
$Zn^{2+} + 2Ac^- = Zn(Ac)_2$	50–300	14
$Zn^{2+} + 3Ac^- = Zn(Ac)_3^-$	50–300	14
Oxalate ($Ox^{2-} = C_2O_4^{2-}$)		
$H^+ + HOx^- = H_2Ox$	25–125	15
$H^+ + Ox^{2-} = HOx^-$	25–175	15
$Al^{3+} + Ox^{2-} = AlOx^+$	25–125	16
$Al^{3+} + 3Ox^{2-} = Al(Ox)_3^{3-}$	80	17
$Pb^{2+} + Ox^{2-} = PbOx$	25–85	12
$Pb^{2+} + 2Ox^{2-} = Pb(Ox)_2^{2-}$	25–85	12
Malonate ($Ma^{2-} = C_3H_2O_4^{2-}$)		
$H^+ + HMa^- = H_2Ma$	25–100	18
$H^+ + Ma^{2-} = HMa^-$	25–100	18

[a] 1, Additonal data near 25 °C are also cited by Martell and Smith (1977, 1982) and Smith and Martell (1989); 2, Ellis (1963); 3, Lown et al. (1970); 4, Fisher and Barnes (1972); 5, Oscarson et al. (1988); 6, Mesmer et al. (1989); 7, Fein (1991a); 8, Seewald and Seyfried (1991); 9, Semmler et al. (1990); 10, Drummond et al. (1989); 11, Palmer and Drummond (1988); 12, Hennet et al. (1988a); 13, Giordano (1989); 14, Giordano and Drummond (1991); 15, Kettler et al. (1991); 16, Thyne et al. (1992); 17, Fein (1991b); 18, Kettler et al. (1992).

4 Ore Solution Models

A variety of geochemical models have been developed to describe the genesis of those deposits typically associated with organic matter. In many of these models, metal transport is dominantly by some mechanism involving inorganic complexes, whereas metal deposition is caused by epigenetic or indigenous organic matter. During the past 15 years, efforts have been initiated by a small number of investigators to study the importance of organic matter as a means of metal transport in hydrothermal ore fluids (Giordano and Barnes 1981; Giordano 1985, 1990; Drummond and Palmer 1986; Manning 1986; Kharaka et al. 1987; Hennet et al. 1988a; Gize and Barnes 1989; Rose 1989). Progress, however, has been slow in developing satisfactory ore fluid models that involve metal-organic complexing. The principal causes for this slow progress can be attributed to a lack of information on the following: the chemical and physical constraints of ore-forming environments, especially pH and oxidation state; the nature and concentration of organic ligands in ore-forming environments; and the thermodynamic properties of pertinent metal-organic complexes. In this section, the results are given for metal and ligand speciation calculations for three proposed Mississippi Valley-type (MVT) ore fluids and a model composite ore fluid for red bed-related base metal (RBRBM) deposits. It is now widely accepted that these two deposit types are related to sedimentary basinal processes and that their ore fluids were probably evolved sedimentary basin brines (Sverjensky 1989). Potential organic ligands in these ore fluids are polar organic compounds, which have been detected in modern oil-field brines (Table 2), and those which may have fractionated into such brines during the early to middle stages of basin diagenesis (Fig. 2). Based solely on analyses of oil-field brines and the available thermodynamic data, the most likely organic complexing agents in MVT and RBRBM ore fluids are the carboxylic acid anions acetate, oxalate, malonate, and succinate. Although monohydroxyphenols have been detected in oil-field brines (Kharaka et al. 1986; Fisher and Boles 1990), they do not form strong complexes with metals. Dihydroxyphenols can form strong complexes with metals and, although they have not been detected in oil-field brines, Surdam et al. (1984) suggested that aluminum catechol complexes may be responsible for significant aluminum mobilization during diagenesis. Certain thiols and mercaptocarboxylate ligands form strong metal-organic complexes, however, to my knowledge specific organosulfur compounds have not been identified in oil-field brines.

Giordano (1990) calculated concentrations of Ca, Mg, Na, Pb, Zn, Fe, and Al as acetate, oxalate, malonate, succinate, and catechol complexes in three reconstructed MVT ore solutions and a model oil-field brine at 100°C. In Table 5, results are presented for revised speciation calculations for the three MVT ore fluids and calculated speciation results for a model RBRBM ore fluid. These model results are based in part on recently published data at elevated temperatures for Ca, Mg, and Al complexes and high-temperature constants for ionization reactions of oxalic and malonic acids (Table 4). In the MVT model proposed by Anderson (1975; model 1, Table 5), the ore fluid is moderately oxidized and falls well above the sulfide-sulfate boundary in $\log a_{o2}$ – pH space at 100°C. It is slightly acid and contains 10^{-2} molal

Table 5. Calculated concentrations of lead and zinc in proposed Mississippi Valley-type ore fluids (models 1, 2, and 3) and a composite ore fluid for red-bed base metal deposits (model 4). Concentrations are in ppm

Complex	Model 1 Anderson (1975)[a]	Model 2 Sverjensky (1984)[b]	Model 3 Giordano and Barnes (1981)[c]	Model 4 Composite[d]
Total lead	79.70	0.45	2.2×10^{-3}	29.30
Lead chloride	76.70	0.43	1.1×10^{-8}	22.40
Lead hydroxide	1.01	3.9×10^{-3}	4.8×10^{-9}	2.09
Lead bisulfide	3.2×10^{-10}	2.3×10^{-4}	2.2×10^{-3}	8.0×10^{-11}
Lead acetate	1.10	5.2×10^{-3}	2.9×10^{-10}	2.45
Lead oxalate	0.65	2.9×10^{-3}	2.0×10^{-10}	2.03
Lead malonate	0.08	3.9×10^{-4}	1.6×10^{-11}	0.09
Lead succinate	3.7×10^{-4}	3.4×10^{-6}	6.2×10^{-14}	5.2×10^{-4}
Lead catechol	3.4×10^{-6}	6.9×10^{-11}	3.4×10^{-13}	9.5×10^{-6}
Total zinc	531.0	2.68	3.0×10^{-3}	186.3
Zinc chloride	519.0	2.66	7.7×10^{-8}	160.6
Zinc hydroxide	4.8	9.3×10^{-3}	2.3×10^{-8}	9.89
Zinc bisulfide	2.7×10^{-10}	6.2×10^{-5}	3.0×10^{-3}	6.7×10^{-11}
Zinc acetate	3.35	6.3×10^{-3}	8.4×10^{-10}	7.03
Zinc oxalate	1.40	6.6×10^{-3}	4.7×10^{-10}	5.07
Zinc malonate	1.08	1.0×10^{-3}	2.4×10^{-10}	2.04
Zinc succinate	2.0×10^{-3}	3.0×10^{-6}	3.7×10^{-13}	3.9×10^{-3}
Zinc catechol	3.1×10^{-4}	3.0×10^{-9}	4.7×10^{-11}	1.3×10^{-3}

[a] Total sulfur = 10^{-2} molal, pH = 5.7, $\log a_{o2}$ = -50, 100 °C
[b] Total sulfur = 10^{-5} molal, pH = 4.5, $\log a_{o2}$ = -50.4, 125 °C
[c] Total sulfur = 10^{-2} molal, pH = 7.2, $\log a_{o2}$ = -55, 100 °C
[d] Total sulfur = 10^{-2} molal, pH = 6.0, $\log a_{o2}$ = -50, 100 °C: the composition of this model ore fluid is a composite based on parameters proposed by Rose (1976, 1989), Sverjensky (1987, 1989), and Branam and Ripley (1990).

total sulfur, principally in the form of SO_4^{2-}, with reduced sulfur concentrations well below 10^{-5} molal. The MVT ore solution proposed by Giordano and Barnes (1981; model 3, Table 5) is reduced and falls well below the sulfide-sulfate boundary at 100 °C. It is slightly alkaline and contains 10^{-2} molal total sulfur, principally as bisulfide. In the MVT model proposed by Sverjensky (1984; model 2, Table 5), the ore solution is also reduced, with $\log a_{o2}$ – pH conditions just below the sulfide-sulfate boundary at 125 °C. The solution is moderately acid and contains 10^{-5} molal total sulfur, principally as H_2S. The composite ore fluid proposed here for red bed-related base metal deposits (model 4, Table 5) is similar to the MVT model proposed by Anderson (1975). It is moderately oxidized and falls well above the sulfide-sulfate boundary at 100 °C. It is a bit less acid than the Anderson (1975) model and contains 10^{-2} molal total sulfur, principally in the form of SO_4^{2-}, with reduced sulfur concentrations well below 10^{-5} molal. The model parameters used for this composite ore fluid are based on those proposed by

Rose (1976, 1989), Sverjensky (1987, 1989), and Branam and Ripley (1990) for sediment-hosted copper-rich deposits.

For the three MVT models, total chloride concentrations are near 3 molal, whereas in the RBRBM model total chloride is 2 molal. For the four ore-fluid models, several mineral saturation constraints were applied. These are listed in Table 6 together with the concentrations of organic ligands used in these models. The concentrations of acetate, oxalate, malonate, and succinate are based on maximum values reported for basinal brines (Table 2). The concentration of total catechol (0.01 molal) is an arbitrarily chosen value that is in the concentration range reported for malonate, but is well below catechol solubility in water (4 molal at 25 °C). The total chloride to total acetate ratio is about 30:1 for the three MVT models and 20:1 for the RBRBM composite fluid.

The results presented in Table 5 shed light on several aspects of lead and zinc transport in the four model ore fluids. If the minimum concentration of lead and zinc required to form an ore deposit is 1 to 10 ppm (Barnes 1979; Anderson 1983), then in model 3 proposed by Giordano and Barnes (1981) sufficient metal cannot be transported by any of the listed complexes. Barely enough lead and zinc are transported in the MVT fluid proposed by Sverjensky (1984), predominantly as choride complexes. In the MVT ore fluid proposed by Anderson (1975) and the composite RBRBM fluid, more than sufficient zinc and lead can be transported, predominantly as chloride complexes. Note that in these model ore fluids, significant amounts of lead and zinc are transported as acetate, oxalate, and malonate complexes. In model 2, chloride complexes are also dominant and total carboxylate complexes account for the second largest fraction of lead and zinc. In model 3, bisulfide complexes are by far the dominant species, with all other listed complexes accounting for many orders of magnitude less metal in solution. In all four models, catechol is unable to transport significant amounts of lead and zinc. This follows from the law of mass action, which favors H^+ in the competition for catechol ions within the given pH ranges of the ore fluids.

The competition among cations for organic ligands is an important factor in controlling the speciation in these model ore fluids. To illustrate the nature of this competition, speciation of Na, Ca, Mg, Fe, Al, Pb, and Zn and the ligands listed in Table 6 is shown in Tables 7 and 8 for the model RBRBM ore fluid (model 4, Table 5). Calculated chloride and organic ligand speciation for this ore fluid is given in Table 7 as percent of ligand bound in the indicated species or complex. In this oxidized and slightly acid solution, the ligands acetate, oxalate, malonate, and succinate are present primarily as free ions, singly protonated species, and complexes of Na, Ca, and Mg. Although carboxylate complexes of Pb, Zn, Al, and Fe have a greater thermodynamic stability than carboxylate complexes of Na, Mg, and Ca, the free ion activities of the former group of metals are too low to allow significant competition with the weakly complexing, but more abundant, ions of Na, Mg, and Ca. Catechol forms very stable metal-organic complexes

Table 6. Mineral saturation constraints and organic ligand concentrations for ore-fluid speciation models

Saturation constraints	Organic ligands
Galena	Acetate (0.1 molal)
Sphalerite	Oxalate (0.005 molal)
Pyrite[a]	Malonate (0.025 molal)
Hematite[b]	Succinate (0.0005 molal)
Calcite	Catechol (0.01 molal)
Quartz	
Potassium feldspar	

[a] MVT models 2 (Sverjensky 1984) and 3 (Giordano and Barnes 1981) (Table 5).
[b] MVT model 1 (Anderson 1975) and RBRBM model 4 (Table 5).

(Martell and Smith 1977), however, at the model pH of 6, H^+ is the dominant cation competitor, and almost 100% of the catechol ligand is in the form of the doubly protonated species. Because of the low activity of ionized catechol, only small amounts of metal-catechol complexes are formed.

Calculated metal speciation for the composite RBRBM fluid is given in Table 8 as percent metal bound in the indicated species. Although presented differently, the results in Table 8 for lead and zinc are similar to those given in Table 5 for model 4; more than 75% of the lead and zinc is in chloride complexes and a significant 15% of these metals is bound in carboxylate complexes. Calcium and magnesium are present primarily as free ions or in chloride complexes, but approximately 12 to 14% of each of these metals is present in acetate complexes, and between 4 and 12% is bound in other carboxylate complexes. Iron is present primarily as a free ion (18%), in chloride complexes (24%), and hydroxide complexes (44%). Similar to calcium and magnesium, about 9% of the iron is in acetate complexes and about 5% is bound in other carboxylate species. Of all the metals considered in this model, aluminum is the most strongly complexed with organic ligands. Of the total dissolved aluminum, 44 and 53% is accounted for by aluminum-oxalate complexes and hydrolyzed species, respectively. Significant amounts of aluminum are also bound in acetate and catechol complexes.

The results presented in Tables 5, 7, and 8 are in agreement with other studies investigating the significance of metal-organic complexing in ore fluids. For the same MVT ore fluids considered in this section, Giordano (1985) calculated lead and zinc speciation involving nine organic ligands (acetate, propionate, n-butyrate, phthalate, oxalate, tartronate, malate, D-tartrate, and salicylate). Of these ligands, acetate and oxalate were found to be the most effective organic transporting agents by several orders of magnitude. Giordano (1985) used only 25 °C stability constants for zinc and

Table 7. Calculated ligand speciation in model ore fluid for red bed-related base metal deposits[a]

Ligand	Total molality	Percent ligand bound in indicated species[b]									
		L	HL	H₂L	Na	Ca	Mg	Pb	Zn	Fe	Al
Chloride	2.0	88.28	NC	NC	10.93	0.25	0.45	0.02	0.38	<0.01	NC
Acetate	0.1	57.73	3.20	NC	23.78	9.63	6.06	0.02	0.18	0.01	<0.01
Oxalate	0.005	29.16	0.30	<0.01	NC	29.80	39.00	0.24	2.28	0.01	<0.01
Malonate	0.025	37.00	8.4	0.01	33.76	8.40	13.20	0.01	0.15	0.01	NC
Succinate	0.0005	60.60	4.74	0.04	23.80	7.20	4.06	<0.01	<0.01	<0.01	NC
Catechol	0.01	<0.01	0.07	99.93	NC	<0.01	<0.01	<0.01	<0.01	<0.01	<0.01

[a] Parameters for composite ore fluid are based on three proposed models (Rose 1976, 1989; Sverjensky 1987, 1989; Branam and Ripley 1990) and include: $T = 100°C$, $pH = 6$, $\log a_{o2} = -50$, total sulfur $= 10^{-2.0}$ molal, total carbon $= 10^{-2.4}$ molal.

[b] Abbreviations: L = nonprotonated ligand; HL = singly protonated ligand; H₂L = doubly protonated ligand; Na, Ca, Mg, Pb, Zn, Fe, and Al refer to metal-organic complexes; NC indicates not calculated.

Table 8. Calculated metal speciation in model ore fluid for red bed-related base metal deposits[a]

Metal	Total (ppm)	Percent metal in free ion and indicated complex[b,c]							
		Free ion	Chloride	Hydroxide	Acetate	Oxalate	Malonate	Succinate	Catechol
Na	3.82×10^4	84.33	13.12	<0.01	1.42	NC	0.50	0.01	NC
Ca	3.35×10^3	76.46	5.93	<0.01	11.51	1.78	2.51	0.05	<0.01
Mg	1.05×10^3	50.00	21.16	<0.01	14.09	4.69	7.67	0.04	<0.01
Pb	2.94×10^1	0.51	76.33	7.10	8.31	6.89	0.51	<0.01	0.01
Zn	1.86×10^2	0.83	86.32	5.26	3.77	2.72	1.10	<0.01	<0.01
Fe	2.30×10^0	17.53	23.85	43.83	9.36	1.36	4.04	<0.01	<0.01
Al	2.22×10^{-3}	<0.01	NC	52.49	0.22	44.19	NC	NC	3.06

[a] Parameters for composite ore fluid are based on three proposed models (Rose 1976, 1989; Sverjensky 1987, 1989; Branam and Ripley 1990) and include: $T = 100°C$, $pH = 6$, $\log a_{o2} = -50$, total sulfur $= 10^{-2.0}$ molal, total carbon $= 10^{-2.4}$ molal.

[b] Total chloride $= 2.0$ molal, total acetate $= 0.1$ molal, total oxalate $= 0.005$ molal, total succinate $= 0.0005$ molal, total malonate $= 0.025$ molal, total catechol $= 0.01$ molal.

[c] NC indicates not calculated.

lead organic complexes and obtained concentrations of lead and zinc in acetate and oxalate complexes several orders of magnitude lower than those reported in Table 5. Kharaka et al. (1987) and Lundegard and Kharaka (1990) calculated metal/organic ligand speciation in metal-rich, oil-field brines from central Mississippi. These brines are among the most metal-enriched subsurface waters documented (Carpenter et al. 1974; Saunders and Swann 1990) and may represent potential ore fluids. Their model calculations indicate that metal-organic complexes are of minor importance in the speciation of lead and zinc but that iron, calcium, and possibly aluminum form important metal-acetate and metal-oxalate complexes in these waters and ore fluids with similar chemistry.

The calculated speciations presented in Tables 5, 7, and 8 should be considered with caution. More reliable estimates of speciation can be made when additional high-temperature thermodynamic data for metal-organic complexes and additional information regarding pertinent organic ligands and their concentrations in ore fluids are available. As a result of these deficiencies, the calculated concentrations reported here should be considered as order of magnitude estimates of actual concentrations, which may have existed in MVT and RBRBM ore fluids.

5   Conclusion

Although the ore solution models described in this chapter are based on a less than ideal thermodynamic data base, the principal deficiency is clearly the lack of well-constrained geochemical parameters for the tested ore-forming systems. Efforts to successfully evaluate metal-organic complexing as an ore transport mechanism must not neglect the importance of well-constrained geochemical parameters and the procurement of reliable thermodynamic data. From the results presented in this chapter, some important inferences can be made regarding metal transport by organic complexes in ore fluids of sedimentary origin. The results presented in Table 5 clearly show that significant amounts of zinc and lead cannot be mobilized as metal-organic complexes involving acetate or other carboxylate ligands in ore fluids containing greater than 10^{-5} molal reduced sulfur as hydrogen sulfide or bisulfide. However, in reduced ore fluids sufficient metal and reduced sulfur may be transported as complexes involving organosulfur ligands (e.g., sulfur-bearing amino acids, mercaptocarboxylic acids, thiols, sulfides). On the other hand, significant quantities of dissolved metal can be transported by carboxylate complexes in oxidized ore fluids containing less than 10^{-9} molal reduced sulfur (models 1 and 4, Table 5).

If the organic ligand concentrations in these model ore fluids are increased by a factor of two or three, a significant shift in favor of organic complexing relative to chloride complexing would be observed. Giordano and Drummond (1991) and Hennet et al. (1988a) show for zinc and lead that acetate complexing is dominant relative to chloride complexing in solutions having [chloride ion]/[acetate ion] concentration ratios of less than about 10 at temperatures between 25 and 300 °C. Lundegard and Kharaka (1990) and Drummond and Palmer (1986) present evidence from brine analyses and theoretical models (designed to simulate the generation and destruction of organic acids in basinal water), which suggests that concentrations two to three times those given in Table 2 are not unreasonable for primary organic acid concentrations. Here, primary concentration is defined as the concentration that exists before significant destruction of the organic acid takes place (Lundegard and Kharaka 1990). Basinal fluids having such high organic ligand concentrations are probably rare but may exist as pore water in metal-rich black shales or petroleum source rocks (Lundegard and Kharaka 1990). Extraction of metals into these organic-rich pore waters, either before or after expulsion into more permeable units, may be the initial stage in the generation of metal-rich ore fluids responsible for epigenetic base metal mineralization in sedimentary basins. Furthermore, oxidized ore fluids similar in chemistry to the RBRBM model solution can now be linked to deposits that have a strong syngenetic character. For example, there is a sufficient body of evidence to infer that most of the lead, zinc, and copper in the highest grade zones of the Kupferschiefer was introduced during late stage diagenesis by oxidized metal-rich brines at temperatures near 100 °C (Jowett 1986; Jowett et al. 1987; Vaughan et al. 1989). And recently, Coveney et al. (1987) and Coveney and Glascock (1989) proposed a genetic link between the mineralization of certain Pennsylvanian black shales of the US midcontinent and basinal brines, possibly MVT ore fluids.

The MVT and RBRBM models evaluated in this chapter show that metal-organic complexes may be important in transporting lead, zinc, and other metals in ore fluids responsible for at least two major classes of deposits listed in Table 1. It is hoped that these results will encourage investigators to evaluate the role metal-organic complexing (including complexes of Cu, Ag, Au, Ni, V, Mo, U, and platinum-group metals) in the formation of all deposits with a clear genetic link to organic processes.

References

Aiken GR, McKnight DM, Wershaw RL (1985) Humic substances in soil, sediment, and water. Wiley, New York, 392 pp

Anderson GM (1975) Precipitation of Mississippi Valley-type ores. Econ Geol 70: 937–942

Anderson GM (1983) Some geochemical aspects of sulfide precipitation in carbonate rocks. In: Kisvarsanyi G, Grant SK, Pratt WP, Koenig JW (eds) Int Conf Mississippi Valley type lead-zinc deposits, Proc vol. Rolla, University of Missouri-Rolla, pp 61–76

Anderson GM, Castet S, Schott J, Mesmer RE (1991) The density model for estimation of thermodynamic parameters of reactions at high temperatures and pressures. Geochim Cosmochim Acta 55: 1769–1779

Bada JL, Zhao M, Miller SL (1991) Alanine stability in aqueous solutions at 250 °C. Geol Soc Am Abstr Programs 23(5): A25

Baker WE (1978) The role of humic acid in the transport of gold. Geochim Cosmochim Acta 42: 645–649

Bannikova LA, Shirinskiy V G (1988) The hydrocarbons in hydrothermal ore-deposit bitumoid. Geochem Int 25: 58–66

Barnes HL (1979) Solubilities of ore minerals. In: Barnes HL (ed) Geochemistry of hydrothermal ore deposits, 2nd edn. Wiley, New York, pp 405–461

Barnes HL (1983) Ore-deposition reactions in Mississippi Valley-type deposits. In: Kisvarsany G, Grant SK, Pratt WP, Koenig JW (eds) Int Conf Mississippi Valley type lead-zinc deposits, Proc vol. Rolla, University of Missouri-Rolla, pp 75–85

Barnes HL, Czamanske GK (1967) Solubilities and transport of ore minerals. In: Barnes HL (ed) Geochemistry of hydrothermal ore deposits. Holt, Rinehart, and Winston, New York, pp 334–381

Barth T (1987) Quantitative determination of volatile carboxylic acids in formation waters by isotachophoresis. Anal Chem 59: 2232–2237

Barton PB (1967) Possible role of organic matter in the precipitation of the Mississippi Valley ores. Econ Geol Monogr 3: 371–378

Barton PB (1982) The many roles of organic matter in the genesis of mineral deposits. Geol Soc Am Abstr Programs 14(7): 440

Bazylinski DA, Farrington JW, Jannasch HW (1988) Hydrocarbons in surface sediments from a Guaymas Basin hydrothermal vent site. Org Geochem 12: 547–558

Bell JL (1991) Acetate decomposition in hydrothermal solutions. PhD Thesis, The Pennsylvania State University, University Park, 228 pp

Bernhardt G, Lundemann H D, Jaenicke R, Konig H, Stetter KO (1984) Biomolecules are unstable under "black smoker" conditions. Naturwissenschaften 71: 583–586

Blumer M (1975) Curtisite, idrialite, and pendletonite, polycyclic aromatic hydrocarbon minerals: their composition and origin. Chem Geol 16: 245–256

Boles JS (1986) Aqueous thermal degradation of naturally occurring aromatic organic acids and the synthetic chelating agent disodium EDTA. PhD Thesis, Princeton University, Princeton, 137 pp

Boles JS, Crerar DA, Grissom G, Key T (1988) Aqueous thermal degradation of gallic acid. Geochim Cosmochim Acta 52: 341–344

Boyle RW, Brown AC, Jefferson CW, Jowett EC, Kirkham RV (1989) Sediment-hosted stratiform copper deposits. Geol Assoc Can Spec Pap 36, 710 pp

Branam TD, Ripley EM (1990) Genesis of sediment-hosted copper mineralization in south-central Kansas: sulfur/carbon and sulfur isotope systematics. Econ Geol 85: 601–621

Buffle J (1988) Complexation reactions in aquatic systems. Wiley, New York, 692 pp

Capuano RM (1990) Hydrochemical constraints on fluid-mineral equilibria during compaction diagenesis of kerogen-rich geopressured sediments. Geochim Cosmochim Acta 54: 1283–1299

Carpenter AB, Trout ML, Pickett EE (1974) Preliminary report on the origin and chemical evolution of lead- and zinc-rich oil field brines in central Mississippi. Econ Geol 69: 1191–1206

Coel RJ, Crock JG, Kyle JR (1991) Biogeochemical studies of gold in a placer deposit, Livengood, Alaska. US Geol Surv Open-File Rep 91–142, 51 pp

Coveney RM, Glascock MD (1989) A review of the origins of metal-rich Pennsylvanian black shales, central USA, with an inferred role for basinal brines. Appl Geochem 4: 347–367

Coveney RM, Leventhal JS, Glascock MD, Hatch JR (1987) Origins of metals and organic matter in the Mecca Quarry Shale Member and stratigraphically equivalent beds across the midwest. Econ Geol 82: 915–933

Crossey LJ (1991) Thermal degradation of aqueous oxalate species. Geochim Cosmochim Acta 55: 1515–1527

Dean WE (1986) Organics and ore deposits. Proc Denver Region Exploration Geologists Society, Wheat Ridge, Co, Symp on Organics in ore deposits, 218 pp

Degens ET, Hunt JM, Reuter JH, Reed WE (1964) Data on the distribution of amino acids and oxygen isotopes in petroleum brine waters of various geologic ages. Sedimentology 3: 199–225

Drummond SE, Palmer DA (1986) Thermal decarboxylation of acetate. Part II. boundary conditions for the role of acetate in the primary migration of natural gas and the transportation of metals in hydrothermal systems. Geochim Cosmochim Acta 50: 825–833

Drummond SE, Palmer DA, Wesolowski DJ, Giordano TH (1989) Hydrothermal transportation of metals via acetate complexes. 28th Int Geol Congr Abstr Washington DC, vol. 1, p 420

Ehrlich HL (1986) Interactions of heavy metals and microorganisms. In: Carlisle D, Berry WL, Kaplan IR, Watterson JR (eds) Mineral exploration: biological systems and organic matter. Prentice-Hall, Englewood Cliffs, pp 221–237

Ellis AJ (1963) The ionization of acetic, propionic, n-butyric, and benzoic acid in water, from conductance measurements up to 225 °C. J Chem Soc 59: 2299–2310

Fein JB (1991a) Experimental study of aluminum-, calcium-, and magnesium-acetate complexing at 80 °C. Geochim Cosmochim Acta 55: 955–964

Fein JB (1991b) Experimental study of aluminum-oxalate complexing at 80 °C: implications for the formation of secondary porosity within sedimentary reservoirs. Geology 19: 1037–1040

Fisher JR, Barnes HL (1972) The ion-product constant of water to 350 °C. J Phys Chem 76: 90–99

Fisher JB, Boles JR (1990) Water-rock interaction in Tertiary sandstones, San Joaquin Basin, California, USA: diagenetic controls on water composition. Chem Geol 82: 83–101

Fitch A, Stevenson FJ (1983) Stability constants of metal-organic matter complexes: theoretical aspects and mathematical models. In: Theophrastus SA (ed) Significance of trace elements in solving petrogenetic problems. National Technical University of Athens, Athens, pp 645–669

Germanov AI (1965) Geochemical significance of organic matter in the hydrothermal process. Geochem Int 2: 643–652

Germanov AI, Bannikova LA (1972) Change in organic matter of sedimentary rocks in the hydrothermal process of sulfide concentration. Dokl Akad Nauk SSSR 203: 1180–1182

Germanov AI, Mel' Kanovitskaga SG (1975) Organic acids in hydrothermal ores of polymetallic deposits in groundwaters. Dokl Akad Nauk SSSR 225: 182–184

Gill GA, Bruland KW (1990) Mercury speciation in surface freshwater systems in California and other areas. Environ Sci Technol 24: 1392–1400

Giordano TH (1985) A preliminary evaluation of organic ligands and metal-organic complexing in Mississippi Valley-type ore solutions. Econ Geol 80: 96–106

Giordano TH (1989) Anglesite (PbSO$_4$) solubility in acetate solutions: the determination of stability constants for lead acetate complexes to 85 °C. Geochim Cosmochim Acta 53: 359–366

Giordano TH (1990) Organic ligands and metal-organic complexing in ore fluids of sedimentary origin. US Geol Surv Circ 1058: 31–41

Giordano TH, Barnes HL (1981) Lead transport in Mississippi Valley-type ore solutions. Econ Geol 76: 2200–2211

Giordano TH, Drummond SE (1991) The potentiometric determination of stability constants for zinc acetate complexes in aqueous solutions to 295 °C. Geochim Cosmochim Acta 55: 2401–2415

Gize AP (1984) The organic geochemistry of three Mississippi Valley-type ore deposits: PhD Thesis, The Pennsylvania State University, University Park, 350 pp

Gize AP (1986) The development of a thermal mesophase in bitumens from high temperature ore deposits. In: Dean WE (ed) Organics and ore deposits. Proc Denver Region Exploration Geologists Society, Wheat Ridge, CO, Symp on Organics in ore deposits, pp 137–150

Gize AP, Barnes HL (1987) The organic geochemistry of two Mississippi Valley-type lead-zinc deposits. Econ Geol 82: 457–470

Gize AP, Barnes HL (1989) Organic processes in Mississippi Valley-type ore genesis. 28th Int Geol Congr Abstr Washington DC, vol 1, pp 557–558

Grauch RI, Huyck HLO (1990) Metalliferous black shales and related ore deposits. Proc, 1989 US Working Group Meet, Int Geol Correlation Program Proj 254. US Geol Surv Circ 1058, Washington DC, 85 pp

Gustafson LB, Williams N (1981) Sediment-hosted stratiform deposits of copper, lead, and zinc. In: Skinner BJ (ed) Economic geology. 75th Anniversary volume. Society of Economic Geologists, El Paso, TX, pp 139–178

Haberstroh PR, Karl DM (1989) Dissolved free amino acids in hydrothermal vent habitats of the Guaymas basin. Geochim Cosmochim Acta 53: 2937–2945

Hall GA (1949) The kinetics of the decomposition of malonic acid in aqueous solution. J Am Chem Soc 71: 2691–2693

Hansley PL, Spirakis CS (1992) Organic matter diagenesis as the key to a unifying theory for the genesis of tabular uranium-vanadium deposits in the Morrison Formation, Colorado Plateau. Econ Geol 87: 352–365

Harrison WJ, Thyne GD (1992) Predictions of diagenetic reactions in the presence of organic acids. Geochim Cosmochim Acta 56: 565–586

Hausen DM, Park WC (1986) Observations on the association of gold mineralization with organic matter in Carlin-type ores. In: Dean WE (ed) Organics and ore deposits, Proc Denver Region Exploration Geologists Society, Wheat Ridge, CO, Symp on Organics and ore deposits, pp 119–136

Helgeson HC (1967) Thermodynamics of complex dissociation in aqueous solution at elevated temperatures. J Phys Chem 71: 3121–3136

Helgeson HC (1969) Thermodynamics of hydrothermal systems at elevated temperatures and pressures. Am J Sci 267: 729–804

Helgeson HC (1991) Organic/inorganic reactions in metamorphic processes. Can Mineral 29: 707–739

Hennet RJC, Crerar DA, Schwartz J (1988a) Organic complexes in hydrothermal systems. Econ Geol 83: 742–764

Hennet RJC, Crerar DA, Schwartz J (1988b) The effect of carbon dioxide partial pressure on metal transport in low-temperature hydrothermal systems. Chem Geol 69: 321–330

Henry AL, Anderson GM, Heroux Y (1992) Alteration of organic matter in the viburnum trend lead-zinc district of southeast Missouri. Econ Geol 87: 288–309

Heroux Y, Michoux D, Desjardins M, Sangster D (1989) Petrographic et geochimie des matieres organiques des sequences plombo – zinciferes d'age Carbonifere, Bassin Salmon River, Nouvelle – Ecosse, Canada. Org Geochem 14: 253–268

Ho ES, Meyers PA, Mauk JL (1990) Organic geochemical study of mineralization in the Keweenawan Nonesuch Formation at White Pine, Michigan. Org Geochem 16: 229–234

Hulen JB (1991) Assessing the role of active and ancient geothermal systems in evolution of oil reservoirs in the Basin and Range province, eastern Nevada. In: Sedimentary basin geochemistry and fluid/rock interactions. Worksh Proc Vol. The University of Oklahoma, Norman, pp 8–17

Jackson KS, Jonasson IR, Skippen GB (1978) The nature of metals-sediment-water interactions in freshwater bodies, with emphasis on the role of organic matter. Earth Sci Rev 14: 97–146

Jowett EC (1986) Genesis of Kupferschiefer Cu-Ag deposits by convective flow of Rotliegendes brines during Triassic rifting. Econ Geol 81: 1823–1837

Jowett EC, Rydzewski A, Jowett RJ (1987) The Kupferschiefer Cu-Ag ore deposits in Poland – a re-appraisal of the evidence of their origin and presentation of a new genetic model. Can J Earth Sci 24: 2016–2037

Kawka OEM, Simoneit BRT (1990) Hydrothermal generation of aromatic hydrocarbons in petroleum and subsurface sediments of Guaymas basin: traces of high-temperature sedimentary processes. Geol Soc Am Abstr Programs 2:A33

Kesler SE, Kettler RM, Meyers PA, Dunham KW, Russell N, Seaward M, McCurdy K (1986) Relation between organic material and precious metal mineralization in the Moore ore body, Pueblo Viejo, Dominican Republic. In: Dean WE (ed) Organics and ore deposits, Proc Denver Region Exploration Geologists Society, Wheat Ridge, CO, Symp on Organics and ore deposits, pp 105–110

Kettler RM, Waldo GS, Penner-Hahu JE, Meyers PA, Kesler SE (1990) Sulfidation of organic matter associated with gold mineralization, Pueblo Viejo, Dominican Republic. Appl Geochem 5: 237–248

Kettler RM, Palmer DA, Wesolowski DJ (1991) Dissociation quotients of oxalic acid in aqueous sodium chloride media to 175 °C. J Solution Chem 20: 905–927

Kettler RM, Wesolowski DJ, Palmer DA (1992) Dissociation quotients of malonic acid in aqueous sodium chloride media to 100 °C. J Solution Chem 21: 883–900

Kharaka YK, Lico MS, Wright VA, Carothers WW (1979) Geochemistry of formation waters from Pleasant Bayou No. 2 well and adjacent areas in coastal Texas. Proc 4th US Gulf Coast Geopressured-Geothermal Energy Conf, Research and Development, University of Texas at Austin, pp 168–199

Kharaka YK, Carothers WW, Rosenbauer RJ (1983) Thermal decarboxylation of acetic acid: implications for origin of natural gas. Geochim Cosmochim Acta 47: 397–402

Kharaka YK, Law LM, Carothers WW, Goerlitz DF (1986) Role of organic species dissolved in formation waters from sedimentary basins in mineral diagenesis. In: Gautier DL (ed) Roles of organic matter in sedimentary diagenesis. Soc Econ Paleontol Mineral Spec Publ 38, pp 111–122

Kharaka YK, Maest AS, Carothers WW, Law LM, Lamothe PJ, Fries TL (1987) Geochemistry of metal-rich brines from central Mississippi Salt Dome basin, USA. Appl Geochem 2: 543–561

Kiene RP, Taylor BF (1989) Metabolism of acrylate and 3-mercaptopropionate. In: Saltzman ES, Cooper WJ (eds) Biogenic sulfur in the environment. Am Chem Soc Symp Ser 393: 222–230

Kotova AV, Bakirova SF, Yag'Yayera SM, Leonov ID (1987) Effects of hydrogen sulfide on petroleum porphyrin complexes. Geochem Int 24: 111–116

Krauskopf KB (1955) Sedimentary deposits of rare metals. In: Bateman AM (ed) Economic geology, 50th Anniversary volume. Society of Economic Geologists, El Paso, TX, pp 411–463

Kribek B (1989) The role of organic matter in the metallogeny of the Bohamian Massif. Econ Geol 84: 1525–1540

Kribek B, Kaigl J, Oruzinsky V (1977) Characteristics of di and trivalent metal-humic acid complexes on the basis of their molecular-weight distribution. Chem Geol 19: 73–81

Landais P, Meyer AJ (1989) Temperature evolution assessment and indirect dating of mineralization in a sedimentary ore deposit: organic matter, fluid inclusions, fission track, and computer modeling. 28th Int Geol Cong Abstr, Washington DC, vol 2, 256

Langmuir D (1979) Techniques of estimating thermodynamic properties for some aqueous complexes of geochemical interests. In: Jenne EA (ed) Chemical modeling in aqueous systems. Am Chem Soc Symp Ser 93, pp 353–387

Laznicka P (1985) The geological association of coal and metallic ores – a review. In: Wolf KH (ed) Handbook of strata-bound and stratiform ore deposits, vol 13. Elsevier, Amsterdam, pp 1–71

Leventhal JS (1986) Roles of organic matter in ore deposits. In: Dean WE (ed) Organics and ore deposits. Proc Denver Region Exploration Geologists Society, Wheat Ridge, CO, Symp on Organics and ore deposits, pp 7–20

Leventhal JS (1990) Organic matter and thermochemical sulfate reduction in the viburnum trend, southeast Missouri. Econ Geol 85: 622–632

Levitskiy VV, Vikulova LP, Demin BG, Popivnyak IV (1982) Comparative analysis of gold-carbon sulfide-quartz ores and organometallic compounds. Dokl Akad Nauk SSSR 255: 240–243

Lewan MD (1984) Factors controlling the proportionality of vanadium and nickel in crude oils. Geochim Cosmochim Acta 48: 2231–2238

Lewan MD, Maynard JB (1982) Factors controlling enrichment of vanadium and nickel in the bitumen of organic sedimentary rocks. Geochim Cosmochim Acta 46: 2547–2560

Lindsay WT (1980) Estimation of concentration quotients for ionic equilibria in high temperature water: the model substance approach. In: Proc Int Water Conf, Engineering Society of Western Pennsylvania, vol 41, pp 284–294

Lipiner G, Willner I, Aizenshtat Z (1988) Correlation between geochemical environments and controlling factors in the metallation of porphyrins. Org Geochem 13: 747–756

Lown DA, Thirsk HR, Wyane-Jones L (1970) Temperature and pressure dependence of the volume of ionization of acetic acid in water from 25 to 225 °C and 1 to 3000 bars. Trans Faraday Soc 66: 51–73

Lundegard PD, Kharaka YK (1990) Geochemistry of organic acids in subsurface waters. In: Melchior DC, Bassett RL (eds) Chemical modeling of aqueous systems II. Am Chem Soc Symp Ser 416, pp 170–189

MacGowan DB, Surdam RC (1988) Difunctional carboxylic acid anions in oilfield waters. Org Geochem 12: 245–259

MacGowan DB, Surdam RC (1990a) Importance of organic-inorganic reactions to modeling water-rock interactions during progressive clastic diagenesis. In: Melchior DC, Bassett RL (eds) Chemical modeling of aqueous systems II. Am Chem Soc Symp Ser 416, pp 494–507

MacGowan DB, Surdam RC (1990b) Carboxylic acid anions in formation waters, San Joaquin Basin and Louisiana Gulf Coast, USA – implications for clastic diagenesis. Appl Geochem 5: 687–701

Macqueen RW (1980) Geochemistry of organic matter in ore deposits: introduction to Carnegie Conference. Carnegie Institution of Washington Conf Geochemistry of organic matter in ore deposits, Extended Abstracts and Bibliographies of Participants, pp 90–93

Macqueen RW (1986) Origin of Mississippi Valley-type lead zinc ores by organic matter-sulfate reactions: the Pine Point example. In: Dean WE (ed) Organics and ore deposits, Proc Denver Region Exploration Geologists Society, Wheat Ridge, CO, Symp on Organics and ore deposits, pp 151–158

Macqueen RW, Powell TG (1983) Organic geochemistry of the Pine Point lead-zinc ore field and region, Northwest Territories, Canada. Econ Geol 78: 1–25

Manning DAC (1986) Assessment of the role of organic matter in ore transport processes in low-temperature base-metal systems. Trans Inst Mining Metallurgy Sec B 95: B195–B200

Marshall WL (1970) Complete equilibrium constants, electrolyte equilibria, and reaction rates. J Phys Chem 74: 346–355

Martell AE, Smith RM (1974) Critical stability constants, vol 1. Amino acids. Plenum Press, New York, 469 pp

Martell AE, Smith RM (1977) Critical stability constants, vol 3. Other organic ligands. Plenum Press, New York, 495 pp

Martell AE, Smith RM (1982) Critical stability constants, vol 5. First supplement. Plenum Press, New York, 604 pp

Maynard JB (1991a) Uranium: syngenetic to diagenetic deposits in foreland basins. In: Force ER, Eidel JJ, Maynard JB (eds) Sedimentary and diagenetic mineral deposits: a basin analysis approach to exploration. Society of Economic Geologists, El Paso, TX, Rev Econ Geol 5: 187–197

Maynard JB (1991b) Copper: product of diagenesis in rifted basins. In: Force ER, Eidel JJ, Maynard JB (eds) Sedimentary and diagenetic mineral deposits: a basin analysis approach to exploration. Soc Econ Geol Rev Econ Geol 5, pp 199–207

McLimans R (1977) Sphalerite stratigraphy, stable isotope studies and fluid inclusion studies of the upper Mississippi Valley lead-zinc district. PhD Thesis, The Pennsylvania State University, University Park, 175 pp

Mesmer RE, Marshall WL, Palmer DA, Simenson JM, Holmes HF (1988) Thermodynamics of aqueous association and ionization reactions at high temperatures and pressures. J Solution Chem 17: 699–718

Mesmer RE, Patterson CS, Busey RH, Holmes HF (1989) Ionization of acetic acid in NaCl (aq) media: a potentiometric study to 573 K. J Phys Chem 93: 7483–7490

Miller SL, Bada JL (1988) Submarine hot springs and the origin of life. Nature 334: 609–611

Morrison RT, Boyd RN (1966) Organic chemistry. Allyn and Bacon, Boston, 1204 pp

Mossman DJ, Dyer BD (1985) The geochemistry of Witwatersrand-type gold deposits and the possible influence of ancient prokaryotic communities on gold dissolution and precipitation. Precambrian Res 30: 303–319

Nash JT, Granger HC, Adams SS (1981) Geology and concepts of genesis of important types of uranium deposits. In: Skinner BJ (ed) Economic geology. 75th Anniversary volume. Society of Economic Geologists, El Paso, TX, pp 63–116

Nissenbaum A, Swaine DJ (1976) Organic matter-metal interactions in recent sediments: the role of humic substances. Geochim Cosmochim Acta 40: 809–816

Oscarson JL, Gillespie SE, Christensen JJ, Izatt RM, Brown PR (1988) Thermodynamic quantities for the interaction of H^+ and Na^+ with $C_2H_3O_2^-$ and Cl^- in aqueous solutions from 275 to 320 °C. J Solution Chem 17: 865–885

Palmer DA, Drummond SE (1986) Thermal decarboxylation of acetate. Part I. The kinetics and mechanism of reaction in aqueous solution. Geochim Cosmochim Acta 50: 813–823

Palmer DA, Drummond SE (1988) Potentiometric determination of the molal formation constants of ferrous acetate complexes in aqueous solutions to high temperatures. J Phys Chem 92: 6795–6800

Perdue EM (1985) Acidic functional groups of humic substances. In: Aiken GR, McKnight DM, Wershaw RL (eds) Humic substances in soil, sediment, and water. Wiley, New York, pp 493–526

Perdue EM, Lytle CR (1986) Chemical equilibrium modeling of metal complexation by humic substances. In: Carlisle D, Berry WL, Kaplan IR, Watterson JR (eds) Minerals exploration: biological systems and organic matter. Prentice-Hall, Englewood Cliffs, pp 428–444

Perrin DD (1979) Stability constants of metal-ion complexes part B, organic ligands. IUPAC chemical data series 22. Pergamon Press, Oxford, 1263 pp

Price PE, Kyle JR (1986) Genesis of Salt Dome hosted metallic sulfide deposits: the role of hydrocarbons and related fluids. In: Dean WE (ed) Organics and ore deposits. Proc

Denver Region Exploration Geologists Society, Wheat Ridge, CO, Symp on Organics and ore deposits, pp 171–182

Radtke AS, Scheiner BJ (1970) Studies of hydrothermal gold deposition (1). Carlin gold deposit, Nevada: the role of carbonaceous materials in gold deposition. Econ Geol 65: 87–102

Radtke AS, Rye RO, Dickson FW (1980) Geology and stable isotope studies of the Carlin gold deposit, Nevada. Am Assoc Pet Geol Bull 75: 641–672

Rapp JB (1976) Amino acids and gases in some springs and an oil field in California. J Res US Geol Surv 4: 227–232

Rashid MA (1985) Geochemistry of marine humic compounds. Springer, Berlin Heidelberg New York, 300 pp

Reuter JH, Perdue EM (1977) Importance of heavy metal-organic matter interactions in natural waters. Geochim Cosmochim Acta 41: 325–334

Ripley EM, Shaffer NR, Gilstrap MS (1990) Distribution and geochemical characteristics of metal enrichment in the New Albany Shale (Devonian-Mississippi), Indiana. Econ Geol 85: 1790–1807

Roedder E (1976) Fluid-inclusion evidence on the genesis of ores in sedimentary and volcanic rocks. In: Wolf KH (ed) Handbook of strata-bound and stratiform ore deposits, vol 2. Elsevier, New York, pp 67–110

Roedder E (1979) Fluid inclusions as samples of ore fluids. In: Barnes HL (ed) Geochemistry of hydrothermal ore deposits. Wiley, New York, pp 684–737

Roedder E (1984) Fluid inclusions. Mineralogical Society of America, Washington DC. Rev Mineral 12: 644 pp

Rose AW (1976) The effect of cuprous chloride complexes in the origin of red-bed copper and related deposits. Econ Geol 71: 1036–1048

Rose AW (1989) Mobility of copper and other heavy metals in sedimentary environments. In: Boyle RW, Brown AC, Jefferson CW, Jowett EC, Kirkham RV (eds) Sediment-hosted stratiform copper deposits. Geol Assoc Can Spec Pap 36: 97–110

Saunders JA, Swann CT (1990) Trace-metal content of Mississippi oil field brines. J Geochem Explor 37: 171–183

Saxby JD (1972) Organic matter in Red Sea sediments. Chem Geol 9: 233–240

Saxby JD (1976) The significance of organic matter in ore genesis. In: Wolf KH (ed) Handbook of strata-bound and stratiform ore deposits, vol 1. Elsevier, New York, pp 111–133

Schnitzer M, Khan SU (1978) Soil organic matter. Elsevier, New York, 319 pp

Schultz RB (1991) Metalliferous black shales: accumulation of carbon and metals in cratonic basins. In: Force ER, Eidel JJ, Maynard JB (eds) Sedimentary and diagenetic mineral deposits: a basin analysis approach to exploration. Society of Economic Geologists, El Paso, TX. Rev Econ Geol 5: 171–176

Seewald JS, Seyfried WE (1991) Experimental determination of portlandite solubility in H_2O and acetate solutions at 100–350°C and 500 bars: constraints on calcium hydroxide and calcium acetate complex stability. Geochim Cosmochim Acta 55: 659–669

Semmler J, Irish DE, Ozeki T (1990) Vibrational spectral studies of solutions at elevated temperatures and pressues. 12. Magnesium acetate. Geochim Cosmochim Acta 54: 947–954

Severson RC, Crock JG, McConnell BM (1986) Processes in the formation of crystalline gold in placers. In: Dean WE (ed) Organics and ore deposits, Proc Denver Region Exploration Geologists Society, Wheat Ridge, CO, Symp on Organics and ore deposits, pp 69–80

Shock EL (1988) Organic acid metastability in sedimentary basins. Geology 16: 886–890

Shock EL (1989) Corrections to "Organic acid metastability in sedimentary basins". Geology 17: 572–573

Shock EL (1990) Do amino acids equilibrate in hydrothermal fluids? Geochim Cosmochim Acta 54: 1185–1189

Simoneit BRT (ed) (1990) Organic matter in hydrothermal systems. Appl Geochem 5: 1–548

Simoneit BRT, Lonsdale PF (1982) Hydrothermal petroleum in mineralized mounds at the seabed of Guaymas basin. Nature 295: 198–202

Skinner BJ (1967) Precipitation of Mississippi Valley type ores: a possible mechanism. Econ Geol Monogr 3: 363–370

Smith JW (1986) The contrasting effect of a carbonaceous host within the Amethyst silver system at Creede, Colorado. In: Dean WE (ed) Organics and ore deposits. Proc Denver Region Exploration Geologists Society, Wheat Ridge, CO, Symp on Organics and ore deposits, pp 111–114

Smith RM, Martell AE (1975) Critical stability constants, vol 2. Amines. Plenum Press, New York, 415 pp

Smith RM, Martell AE (1989) Critical stability constants, vol 6. Second supplement. Plenum Press, New York, 643 pp

Sokolova MT, Karyakin AV, Yefimova NF, Kremneva MA (1980) Dispersed organic matter in hydrothermal mineral formation. Geochem Int 16: 57–64

Stevenson FJ (1983) Trace metal-organic matter interactions in geologic environments. In: Theophrastus SA (ed) Significance of trace elements in solving petrogenetic problems. National Technical University of Athens, Athens, pp 671–691

Stumm W, Morgan JJ (1981) Aquatic chemistry. Wiley, New York, 780 pp

Surdam RC, Boese SW, Crossey LJ (1984) The chemistry of secondary porosity. In: McDonald DA, Surdam RC (eds) Clastic diagenesis. Am Assoc Pet Geol Mem 37: 127–151

Sverjensky DA (1984) Oil field brines as ore-forming solutions. Econ Geol 79: 23–37

Sverjensky DA (1986) Genesis of Mississippi Valley-type lead-zinc deposits. Annu Rev Earth Planet Sci 14: 177–199

Sverjensky DA (1987) The role of migrating oil field brines in the formation of sediment-hosted Cu-rich deposits. Econ Geol 82: 1130–1141

Sverjensky DA (1989) Chemical evolution of basinal brines that formed sediment-hosted Cu-Pb-Zn deposits. In: Boyle RW, Brown AC, Jefferson CW, Jowett EC, Kirkham RV (eds) Sediment-hosted stratiform copper deposits. Geol Assoc Can Spec Pap 36: 127–134

Thurman EM (1985) Organic geochemistry of natural waters. Nijhoff, Boston, 497 pp

Thyne GD, Harrison WJ, Alloway MD (1992) Experimental study of the stability of the Al-oxalate complexation at 100 °C and calculations of the effects of complexation on clastic diagenesis. In: Kharaka YK, Maest AS (eds) Water-rock interaction, vol 1. Balkema, Rotterdam, pp 353–357

Tissot BP, Welte DH (1984) Petroleum formation and occurrence. Springer, Berlin Heidelberg New York, 699 pp

Tourtelot HA (1979) Black shale-its deposition and diagenesis. Clays Clay Minerals 27: 313–321

Trudinger PA (1976) Microbiological processes in relation to ore genesis. In: Wolf KH (ed) Handbook of strata-bound and stratiform ore deposits, vol 2. Elsevier, New York, pp 135–190

Trudinger PA (1980) Biological factors in mineral genesis. Carnegie Institution of Washington Conf on the Geochemistry of organic matter in ore deposits, extended abstracts and bibliographies of participants, Washington DC, pp 151–155

Trudinger PA, Chambers LA, Smith JW (1985) Low-temperature sulfate reduction: biological versus abiological. Can J Earth Sci 22: 1910–1918

Turner-Peterson CE, Fishman NS (1986) Geologic synthesis and genetic models for uranium mineralization in the Morrison Formation, Grants Uranium Region, New

Mexico. In: Turner-Peterson CE, Santos ES, Fishman NS (eds) A basin analysis case study: the Morrison Formation, Grants Uranium Region, New Mexico. Am Assoc Pet Geol, Tulsa, OK. Stud Geol Ser 22: 357–388

Turner-Peterson CE, Fishman NS, Hatcher PG, Spiker EC (1986) Origin of organic matter in sandstone uranium deposits of the Morrison Formation, New Mexico: geologic and chemical constraints. In: Dean WE (ed) Organics and ore deposits. Proc Denver Region of Exploration Geologists, Wheat Ridge, CO, Symp on Organics and ore deposits. pp 185–195

Vairavamurthy A, Mopper K (1989) Mechanistic studies of organosulfur (thiol) formation in coastal marine sediments. In: Saltzman ES, Cooper WJ (eds) Biogenic sulfur in the environment. Am Chem Soc Symp Ser 393, Washington DC, pp 231–242

Varshal GM, Velynkhanova T, Baranova N (1984) Geochemical role of gold (III) fulvate complexes. Geochem Int 21: 139–145

Vaughan DJ, Sweeney M, Diedel GFR, Haranczyk C (1989) The Kupferschiefer – an overview with an appraisal of different types of mineralization. Econ Geol 84: 1003–1027

Veitch JD, McLeroy DG (1972) Organic mobilization of ore metals in low-temperature carbonate environments. Geol Soc Am Abstr Programs 7(4): 110–111

Vine JD, Tourtelot EB (1970) Geochemistry of black shale deposits – a summary report. Econ Geol 65: 253–272

Vlassopoulous D, Wood SA, Mucci A (1990) Gold speciation in natural waters. II. The importance of organic complexing – experiments with some simple model ligands. Geochim Cosmochim Acta 54: 1575–1586

Weast RC (1977) Handbook of chemistry and physics, 58th edn. CRC Press, Cleveland, 2348 pp

Welhan JA, Lupton JE (1987) Light hydrocarbon gases in Guaymas basin hydrothermal fluids: thermogenic versus abiogenic origin. Am Assoc Pet Geol Bull 71: 215–223

Westbroek P, DeJong EW (1983) Biomineralization and biological metal accumulation. Reidel, Boston, 533 pp

White RH (1984) Hydrolytic stability of biomolecules at high temperatures and its implication for life at 250 °C. Nature 310: 430–432

Willingham TO, Nagy B, Nagy LA, Krinsley DH, Mossman DJ (1985) Uranium-bearing stratiform organic matter in paleoplacers of the lower Huronian Supergroup, Elliot Lake – Blind River region, Canada. Can J Earth Sci 22: 1930–1944

Wood SA (1990) The interaction of dissolved Pt with fulvic acid and simple organic acid analogues in aqueous solutions. Can Mineral 28: 665–673

Wood SA (1991) The interaction of Pd^{2+} with fulvic acid and simple organic acids – solubility and spectroscopic studies. Geol Soc Am Abstr Programs 23(4): A214

Zezin RB, Sokolova MN (1968) Macroscopic occurrences of carbonaceous matter in hydrothermal deposits of the Khibiny pluton. Dokl Akad Nauk SSSR 177: 217–221

Zumberge JE, Sigleo AC, Nagy B (1978) Molecular and elemental analyses of the carbonaceous matter in the gold and uranium bearing Vaal Reef carbon seams, Witwatersrand sequence. Minerals Sci Eng 10: 223–246

Chapter 12   Geochemical Models of Rock-Water Interactions in the Presence of Organic Acids

Wendy J. Harrison[1] and Geoffrey D. Thyne[2]

Summary

Theoretical analysis of the role of organic acids (OA) and acid anions (OAA) in sedimentary environments is an effective way of evaluating the importance of inorganic-organic interactions in rock-water systems. Geochemical modeling provides insight into the role of these organic constituents in aqueous sedimentary environments by *defining the boundary conditions under which such interactions may be significant.* In many cases, carefully designed geochemical models allow conflicts to be resolved among differing experimental results and among various working hypotheses. The approach can be limited by lack of understanding of the processes being modeled, lack of testable working hypotheses, or lack of either experimental or field observations against which to calibrate model results. Interpretation of results from chemical models always needs to accommodate uncertainties in the supporting thermodynamic data.

Geochemical models of rock-water interactions in the presence of organic acids, using a combination of experimental and estimated thermodynamic data, result in several important predictions: first, trivalent cation-difunctional acid anion complexes are the most stable, whereas monovalent cation-monofunctional acid anion complexes are the least stable. Second, the stability of all complexes is pH-dependent and, whereas aluminum oxalate dominates the species distribution of aluminum under acidic conditions, in alkaline waters, inorganic aluminum species are predominant. Calculations using reaction path models reveal additional constraints on the role of OAA in geologic processes. The roles that OA and OAA play in modifying mineral solubility equilibria depend, among others, on pH, acid composition, competing inorganic equilibria with CO_2, and temperature. Considerations of simple system geochemistry thus cannot be readily used to predict the effect the OA/OAA may have in modifying rock-water interactions. Indeed,

---

[1] Department of Geology and Geological Engineering, Colorado School of Mines, Golden, Colorado 80401, USA
[2] Present address, Department of Physics and Geology, California State University at Bakersfield, Bakersfield, California 93311, USA

in some circumstances OAA-bearing waters are less effective at producing porosity in an arkosic sandstone than are OAA-free waters.

An overview of the role of OA in sedimentary processes is that they contribute to overall patterns of fluid-rock interaction, but appear unlikely to dominate such reactions except in restricted geochemical environments where concentrations are in excess of typical values. Such environments might include wetlands, gasoline-contaminated groundwaters, within organic-rich shales, and within sandstones immediately adjacent to such rocks.

1 Introduction

The geochemical role of dissolved organic acid anions (OAA) in ground-waters has attracted considerable attention in the last decade. Antweiler and Drever (1983), Thurman (1985), Evans (1988), and Bennett et al. (1991) observed that interactions among inorganic and organic constituents in soils could account for observed geochemical signatures in near-surface aqueous environments. Field evidence has been used to demonstrate that petroleum-contaminated groundwaters have dissolved quartz (Bennett and Seigel 1987), implying that the solubility of quartz is enhanced by organic complex for-mation at low temperatures (Bennett et al. 1988; Marley et al. 1989) and experiments demonstrating increases in quartz solubility in the presence of citric and oxalic acids appear to confirm field observations (Bennett 1991). Surdam and coworkers (Crossey et al. 1984; Surdam et al. 1984, 1989) proposed that complexing of Al^{3+} by oxalic and acetic acid anions, generated during thermal maturation of kerogen, could enhance plagioclase solubility and account for the formation of kaolinite distant from feldspar dissolution sites thereby reconciling observed diagenetic sequences in clastic sedimen-tary rocks. Studies of uranium mineralization in sandstones (Adams et al. 1978) provide additional textural evidence for the association of organic compounds with feldspar dissolution and quartz and kaolinite precipitation.

Undoubtedly, one of the major stimulants of interest in dissolved organic acids has been the role that these acids might play in creating secondary porosity in clastic rocks and in accounting for precipitation of cements distant from the implied source mineral dissolution sites (Crossey et al. 1984; Surdam et al. 1984, 1989). The particular appeal of this hypothesis appears to be that it provides a mechanism for circumventing the so-called water volume problem wherein amounts of diagenetic products in sedimen-tary rocks require unrealistically large volumes of fluid flowing in the sub-surface (Bjørlykke 1979, 1984; Blatt 1979; Land 1984; Land et al. 1987). Unfortunately, several observational, experimental, and theoretical uncer-tainties arise on closer examination of the chemical mechanisms that might be involved.

Analyses of formation waters reveal the presence of a considerable number of different organic acids, the most abundant being monofunctional acetic acid (maximum of about $10\,000\,mg\,kg^{-1}$) and difunctional oxalic and malonic acids (maxima of about 500 and $2550\,mg\,kg^{-1}$, respectively; Carothers and Kharaka 1978; Fisher 1987; MacGowan and Surdam 1988). Although these upper limits are comparable to concentrations used in experiments where aqueous aluminum concentrations were observed to be in excess of normal amounts determined by equilibrium with aluminous solids (Surdam et al. 1984), it is more common for measured organic acid concentrations to be much lower. Controversy exists over the interpretation of some experiments demonstrating the role of organic acids in feldspar solubility (Surdam et al. 1984; Stoessell and Pittman 1991). Other studies do provide convincing evidence that feldspar solubilities are modified in the presence of organic acids (Huang and Keller 1970; Bevan and Savage 1989; Hajash et al. 1989; Manning et al. 1991), although elevated concentrations of Si and Al over expected inorganic equilibria may represent an enhancement to reaction rates through a modification to the surface dissolution mechanism (Bevan and Savage 1989; Manning et al. 1991).

Theoretical analysis of the role of OAA in sedimentary environments is an effective way of evaluating the importance of inorganic-organic interactions. Geochemical modeling provides insight into the role of OAA in aqueous sedimentary environments by *defining the boundary conditions under which such interactions may be significant.* As with field and experimental approaches, chemical models sometimes provide equivocal evidence concerning the role of organic acids in fluid-rock interactions. Results can be limited by lack of understanding of the processes being modeled, lack of testable working hypotheses, or lack of either experimental or field observations against which to calibrate model results. In the case of OAA and their role in modifying mineral equilibria none of these criteria are limiting! Numerous published studies have presented hypotheses, experiments, and field measurements. The numerical approach is always limited by the availability and quality of supporting thermodynamic data, however.

2 Overview of Chemical Modeling

2.1 Model Definitions

The use of the term chemical modeling is somewhat loosely applied with resulting confusion in meaning. In this chapter we adhere to strict definitions as follows: as conventionally used in aqueous chemistry a *chemical model* is a numerical algorithm designed to calculate aqueous species distributions and mineral saturation states in waters. One of the fundamental principles of such calculations is that competition among cations and anions can be evaluated and a distribution of an element among many aqueous species can be obtained. Following from the original application to the speciation of seawater by Garrels and Thompson (1962), computer software has been written by more than ten authors to perform these specific calculations (for an excellent review of chemical models, see Nordstrom et al. 1979). Individual computer packages vary in options that may be applied to the basic speciation and saturation state calculations, and in the source, reliability, and completeness of the supporting thermodynamic data base. Most computer codes allow selection from at least two aqueous ion activity models, one or more adsorption equations, and permit redox disequilibria. Most also accommodate organic acids and other organic compounds, although the completeness of these compounds and the reliability of the thermodynamic data are highly variable. Some codes include calculations of isotopic fractionation (Bassett and Melchior 1990). The increasing sophistication of such computer algorithms allows for improved understanding of geochemical processes in a wide variety of aqueous environments.

A *reaction path model* advances from this point by allowing predictions about the chemical equilibria in a system in which mass transfer is allowed among phases in response to changes in system composition or in physical constraints. Using software codes such as EQ3/6 (Wolery 1979, 1983), CHILLER (Reed 1983), PHREEQE (Parkhurst et al. 1980) forward predictions can be made about rock-fluid-gas interactions in a wide range of geochemical environments (Table 1; modified from Harrison 1990). The same options found in the chemical speciation models are implemented as well as precipitation and dissolution rate expressions, mineral solid solution expressions, and reaction options (open system, closed system, flow-through, etc.). These calculations are all made without being tied to formal equations for fluid flow through porous media, however, and thus some uncertainties arise in the application of reaction path results to systems where time-dependent modifications to permeability occur which may in turn cause changes in the nature of the chemical reactions.

A *coupled reaction-transport model* is an algorithm in which the feedbacks among chemical reaction and fluid flow through a porous medium are explicitly accounted for through time. Clearly, this set of equations most accurately describes the chemical evolution of a dynamic fluid-rock system. Two principal disadvantages in this approach are, first, that most of the software capable of simulating coupled flow reaction is developmental and, second, in combining two significant numerical problems, simplifications of both the flow and chemical components are needed. As a consequence, applications of flow-reaction calculations to geochemical problems are relatively few (e.g., Chen et al. 1990; Meshri and Walker 1990; Moore and Ortoleva 1990; Nagy et al. 1990).

Further discussion of individual computer codes and their options and limitations is beyond the scope of this chapter. The reader is referred to publications of conferences on Chemical Modeling of Aqueous Systems,

Table 1. Possible reaction path models for fluid/rock interactions in sedimentary basins. (Harrison 1990)

Chemical system	Model	Reaction path variables	Examples of applications
A. *Fluid*	(i) Closed system equilibrium	Pressure and/or temperature	Species distribution; saturation calculations
	(ii) Open system flow-through	Pressure and/or temperature	Evolution of ore-forming solutions, well-bore scale precipitation
B. *Fluid + rock*	(i) Closed system equilibrium	Pressure and/or temperature with rock: fluid ratio	Burial diagenesis of hydrologically isolated sediments
	(ii) Closed system with kinetic constraints	As B(i), with time	Interpretation of laboratory experiments
	(iii) Open system flow-through	As B(i)	Leachate generation; formation of hydrothermal ores and alteration haloes
	(iv) Open system flow-through or flush, with kinetic constraints	As B(i), with time	Experimental column leaching studies; leachate generation
C. *Fluid A + fluid B*	(i) Closed system equilibrium	Fluid A:B ratio	Mixing of contaminant with water, mixing in coastal aquifers
D. *Fluid A + rock, mixed with fluid B*	(i) Open system flow-through or flush with/without kinetic constraints	Fluid A:B ratio $+/-$ B(i), time	Mixing zone diagenesis; unconformity diagenesis; steam, CO_2, or alkali injection for enhanced oil recovery

sponsored by the American Chemical Society in 1978 (Jenne 1979) and in 1988 (Bassett and Melchior 1990) for a more comprehensive treatment.

2.2   Review of Good Modeling Habits

Judicious use of chemical models of all three types is called for at all times. It is particularly appropriate to use chemical models as part of an integrated approach to solving a given geochemical problem, using the model results to provide insight into geochemical and hydrologic processes, to set boundary limits for any given process, and to evaluate multiple working hypotheses.

Although calculations are made to high precision (as many as 12 figures after the decimal place) and internal tolerances for the solutions to algebraic matrices and to differential equations may be in the fifteenth decimal place, calculated parameters are never this accurate and, indeed, can easily be quite wrong. Any chemical, reaction path, or coupled model should be viewed as a *tool* permitting ready solution to a complex set of mathematical expressions. As with all tools, their appropriate use on a problem is needed and particularly pertinent to fast, complex, computer tools is the self-explanatory expression "Garbage in, garbage out"!

A principal uncertainty with chemical models lies in the quality and completeness of the supporting thermodynamic or kinetic data. Although several groups have made pioneering attempts to produce internally consistent thermodynamic data bases (Helgeson et al. 1978; Nordstrom et al. 1990; NEA/CODATA; SUPCRT91, Johnson et al. in press), many of the original experiments, measuring for example, intra-aqueous reaction constants, date to the 1920s and 1930s (Helgeson 1969; Helgeson et al. 1978). In any data base, there are experimentally measured, estimated, and extra-polated thermodynamic quantities all of variable accuracy. A realistic approach to the use of chemical calculations is thus to perform sensitivity analyses wherein critical reactions and predictions are reevaluated using a range of possible stability constants, or possible temperature extrapolations, depending on the problem of interest. All users have to be personally responsible for assuring data base completeness for a given problem and adding missing quantities if appropriate.

Uncertainty also arises in the accuracy of the measured chemical or physical data (field or experimental) that may be used as model input for a particular problem: for example, measurements of pressure, temperature, alkalinity, pH, and Eh, petrographic descriptions, and mineral chemistries of phases all have uncertainty associated with them. Analytical incompleteness is also a concern if missing compositions must then be estimated. Additional uncertainty arises when formulae and supporting parameters are extrapolated beyond their range of applicability. The use of the Debye-Huckel expression to calculate ion activity coefficients at ionic strengths greater than 1 molal provides an example.

2.3 Assumptions in Chemical Models of Inorganic-Organic Interactions

Several specific assumptions are made in the prediction of organic acid interactions with minerals and waters. We set these out prior to discussion of results so that the limitations of the theoretical approach are clear to the reader.

2.3.1 Data Base

In both chemical and reaction path calculations some uncertainty arises because of the incompleteness of the supporting thermodynamic data, notably stability constants for the cation-organic acid anion complexes. Shock and Helgeson (1990) derived partial molal thermodynamic properties of an extensive set of organic compounds from which stability constants can be calculated. Using thermodynamic estimation methods, the temperature dependencies of organic compounds can be derived from heat capacities. Harrison and Thyne (1992) compiled aqueous complex stability constants for seven mono- and difunctional organic acids with as many as 12 cations. They also used theoretical methods for extrapolating these constants to elevated temperatures (DQUANT, Helgeson 1969, 1970; Similar Slope Method, Kharaka et al. 1988). A few data come from direct experimental measurements, but these are relatively sparse particularly at temperatures above 25 °C. Several experimental studies of cation-organic acid complexation have been published recently dealing with selected cations and acids (Giordano and Drummond 1987, 1991; Hennett et al. 1988; Palmer and Drummond 1988; Giordano 1989; Fein 1991a,b; Seewald and Seyfried 1991).

A combination of uncertainty in thermodynamic data accuracy at elevated temperatures and overall lack of completeness with respect to organic compound class and interaction types (complex formation, adsorption, solid precipitate) poses some real limitations in the use of chemical and reaction path models. Thus, a cautionary note is that to compare results of different numerical studies it is necessary to know whether similar thermodynamic data bases were used. As we will show below, apparently discrepant conclusions drawn from modeling studies probably result from the use of more or less complete thermodynamic data. Of concern in the investigation of organic acid interactions in silicate rocks is the complete lack of any equilibrium measurements of silica-organic acid anion interactions, although dissolution rate measurements have been made (Bennett et al. 1988; Bennett 1991).

2.3.2 Metastability

Shock (1988, 1990) has shown that organic compounds are in metastable equilibrium in most aqueous sedimentary and hydrothermal systems. Stable phases should be CH_4, CO_2, C, and H_2O (Fig. 2 in Shock 1988). Thus, the natural occurrences of organic compounds imply substantial kinetic inhibition of their breakdown to these products, although individual acids and acid anions appear to be in chemical equilibrium. Experimental studies of acetic acid and oxalic acid decarboxylation rates confirm that these acids may persist for hundreds to ten thousands of years, the longer times clearly being geologically significant periods (Kharaka et al. 1983; Drummond and

Palmer 1986a,b; Crossey 1991; Bell and Palmer, Chap. 9, this Vol.). Conversely, experimental studies by Thyne and Harrison (unpubl. data) reveal half-lives of less than 10 days for malonic acid at temperatures over 100 °C. In a chemical or reaction path model, where an equilibrium condition is being calculated, the decarboxylation reactions for the organic acids have to be disabled (suppressed) and thus there is no feedback between organic and inorganic carbon in these calculations nor is there any redox control imposed by the organic acids (Rose and Jackson 1989; Harrison and Thyne 1992, and see also Bell and Palmer, Chap. 9 and Shock, Chap. 10, this Vol.).

2.3.3 Organic Solids

Most chemical and reaction path models currently account only for the aqueous carboxylic acids and their cation complexes, amino acids, some liquid hydrocarbons, alcohols, and certain other compounds entered into the data base for project-specific purposes. Adsorption of trace metals onto solid humic substances, for example, requires the user to create a fictitious solid and use an empirical adsorption coefficient. Scattered reports of carboxylic acid solids such as calcium oxalate (Marlowe 1970; Naumov et al. 1971; Galimov et al. 1975; Graustein et al. 1977; Campbell and Roberts 1986) emphasize the necessity to perform sensitivity analyses on the formation of such solids and indicate another area of uncertainty in the interpretation of chemical and reaction path model results.

2.3.4 Limitations on Model Results

Given these three areas of uncertainty chemical models need to be used with caution. The integration of organic and inorganic geochemistry into a single computer code is in its infancy and as the application of these calculations moves increasingly into the field of environmental contamination, improvements to our predictive ability are inevitable. Sensible use and interpretation are in order.

3 Previous Work

Relatively little published work exists in the field of modeling interactions between organic acids and minerals in aqueous systems. Most of the published studies include chemical model calculations as part of an integrated experimental/field/theoretical analysis of a specific geochemical problem. We can break this work into three major applications:

1. Chemical models of aqueous speciation and mineral saturations in natural groundwaters as these predictions apply to problems in geological controls on surface and groundwater geochemistry, soil geochemistry, sediment diagenesis, and the transport and deposition of ore-forming elements.
2. Chemical models of aqueous speciation and mineral saturations in contaminated surface and groundwaters.
3. Reaction path models of the role of organic acids in sediment diagenesis.

Predictions of aqueous speciation and mineral saturations in natural surface and groundwaters in which metal-organo interactions are accounted for, have been made for a wide range of applications, including assessment of bioavailability and toxicity of metals in river, estuarine, and marine waters (Holm et al. 1979; Sunda and Hanson 1979; Holm and Curtiss 1990), controls on pore water chemistry and element availability in soils (Antweiler and Drever 1983; Bilinksi et al. 1986; Evans 1988; Lovgren 1991), behavior of radioactive nuclides in groundwaters (Cleveland 1979), groundwater pH, Eh, and alkalinity, and solubilities of aluminosilicate and carbonate minerals in sandstone aquifers and reservoirs (Drez and Harrison 1985; Kharaka et al. 1986; Lundegard and Kharaka 1990; Fein 1991a,b; Harrison and Thyne 1992), and the transport and deposition of ore-forming metals such as Pb, Zn, and Fe (Giordano 1985; Hennett et al. 1988; Giordano and Drummond 1991; Seewald and Seyfried 1991).

Predictions of the fate of organic compounds in contaminated groundwaters are usually made by including an empirically determined "retardation factor" or distribution (partition) coefficient in a hydrologic model (Freeze and Cherry 1979). This variable describes the removal of the compound from water due to all processes without the necessity of distinguishing among them. Included in the retardation factor is removal due to adsorption-desorption reactions on all mineral and organic surfaces, and thus the retardation factor is sediment-specific. Such studies do not routinely include the role that the dissolved organic compounds play in modifying the inorganic constituents of the rock-water system, although field studies show that these interactions do occur (Bennett and Seigel 1987; Bennett and Casey, Chap. 7, this Vol.). We do not cover this aspect of modeling inorganic-organic acid interactions further in this chapter because of the simplicity of the chemical approach. We note, however, that improved capabilities of coupled fluid flow-chemical reaction codes will make a significant contribution to the prediction of the fate of organic compounds in surface and groundwater systems. These is currently no published work utilizing formally coupled flow-chemical reaction simulations of interactions between minerals and waters containing organic compounds known to these authors.

Reaction path calculations of fluid-rock interactions in which organic acids are specifically included in the thermodynamic data base appear to be presently limited to a study of the buffering capacity of carbon dioxide in the presence of organic acids (Lundegard and Land 1989; Lundegard and

Kharaka 1990) and the role that organic acids play in sandstone diagenesis (Harrison and Thyne 1992).

4 Chemical Models of Aqueous Speciation in Organic Acid-Bearing Waters

4.1 Speciation of Acids and Acid Anions

Values of pKa for several of the commonly reported acids in natural groundwaters are provided in Table 2. These values indicate that the organic acid anions (OAA) are typically stable above pH values of greater than 4 to 5 at 25 °C (Fig. 1). The exceptions are oxalic acid and malonic acid, which have two hydrogen ions. Oxalic and malonic acids lose the first hydrogen ion at pH values less than 3 so that there is always some dissociation of these acids

Table 2. Commonly reported carboxylic acids in sedimentary systems. (Harrison and Thyne 1992)

Name	Formula	Max. reported conc. (mg/kg)[a]	pka[b] at 25 °C	pka[b] at 100 °C
Monocarboxylic acids:				
Formic (methanoic)	$HCOOH$	62.6	-3.75	-3.91
Acetic (ethanoic)	CH_3-COOH	10 000.0	-4.76	-4.98
Propionic (propanoic)	CH_3-CH_2-COOH	4400.0	-4.88	-5.11
Dicarboxlylic acids:				
Oxalic (ethandioic)	$HOOC-COOH$	494.0	$-0.67;\ -4.27$	$-0.86;\ -4.77$
Malonic (propanedioic)	$HOOC-CH_2-COOH$	2540.0	$-2.85;\ -5.71$	$-2.95;\ -6.14$
Succinic (butanedioic)	$HOOC-(CH_2)_2-COOH$	63.0	$-4.206;\ -5.65$	$-4.25;\ -5.89$
Aromatic acids:				
Salicylic (2-hydroxybenzoic)		65.0[c]	$-2.97;\ -13.75$	$-2.95;\ -12.79$
Carbonic Acid[d]:	H_2CO_3		$-6.3;\ -10.33$	$-6.43;\ -10.16$

[a] MacGowan and Surdam (1988).
[b] Martell and Smith (1977); Harrison and Thyne (1992).
[c] MacGowan and Surdam (unpubl.).
[d] Kharaka et al. (1988).

Fig. 1. Acid-acid anion speciation as a function of pH at 105 °C (**A**), 60 °C (**B**), and 25 °C (**C**). Calculations are for a water in equilibrium with kaolinite, Strong partitioning of aluminum to Al-oxalate causes the oxalic acid activity to decrease at pH < 5. In the absence of aluminum, the oxalic acid would be predominant under these pH conditions. Calculations did not include the partially ionized species $HC_2O_4^-$

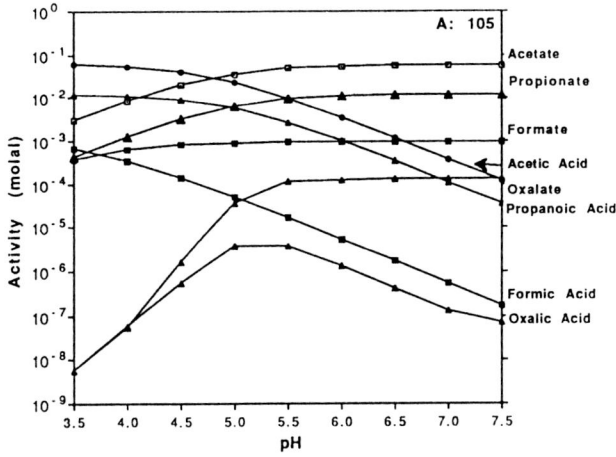

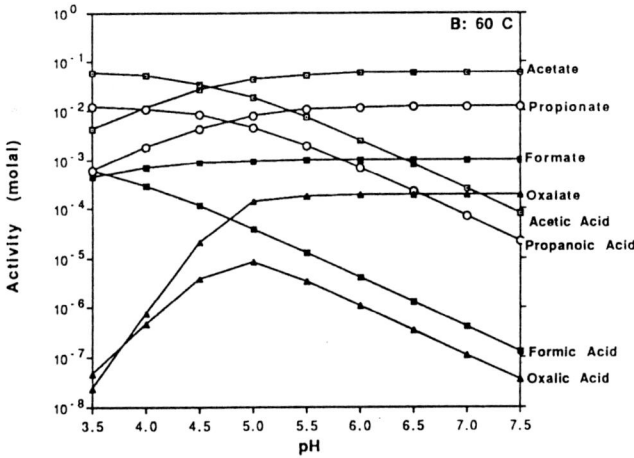

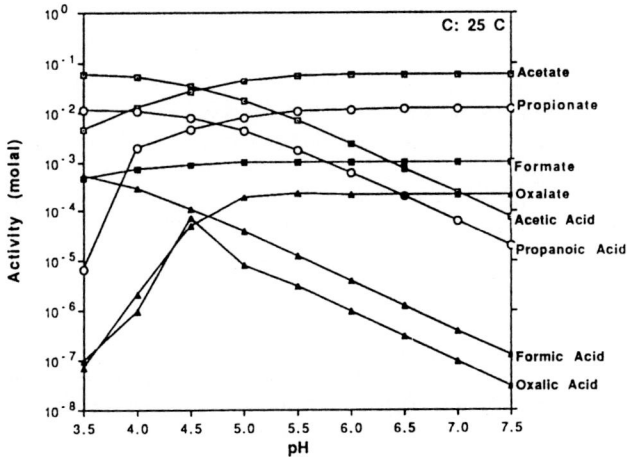

to the acid anions in most naturally occurring groundwaters. Values of pKa become smaller as temperature increases, i.e., the acid-acid anion equilibrium shifts to more alkaline pH as temperature increases. For temperatures in the range 0–250 °C the pH shift is about one unit (Kharaka et al. 1986; Shock and Helgeson 1990; Harrison and Thyne 1992).

The presence of dissolved organic acids and the calculation of the species distribution of the acids and acid anions play an important role in the interpretation of alkalinity and in the reconstruction of subsurface pH from analytical data. Kharaka et al. (1986) report that organic acid anions may contribute between 10 to 90% of an alkalinity measurement. If alkalinity is used to determine dissolved carbonate, a substantial error may result if organic acid contributions are not taken into account. Kharaka et al. (1986) report that in the reconstruction of subsurface pH calculated values may vary by as much as 2 pH units, depending on whether or not organic acids are included in the calculation. As the analysis of organic acids in subsurface waters becomes more routine (MacGowan and Surdam, Chap. 2, this Vol.) their contribution to pH and alkalinity can be evaluated; however, utilization of large water analysis data bases for some problems in sedimentary geochemistry needs to be undertaken with caution because older water analyses will not include organic acid measurements. Lundegard and Kharaka cover these issues in considerable detail (Chap. 3, this Vol.). Kharaka et al. (1983) also note that the presence of organic acids may also control the oxidation potential of the water. Organic acids are involved in the reduction of sulfate to sulfide according to the reaction:

$$SO_4^{2-} + CH_3COOH = HS^- + 2HCO_3^- + H^+, \qquad (1)$$

and may also cause the reduction of ferric to ferrous iron (Surdam et al. 1984; Kharaka et al. 1986) according to the reaction:

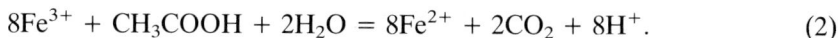

$$8Fe^{3+} + CH_3COOH + 2H_2O = 8Fe^{2+} + 2CO_2 + 8H^+. \qquad (2)$$

Redox disequilibrium exists among organic acids and CH_4, so organic acids and acid anions are not linked by intra-aqueous reactions to either carbon gases or to inorganic carbon species in chemical models where the geochemical consequences of organic acids are being evaluated[3]. Chemical model predictions of the redox role of organic acids are currently limited by lack of general understanding of the processes and reactions involved.

---

[3] This may be accomplished by either setting values of the dissociation constants that inhibit the breakdown of the acids and acid anions to bicarbonate or by defining carbon in organic acids and acid anions as a "new element". The calculated result is the same in either case.

4.2   Cation-Acid Anion Complexes

The stability of all cation-acid anion complexes [CA(OAA)] is described by a reaction of the general form:

CA(OAA) = CA + (OAA).

Equilibrium constants for these reactions will determine the relative distributions of a single OAA among all the available cations, and, for a single cation, the relative importance of the OAA complex among free ion, ion pairs, and inorganic complexes. In general, the monovalent cations are only weakly associated with any of the organic acid anions, the divalent cations and monofunctional acid anions form strong complexes, and the trivalent cations and difunctional acid anions form still stronger complexes. Assessment of the temperature dependency of these equilibrium constants by Shock and Helgeson (1990) and Harrison and Thyne (1992) as well as selected temperature-dependent experimental determinations of the stability constants (see Sect. 4.1) shows a slight to moderate increase in complex stability at temperatures above 25 °C with the exception of the aluminum malonate and oxalate complexes, which are discussed below.

Table 3. Water composition used in species distribution calculations. (Harrison and Thyne 1992)

Species	Concentration (mg/kg)
Na^+	9 300
K^+	200
Ca^{2+}	790[a] or calcite equilibrium
Mg^{2+}	21
Sr^{2+}	49
Sio_2	Quartz saturation[a]
Fe^{3+}	1.9
Al^{3+}	0.5×10^{-7} molal or kaolinite equilibrium[a]
Mn^{2+}	0.70
Cl^-	14 000
SO_4^{2-}	40
HCO_3^-	96
$CHOO^-$	50
CH_3COO^-	3 730
$CH_3CH_2COO^-$	896
$C_2O_4^{2-}$	40
pH	6.9 (values of 3.5 to 7.5)
T (°C)	105 (values of 25–105 °C)

North Coles Levee Well 14–31, (Boles 1987). Organic acids based on MacGowan and Surdam (1988).
[a] Multiple values or constraints indicate assumptions made in subsequent distribution of species calculations.

In the following paragraphs we show aqueous speciation calculations for a typical low salinity subsurface water in the temperaure range 25–105 °C (Table 3). An important feature of chemical species distribution models is that competing effects among all cations and anions in the calculation are included as determined by the values of the dissociation constants. In the subsequent sections we show results for a water that contains nine major cations and seven anions and all related aqueous complexes. *Competition among cations and anions, such as Ca and Al for oxalate, is specifically accounted for in all chemical speciation models.* The calculations show certain features of the role of organic ligands but are by no means the only possible distributions. Aqueous speciation is strongly dependent upon both the chemical and physical characteristics of the system and the reader is encouraged to make additional calculations for water chemistries that are different from the one we have used here.

4.2.1 Monovalent Cations

Monovalent cations such as Na^+ and K^+ form weak complexes with the mono- and difunctional organic acid anions. Stability constants are typically in the range of log K = −1.1 to −1.5 (for the dissociation reaction; Martell and Smith 1977). In any aqueous speciation model a small percentage (less than about 10%) of the available organic acid anion will form monovalent cation complexes. These complexes should be included in chemical models for completeness. A small error will result if the complexes are ignored and this may or may not be of concern depending on the application. Kharaka et al. (1986) report that the Na and K salicylate complexes are more stable, however, and inclusion of these species will be necessary in waters having substantial salicylic acid concentrations.

4.2.2 Divalent Cations

In general, divalent cations form stronger complexes with the monofunctional acid anions than with the difunctional acid anions and aromatic compounds. Typical speciation calculation results are shown in Figs. 2, 3, and 4. In all cases, the divalent ion is the predominant species at low

Fig. 2. Calcium species distribution as a function of pH at 105 °C (**A**), 60 °C (**B**), and 25 °C (**C**). The divalent free ion dominates the species distribution at all temperatures and all values of pH. In these calculations, calcium oxalate formation is suppressed. In the presence of the solid, the species distribution remains similar, however, values of calcium and oxalate activites are much lower as calcium is removed from solution by the solid formation (Harrison and Thyne 1992)

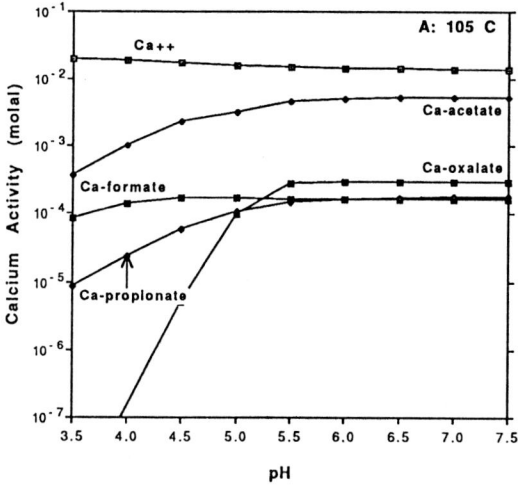

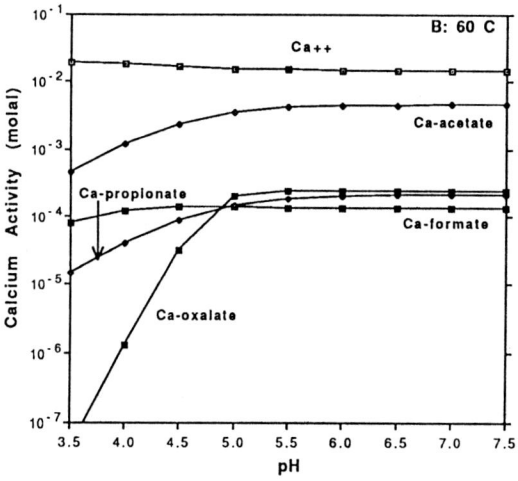

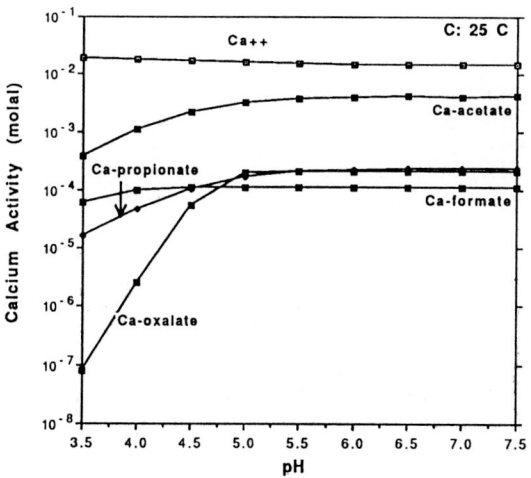

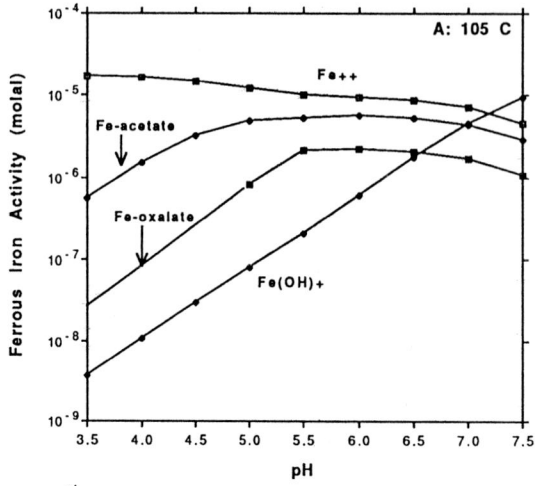

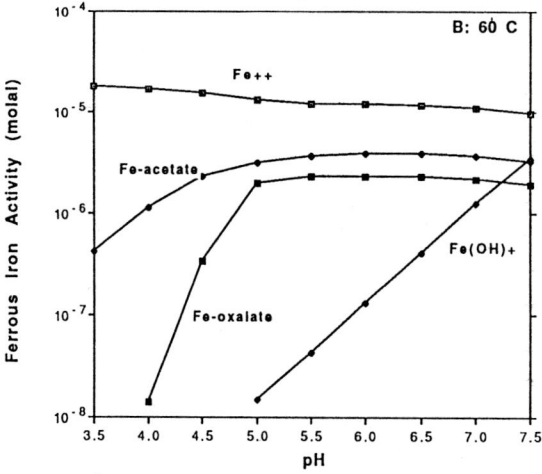

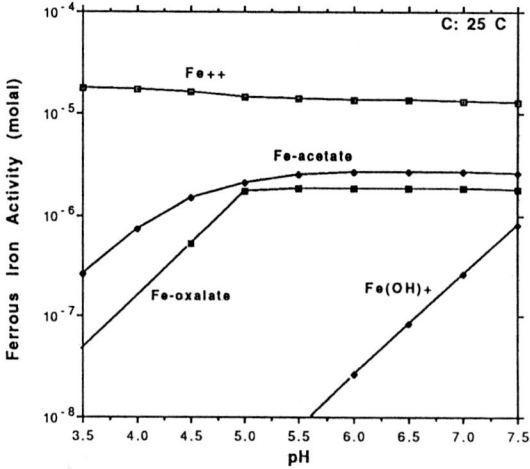

temperatures over the pH interval shown. As temperature increases the stability of the acetate complexes increases but only in the case of Mg-acetate does it actually become predominant. Thyne (1992) shows the malonate complexes being predominant in alkaline solutions, however, we noted previously that this acid appears to have a short half-life above 100 °C and may only be important in near-surface aqueous environments. In waters containing chloride, carbonate, or sulfate in higher abundances than the analysis shown in Table 3, cation complexes with these inorganic anions will be present. In waters more alkaline than those shown in Figs. 2, 3, and 4, the hydroxide complexes will become dominant (see also Seewald and Seyfried 1991). Of the monofunctional acids only the acetate anion contributes significantly to the cation complex distribution. Formate, propionate, and citrate do not form very strong complexes with divalent cations. The difunctional acids are likewise not important complex formers with the divalent cations.

In considering the speciation of waters containing calcium, it is appropriate to discuss the role of the solid phase, calcium oxalate (whewellite – the monohydrate, and weddelite – the dihydrate). These minerals have been reported, occasionally, from several geochemical environments (Marlowe 1970; Naumov et al. 1971; Graustein et al. 1977; Campbell and Roberts 1986) and may actually be quite common, but not frequently observed by petrographers. The solubility of calcium oxalate is low and increases slightly with temperature ($0.67 \, \text{mg} \, \text{kg}^{-1}$ at 13 °C and $14 \, \text{mg} \, \text{kg}^{-1}$ at 95 °C; CRC Handbook). The formation of a solid will fix the concentrations of calcium and oxalate anions at values around $1-10 \, \text{mg} \, \text{kg}^{-1}$ between 15 and 100 °C, depending on solution composition (Harrison and Thyne 1992). Although calcium concentrations could be fixed at higher amounts by alternative inorganic equilibria, the measurement of oxalate concentrations up to $500 \, \text{mg} \, \text{kg}^{-1}$ (MacGowan and Surdam 1988) implies inhibition of the solid precipitation or an acid generation rate faster than the precipitation rate of the solid. If oxalate concentrations are limited, the formation of the cation-OAA complexes will also be limited. Meshri (1986) has shown that other organic acid solids are highly soluble and not likely to form readily.

Fig. 3. Ferrous iron species distribution as a function of pH at 105 °C (**A**), 60 °C (**B**), and 25 °C (**C**). The divalent free ion dominates the species distribution at all temperatures and values of pH except at 105 °C and pH values of 7 and greater. Ferrous iron-OAA complexes become proportionally more important at higher temperatures and in more alkaline waters but never exceed more than about 30% of the dissolved iron. Absence of the iron chloride complexes is a consequence of the water composition used for modeling purposes (Harrison and Thyne 1992)

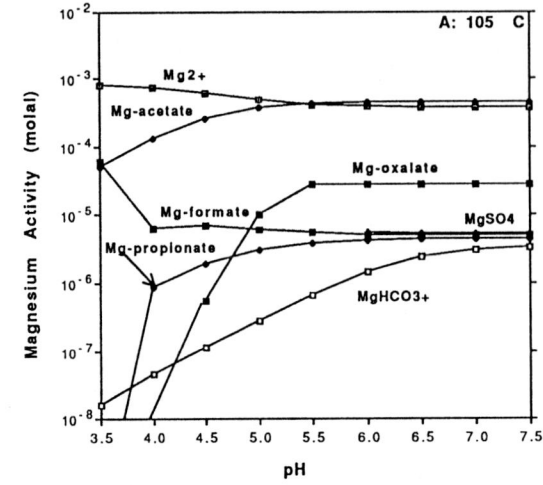

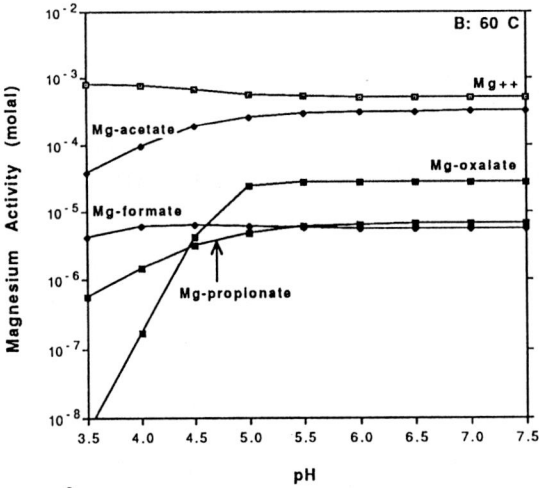

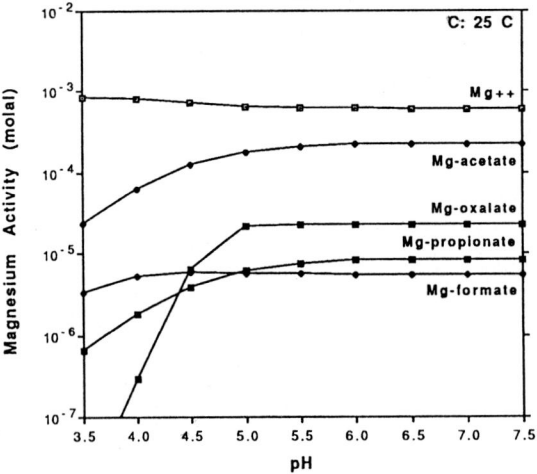

4.2.3 Trivalent Cations

These form stronger bonds with difunctional acid anions, such as malonate, oxalate, and succinate, and because the difunctional acids are significantly dissociated in most normal waters (Table 2; also Lind and Hem 1975), the difunctional acid anion complexes will have a wider stability range with respect to pH than the monofunctional acid anion complexes. In the trivalent cation group, the behavior of aluminum is of particular interest because of the dominance of aluminosilicate minerals in most geological environments. Metals such as Ti and Fe are important minor elements whose geochemical behavior is modified by the presence of organic acids (see experimental studies involving titanium vessels by Stoessell and Pittman 1991).

Speciation calculations shown in Fig. 5 reveal that the Al-oxalate complex can dominate the geochemistry of aluminum in acidic waters at all temperatures. The effect of this strong affinity between Al and oxalate is to modify the solubility of the aluminosilicates by several orders of magnitude as illustrated by the halloysite solubility curves in Fig. 6A. Similar, although reduced, modifications appear to be possible in the presence of malonate (Fig. 6B) and salicylate. Considerable uncertainty exists in the thermodynamic projection of the stability constants for Al-oxalate and Al-malonate from 25 to 200 °C (Harrison and Thyne 1992). Experimental measurements (Fein 1991b; Thyne and Harrison 1991; Thyne et al. 1992) suggest that there may not be a substantial increase in complex stability at elevated temperatures and thus the speciation calculations shown in Fig. 5 are an optimistic scenario at temperatures above about 50 °C. Nevertheless, using the more reliable low temperature data it remains apparent that the difunctional acids do control the behavior of aluminum in the pH range 4–6.

The effect of the formation of Al-oxalate and/or malonate complexes is to increase the amount of dissolved aluminum in a water perhaps by as much as 2 to 3 orders of magnitude according to Figs. 5 and 6. Chemical model predictions are in accordance with studies of soil zone geochemistry (Lind and Hem 1975; Antweiler and Drever 1983) where elevated aluminum concentrations are measured in the presence of OAA. Conversely, reports of elevated Al in petroleum reservoir formation waters or in other deep brines are rare (Hem 1985; L.S. Land, pers. comm. to W.J.H. 1990). Normal Al concentrations may indicate lack of OAA in these waters or

---

Fig. 4. Magnesium species distribution as a function of pH at 105 °C (**A**), 60 °C (**B**), and 25 °C (**C**). The divalent free ion dominates the species distribution at temperatures of 25 and 60 °C and all values of pH. Magnesium-OAA complexes become proportionally more important at higher temperatures and in more alkaline waters. At 105 °C, magnesium acetate and Mg^{2+} activities become equal at pH values greater than 5.5 (Harrison and Thyne 1992)

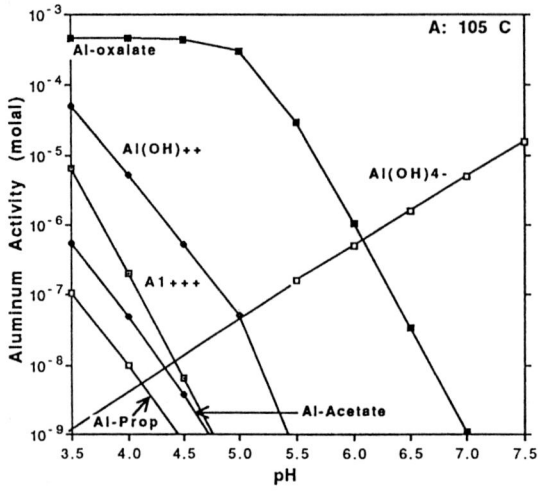

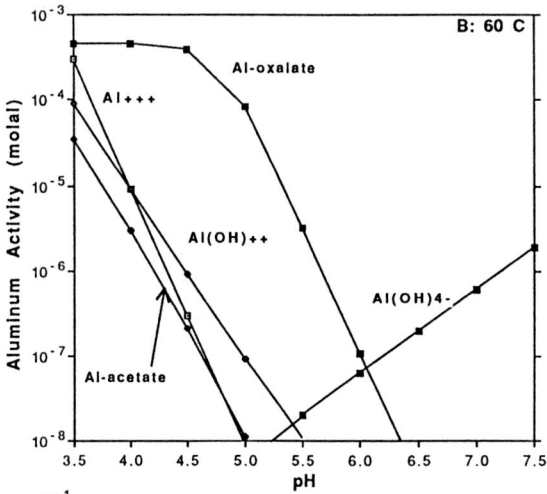

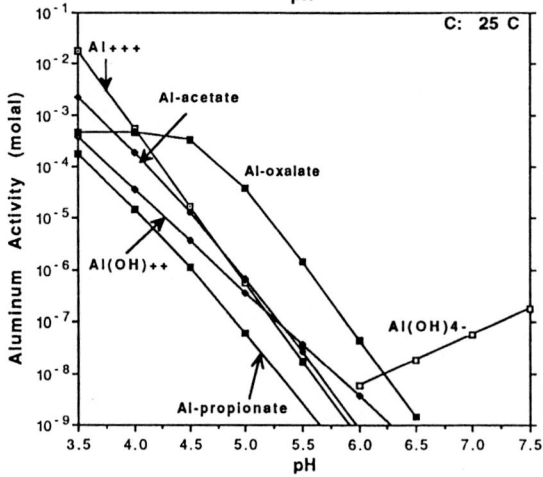

destabilization of the Al-OAA complex and deposition of excess aluminum as borehole scale. Reports of aluminous scale-forming minerals are likewise uncommon. However, if the complex is destabilized in the near well-bore environment and Al deposition occurs in the formation as an amorphous phase or as a clay mineral, it would be almost impossible to demonstrate unequivocally that subsurface pore waters had indeed contained the Al-OAA complex.

4.2.4   Trace Metals (Cu, As, Zn, Pb, and others)

Several studies (Cleveland 1979; Holm et al. 1979; Sunda and Hanson 1979; Giordano and Barnes 1981; Giordano 1985; Lovgren 1991) have shown that minor elements may be complexed by organic acid anions and that such complexes play an important role in the contaminant geochemistry and bioavailability of metals such as arsenic, copper, zinc, lead, and plutonium as well as in the transport and deposition of metals in ore-forming environments. In general, observations about minor and trace element species distributions are similar to results discussed above for the major elements. Chemical models of aqueous systems must include appropriate complexes if they are expected to be accurate. The prediction of the behavior of minor elements is generally limited by lack of thermodynamic data, however, estimation techniques (Langmuir 1979; Shock and Helgeson 1990; Harrison and Thyne 1992) allow approximations to be made. We do not elaborate on the behavior of metals in ore-forming environments as Giordano addresses this subject in Chapter 11.

4.2.5   Role of Organic Acid Concentration

Calculations shown in Figs. 2–5 were made assuming only mono-ligand complexes with stable cations. Di- and tri-ligand complexes have been observed using spectroscopy techniques (Caminiti et al. 1984; Bilinski et al. 1986; Palmer and Drummond 1988) and have been interpreted from ex-

---

Fig. 5. Aluminum species distribution as a function of pH at 105 °C (**A**), 60 °C (**B**), and 25 °C (**C**). The expansion of the stability field for the aluminum-oxalate complex at elevated temperatures is a consequence of the temperature-dependent stability constants for that aqueous species. Aluminum acetate and propionate are relatively minor components of the total dissolved aluminum. In our study, the presence of oxalate at 25 °C increases the total dissolved aluminum by almost two orders of magnitude at pH 5.5 but plays no role at values of pH outside the range 4–6.5. A greater increase occurs at higher temperatures, however, more uncertainty exists in these data (Harrison and Thyne 1992)

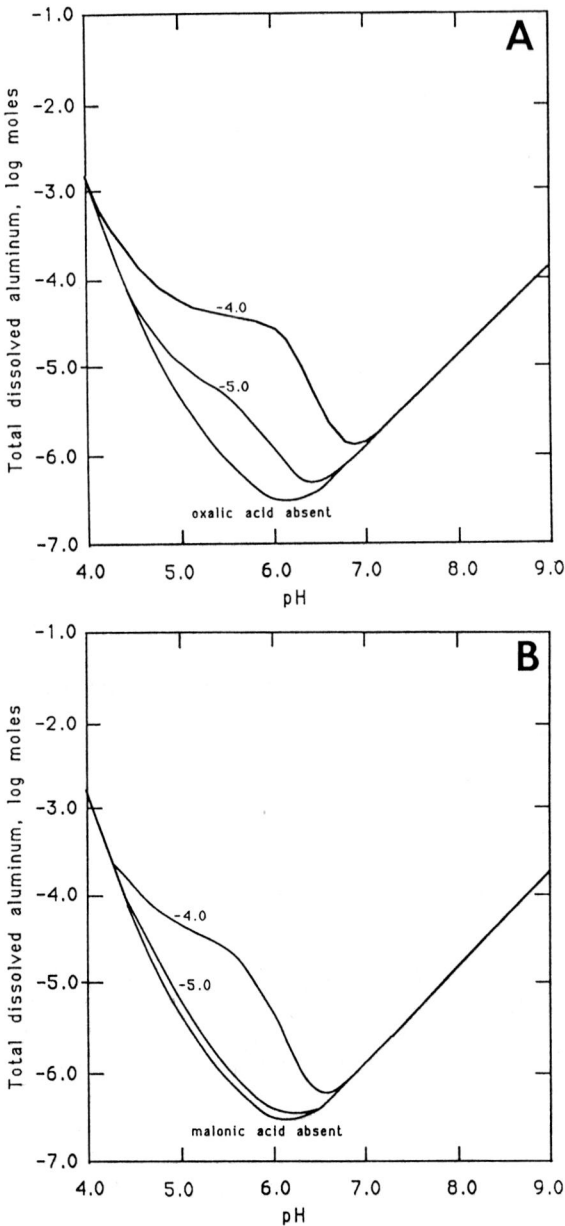

Fig. 6. Effect of oxalate (**A**) and malonate (**B**) complexes on the equilibrium solubility of halloysite $[Al_2Si_2O_5(OH)_4]$. Dissolved silica is $20\,mg\,kg^{-1}$ as SiO_2 (*aq*). A diagram for salicylic acid looks similar to that for malonic acid but the solubility increase is reduced. (After Lind and Hem 1975)

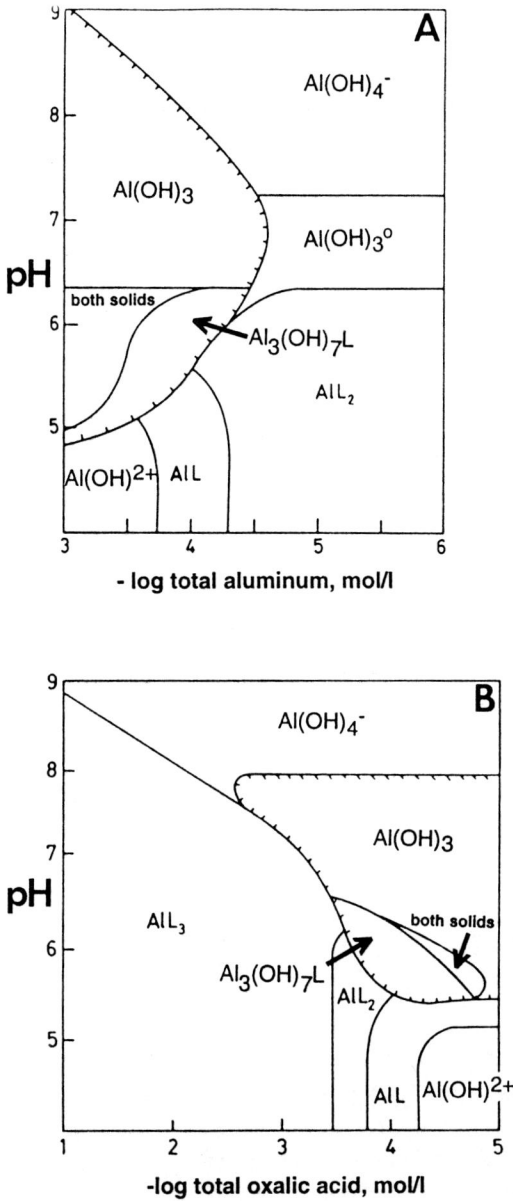

Fig. 7. Stability of aluminum complexes as a function of dissolved aluminum concentration (**A**) and oxalic acid concentration (**B**). Mono-ligand species are stable at aluminum concentrations of less than $1.3 \, mg \, kg^{-1}$ at an oxalic acid concentration of $9 \, mg \, kg^{-1}$ (**A**). Di- and tri-ligand complexes are only stable if the oxalic acid concentration exceeds about $50 \, mg \, kg^{-1}$ and the aluminum concentration is $260 \, mg \, kg^{-1}$ (**B**). (After Bilinski et al. 1986)

periments on aluminum oxalate (Fein 1991b). The appearance of these higher ligand complexes is dependent on the concentration of the acid anion as well as on the cation concentration. Figure 7 shows the stability range for aluminum oxalate complexes determined by Bilinski et al. (1986). Clearly, at acid anion concentrations greater than about 0.1 mmol oxalic acid and 10 mmol acetic acid, the di- and tri-ligands must be included in the speciation calculations. These concentration limits appear to be shifted to higher values as temperature increases based on similar data for Zn-acetate (Giordano and Drummond 1991), i.e. the mono-ligand species have an expanded stability field at elevated temperature. Calculations by Bilinski et al. (1986) show that the di-ligand oxalate complex may be the dominant aluminum species at low pH, whereas Evans (1988) shows the mono-salicylate aluminum complex accounts for more than 60% of the dissolved aluminum in certain soil waters. The differences among these studies with respect to ligand number and complexing ability largely reflect the completeness, or lack thereof, of the thermodynamic data used in the chemical models and point to the need to evaluate this parameter when comparing apparently discrepant results.

5 Chemical Models of Mineral Stabilities in the Presence of OA-Bearing Waters

Saturation index calculations made as part of a species distribution problem allow an assessment to be made of the effect of organic acids on the likely state of heterogeneous equilibria in an aqueous system (see Drever 1988, for discussion and definitions). By comparing saturation indices for minerals in systematically different waters we can predict the likely behavior of these minerals in the presence of organic acids. The predictions about mineral stability vary with the precise constraints that are placed on the calculations, in particular whether the cations are constrained to be in equilibrium with a mineral phase or set as a total concentration, the temperature, the partial pressure of CO_2, and the anionic composition of the water. Conclusions that differ from those presented here may be possible, nevertheless, some consistent trends emerge that are related to observations made in the preceding section about speciation.

5.1 Role of Increasing Organic Acid Concentration

Figures 8 and 9 show saturation states as a function of OAA concentration calculated by Harrison and Thyne (1992). Conditions are identical with respect to bulk starting composition, temperature, and pH (5.5 or 7.5), but different constraining equilibria have been used for aluminum. Silica con-

A: T=105, pH=5.5, kaolinite and quartz equilibria

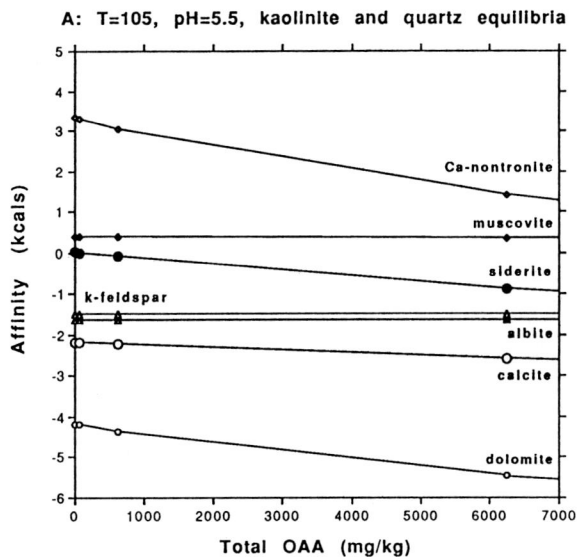

B: T=105, pH=7.5, kaolinite and quartz equilibria

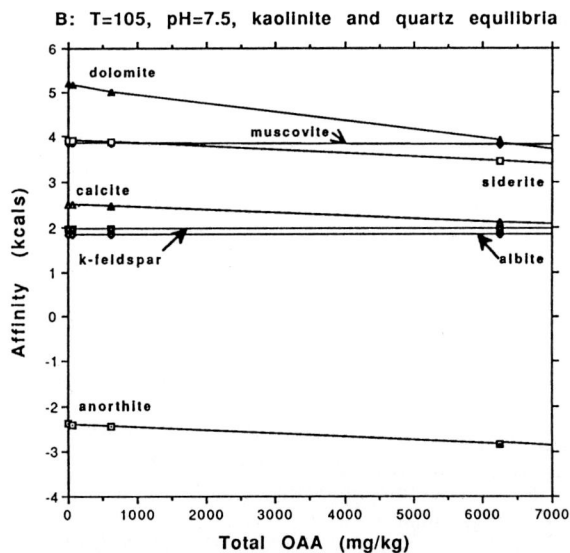

Fig. 8. Mineral affinities (log[A = $-2.303RT(SI)$]; SI = saturation index) as a function of OAA concentration. In these figures the sum of acetate + oxalate is plotted as the abscissa. Calculations were made using a combination of acetate:oxalate of 25:1 and the water analysis in Table 3. (Harrison and Thyne 1992)

380 W.J. Harrison and G.D. Thyne

A: T=105, pH=5.5, aAl=0.5e-7, quartz equilibrium

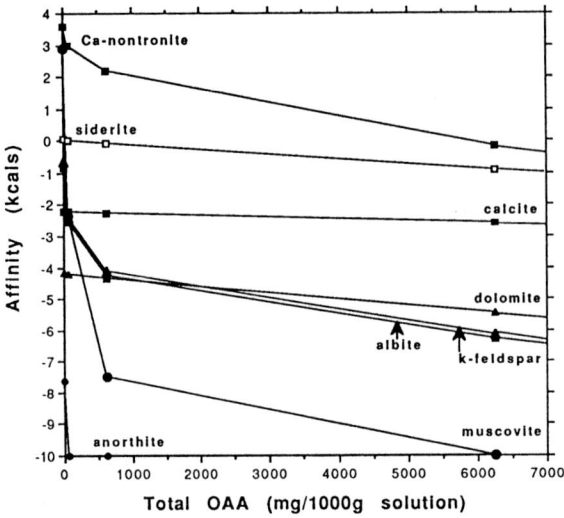

B: T=105 C, pH=7.5, aAl=0.5e-7, quartz equilibrium

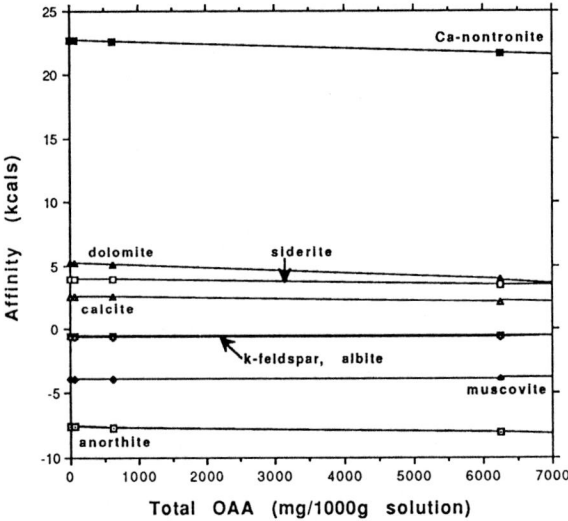

Fig. 9. Mineral affinities (log[A = −2.303RT(SI)]; SI = saturation index) as a function of OAA concentration. In these figures the sum of acetate + oxalate is plotted as the abscissa. Calculations were made using a combination of acetate:oxalate of 25:1 and the water analysis in Table 3. (Harrison and Thyne 1992)

centrations close to equilibrium with quartz are normal in most sedimentary pore waters and because we lack thermodynamic data for silica-OAA complexes this constraint has little significance in these calculations. In Fig. 8 the dissolved aluminum concentration is determined by equilibrium with kaolinite. As pH and organic acid concentrations are changed the total amount of aluminum in solution also changes. In Fig. 9 the amount of aluminum is fixed at a constant concentration of 0.5×10^{-7} molal. The choice of constraints thus leads to one condition where aqueous aluminum concentrations are fixed (representing equilibria in rocks where no aluminous phase is present and pore water composition is externally controlled) and one where aqueous aluminum concentrations may vary (representing equilibria in sandstones having one or more aluminous phases). Although other minerals could be chosen instead of kaolinite (e.g., muscovite, k-feldspar), conclusions drawn below would be unchanged. Absolute values of affinity[4] are not as significant as the direction in which affinity changes as OAA concentration increases: any decrease in affinity implies that a mineral is less stable (i.e., more likely to dissolve), whereas no change implies that the mineral stability is not affected by OAA and an increase in affinity (not seen in these figures) implies an increase in mineral stability (i.e., more likely to precipitate).

Overall, most minerals decrease in stability in the presence of OAA but the extent of this destabilization depends on pH. At low pH the free divalent cations and the Al-oxalate and malonate complexes are predominant. Increasing OA concentrations cause destabilization of the aluminosilicates as inorganic Al^{3+} and Al-hydroxide complexes are converted to Al-oxalate. Minor destabilization of the carbonates occurs because some of the divalent cations are partitioned into the acid anion complexes. At more alkaline values of pH, however, the Al-oxalate complex is not stable, thus increasing OA concentrations does not modify the stability of the aluminosilicates albite and K-feldspar (Na- and K-OAA complexes are insignificant). The increased stability of the divalent cation-OAA complexes causes Ca-, Fe- and Mg-bearing clays as well as Ca-plagioclase (either anorthite or anorthite component in plagioclase) to become less stable even though the Al-oxalate complex is not present. Carbonate minerals will also become less stable due to the reduction in the free ion activities of Mg, Ca, and Fe. If calcite activity is independently fixed by equilibrium with calcite, even anorthite can be stabilized as OAA concentrations increase. Interpretation of Figs. 8 and 9 suggests that any of the following general trends can apply when OAA concentrations increase in the pore waters of sandstones: (1) K-feldspar and albite remain stable, anorthite dissolves; (2) all feldspars and

---

[4] Following Wolery (1983), we utilize chemical affinity (A: units of kcal) as a measure of mineral saturation state, which is defined as $A = 2.303RT(SI)$ where SI is the conventional saturation index (see Drever 1988).

clays remain stable; (3) all feldspars and clays become unstable; and (4) carbonates can be slightly to very unstable. It appears to be difficult to cause the precipitation of carbonates except under some conditions of combined increases in OAA with increases in pCO_2 (Thyne 1992).

Lack of data for the formation of Si-OAA complexes makes the prediction of the behavior of quartz difficult. Dissolution experiments by Bennett et al. (1988) and Bennett (1991) as well as field studies (Bennett and Seigel 1987) show that the dissolution rate of quartz is modified in the presence of some acids but thermodynamic data cannot be retrieved from these experiments for use in chemical models. If indeed a stable Si-OAA complex forms in sedimentary environments, then the stabilities of quartz as well as all the silicate minerals will be reduced further than the destabilization shown in Figs. 8 and 9 (see also Bennett and Casey, Chap. 7, this Vol.).

5.2   Interactions Among Organic Acids, CO_2, and Minerals

The geochemical role of OA in combination with CO_2 is central to the problem of predicting diagenesis in sedimentary basins. Both are the products of the thermal decarboxylation of kerogen and extensive literature exists on the importance of both OA and CO_2 in generating secondary porosity through dissolution of detrital aluminosilicate grains and earlier-formed carbonate cements during burial (McDonald and Surdam 1984 and references therein; Gautier 1986 and references therein; Surdam et al. 1989 and references therein; Surdam and Yin, Chap. 13 and Giles et al., Chap. 14, this Vol.). Conventionally, the role of increasing partial pressure of CO_2 is to cause dissolution of aluminosilicate and carbonate minerals. Such dissolution results from a decrease in water pH due to the formation and partial dissociation of carbonic acid according to the standard reactions:

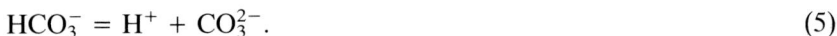

$$CO_2 + H_2O = H_2CO_3 \tag{3}$$

$$H_2CO_3 = H^+ + HCO_3^- \tag{4}$$

$$HCO_3^- = H^+ + CO_3^{2-}. \tag{5}$$

Likewise reactions among OA and OAA will also liberate (or consume) hydrogen ions:

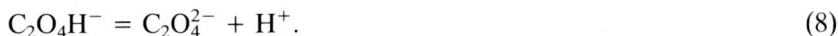

$$CH_3COOH = CH_3COO^- + H^+ \tag{6}$$

$$C_2O_4H_2 = C_2O_4H^- + H^+ \tag{7}$$

$$C_2O_4H^- = C_2O_4^{2-} + H^+. \tag{8}$$

The differences among pKa values for carbonic acids and organic acids (Table 2) result in the inorganic and organic acids having a different ability to buffer water pH (Kharaka et al. 1986; Lundegard and Kharaka, Chap. 3 this Vol.). Lundegard and Land (1989) and Thyne (1992) have configured

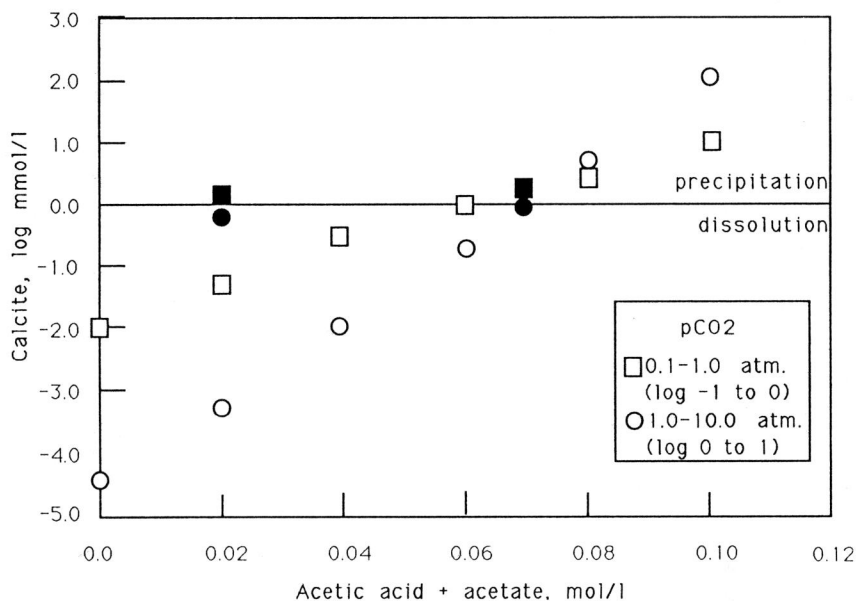

Fig. 10. Effect of increasing pCO_2 on the stability of calcite in the presence of acetic acid. As the acetic acid concentration increases, the buffering capacity of the water increases with respect to the water's ability to consume protons and resist pH change. Thus, at a critical acetic acid concentration (about $0.07\,\mathrm{mol\,l^{-1}}$; $4200\,\mathrm{mg\,kg^{-1}}$) increases in acidity caused by dissolving CO_2 in the water can be offset by the consumption of H^+ by the acetic acid-acetate reaction. At lower acetic acid concentrations calcite dissolves, whereas at higher aceteic acid concnetrations calcite may precipitate (Lundegard and Land 1989). *Solid symbols* are data from Thyne (1992) as shown in Fig. 11. Discrepant predictions between the two studies at the lower acid concentration can be reconciled by a higher concentration of cation-acid anion complexes in Thyne's study. These complexes release cations that cause carbonate mineral precipitiation. The absolute amount of solid formation cannot be calculated from Thyne's data

chemical models to assess the relative roles of OA and CO_2 in modifying carbonate stabilities. Their results are partically contradictory on first inspection.

Figure 10 shows the effect of increasing partial pressure of CO_2 at various OAA concentrations on the stability of calcite (modified from Lundegard and Land 1989). For geologically likely acetic acid concentrations, increasing pCO_2 causes calcite dissolution, as the ability of the acetic acid-acetate reaction to consume protons is exceeded. At very high acetic acid concentrations (above about $3600\,\mathrm{mg/l}$: 0.6 to $0.7\,\mathrm{mol/kg}$ in Fig. 10) the acetic acid-acetate buffer can consume all the protons produced as pCO_2 increases. Subsequently, calcite precipitation occurs because of increases in dissolved carbonate and bicarbonate species [reactions (3)–(5)] and in free calcium ions. The latter are released by the increased hydrogen ion concentration [reactions (4) and (5)] which drives Ca-acetate breakdown and the conversion of acetate to acetic acid:

$$CaCH_3COO^+ + H^+ = Ca^{2+} + CH_3COOH. \tag{9}$$

Figure 11 shows results from Thyne (1992). Thyne (1992) has configured the problem differently and predicts initial stabilization of carbonates (values of $\log Q/K$ become more positive as $\log pCO_2$ increases from -1 to about 0; Fig. 11): Thyne's explanation is that protons generated by reactions (3)–(5)

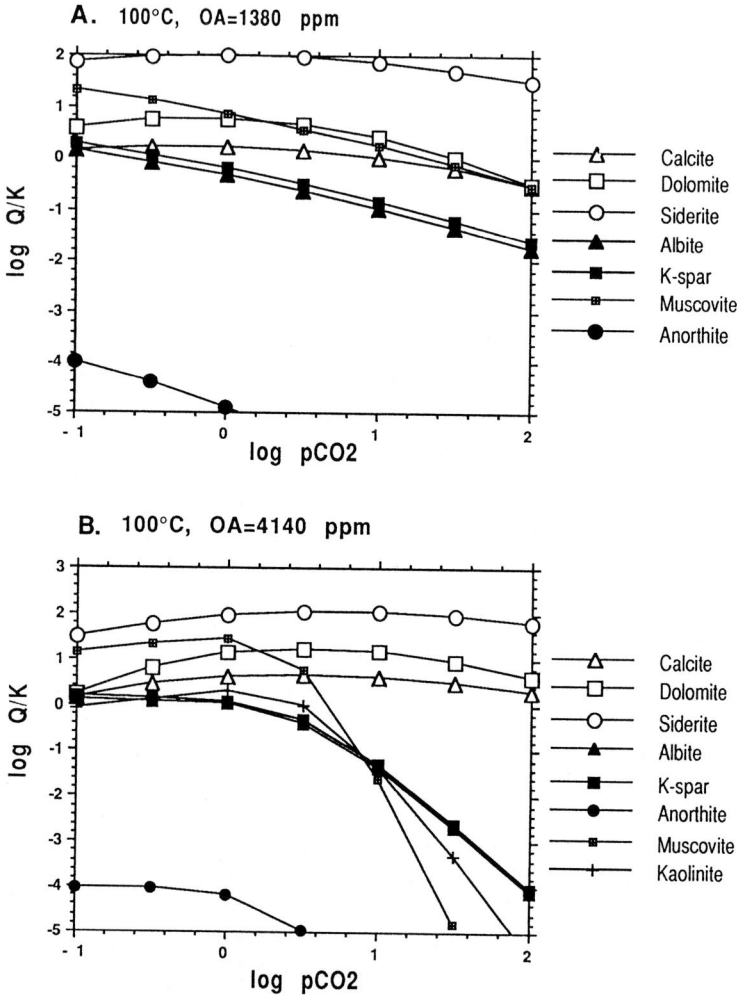

Fig. 11A,B. Effect of increasing pCO_2 on the saturation indices of several aluminosilicate and carbonate minerals. **A** $1380\,mg\,kg^{-1}$ total OAA ($2000\,mEq\,kg^{-1}$ protons) and **B** $4140\,mg\,kg^{-1}$ total OAA ($7000\,mEq\,kg^{-1}$ protons) (Thyne 1992). Increased stability occurs for carbonate minerals as pCO_2 is increased from log 0 to 1 bar (1–10 bar) and decreased stability occurs for carbonates when pCO_2 increases to more than log 1 bar (10 bar)

are consumed by the acid-acid anion reaction which in turn drives the breakdown of cation-acid complexes to replenish the acid anions. This reaction also releases free cations causing the solubility product of the carbonate minerals to be exceeded, resulting in precipitation. Once the capacity of the acid-acid anion reaction to buffer pH is exhausted, further increase in pCO_2 causes carbonate dissolution according to conventional expectations.

The two studies are not inconsistent but emphasize, firstly, the need to undertake chemical modeling with careful problem-posing so that the calculations address the geochemical question of interest and, secondly, the need to pay attention to differences in the way the models have been used. For example, Thyne (1992) uses an OA mixture having both mono- and difunctional acids. The difunctional acids have considerably lower values of pKa (Table 2) than the monofunctional acetic acid used by Lundegard and Land (1989); thus, Thyne's (1992) water is capable of consuming considerably more protons in the OA-OAA buffer than Lundegard and Land's (1989) water for a comparable acetic acid concentration. If Thyne's total acid concentrations are converted to milliequivalents of protons and are plotted in a similar form to that used by Lundegard and Land (1989) in Fig. 10, it is apparent that although Thyne's higher concentration model calculations ($7000\,mEq\,H^+\,kg^{-1}$ and Fig. 11A) produce results that are similar, the lower concentration ($2000\,mEq\,H^+\,kg^{-1}$ and Fig. 11B) calculations do not. Thyne (1992) also uses a water composition that includes several cations compared with the simple system (calcite-CO_2-H_2O-CH_3COOH) used by Lundegard and Land (1989). Magnesium, iron, and aluminum in Thyne's system also form complexes with the acid anions and contribute to the buffer capacity of the system. Cations released during the breakdown of these complexes are available for precipitation of carbonate minerals. Clearly, accurate prediction of the stabilities of carbonate minerals in aqueous systems must consider the roles of organic acids, acid anions, cation competition, and inorganic carbonate equilibria.

In comparing these two studies, some notes of caution are appropriate: absolute values of affinity or saturation index should not be compared between these two studies, nor should inferences about the relative importances of CO_2 and OA in modifying carbonate and silicate vs. carbonate equilibria be made. This entire issue can only be assessed accurately by using reaction path calculations: chemical models can only be used to reveal trends, not absolute amounts of material transfer.

6   Reaction Path Calculations of the Role of OAA in Fluid-Rock Reactions

Reaction path calculations have been used by Harrison and Thyne (1992) to assess the potential implications that OAA might have for mass transfer processes in the subsurface. Closed system (titration) reaction path models, with and without organic acids, using the input data in Table 4 are shown in Fig. 12. These calculations have been configured to investigate the reactions that may occur between an arkosic sandstone and organic acid-rich water being expelled from a kaolinitic shale. In these calculations, the progressive reaction of the hypothetical arkosic sandstone represents an increasing ratio of rock:water equilibration as the reaction progress variable, Zi, increases towards one.[5] No dramatic differences result in the presence of OAA ($3158\,mg\,kg^{-1}$ acetate and $58\,mg\,kg^{-1}$ oxalate), although differences in mineral stability fields are observed as a result of cation complex formation, which causes undersaturation of dolomite. The appearance of all other minerals is suppressed to higher amounts of rock reaction in the presence of OAA because of the capability of the fluid to complex cations thereby reducing the free ion activities: for example, the water first saturates with Ca-saponite at $\log Zi = -3.4$ in the absence of OAA, whereas the water does not saturate with the same phase until $\log Zi = -2.1$ in the presence of OAA. The OAA-bearing waters precipitate more kaolinite in the early parts of the reaction path. This is a consequence of the way the problem has been posed: the water is assumed to be in equilibrium with kaolinite when expelled from the shale and thus will contain more dissolved Al than the OAA-free water.

One of the significant uncertainties in applying these reaction path calculation results to mass transfer processes in nature is the ratio of the reactive surface area to the total surface area, which is needed to calculate the extent of reaction that will occur in any given time increment. This ratio is known as the effective surface area and may vary from 10^{-2} to 10^{-5} (Helgeson et al. 1984; Bruton 1986; White and Peterson 1990; Harrison and Tempel 1992; Harrison and Thyne 1992). By using a range of effective

---

[5] Reaction progress (Helgeson 1970) is a measure of the extent to which an irreversible reaction has proceeded. Conventionally, the reaction progress variable is used to measure the molar transfer of material during an irreversible reaction, such as the moles of feldspar titrated into a mass of water. In a complex rock-water system, the reaction progress variable describes an array of simple irreversible reactions (Wolery and Daveler 1992) as the rock is progressively equilibrated with the water. In this case the reaction progress variable is a direct monitor of the total amount of rock material titrated into the water. Reaction progress may also describe the redistribution of material that occurs during the closed system heating or cooling of a system in which case the reaction progress variable is a direct monitor of temperature change. In this chapter, Zi has a value of between 0 (no reaction has occurred) and 1 (complete reaction has occurred). The total amount of material to be transferred is given in Table 4 and thus the actual amount of reaction is determined by the product to Zi and the total moles of reactant minerals. In Fig. 12 lines show the behavior of phases as the amount of rock-water reaction increases. Phases may continuously precipitate, such as muscovite, or precipitate then dissolve, such as kaolinite and dolomite, or the water may become saturated with a phase whereupon no further reaction occurs, such as quartz.

Table 4. Water and sandstone compositions used in reaction path calculations. (Harrison and Thyne 1992)

Water composition		Sandstone composition			
Species	Concentration (mg kg^{-1})	Mineral	(mol)	(cm^3)^a	(vol%)
Na$^+$	9 300	Quartz	68.7	1558.67	31.82
K$^+$	50	K-feldspar	9.9	1077.81	22.00
Ca^{2+}	790	Albite	6.9	690.48	14.09
Mg^{2+}	21	Anorthite	3.1	312.45	6.38
Sr^{2+}	49	Calcite	7.0	258.54	5.28
SiO$_2$	Quartz saturation	Pore space^b		1000.00	20.41
Fe^{3+}	1.9			4897.95	100.00
Al^{3+}	Kaolinite saturation				
Mn^{2+}	0.70				
Cl$^-$	14 000				
SO$_4^{2-}$	40				
HCO$_3^-$	96				
CH$_3$COO$^-$	3 152				
C$_2$O$_4^{2-}$	58				
pH	5.5				
T (°C)	105				

^{a} Mineral volumes calculated assuming 20% pore space occupies 1000 cm^3.
^{b} Volume of 1 kg of fluid used in EQ3/6 calculations: assumed to be equivalent to about 20% porosity.

surface areas we can, however, look at the likely range of effects that the OA may play in fluid-rock reactions during sediment diagenesis. Diagenetic consequences of OAA-rich fluids are evaluated for three different amounts of reactant per pore volume of fluid: 0.0255 g/pore volume (Zi = log −4.5), 0.255g/pore volume (Zi = log −3.5), and 2.55 g/pore volume (Zi = log −2.5), representing effective surface areas of $10^{-4.5}$, $10^{-3.5}$, and $10^{-2.5}$, respectively. Tables 4 and 5 list all the assumptions made in translating these reaction path calculations to diagenetic predictions.

The starting mineral and pore space volumes shown in Table 4 are representative of a hypothetical arkose. We assume the pore volume is 1000 cm^3 because this is the amount of solution used in the EQ3/6 calculations. The rock volume is accordingly calculated to allow 1000 cm^3 to be equal to about 20% porosity. The amounts of rock reaction per 1000 cm^3 (approximately 1 kg) taken from Fig. 12 are converted to volumes (using molar volumes listed by Robie et al. 1978) and a modified rock composition is calculated based upon the amounts of dissolved starting minerals and new products. These changes, typically very small for reaction of rock with 1000 cm^3 of water, are multiplied by 500 to predict the volumetric change after 500 pore volumes of fluid. These volumes are provided in Table 5.

grams rock/1000g fluid

A: OAA-absent

grams rock/1000g fluid

B: OAA-present

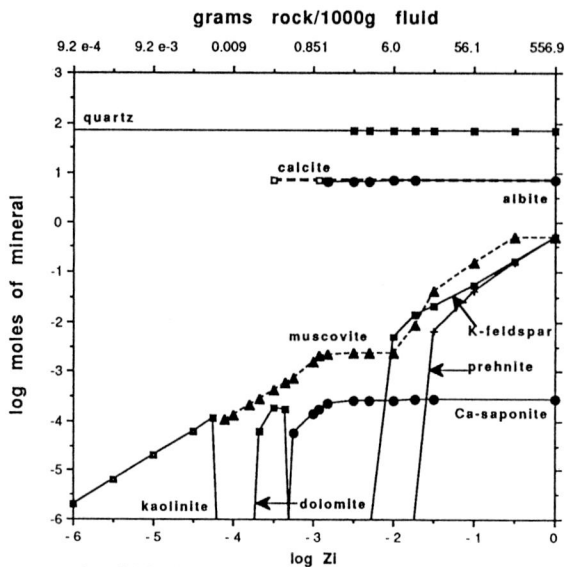

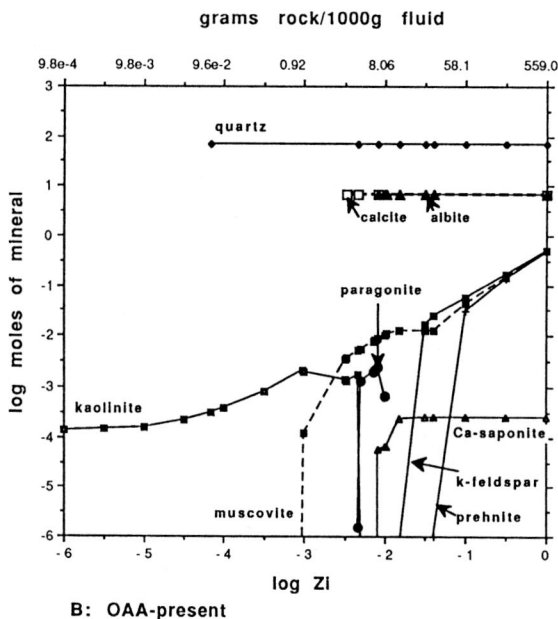

Fig. 12A,B. Reaction path calculations of sandstone-pore water interactions at 105 °C. **A** OAA absent; **B** OAA present in pore water (3120 mg kg^{-1} acetate; 58 mg kg^{-1} oxalate; other input data summarized in Table 4). Dolomite is destabilized by the presence of OAA because of the ability of these anions to complex calcium and magnesium, thereby reducing the free ion activities. The final mineral assemblage, which represents reaction of a single pore volume of fluid with an arkose having 20% porosity, is not reported from studies of sandstone diagensis, implying that this much mineral reaction is unrealistic. Calculations of diagenetic alteration patterns (Table 5; Fig. 13) utilized amounts of reaction at log Zi = −4.5, −3.5, and −2.5. (Harrison and Thyne 1992)

Table 5. Modal mineralogy of a sandstone after 500 pore volumes of fluid flow

Mineral	Initial vol%	Sandstone mineralogy (vol%) After 500 pore volumes					
Effective surface area		$10^{-4.5}$		$10^{-3.5}$		$10^{-2.5}$	
Rock dissolved[a]		0.0255 g/kg		0.255 g/kg		2.55 g/kg	
		No OAA	OAA	No OAA	OAA	No OAA	OAA
Quartz	31.82	31.82	31.75	31.82	31.82	31.82	34.58
K-feldspar	22.00	21.97	21.97	21.65	21.65	18.53	18.53
Albite	14.09	14.06	14.06	13.77	13.77	16.77	17.32
Anorthite	6.38	6.35	6.35	6.05	6.05	3.12	3.12
Calcite	5.28	5.26	5.26	5.16	5.16	5.77	3.47
Kaolinite	–	0.02	0.22	–	0.84	–	1.41
Muscovite	–	–	–	0.6	–	3.85	4.97
Smectite	–	–	–	0.01	–	0.4	–
Dolomite	–	–	–	0.1	–	–	–
Pore space	20.43	20.48	20.39	20.84	20.71	19.74	16.60
Leached Porosity	–	0.11	0.18	1.12	1.12	6.74	8.54
Primary Porosity	20.43	20.43	20.43	20.43	20.43	20.43	20.43
Authigenic minerals	–	0.06	0.22	0.71	0.84	7.42	12.37

[a] Per pore volume assuming one pore volume is 1000 cm³ and at the starting condition is about 20% porosity.

Figure 13 shows the potential impact the OAA may have on sandstone porosity for an effective surface area of $10^{-2.5}$. If the kaolinite is precipitated adjacent to the sites of feldspar dissolution, the net result will be porosity loss because of the higher molar volumes of the clays relative to the feldspars (Robie et al. 1978). In these calculations the OAA play a more important role, volumetrically, in causing calcite dissolution than they do in aluminosilicate reactions. As the effective surface area decreases, and the amount of rock reaction per pore volume of fluid decreases, the diagenetic modifications become volumetrically less important (Table 5). For low amounts of rock reaction per pore volume, diagenetic modification can only be significant if enormous fluid volumes are available: for example, as much as 5000 pore volumes would be needed to create 2.2% kaolinite and about 1% leached grain porosity when OAA are present at effective surface areas of $10^{-4.5}$. As the amount of rock reaction per pore volume increases, the modifications to porosity and authigenic mineral assemblage do become more significant: however, conceptually increasing the amount of rock reaction implies a longer residence time for fluid in pore space and thus slower moving fluid and, for a given time increment, fewer pore volumes of fluid in total. Fluid volumes used in Fig. 13 and Table 5 appear reasonable based on paleohydrologic analysis (Harrison and Summa 1991; Harrison and Tempel 1992).

Clearly, these observations are based upon a specific set of reaction path calculations: many others could be configured and different results obtained.

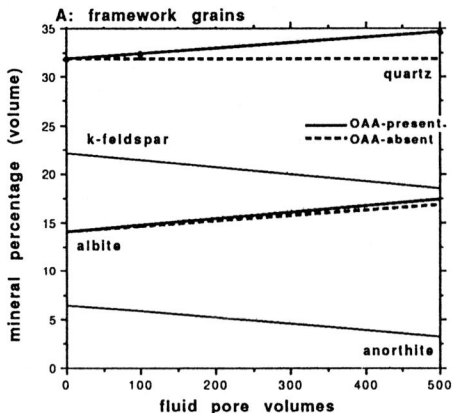

A: framework grains

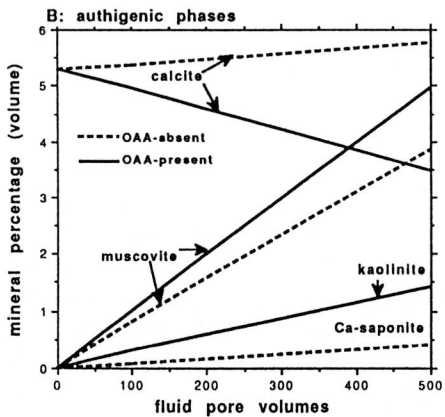

B: authigenic phases

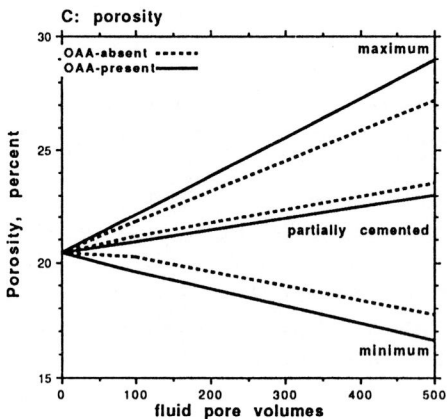

C: porosity

Changing the amount of acid, and the assumptions about the chemical state of the fluid expelled from an adjacent shale, as well as changing the temperature conditions and gas partial pressures will result in variable relationships among feldspars, carbonates, and clays. However, it is difficult to anticipate that the mass transfer implications could be substantially different. *A reasonable conclusion to be drawn is that the role of OAA may be to modify mineral-water phase equilibria but not to provide a substantial means by which porosity enhancement may occur in sedimentary rocks.* These conclusions appear to be in agreement with arguments made by Giles and Marshall (1986) and Giles et al. (Chap. 14, this Vol.), but not with those reached by Fein (1991b) who measured the stability of Al-oxalate. Although Fein (1991b) finds that the Al-oxalate complex can modify the solubility of aluminosilicates, Fein's proposed increase in mass transfer is not so straightforward and is limited by the simplicity of the chemical system selected. Reaction path calculations allow more accurate predictions of mass transfer effects, however, a reliable assessment of the role of OAA in porosity enhancement in the subsurface may not be achieved without formal coupling of the equations for fluid flow with those for chemical reaction in aqueous systems.

7 Conclusions

Chemical models using a combination of experimental and estimated thermodynamic data result in several important predictions: (1) monovalent cations form very weak complexes with organic acids and such complexes typically account for less than 10% of the dissolved element; (2) divalent cations form only weak complexes with difunctional acids but stronger complexes with monofunctional acids; (3) divalent metal cation-OAA complexes are stable in alkaline waters but even under such conditions rarely

---

Fig. 13A–C. Hypothetical changes in mineral abundances assuming mineral reaction sequences as defined in the reaction path calculations (Fig. 12). *Dashed lines* distinguish OAA-absent reactions. Amounts of albite, K-feldspar, and anorthite reacted are similar in both cases; however, more quartz is precipitated when OAA are present (**A**). Significant differences in authigenic mineralogies are shown in **B**: OAA-bearing fluids dissolve calcite and precipitate kaolinite, whereas OAA-free fluids precipitate calcite and smectite (Ca-saponite). Both fluids precipitate muscovite (proxy for illite). **C** The changes that might result in sandstone porosity for three different scenarios. Maximum porosity changes are calculated assuming all authigenic phases are removed from the pore space, minimum porosities are calculated assuming that all authigenic phases remain adjacent to the reacting framework grains. An intermediate scenario (*partially cemented*) represents the porosities calculated if the authigenic clays do not precipitate but quartz and calcite are allowed to form as overgrowths or poikilotopic cements (Harrison and Thyne 1992)

account for more than 60% of the total dissolved cation; (4) trivalent cations form stronger complexes with difunctional than monofunctional acids: (5) aluminum acetate complexes are of minor importance, whereas alumnum oxalate dominates the species distribution of aluminum under acidic conditions; (6) in waters having pH values more alkaline than about 6.5, inorganic aluminum species are predominant at high temperatures. The stability of aluminum malonate is uncertain because decarboxylation rates for the malonate anion are poorly known; (7) OAA are ineffective at neutral to alkaline pH in modifying stabilities of aluminosilicate minerals, whereas these anions have variable effectiveness over a wide range of pH in modifying carbonate mineral stabilities; (8) prediction of carbonate mineral stabilities needs to include the role of CO_2 and OAA; and (9) predictions of the diagenetic alteration of sandstones by OAA-bearing waters show that in some circumstances OAA-bearing waters are less effective at producing porosity in an arkosic sandstone than are OAA-free waters.

An overview of the role of organic acids in sedimentary processes is that they contribute to overall patterns of fluid-rock interaction but appear unlikely to dominate such reactions except in restricted geochemical environments where concentrations are in excess of typical values. Such environments might include wetlands, gasoline-contaminated groundwaters, and organic-rich shales, and adjacent sandstones.

References

Adams SS, Curtis HS, Hafen PL, Salek-Nejad H (1978) Interpretation of postdepositional processes related to the formation and destruction of the Jackpile-Paguate uranium deposit, northwest New Mexico. Econ Geol 73: 1635–1654

Antweiler RC, Drever JI (1983) The weathering of a Late Tertiary volcanic ash: the importance of organic solutes. Geochim Cosmochim Acta 47: 623–629

Bassett RL, Melchior DC (1990) Chemical modeling of aqueous systems: an overview. In: Melchior DC, Bassett RL (eds) Chemical modeling of aqueous systems II. Am Chem Soc Symp Ser 416, Washington DC, pp 1–15

Bennett P (1991) Quartz dissolution in organic-rich solutions. Geochim Cosmochim Acta 55: 1781–1799

Bennett PC, Siegel DI (1987) Increased solubility of quartz in water due to complexing by organic compounds. Nature 326: 684–686

Bennett PC, Melcer ME, Siegel DI, Hassett JP (1988) The dissolution of quartz in dilute solutions of organic acids at 25 °C. Geochim Cosmochim Acta 52: 1521–1530

Bennett PC, Siegel DI, Hill BM, Glasser PH (1991) Fate of silicate minerals in a peat bog. Geology 19: 328–331

Bevan J, Savage D (1989) The effect of organic acids on the dissolution of K-feldspar under conditions relevant to burial diagenesis. Mineral Mag 53: 415–425

Bilinski H, Horvath L, Ingri N, Sjoberg S (1986) Equilibrium aluminium hydroxo-oxalate phases during initial clay formation; H^+-Al^+-oxalic acid-Na^+ system. Geochim Cosmochim Acta 50: 1911–1922

Bjørlykke K (1979) Cementation of sandstones. J Sediment Pet 49: 1358–1359

Bjørlykke K (1984) Formation of secondary porosity: how important is it? In: McDonald DA, Surdam RC (eds) Clastic diagenesis. Am Assoc Pet Geol Mem 37, pp 277–289

Blatt H (1979) Diagenetic processes in sandstones. In: Scholle PA, Schluger PR (eds) Aspects of diagenesis. Soc Econ Paleontol Mineral Spec Publ 26, pp 141–157

Boles JR (1987) Six million year diagenetic history, North Coles Levee, San Joaquin Basin, California. In: Marshall J D (ed) Diagenesis of sedimentary sequences. Geol Soc Spec Publ 36: 191–200

Bruton CJ (1986) Predicting mineral dissolution and precipitation during burial: synthetic diagenetic sequences. In: Proc Worksh Geochemical modeling. Lawrence Livermore National Laboratory Publ CONF-8609134, Livermore, CA, pp 111–120

Caminiti P, Cucca P, Monduzzi M, Suba G (1984) Divalent metal acetate complexes in concentrated aqueous solutions: an X-ray diffraction and NMR spectroscopy study. J Chem Phys 81: 543–551

Campbell TJ, Roberts WL (1986) Whewellite from South Dakota and a review of other North American localities. Mineral Rec 17: 131–133

Carothers WW, Kharaka YK (1978) Aliphatic acid anions in oil-field waters – implications for origin of natural gas. Am Assoc Pet Geol Bull 62: 2441–2453

Chen W, Ghaith A, Park A, Ortoleva PJ (1990) Diagenesis through coupled processes: modeling approach, self-organization, and implications for exploration. In: Meshri I (ed) Prediction of reservoir quality through chemical modeling. Am Assoc Pet Geol Mem 49, pp 103–131

Cleveland JM (1979) Critical review of plutonium equilibria of environmental concern. In: Melchior DC, Bassett RL (eds) Chemical modeling of aqueous systems II. Am Chem Soc Symp Ser 416, Washington DC, pp 321–338

Crossey LJ (1991) Thermal degradation of aqueous oxalate species. Geochim Cosmochim Acta 55: 1515–1529

Crossey LJ, Frost BR, Surdam RC (1984) Secondary porosity in laumontite-bearing sandstones. In: McDonald D A, Surdam R C (eds) Clastic diagenesis. Am Assoc Pet Geol Mem 37, pp 225–239

Drever JI (1988) The geochemistry of natural waters, 2nd edn. Prentice-Hall, Englewood Cliffs, 437 pp

Drez PE, Harrison WJ (1985) Do organic acids play a role in diagenesis? In: Am Assoc Pet Geol Res Conf, Prediction of reservoir quality through chemical modeling, Park City, Utah (Abstr). Am Assoc Pet Geol, Tulsa, OK

Drummond SE, Palmer DA (1986a) Thermal decarboxylation of acetate. Part 1. The kinetics and mechanism of reaction in aqueous solution. Geochim Cosmochim Actea 50: 813–823

Drummond SE, Palmer DA (1986b) Thermal decarboxylation of acetate. Part 2. Boundary conditions for the role of acetate in the primary migration of natural gas and the transportation of metals in hydrothermal systems. Geochim Cosmochim Acta 50: 825–833

Evans LJ (1988) Some aspects of the chemistry of aluminum in podzolic soils. Commun Soil Sci Plant Anal 19: 793–803

Fein JB (1991a) Experimental study of aluminum-, calcium- and magnesium-acetate complexing at 80 °C. Geochim Cosmochim Acta 55: 955–964

Fein JB (1991b) Experimental study of aluminum-oxalate complexing at 80 °C implications for the formation of secondary porosity within sedimentary reservoirs. Geology 1037–1040

Fisher JB (1987) Distribution and occurrence of aliphatic acid anions in deep subsurface waters. Geochim Cosmochim Acta 51: 2459–2468

Freeze RA, Cherry JA (1979) Groundwater. Prentice-Hall, Englewood Cliffs, 604 pp

Galimov EM, Tugarinov AI, Nikitin AA (1975) On the origin of whewellite in a hydrothermal uranium deposit. Geochem Int 1975: 31–37 (Translation from Geokhimiya 676–683)

Garrels RM, Thompson ME (1962) A chemical model for sea water at 25 °C and one atmosphere total pressure. Am J Sci 260: 57–66

Gautier DL (ed) (1986) Roles of organic matter in sediment diagenesis. Soc Econ Paleontol Mineral Spec Publ 38, 204 pp

Giles MR, Marshall JD (1986) Constraints on the development of secondary porosity in the subsurface: re-evaluation of processes. Mar Pet Geol 3: 243–256

Giodarno TH (1985) A preliminary evaluation of organic ligands and metal organic complexing in Mississippi Valley-type ore solutions. Econ Geol 80: 96–106

Giordano TH (1989) Anglesite (PbSO₄) solubility in acetate solutions: the determination of stability constants for lead acetate complexes to 85 °C. Geochim Cosmochim Acta 53: 359–366

Giordano TH, Barnes HL (1981) Lead transport in Mississippi Valley-type ore solutions. Econ Geol 76: 2200–2211

Giordano TH, Drummond SE (1987) Zinc acetate complexing in hydrothermal solutions. Geol Soc Am Abstr Program 19: 677

Giordano TH, Drummond SE (1991) The potentiometric determination of stability constants for zinc acetate complexes in aqueous solutions to 295 °C. Geochim Cosmochim Acta 55: 2410–2417

Graustein WC, Cromack J Jr, Sollins P (1977) Calcium oxalate: occurrence in soils and effect on nutrient and geochemical cycles. Science 198: 1252–1254

Hajash A, Mahoney AJ, Elias BP (1989) Role of carboxylic acids in the dissolution of silicate sands: an experimental study at 100 °C and 345 bars. Geol Soc Am Annu Meet St. Louis, Abstr Program 21: A49

Harrison WJ (1990) Modeling fluid/rock interactions in sedimentary basins. In: Cross TA (ed) Quantitative dynamic stratigraphy. Prentice-Hall, Englewood Cliffs, pp 195–231

Harrison WJ, Summa LL (1991) Paleohydrology of the Gulf of Mexico basin. Am J Sci 291: 109–176

Harrison WJ, Tempel RN (1993) Diagenetic pathways in sedimentary basins. In: Robinson A, Horbury A (eds) paleohydrology and diagenesis in sedimentary basins. Am Assoc Pet Geol Stud Geol 36: 69–86

Harrison WJ, Thyne GD (1992) Predictions of diagenesis in the presence of organic acids. Geochim Cosmochim Acta 56: 565–586

Helgeson HC (1969) Thermodynamics of hydrothermal systems at elevated temperatures and pressures. Am J Sci 267: 729–804

Helgeson HG (1970) Calculation of mass transfer in geochemical processes involving aqueous solutions. Geochim Cosmochim Acta 34: 569–592

Helgeson HG, Delany JM, Nesbitt HW, Bird DK (1978) Summary and critique of the thermodynamic properties of the rock-forming minerals. Am J Sci 278A: 1–229

Helgeson HG, Murphy WM, Aargaard P (1984) Thermodynamic and kinetic constraints on reaction rates among minerals and aqueous solutions II. Rate constants, effective surface area and the hydrolysis of feldspar. Geochim Cosmochim Acta 48: 2405–2433

Hem JD (1985) Study and interpretation of the chemical characteristics of natural water. US Geol Surv Water Supply Pap 2254, 263 pp

Hennett RJ-C, Crerar DA, Schwartz J (1988) Organic complexes in hydrothermal systems. Econ Geol 83: 742–764

Holm TR, Curtiss CD III (1990) Copper complexation by natural organic matter in groundwater. In: Melchior DC, Bassett RL (eds) Chemical modeling of aqueous systems II. Am Chem Soc Symp Ser 416, Washington DC, pp 508–518

Holm TR, Anderson MA, Iverson DG, Stanforth RS (1979) Heterogeneous interactions of arsenic in aquatic systems. In: Jenne EA (ed) Chemical modeling of aqueous systems I. Am Chem Soc Symp Ser 93, Washington DC, pp 711–736

Huang WH, Keller WD (1970) Dissolution of rock-forming silicates in organic acids: simulated first stage weathering of fresh mineral surfaces. Am Mineral 55: 2076–2094

Jenne EA (1979) Chemical modeling-goals, problems, approaches and priorities. In: Jenne EA (ed) Chemical modeling of aqueous systems I. Am Chem Soc Symp Ser 93, Washington DC, pp 3–24

Johnson JW, Lundeen SR, Chamberlain SC, Thermodynamic databases for the EQ3/6.3245.1090 Software Package: UCRL-XXXXX, Lawrence Livermore National Laboratory, Livermore, CA (in press)

Kharaka YK, Carothers WW, Rosenbauer RJ (1983) Thermal decarboxyaltion of acetic acid: implications for origin of natural gas. Geochim Cosmochim Acta 47: 397–402

Kharaka YK, Law LM, Carothers WW, Goelitz DF (1986) Role of organic species in formation waters from sedimentary basins in mineral diagenesis. In: Gautier D L (ed) Roles of organic matter in sediment diagenesis. Soc Econ Paleontol Mineral Spec Publ 38, pp 111–123

Kharaka YK, Gunter WD, Aggerwaal PK, Perkins EH, DeBraal JD (1988) SOLMINEQ.88: a computer model for geochemical modeling of rock-water interactions. US Geol Surv Water Resources Investigations Rep 88-4227, 200 pp

Land LS (1984) Diagenesis of Frio Sandstones, Texas Gulf Coast: a regional isotopic study. In: McDonald DA, Surdam RC (eds) Clastic diagenesis. Am Assoc Pet Geol Mem 37, pp 37–62

Land LS, Milliken KL, McBride EF (1987) Diagenetic evolution of Cenozoic sandstones, Gulf of Mexico Sedimentary Basin. Sediment Geol 50: 195–225

Langmuir D (1979) Techniques for estimating thermodynamic properties of some aqueous complexes of geochemical interest. In: Jenne EA (ed) Chemical modeling of aqueous systems I. Am Chem Soc Symp Ser 93, Washington DC, pp 353–387

Lind CJ, Hem JD (1975) Effects of organic solute on chemical reactions of aluminum. US Geol Surv Water Supply Pap 1827-G, pp G1–G83

Lovgren L (1991) Complexation reactions of phthalic acid and aluminum (III) with the surface of goethite. Geochim Cosmochim Acta 55: 3639–3646

Lundegard PD, Kharaka YK (1990) Geochemistry of organic acids in subsurface waters: field data, experimental data and models. In: Melchior DC, Bassett RL (eds) Chemical modeling of aqueous systems II. Am Chem Soc Symp Ser 416, Washington DC, pp 169–189

Lundegard PD, Land LS (1989) Carbonate equilibria and pH buffering by organic acids-response to changes in pCO_2. Chem Geol 74: 277–287

MacGowan DB, Surdam RC (1988) Difunctional carboxylic acid anion in oil-field waters. Org Geochem 12: 245–259

Manning DAC, Rae EIC, Small JS (1991) An exploratory study of acetate decomposition and dissolution of quartz and Pb-rich potassium feldspar at 150°C, 50 mPa (500 bars). Mineral Mag 55: 183–195

Marley NA, Bennett P, Janecky DR, Gaffney JS (1989) Spectroscopic evidence for organic diacid complexation with dissolved silica in aqueous systems. I. Oxalic acid. Org Geochem 14: 525–528

Marlowe JI (1970) Weidellite in bottom sediment from the St. Lawrence and Saguenay Rivers. J Sediment Pet 40: 499–506

Martell AE, Smith RM (1977) Critical stability constants, vol 3. Other organic ligands. Plenum Press, New York, 495 pp

McDonald DA, Surdam RC (eds) (1984) Clastic diagenesis. Am Assoc Pet Geol Mem 37, 434 pp

Meshri ID (1986) On the reactivity of carbonic and organic acids and generation of secondary porosity. In: Gautier DL (ed) Roles of organic matter in sediment diagenesis. Soc Econ Paleontol Mineral Spec Publ 38, pp 123–129

Meshri ID, Walker JM (1990) A study of rock-water interaction and simulation of diagenesis in the Upper Almond Sandstones of the Red Desert and Washakie Basins, Wyoming. In: Meshri ID, Ortoleva PJ (eds) Prediction of reservoir quality through chemical modeling. Am Assoc Pet Geol Mem 49, pp 55–70

Moore CH, Ortoleva PJ (1990) Effects of fluid and rock compositions on diagenesis: a modeling approach. In: Meshri ID, Ortoleva PJ (eds) Prediction of reservoir quality through chemical modeling. Am Assoc Pet Geol Mem 49, pp 131–146

Nagy KL, Steefel CI, Blum AE, Lasaga AC (1990) Dissolution and precipiation kinetics of kaolinite: initial results at 80 °C with application to porosity evolution in a sandstone. In: Meshri ID, Ortoleva PJ (eds) Prediction of reservoir quality through chemical modeling. Am Assoc Pet Geol Mem 49, pp 85–102

Naumov GB, Nikitin AA, Naumov VB (1971) The origin of hydrothermal whewellite from fluorite veins in Transbaykalia. Geochem Int 1971: 107–112 (Translated from Geokhimiya 2: 180–186)

Nordstrom DK, Plummer LN, Wigley TML, Wolery TJ et al. (1979) A comparison of computerized chemical models for equilibrium calculations in aqueous systems. In: Jenne EA (ed) Chemical modeling of aqueous systems I. Am Chem Soc Symp Ser 93, Washington DC, pp 857–892

Nordstrom DK, Plummer LN, Langmuir D, Busenburg E, May HM, Jones BF, Parkhurst DL (1990) Revised chemical equilibrium data for major water-mineral reactions and their limitations. In: Melchior DC, Bassett RL (eds) Chemical modeling of aqueous systems II. Am Chem Soc Symp Ser 416, Washington DC, pp 398–413

Palmer DA, Drummond SE (1988) Potentiometric determination of the molal formation constants of ferrous acetate complexes in aqueous solutions to high temperatures. J Phys Chem 92: 6795–6800

Parkhurst DL, Thorstenson DC, Plummer NL (1980) PHREEQE – a computer program for geochemical calculations. US Geol Surv Water Resources Investigation 80-96, 210 pp

Reed MH (1983) Calculation of multicomponent chemical equilibria and reaction processes in systems involving minerals, gases and an aqueous phase. Geochim Cosmochim Acta 46: 513–528

Robie HA, Hemingway BS, Fisher JR (1978) Thermodynamic properties of minerals and related substances at 298.15 K and 1 bar (10^5 Pascals) pressure and at higher temperatures. US Geol Surv Bull 1452, 456 pp

Rose NM, Jackson KJ (1989) Computation of coupled aqueous organic-inorganic equilibria in hydrothermal systems. Geol Soc Am Annu Meet St. Louis, Abstr Program 21: A49

Seewald JS, Seyfried WE Jr (1991) Experimental determination of portlandite solubility in H_2O and acetate solutions at 100–350 °C and 500 bars: constraints on calcium hydroxide and calcium acetate complex stability. Geochim Cosmochim Acta 55: 647–658

Shock EL (1988) Organic acid metastability in sedimentary basins. Geology 16: 886–890

Shock EL (1990) Do amino acids equilibrate in hydrothermal fluids? Geochim Cosmochim Acta 54: 1185

Shock EL, Helgeson HC (1990) Calculation of the thermodynamic and transport properties of aqueous species at high pressures and temperatures: standard partial molal properies of organic species. Geochim Cosmochim Acta 54: 915–945

Stoessell RK, Pittman ED (1991) Secondary porosity revisited: the chemistry of feldspar dissolution by carboxylic acids and anions. Am Assoc Pet Geol Bull 74: 1795–1805

Sunda WG, Hanson PJ (1979) Chemical speciation of copper in river water: effect of total copper, pH, carbonate, and dissolved organic matter. In: Jenne EA (ed) Chemical modeling of aqueous systems I. Am Chem Soc Symp Ser 93, Washington DC, pp 147–180

Surdam RC, Boese SW, Crossey LJ (1984) The chemistry of secondary porosity in clastic diagenesis. In: McDonald D A, Surdam R C (eds) Clastic diagenesis. Am Assoc Pet Geol Mem 37, pp 127–150

Surdam RC, Crossey LJ, Hagan ES, Heasler HP (1989) Organic-inorganic interactions and sandstone diagenesis. Am Assoc Pet Geol Bull 73: 1–23

Thurman EM (1985) Organic Geochemistry of Natural Waters. Nijhoff Junk Dordrecht, 497 pp

Thyne GD (1992) The early diagenesis of the Cardium Sandstone Alberta and calculations of the effect of organic acids on clastic diagenesis. PhD Thesis, University of Wyoming, Laramie, Wyoming, 186 pp

Thyne GD, Harrison WJ (1991) Stability of aluminum oxalate aqueous complexes from 25–150 °C. Annu Meet Geol Soc Am Boulder, CO, Abstr Program, vol 23, San Diego, p A212

Thyne GD, Harrison WJ, Alloway MD (1992) Experimental study of the stability of the Al-oxalate complexation at 100 °C and calculation of the effects of the complexation in clastic diagenesis. In: Kharaka YK, Maest AS (eds) Water rock interaction, vol 1. Proc 7th Int Symp Water rock interactions. Balkema, Rotterdam, pp 353–357

White AF, Peterson ML (1990) Role of reactive-surface area characterization in geochemical knietic models. In: Melchior DC, Bassett RL (eds) Chemical modeling of aqueous systems II. Am Chem Soc Symp Ser 416, Washington DC, pp 461–477

Wolery TJ (1979) Calculation of chemical equilibria between aqueous solutions and minerals: the EQ3/6 software package. Lawrence Livermore National Laboratory Rep. UCRL-52658, 41 pp

Wolery TJ (1983) EQ3NR. A computer program for geochemical aqueous speciation-solubility calculations: user's guide and documentation. Lawrence Livermore National Laboratory Rep UCRL-53414, 189 pp

Wolery TJ, Daveler SA (1992) EQ6, a computer program for reaction path modeling of aqueous geochemical systems: theoretical manual, user's guide, and related documentation (version 7.0), MCRL-MA-110662, Pt. IV. Lawrence Livermore National Laboratory, Livermore, CA, 255 pp

Chapter 13   Organic Acids and Carbonate Stability, the Key to Predicting Positive Porosity Anomalies

Ronald C. Surdam[1] and Peigui Yin[1]

Summary

A systematic and sequential set of carbonate reactions characterizes many clastic source/reservoir rock systems during progressive burial. Typically, in its simplest form, this sequence of carbonate reactions with increasing thermal exposure is: (1) formation of early carbonate cements that preserve intergranular volume (IGV); (2) dissolution of early carbonate cements, enhancing porosity and resulting in positive porosity anomalies; (3) formation of late carbonate cements, again preserving IGV; and (4) if temperatures are high enough and if quartz cementation is inhibited, dissolution of late carbonate cements, again enhancing porosity.

If this sequence of carbonate reactions is coupled with parallel organic reactions, including generation and decarboxylation of organic acids and acid anions, a predictive, process-oriented model can be constructed for the carbonate reactions. The model consists of three operations: (1) interpretation of reaction pathways; (2) kinetic modeling of organic reactions; and (3) simulation of rock/water interactions in either time or temperature space. Integrating these three operations allows us to predict zones of carbonate dissolution or optimum porosity enhancement (positive porosity anomalies) in source/reservoir rock systems.

The sandstones of the Latrobe Group in the Gippsland Basin of Australia are characterized by early dolomite and late Fe-magnesite cements. The two cementation events were separated temporally by a significant carbonate dissolution event (early dolomite dissolution) that resulted in a zone of porosity enhancement, a positive porosity anomaly characterized by 30+% porosity and up to 2 darcies permeability, in a present-day depth interval from 1400 to 2600 m. The predictive, process-oriented diagnostic model described in this chapter predicts a significant positive porosity anomaly resulting from dolomite dissolution (dolomite undersaturation) within a present-day depth interval from 1200 to 2800 m. The predicted positive porosity anomaly is based on the assumption that organic acid/acid anions,

---

[1] Department of Geology and Geophysics, University of Wyoming, Laramie, Wyoming 82071, USA

derived from the thermocatalytic cleavage of oxygen-bearing functional groups from kerogen, from the reaction of water with kerogen during maturation, and from redox reactions between hydrocarbons and mineral oxidants, are a definitive component in the alkalinity of the formation fluids at temperatures greater than 80 °C.

1 Introduction

A typical chain, or series, of predictable, progressive mineral reactions occurs during continuous progressive burial in sandstone/shale successions (Fig. 1). Competition between key elements in this progressive reaction chain determines net porosity, and establishes positive or negative anomalies relative to normal compactional porosity loss. The most volumetrically significant of these competitions relative to porosity anomalies developing with increasing depth are (1) early mechanical deformation and grain rotation competing with early carbonate cementation for preservation of intergranular volume (IGV); (2) dissolution of early carbonate cements and feldspar/lithic grains competing with precipitation of kaolinite and quartz for porosity

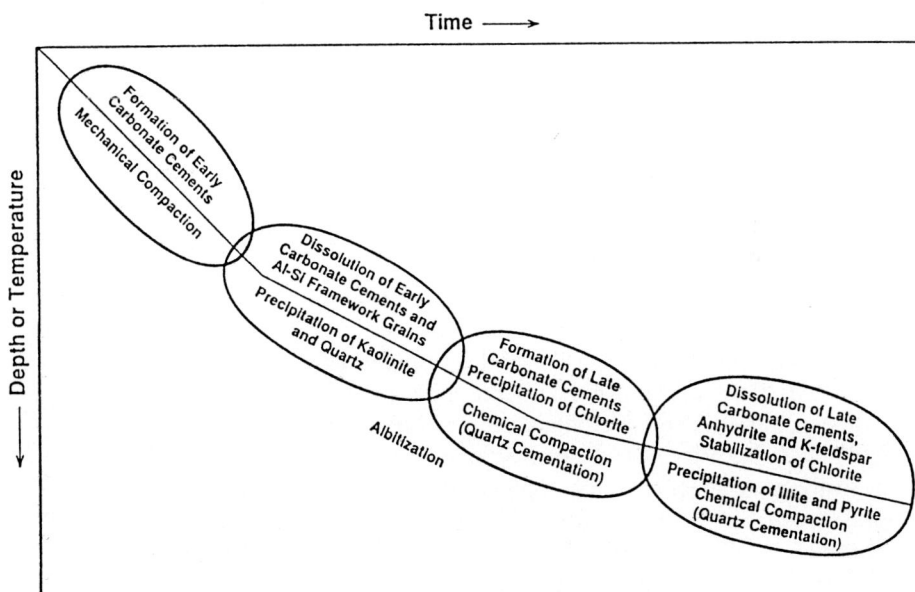

Fig. 1. Schematic diagram of the chain of diagenetic reactions that most significantly affect porosity development. The spatial sequence of these reactions is shown with increasing time and depth or temperature. Note that the individual links in the reaction chain overlap

enhancement; (3) formation of late carbonate cements and chlorite formation competing with quartz cementation and illite formation for preservation of IGV; and (4) dissolution of late carbonate and anhydrite competing with quartz cementation for deep porosity enhancement. Intergranular volume (IGV) as used in this chapter is equivalent to intergranular pore volume (i.e., IGV = intergranular pore space + intergranular cement). The outcome of this series of overlapping diagenetic events determines the porosity/permeability characteristics relative to compactional porosity loss (baseline) through time and through temperature increase. Positive porosity anomalies are primarily a function of carbonate, feldspar, chlorite, anhydrite, and lithic material stability in the various chemical and physical environments of progressive burial.

We suggest that, acting with this series of progressive mineral reactions, there are parallel, serial organic reactions that exert significant control on the inorganic reaction processes by controlling formation water alkalinity and aqueous speciation and thus the inorganic mineral stabilities. These organic reactions govern the mineral stabilities because they (1) determine the alkalinity of the fluid phase, (2) establish pH buffers, and (3) complex metals, allowing key components in the system to become mobile. These three effects are critical in establishing the stability of both carbonates and aluminosilicates at any particular position in time and temperature space.

The most important links volumetrically in the reaction chain with respect to developing zones of enhanced porosity – positive porosity anomalies – are the carbonate cementation/decementation events (Fig. 1). The sequence of carbonate cementation/decementation events shown schematically in Fig. 1 has been noted by many investigators (see especially McBride 1977). Recent work includes Boles (1987), Taylor (1990), Moraes (1989), Yin and Surdam (1985), Jansa and Urrea (1990), and Surdam et al. (1989c,d). Although this list is by no means exhaustive, it does demonstrate that this series of progressive carbonate reactions has been observed in basins around the world by many different investigators.

The formation of carbonate cements typically preserves the intergranular volume (IGV) present at the time of cement formation; with subsequent progressive burial, the IGV will be anomalously high. If the carbonate cement is later destabilized and the carbonate is dissolved, the sandstone will be characterized by enhanced porosity, a positive porosity anomaly. Enhanced porosity, or a positive porosity anomaly, is defined herein as porosity at a specific depth that is greater than the porosity predicted at that depth by a compaction/depth curve for the corresponding sandstone lithology (see Sclater and Christie 1980; Pittman and Larese 1991).

The problems to be addressed in this chapter are: (1) the role that organic acids play in porosity evolution during progressive burial and (2) the predictability in time-temperature space of the sequence of carbonate cementation and decementation events frequently observed in reservoir sandstones. If the carbonate reaction pathway can be systematically predicted, it will reduce exploration risk by allowing the determination of windows of optimum reservoir conditions in sandstone/shale systems.

We will evaluate these problems in a series of steps:

1. We will discuss the processes determining carbonate mineral stability in each of the progressive burial zones (shallow, intermediate, and deep; for a more detailed description of these zones, see Surdam et al. 1989d).
2. We will determine if there are systematic and predictable chemical relationships between the vertically stacked carbonate cementation/decementation zones, and the relationship between organic acids and carbonate mineral stability.
3. We will briefly describe the modeling technology developed by Surdam et al. (1989a,b,c,d) and Surdam and MacGowan (1992) to predict the spatial distribution of carbonate stability in time-temperature space.
4. We will apply and test this predictive technology in a basin where the sequential carbonate reactions have been well documented (Bodard et al. 1984; Yin and Surdam 1985; Yin 1988): the Gippsland Basin, Australia.
5. We will evaluate the relationship between carbonate dissolution events, porosity enhancement, and organic acids.

2 General Principles

2.1 Carbonate Cementation During Shallow Burial (Temperature < 80 °C)

Carbonate cementation of sandstone during shallow burial is a common diagenetic phenomenon. Sandstones with early carbonate cement are typically characterized by relatively high IGV values and by relatively low isotopically determined temperatures of formation. If IGV is plotted versus depth for a sandstone characterized by early carbonate cementation and subsequent burial, the IGV will be greater than that predicted by a normal compaction/depth curve (Sclater and Christie 1980). As a result of early carbonate cement, a significant proportion of the original fabric (porosity) can survive compactional effects. If IGV is to become converted to enhanced porosity, the carbonate must be removed during subsequent burial.

The possibility of wholesale removal of pore-filling carbonate cement from sandstones has been postulated by many authors, most notably Schmidt and McDonald (1979). Ideally, the most significant carbonate cements are those that incompletely fill pore space, or that line pores; for these cements lend strength to the rock, while maintaining some permeability. It has been suggested by many that the potential for enhanced porosity due to decementation is strongly related to the distribution of early cements.

Carbonate cements appear to be volumetrically the most important of the possible early sandstone cements (Berner 1980; Gautier 1985; Curtis and

Coleman 1986, among others). The reason for the high frequency of early carbonate cements in sandstone/shale sequences is that early burial optimizes several organic and inorganic processes whose interaction results in carbonate precipitation.

Volumetrically important early carbonate cements commonly form in association either with the zone of bacterially mediated sulfate reduction or with the zone of microbial methanogenesis, and sometimes with both (Curtis and Coleman 1986). Sulfate reduction typically occurs in fine-grained sediments to a depth of approximately 10 m, but may occur much deeper in coarser-grained sandstones due to the migration of sulfate-charged water into deeper pore-water systems. Bacterially mediated sulfate reduction,

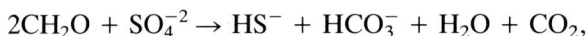

$$2CH_2O + SO_4^{-2} \rightarrow HS^- + HCO_3^- + H_2O + CO_2,$$

is limited by the availability of SO_4^{-2} and organic material. Microbial methanogenesis,

$$2CH_2O \rightarrow CH_4 + CO_2,$$

typically occurs down to a depth of approximately 1000 m.

The introduction of CO_2 into pore water as a result of oxidation of organics, sulfate reduction, or methanogenesis does not result in the precipitation of carbonate minerals. As Curtis (1987) points out, the resulting drop in pH due to CO_2 dissolution and H_2CO_3 dissociation,

$$CO_2 + H_2O \rightarrow H_2CO_3 \rightarrow H^+ + HCO_3^-,$$

will favor carbonate mineral dissolution,

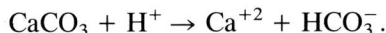

$$CaCO_3 + H^+ \rightarrow Ca^{+2} + HCO_3^-.$$

In sand/shale sequences, significant carbonate mineral precipitation requires a parallel reaction that *consumes hydrogen ions, or produces proton-consuming species* (Curtis 1987). Two such reactions commonly observed at shallow burial are Fe/Mn reduction by organic matter oxidation (Curtis 1987), for example,

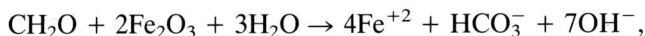

$$CH_2O + 2Fe_2O_3 + 3H_2O \rightarrow 4Fe^{+2} + HCO_3^- + 7OH^-,$$

and silicate hydrolysis (Hay 1963), for example

$$4KAlSi_3O_8 + 22H_2O \rightarrow 4K^+ + 4OH^- + 8H_4SiO_4 + Al_4[Si_4O_{10}](OH)_8,$$

both of which give net hydroxyl production or proton consumption. Thus, provided a Ca^{+2} source is available, in an environment where either sulfate reduction or methanogenesis is accompanied by Fe/Mn reduction or silicate hydrolysis, there is a high probability that a carbonate phase will precipitate. These coupled reactions can be shown as follows:

1. Sulfate reduction:

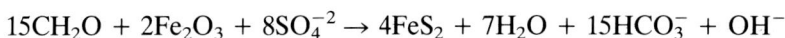

$$15CH_2O + 2Fe_2O_3 + 8SO_4^{-2} \rightarrow 4FeS_2 + 7H_2O + 15HCO_3^- + OH^-$$

or

$$4KAlSi_3O_8 + 2CH_2O + SO_4^{-2} + 3H^+ + 18H_2O$$
$$\rightarrow 2HCO_3^- + HS^- + 8H_4SiO_4 + Al_4[Si_4O_{10}](OH)_8 + 4K^+.$$

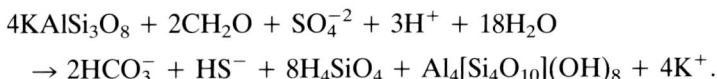

Pyrite precipitation is accompanied by a significant increase in bicarbonate concentration in the pore water. As a result, Fe-poor carbonate (calcite and dolomite) will form with insoluble Fe sulfides.

2. Methanogenesis:

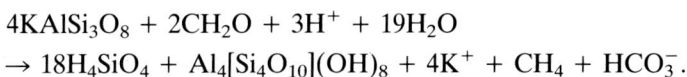

$$7CH_2O + 2Fe_2O_3 \rightarrow 3CH_4 + 4FeCO_3 + H_2O$$

$$4KAlSi_3O_8 + 2CH_2O + 3H^+ + 19H_2O$$
$$\rightarrow 18H_4SiO_4 + Al_4[Si_4O_{10}](OH)_8 + 4K^+ + CH_4 + HCO_3^-.$$

If the sequence enters a zone characterized by microbial methanogenesis, a combination of methanogenesis and Fe reduction is highly conducive to the precipitation of Fe-rich carbonates (Curtis 1987).

Thus, as a sandstone passes through early burial environments characterized by sulfate reduction or methanogenesis, Fe/Mn reduction or silicate hydrolysis, and a source of calcium, there is ample opportunity for the precipitation of carbonate minerals. Sulfate reduction and methanogenesis during shallow burial are inevitable processes if the system contains bacteria and organic material. Likewise, Fe/Mn reduction and silicate hydrolysis are common processes in early burial (Berner 1980). It would be a rare sandstone/shale system that during shallow burial would not have the necessary components for sulfate reduction or methanogenesis and for Fe/ Mn reduction or silicate hydrolysis. Thus, the key to the formation of early carbonate cements typically may be the availability of Ca^{+2} and Mg^{+2}.

Several obvious sources of Ca^{+2} in sandstone/shale systems during shallow burial are: (1) influx of sea or meteoric water; (2) in situ hydrolysis reactions involving Ca-Al silicates; (3) dissolution of intercalated shelly material; and (4) upward migration of pore water from deeper zones of carbonate or Ca-plagioclase dissolution. All of these mechanisms of calcium transport, and perhaps other processes as well, are operative during shallow burial. As for the question of which mechanisms are volumetrically important in the formation of early carbonate cements, studies of the 87/86 Sr contents of early carbonate cements in the San Joaquin Basin of California strongly support the idea of Ca^{+2} imported from a deeper source (plagioclase dissolution) to form early carbonate cements in the Stevens Sandstone (Schultz et al. 1989). In other sandstones, it is probable that local hydrolysis of framework grains or dissolution of shelly material is the chief source of Ca^{+2} for early carbonate cementation. In some cases no carbonate cement will form, but in most sandstones some early carbonate cement occurs. The ideal setting to maximize early carbonate cements is to have organic-rich fine-grained sediments (precursor to source rocks) and coarse-grained sand (precursor to reservoir rocks) in close proximity.

The importance of the spatial relations between fine-grained, organic-rich sediments, sands, and early carbonate cements is neatly illustrated by the recent work of McMahon and Chapelle (1991). McMahon and Chapelle have shown that rates of microbial fermentation exceed rates of respiration (sulfate reduction) in organic-rich aquitards (shale precursors), resulting in a net accumulation of organic acids in pore waters. In contrast, in adjacent aquifers (sandstones) microbial respiration exceeds fermentation, resulting in a net consumption of organic acids. They suggest that this imbalance results in a concentration gradient that drives a flux of organic acids from organic-rich, fine-grained sediment to adjacent coarse-grained sand. Moreover, they conclude that this imbalance links the large pool of sedimentary organic carbon in aquitards (mudrock) to microbial respiration in aquifers (sandstone), and is a significant mechanism driving groundwater chemistry changes in aquifers (i.e., shallow burial diagenesis). This chemical link between organic-rich aquitards and aquifers provides a carbon source for early carbonate cements in sandstones.

During shallow burial in the Black Creek Formation of South Carolina, microbial processes have resulted in the formation of 10 vol% calcite cement in sands (McMahon et al. 1991). This distribution of the calcite cement in the sand appears to be related to grain size and porosity, the cement decreasing away from porous zones or away from the zone of active fluid movement (McMahon et al. 1991). Isotopic data suggest that all the organic carbon in the carbonate cements in the Black Creek Formation sands can be accounted for by microbial processes (McMahon et al. 1991).

2.2   Intermediate Burial (80 to 130 °C)

With continuous and progressive burial, a sandstone enters the intermediate burial zone at approximately 80 °C (zone of intense diagenesis, ~80 to ~110 °C; Table 1). In this interval, a potential hydrocarbon reservoir accommodates important porosity enhancing reactions: early-formed carbonate cements may be dissolved (or later carbonate cements inhibited), and aluminosilicate framework grains (both feldspar and lithic) may be dissolved. Thus, in this zone, porosity can be preserved or significantly enhanced.

Several porosity-destroying reactions also characterize this zone. For example, late ferroan carbonate cements are potential reaction products in this zone, and may significantly occlude porosity (Boles 1978). Also, there is a wide variety of aluminosilicate reaction products (e.g., kaolinite, illite, chlorite, and quartz) that can form in this zone as a result of aluminosilicate dissolution. The imbalance between porosity-enhancing or-preserving reac-

Table 1. Observed sequence of carbonate cementation/dissolution events in clastic units during continuous burial, and suggested causes and effects

Burial[a]	Master process (cause)	Result (effect)
Shallow (sediment/ water interface to ~80 °C	Sulfate reduction/methanogenesis in presence of H^+ consuming or OH^- generating reaction (i.e., Fe/Mn reduction or silicate hydrolysis)	Early carbonate Cement
Intermediate (~80 °C to ~110 °C)	Organic acid anion buildup with relatively low P_{CO_2}	Carbonate dissolution
Intermediate (~110 °C to ~130 °C)	Decarboxylation of organic acid anions (i.e., alkalinity dominated by organic acid anions with elevated P_{CO_2})	Late carbonate Cement
Deep (>130 °C)	Decarboxylation and/or abiotic thermal sulfate reduction	Carbonate dissolution

[a] Suggested temperatures are ±10 °C.

tions and porosity-destroying reactions determines the net porosity gain or loss in a clastic reservoir facies as it moves through the intermediate burial zone. In many clastic reservoir facies, there is no demonstrable *net* gain in porosity due to dissolution. Since in some reservoir facies, mass is conserved in the sandstone interval, some portions of the potential reservoir may undergo extensive dissolution (porosity enhancement), and other portions the precipitation of dissolution products (porosity destruction).

The magnitude of porosity enhancement due to aluminosilicate grain dissolution in a reservoir and source-rock system depends on facies relationships, variations in original composition, formation of subsequent cements, availability of fluid conduits, fluid flux, and the proximity of organic-rich source rocks in hydrologic connection with the reservoir rock. In contrast, carbonate decementation and mass transfer apparently can occur on a scale larger than a specific reservoir and source-rock system (Schultz et al. 1989).

We believe that the probability of significantly enhancing effective porosity (generating a positive porosity anomaly) during burial diagenesis is largely a function of carbonate mineral stability.

2.2.1 Alkalinity in the Zone of Intermediate Burial

Figure 2 suggests that the alkalinity of the fluid phase in the zone of intermediate burial is determined by a very different set of processes from those characterizing the zone of shallow burial. It is suggested that the intermediate burial zone is characterized by the following conditions: (1) the alkalinity of the fluid phase in connate waters typically is dominated by carboxylic acid/acid anions over the 80 to 130 °C (±10 °C) temperature

Porosity

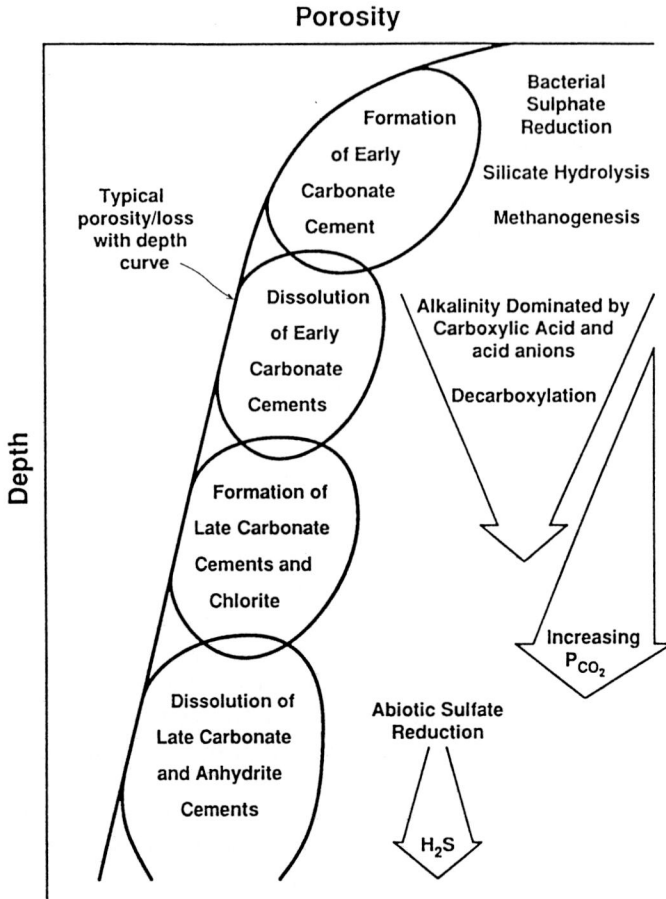

Fig. 2. Schematic diagram showing (1) a typical baseline porosity curve, (2) the chain of mineral reactions that are capable of most significantly chemically enhancing porosity, and (3) the parallel and serial organic and inorganic reactions that control alkalinity and buffer the fluid pH

interval; (2) as the sandstone progresses through this intermediate burial zone, the ratio of CO_2 to carboxylic acid/acid anions increases; (3) as long as carbonate minerals in the sandstone remain in the 80–130 °C temperature interval, the alkalinity of the fluid phase is buffered by the carbonate-organic anion system; and (4) with the addition of hydrocarbons to the fluid phase over the intermediate burial temperature interval, the fluid-flow system evolves from a single-phase (H_2O) to a multiphase (H_2O-oil or H_2O-oil-gas) system, with the result that capillarity dominates the fluid-flow system over this temperature interval. In addition, with the influx of hydrocarbons into the reservoir system, the chemical environment becomes highly reducing.

2.2.2 Aluminosilicate and Aluminosilicate-Carbonate Reactions and pH Buffers

Recently, much attention has been given to "reverse-weathering" reactions (kaolinite + $K^+ \rightarrow$ illite + H^+) and aluminosilicate-carbonate reactions (of the type kaolinite + Fe-carbonate $\rightarrow$ chlorite + calcite + CO_2) as possible pH buffers in formation waters (Hutcheon and Abercrombie 1989; Smith and Ehrenberg 1989; Milliken and Land 1991). There is no question that protons are generated in shales by reverse-weathering reactions such as kaolinite $\rightarrow$ illite, or kaolinite reacting to chlorite,

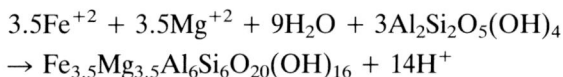

$$3.5Fe^{+2} + 3.5Mg^{+2} + 9H_2O + 3Al_2Si_2O_5(OH)_4$$
$$\rightarrow Fe_{3.5}Mg_{3.5}Al_6Si_6O_{20}(OH)_{16} + 14H^+$$

(Boles and Franks 1979). However, it is our position that although these relatively slow aluminosilicate reactions contribute to the alkalinity, they *do not buffer* the alkalinity until all relatively fast-reacting minerals such as carbonates are eliminated from the fluid/rock *system*. It is the weak acids and bases (dominately carboxylic acid anions, HCO_3^-, and HS^-) in formation waters, coupled with rapidly reacting minerals, such as the carbonate minerals, that buffer formation water pH.

Clay-carbonate reactions also have been suggested as playing an important role in buffering formation-water pH. These reactions typically are of the type kaolinite + Fe-carbonate $\rightarrow$ chlorite + calcite (Smith and Ehrenberg 1989). Muffler and White (1969) suggested a similar reaction for the formation of chlorite in the low-grade metamorphic environment of the Salton Sea trough (dolomite + ankerite + kaolinite + $Fe^{+2} \rightarrow$ chlorite + calcite + CO_2). In the Texas Gulf Coast, Boles and Franks (1979) specifically searched for this type of reaction to explain the chlorite in deeply buried (>200 °C) Wilcox rocks and rejected the idea on the grounds that (1) calcite is restricted to shallowly buried rocks (<2300 m); (2) ankerite is associated with chlorite in deeply buried rocks, there being no apparent net decrease in ankerite and concomitant increase in chlorite; and (3) there is sufficient illite/smectite and kaolinite to account for the formation of chlorite as a result of the reaction

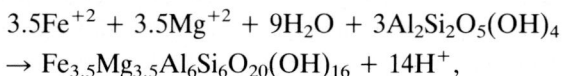

$$3.5Fe^{+2} + 3.5Mg^{+2} + 9H_2O + 3Al_2Si_2O_5(OH)_4$$
$$\rightarrow Fe_{3.5}Mg_{3.5}Al_6Si_6O_{20}(OH)_{16} + 14H^+,$$

with the Fe^{+2} and Mg^{+2} provided by the smectite $\rightarrow$ illite transition. Therefore, although the "reverse-weathering" and clay-carbonate reactions contribute to the alkalinity, they do not typically *buffer* the pH of the fluid in a source-reservoir rock system.

We offer as an alternative the suggestion that in the zone of intermediate burial, the alkalinity of the fluid phase and the buffering of pH typically are determined by organic-inorganic reactions. It is our observation and the observation of others in studying formation water chemistry that over the 80–130 °C temperature interval the alkalinity is dominated by the inter-action of carboxylic acid/acid anions and the carbonate system, and that pH buffering is a function of this interaction. An important exception would occur if there were significant mixing of fluids caused by an influx of meteoric water or hot deep brines. Therefore, we exclude polyhistory basins, and assume that in the rocks we are discussing there has been no significant mixing of connate and meteoric waters after the rocks have reached a depth equivalent to about 80 °C, and that there has been no subsequent mixing of connate waters with hot brines derived from significantly deeper portions of the basin.

The recent experimental work of Reed and Hajash (1992) on the dissolu-tion of granitic sand supports our suggestion that in the zone of intermediate burial, the alkalinity of the fluid phase and buffering of pH are functions of organic-inorganic reactions. Reed and Hajash conclude from their experi-ments that: "although the buffering capacity of granitic sand is apparently large, organic species such as aqueous ammonium acetate/acetic acid in concentrations typical of formation waters are effective pH buffers, at least over the time frame of these experiments."

2.2.3   Carboxylic Acid/Acid Anion Sources

The sources of carboxylic acids and acid anions (organic acid anions, OAA) in the zone of intermediate burial are manifold (MacGowan and Surdam 1990a,b). The three most important processes that result in the generation of OAA are (1) thermal maturation of kerogen in the source rocks (see Surdam and Crossey 1985a,b); (2) redox reactions involving mineral oxidants and kerogen (i.e., clay diagenesis; see Crossey et al. 1986), and (3) redox reactions involving hydrocarbons and mineral oxidants (Surdam et al. in press).

The behavior of various kerogen types during burial and increasing ther-mal exposure indicates that oxygen-bearing functional groups are among the first products to be cleaved from kerogen (Fig. 3). Initially, with increasing thermal exposure, the atomic oxygen/carbon ratio of kerogen changes more sharply than the atomic hydrogen/carbon ratio (Fig. 3). Both oxidation experiments that reconstruct kerogen structures and nuclear magnetic re-sonance (NMR) analyses of kerogen structures demonstrate that a significant proportion of the peripheral oxygen-bearing functional groups that are cleaved from kerogen during diagenesis (prior to or concomitant with

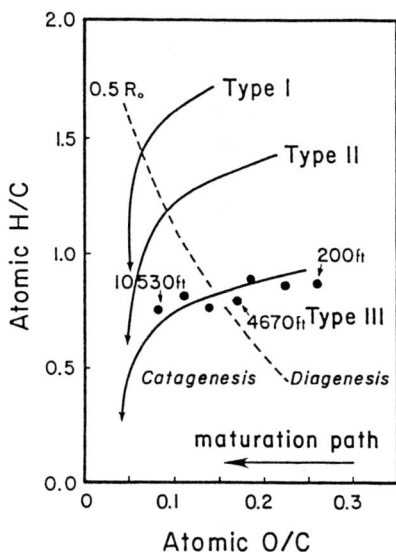

Fig. 3. Van Krevelan diagram outlining maturation pathways of Types I, II, and III kerogens. The *dashed line* represents onset of liquid hydrocarbon generation. Samples are from the Latrobe Group, Gippsland Basin, Australia

hydrocarbon generation) are carboxylic (Surdam et al. 1989a). Thus, thermocatalytic cleaving of oxygen-bearing functional groups from kerogen is an important source of water-soluble OAA. (For a detailed discussion of the importance of thermal maturation of kerogen as a source of OAA, see Surdam et al. 1989d.)

Crossey et al. (1986) examined mineral oxidants as possible agents in the production of carboxylic acids from kerogenous material during the progressive burial of reservoir-source rock systems. In particular, the diagenetic conversion of smectite to illite was quantitatively evaluated as an oxidative agent using Gulf Coast petrologic data, chemical data, and experimental results. The smectite to illite transition is important because the maximum organic acid concentrations in interstitial waters in sandstones spatially overlaps the conversion from random to allevardite-type ordering in the mixed-layer clay minerals in adjacent shales (in the range 80 to 110 °C). Upon conversion from random to allevardite-type ordering in the mixed-layer clay minerals, there is a net release of iron from octahedral sites (Hower et al. 1976; Boles and Franks 1979). Whether the iron is released and then reduced, or reduced within the clay structure itself and released as ferrous iron (Almon 1974), an electron transfer occurs. The organic matter dispersed in the shale is an ideal electron donor for redox reactions, and the most likely oxidation product is a carboxyl group (Almon 1974). Chemical analyses of clays from the Texas Gulf Coast show a decrease in iron content from 7 to 4% at the depth where the mixed-layer clays go through the random to ordered transition (Crossey et al. 1986). Diagenetic minerals containing ferrous iron should show a complementary trend. Thus, it is not surprising that chlorite and ankerite are two commonly observed ferrous minerals in sandstones associated with shales containing ordered mixed-layer clay minerals (Boles and Franks 1979; Moncure et al. 1984).

The data presented by Eglinton et al. (1987) illustrate the importance of redox reactions in generating organic acids. They heated kerogen in contact with limonite (a mineral oxidant) and showed that the presence of the mineral oxidant significantly increased the amount of organic acids produced. It was shown that the amount of carboxylic acids generated in the presence of the mineral oxidant was increased threefold relative to the amount produced by kerogen alone (Eglinton et al. 1987).

Not only does mixed-layer clay diagenesis serve as a potential mineral oxidant, it also provides a significant increase in fluid flux. If the transfer or redistribution of material is to occur, some type of fluid flux is necessary to transport the water-soluble organic acids and their dissolved or chelated load. Redox reactions involving clay minerals and kerogen may be a source of carboxylic acids/acid anions as significant as the thermocatalytic cleaving of oxygen-bearing functional groups from kerogen. It is noteworthy that thermocatalytic cleavage and redox involving mineral oxidants are complementary (both are potential sources of copious quantities of organic acids) and that the processes overlap in time-temperature space.

Redox reactions involving mineral oxidants and hydrocarbons can be a volumetrically important source of OAA (Surdam et al. in press). Geologists have long known that when hydrocarbons invade red sandstones, significant bleaching (iron reduction) takes place. The reaction responsible for the color distribution in the red (oxidized) and white (reduced) zones is the reaction of iron oxides ($\pm$ sulfate) with hydrocarbons. The iron oxides ($\pm$ sulfate) oxidize the hydrocarbon (reductant) to oxygenated organic compounds (OAA), while the Fe_2O_3 (oxidant) is reduced by hydrocarbons to pyrite ($\pm$ chlorite) (see Levandowski et al. 1973). The results of these redox reactions are most obvious in red sandstones associated with hydrocarbon accumulations. However, other reservoir lithologies can be significantly affected, although the results are more subtle. All that is required for significant quantities of OAA to be generated is an oxidant source (hematite, sulfates, etc.) and hydrocarbons (reductant source). Potential mineral oxidants other than hematite and sulfates include Fe^{+3} released during clay diagenesis (Crossey et al. 1986) and Fe^{+3} released during the dissolution of framework grains (e.g., glauconite). Any mineral oxidant that reacts with hydrocarbons in the 80 to 130 °C thermal window can produce volumetrically significant quantities of OAA in the intermediate burial zone (Surdam et al. in press). Most important, the redox reactions of kerogen or hydrocarbons with mineral oxidants to produce OAA are not constrained by the amount of oxygen in the kerogen. For a description of the relationship between redox reactions involving hydrocarbons and mineral oxidants and porosity enhancement, see Surdam et al. (in press).

A fourth process to generate OAA has been suggested recently by Lewan (1992). Lewan, interpreting the differences between kerogen reaction products in anhydrous and hydrous pyrolysis experiments, has suggested that H_2O may act as a source of hydrogen during kerogen maturation. He suggests that free radical sites are frequently terminated by water-derived hydrogen during kerogen maturation. Evidence supporting this suggestion includes mass balance calculations that show that the oxygen content of CO_2 generated under maturation simulations exceeds oxygen loss from the original organic matter: the only source for the excess oxygen is the accompanying water. Lewan's working hypothesis is that oxygen from water dissolved in the organic network of the source rock reacts with existing carbonyl groups to form carboxyl groups and *subsequent* CO_2 through the decarboxylation of the carboxyl groups with increasing temperature. In addition, he proposes that hydrogen released from the reacted water is avaiable to terminate free radical sites.

The work of Lewan (1992) suggests a way to identify the maximum potential amount of OAA that can be derived from kerogen during thermal maturation. For example, the sum of the organic acid anions/acids and CO_2 in a hydrous pyrolysis experiment may be equal to the total OAA generative capacity of a particular source rock. This proposition assumes that all of the CO_2 generated in the experiment results from the decarboxylation of OAA. In our own hydrous pyrolysis experiments, using an experimental design similar to Lewan's and using coal as a hydrocarbon source rock, we generated OAA + CO_2 values between 5 and 17 wt % of the solid (Table 2). Obviously, if water and kerogen react during thermal maturation, as Lewan proposes, there is an additional important process that can generate significant quantities of OAA and that is not constrained by the oxygen content of kerogen.

Table 2. Hydrous pyrolysis experiments on coals

	Almond coal (360 °C) (72 h)	Lance coal (360 °C) (72 h)	Brown coal (360 °C) (144 h)	Indo. coal[a] (360 °C) (72 h)
L. HC. (wt%)	2.11	2.32	4.90	2.48
T. HC. (wt%)	3.84	3.99	6.36	4.05
CO_2 (wt%)	8.50	8.59	2.66	11.87
CAA (wt%)	0.76	0.90	1.12	0.59
Max. CAA (wt%)	12.16	12.41	4.81	16.78

[a] Indo. coal = Indonesian coal; L. HC. = liquid hydrocarbons; T. HC. = total hydrocarbons; CAA = carboxylic acid anions; Max. CAA = maximum potential for generation of carboxylic acid ions (OAA + CO_2).

For source-reservoir rock systems in the 80 to 130 °C temperature interval, four important and overlapping sources of carboxylic acids/acid anions in the fluid phase have been identified. All four processes are capable of generating significant quantities of water-soluble carboxylic acids/acid anions. Thus, it is not surprising that the alkalinity of the fluid phase in source/reservoir rock systems in the 80 to 110 °C (±10 °C) temperature interval typically is dominated by OAA.

On the basis of the oxygen content of kerogen, Lundegard et al. (1984), Bjørlykke (1984), and Giles and Marshall (1986) have questioned thermocatalytic cleaving of kerogen as a source of significant amounts of oxygen-bearing functional groups (organic acids, CO_2, etc.). However, it should be emphasized that thermocatalytic cracking of kerogen is not the only source of OAA to the diagenetic system. It is important to note that whatever the source, significant quantities of OAA are observed in many oil-field waters in the 80 to 130 °C temperature interval.

2.2.4   Carbonate Stability

Our research has shown us that evaluating the stability of carbonate minerals during progressive burial simply in terms of pH and the aqueous species Ca^{+2}, H_2CO_3, HCO_3^-, CO_3^{-2}, and CO_2 is unrealistic. Under that evaluation, alkalinity is attributed solely to these aqueous carbonate species, and carbonate stability in the subsurface is assumed (erroneously) to be solely a function of P_{CO_2} and the concentrations of Ca, Mg, and Fe (e.g., as P_{CO_2} increases, the solubility of calcite increases; Holland and Borcsik 1965).

However, over the temperature range 80 to 110 °C (±10 °C), where organic acid anion concentration is maximum (MacGowan and Surdam 1990a,b), the organic acid anions dominate alkalinity (Willey et al. 1975; Carothers and Kharaka 1978) and significantly affect carbonate stability (Surdam and Crossey 1985a,b; Meshri 1986). The capacity of acetate to buffer pH is much greater than that of any aqueous carbonate species. Over the 80 to 110 °C (±10 °C) temperature range, the pH of the water is buffered by carboxylic acid anions. Acetate is commonly the most abundant organic acid anion in formation waters, and its maximum buffering capacity is at a pH of 5.4 (Martell and Smith 1977; Carothers and Kharaka 1978).

As the formation temperature increases in the zone of intense diagenesis (80 to 130 °C), the concentration of organic acid anions in the formation water increases to the maximum concentration at which the carbonate system is externally buffered (by acetate). At temperatures greater than 100 °C the carboxylic acids, particularly the difunctional species, begin to decarboxylate:

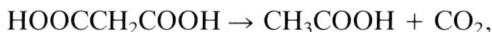

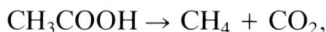

$$HOOCCH_2COOH \rightarrow CH_3COOH + CO_2,$$

$$CH_3COOH \rightarrow CH_4 + CO_2,$$

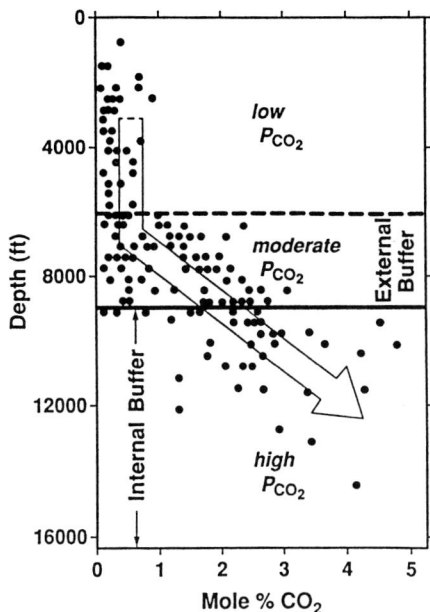

Fig. 4. Mol% CO_2 in gas vs. depth for the Wilcox Formation, Texas Gulf Coast. *Horizontal lines* separate zones considered to have low, moderate, and high P_{CO_2} values. *Large arrow* represents general trend. Internal and external buffers refer to carbonate system: the increasing CO_2 content with depth results from (1) generation of organic acid anions, (2) progressive decarboxylation and carbonate dissolution (external buffers), and (3) reestablishment of the carbonate system as the dominant aspect of alkalinity (internal buffer). (After Franks and Forester 1984)

resulting in increased P_{CO_2}. At some temperature in the interval 120 to 160 °C, the carboxylic acid anions completely decarboxylate and the alkalinity is again dominated by the carbonate system (internally buffered) (Surdam et al. 1984). Implicit in the decarboxylation process is the progressive elevation of formation-water P_{CO_2}. The increasing CO_2 content of produced gases with depth in the Gulf Coast described by Franks and Forester (1984) can be explained by a scenario that includes (1) the generation of organic acid anions; (2) the subsequent, progressive decarboxylation of organic acid anions with concomitant increase in P_{CO_2}; and (3) at temperatures greater than 120 °C, the reestablishment of the carbonate system as the dominant aspect of alkalinity (Fig. 4).

The evolution of formation water alkalinity, then, can be generalized in the following scenario. Over the 80 to 110 °C interval, the carbonate system will be externally buffered and P_{CO_2} will be relatively low. As decarboxylation begins (at about 100 °C), the system remains externally buffered with carboxylic acid anions as P_{CO_2} increases due to decarboxylation and carbonate dissolution. At a temperature somewhere between 120 and 160 °C,

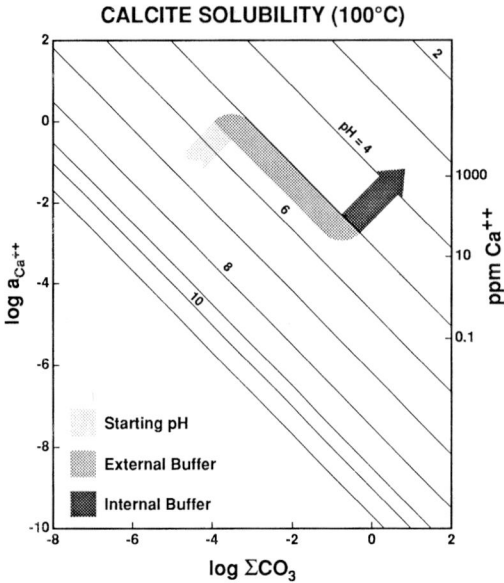

Fig. 5. Log ΣCO_3 vs. calcite solubility (as Ca^{2+} concentration); *stippled pattern* shows the evolution of fluid alkalinity (internal vs. external buffer) with increasing burial. $\Sigma CO_3 = H_2CO_3 + HCO_3^- + CO_3^{-2}$. (Surdam et al. 1984)

the buffering capacity of the carboxylic acid anions will be lost as their concentration is diminished. This scenario is depicted in Fig. 5. Initially, at temperatures greater than 80 °C, the carbonate system is externally buffered by carboxylic acid anions. For example, if the pH is externally buffered at a pH of 5.4, the solution composition is confined to a diagonal line (pH 5.4) in Fig. 5. At this point, carbonate solubility may be high, and carbonate dissolution (or lack of carbonate precipitation) will result. Carbonate dissolution will inevitably result in increasing P_{CO_2}. As the temperature increases (resulting in decarboxylation and elevated P_{CO_2}), but with the system still externally buffered (and thus confined to the pH = 5.4 line), the system will move to decreasing carbonate solubility, and the carbonate minerals will become stable. It is important to note that in Fig. 5, P_{CO_2} increases from left to right along the abscissa, as does ΣCO_3. At the temperature where the carboxylic acid anion concentration is diminished (from 120 to 160 °C), the carbonate system becomes internally buffered and the system will no longer be confined to a constant pH line. As P_{CO_2} increases, the system will move toward lower pH and increasing carbonate solubility (calcite dissolution or lack of precipitation; Fig. 5).

The discussion above is for a rock-fluid system containing carbonate minerals; in the absence of carbonate minerals, the alkalinity and buffering capacity of the fluid system will be controlled by other mechanisms, such as aluminosilicate reactions.

The competition described above between carbonate species and organic acid anions in buffering formation water pH is the general path that the carbonate system will take through the 80 to 130 °C thermal window. Individual cases may vary in detail, depending on such variables as Fe^{+2} and Mg^{+2} concentration and sulfate reduction – hydrocarbon oxidation. Although the Al produced from aluminosilicate dissolution appears to be transported only relatively short distances (perhaps up to hundreds of meters), the Ca produced during dissolution can be transported on the order of 1 km (Schultz et al. 1989). Calcium produced as the result of plagioclase and carbonate dissolution in the intermediate burial zone may be transported up-section, and may serve as a source of Ca^{+2} for carbonates precipitating in the zone of shallow burial (the zone methanogenesis or sulfate reduction). Schultz et al. (1989), working in the San Joaquin Basin, believe that the ultimate source of calcium in the carbonate cements is plagioclase dissolution.

2.2.5 Aluminosilicate Mineral Stability

Aluminosilicate framework grain dissolution is commonly associated with carbonate dissolution events in the temperature interval 80 to 110 °C. Surdam et al. (1984, 1989a,b,c,d) and Surdam and Crossey (1985a,b) have stressed the role of organic acids in elevating aluminum and other metal solubilities in this temperature range. They have suggested that the increased mobility of metals and the increased solubility of aluminosilicate minerals associated with this temperature range are functions of organometallic complexation with organic acids and acid anions.

This suggestion was based in part on experiments originally reported in Surdam et al. (1984). Stoessell and Pittman (1990) have been critical of these experiments (Surdam et al. 1984) and, as a consequence, they conclude that the idea of aluminosilicate framework grain dissolution being a function of organic acids and a mechanism for secondary porosity is flawed. In reaction to this criticism, we have repeated similar experiments to test the original hypothesis.

In summary, our experimental design for the new set of experiments was as follows: (1) the experimental solution was 0.15 M NaCl; 0.075 M oxalic acid adjusted to a pH of 5 with sodium acetate; (2) 5 ml of this solution plus 50 mg of either andesine or microcline (sized to <60 mesh) was heated to 100 °C and agitated for 2 weeks; (3) after 2 weeks, the reactor was cooled to room temperature, and the solution was diluted 1 : 10 and the metal concentrations analyzed by ICP; (4) some of the Teflon reaction vessels contained titanium liners. The experimental results are shown in Table 3. These

Table 3. Experiment on mineral dissolution in the presence of organic acids

Sample	Ti liner	Al ppm	Si ppm	Ti ppm	Ca ppm	Mg ppm	Fe ppm
Andesine white	Yes	56.0	106.1	4.1	0.4	14.6	24.7
Andesine white	Yes	57.7	109.7	3.0	0.3	14.6	24.4
Microcline	Yes	29.7	87.6	7.1	0.6	0.3	4.5
Microcline	Yes	28.9	86.5	8.7	2.6	0.5	4.7
Andesine white	No	57.5	115.2	0.5	0.1	15.0	24.7
Microcline	No	30.9	91.2	0.2	0.8	0.4	6.9
Blank	Yes	0.5	0.0	13.4	0.5	0.0	0.0
Blank	No	0.0	0.0	0.0	0.0	0.0	0.0

experiments strongly support the original hypothesis of Surdam et al. (1984) that organic acids in concentrations comparable to those observed in oil-field waters at a pH of 5 are capable of significantly increasing aluminum mobility, and hence aluminosilicate mineral solubility. This is not a particularly surprising conclusion, for chemists have been discussing how organic acids increase mineral solubilization and metal transport for 150 years (Sprengel 1826; Berzelius 1839; Senfl 1871). For additional documentation of the importance of organic acids to aluminosilicate mineral solubility and porosity evolution due to framework grain dissolution, see the outstanding work of Reed and Hajash (1992).

2.3 Deep Burial (Temperature > 130 °C)

Probably the process most devastating to porosity at temperatures in excess of 130 °C is quartz cementation. Leder and Park (1986) have an empirical approach to predicting quartz cementation resulting from diagenesis during burial. They plot porosity versus depth for relatively clean sandstones from various locations with different thermal regimes. From this data they construct porosity vs. depth curves for clean sandstones susceptible to quartz cementation. These empirical curves reasonably estimate porosity loss due to cementation and compaction for a sandstone undergoing progressive burial. However, there are well-documented cases where significantly more porosity is present in sandstones than the Leder and Park (1986) model would suggest (e.g., the Norphlet and Tuscaloosa formations, in which some samples from greater than 20 000 ft depth contain in excess of 20% porosity). Clearly, diagenetic features not modeled by Leder and Park have inhibited quartz cementation and allowed porosity or integranular volume to remain anomalously high. The problem in predicting enhanced or preserved porosity in sandstones susceptible to quartz cementation is to predict the presence and effect of these unmodeled diagenetic features, the most significant of which are carbonate cement, chlorite rims, early migration of hydrocarbons, and geopressuring. For a comprehensive treatment of quartz cementation, see Houseknecht (1987, 1988) and Houseknecht and Hathon (1987).

As the sandstone/shale system exits the zone of intermediate burial and enters the zone of deep burial ($>130\,°C$), the organic acids have been decarboxylated and the alkalinity of the pore waters is again dominated by HCO_3^-. Siebert (1985) has suggested that the thermal (or abiotic) reduction of sulfate by hydrocarbons begins at approximately $140\,°C$ and ends at $210\,°C$, and that the chemical evolution of many deeply buried and relatively exotic reservoir fluids can be explained by the reduction of sulfate by hydrocarbons and the reaction of the resulting hydrogen sulfide with other minerals. This mechanism is outlined in the seven equations in Table 4. Reactions (1)–(5) in Table 4 show that the reduction of sulfate to H_2S and the oxidation of hydrocarbons to CO_2 in the presence of iron can be an important source of protons in deeply buried diagenetic systems.

Hydrocarbons from the maturation of sedimentary organic matter and sulfate from anhydrite dissolution can undergo oxidation-reduction reactions to form calcite and hydrogen sulfide. This set of reactions becomes kinetically favorable at temperatures above about $140\,°C$ (Siebert 1985), but may proceed at temperatures as low as $100\,°C$ (Machel 1987). If there is an excess of hydrocarbons over sulfate in the sequence and little available Fe, then the reaction series ends, and the reservoir fluid is dominated by hydrocarbons with some hydrogen sulfide. However, if sulfate exceeds hydrocarbons, the resultant formation fluid is dominated by hydrogen sulfide (with elemental sulfur at higher temperature). If Fe is available, the hydrogen sulfide and sulfur react with the Fe to form pyrite and protons. The protons generated from these reactions are then available to participate in calcite dissolution reactions. If Fe is not readily available in the system, then hydrogen sulfide with subordinate carbon dioxide dominates the reservoir fluid. Carbon dioxide can react with feldspars, causing dissolution of the feldspar grains. However, since there are few known aqueous species present at the conditions of temperature and pH under which these reactions occur that are capable of complexing and transporting Al and Si from the site of

Table 4. Clastic chemical reactions based upon hydrocarbon reduction of sulfate. (Siebert 1985)

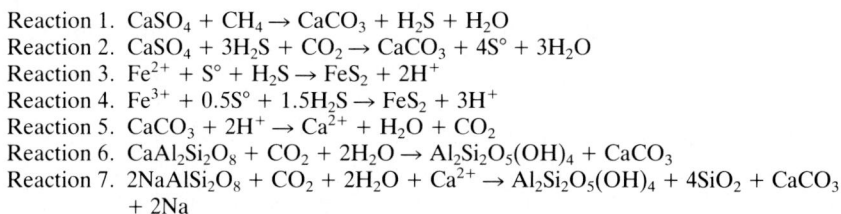

Reaction 1. $CaSO_4 + CH_4 \rightarrow CaCO_3 + H_2S + H_2O$
Reaction 2. $CaSO_4 + 3H_2S + CO_2 \rightarrow CaCO_3 + 4S° + 3H_2O$
Reaction 3. $Fe^{2+} + S° + H_2S \rightarrow FeS_2 + 2H^+$
Reaction 4. $Fe^{3+} + 0.5S° + 1.5H_2S \rightarrow FeS_2 + 3H^+$
Reaction 5. $CaCO_3 + 2H^+ \rightarrow Ca^{2+} + H_2O + CO_2$
Reaction 6. $CaAl_2Si_2O_8 + CO_2 + 2H_2O \rightarrow Al_2Si_2O_5(OH)_4 + CaCO_3$
Reaction 7. $2NaAlSi_2O_8 + CO_2 + 2H_2O + Ca^{2+} \rightarrow Al_2Si_2O_5(OH)_4 + 4SiO_2 + CaCO_3$
$\quad\quad\quad + 2Na$

dissolution, Al and Si are typically precipitated out of the formation fluid at the site of dissolution, frequently as chlorite or illite.

In summary, acidic formation fluids generated by abiotic sulfate reduction (Table 4) can account for the observed additional dissolution of feldspar, carbonate, and sulfate minerals and the formation of such products as "late" illite, chlorite, and pyrite in deeply buried (130 to 210°C) clastic systems. The porosity enhancing and preserving potential of these reactions will be directly dependent upon the availability and spatial distribution of hydrocarbons, sulfate, and iron.

2.4 Systematic Cycling of Carbonate Cementation/Dissolution Events

Figure 6 is a diagram summarizing the previous discussion on carbonate stability. In Fig. 6, we have attempted to list the major processes controlling carbonate stability in temperature-depth space. In this arrangement of processes, two important points are apparent. First, there is a direct relationship

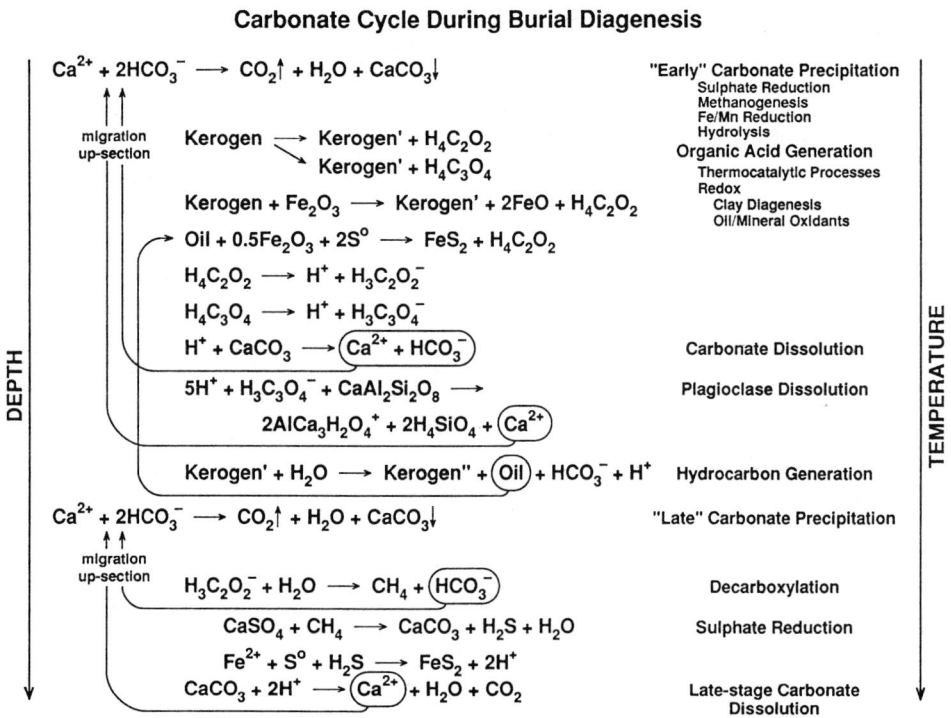

Carbonate Cycle During Burial Diagenesis

$$Ca^{2+} + 2HCO_3^- \longrightarrow CO_2\uparrow + H_2O + CaCO_3\downarrow$$

migration up-section

"Early" Carbonate Precipitation
Sulphate Reduction
Methanogenesis
Fe/Mn Reduction
Hydrolysis

$$Kerogen \longrightarrow Kerogen' + H_4C_2O_2$$
$$Kerogen' + H_4C_3O_4$$

Organic Acid Generation
Thermocatalytic Processes
Redox
Clay Diagenesis
Oil/Mineral Oxidants

$$Kerogen + Fe_2O_3 \longrightarrow Kerogen' + 2FeO + H_4C_2O_2$$

$$Oil + 0.5Fe_2O_3 + 2S^0 \longrightarrow FeS_2 + H_4C_2O_2$$

$$H_4C_2O_2 \longrightarrow H^+ + H_3C_2O_2^-$$

$$H_4C_3O_4 \longrightarrow H^+ + H_3C_3O_4^-$$

$$H^+ + CaCO_3 \longrightarrow \boxed{Ca^{2+} + HCO_3^-}$$ Carbonate Dissolution

$$5H^+ + H_3C_3O_4^- + CaAl_2Si_2O_8 \longrightarrow$$ Plagioclase Dissolution
$$2AlCa_3H_2O_4^+ + 2H_4SiO_4 + \boxed{Ca^{2+}}$$

$$Kerogen' + H_2O \longrightarrow Kerogen'' + \boxed{Oil} + HCO_3^- + H^+$$ Hydrocarbon Generation

$$Ca^{2+} + 2HCO_3^- \longrightarrow CO_2\uparrow + H_2O + CaCO_3\downarrow$$ "Late" Carbonate Precipitation

migration up-section

$$H_3C_2O_2^- + H_2O \longrightarrow CH_4 + \boxed{HCO_3^-}$$ Decarboxylation

$$CaSO_4 + CH_4 \longrightarrow CaCO_3 + H_2S + H_2O$$ Sulphate Reduction

$$Fe^{2+} + S^0 + H_2S \longrightarrow FeS_2 + 2H^+$$

$$CaCO_3 + 2H^+ \longrightarrow \boxed{Ca^{2+}} + H_2O + CO_2$$ Late-stage Carbonate Dissolution

DEPTH   TEMPERATURE

Fig. 6. Carbonate cycle during burial diagenesis: diagram of carbonate reactions with their corresponding diagenetic changes in clastic sediments

between the organic and inorganic reactions. Second, the organic/inorganic reactions are in a series and are sequential. If Fig. 6 correctly approximates the carbonate cementation/decementation events and causative coupled reactions, then the cementation/decementation events are systematically arranged in time-temperature space, and the distribution in time-temperature space of these systematic reactions is predictable. Such prediction can be accomplished using techniques similar to those developed by organic geochemists in the last decade to predict the distribution of hydrocarbon maturation reactions (Surdam and MacGowan 1992).

2.5 Diagenetic Modeling

2.5.1 Reaction Pathways

Sandstone and shale sequences undergo progressive diagenetic transformations from the time of erosion of the source material through deposition, early burial, and burial to a depth corresponding to the onset of metamorphism. The critical factors along the burial path can be represented as reaction divides on a diagenetic pathway diagram (Surdam et al. 1989b). The divides depend upon the presence or absence, or the relative abundance, of critical components. Critical components include sulfate, iron, and organic material (early burial diagenesis); bicarbonate, carboxylic acid anions, and iron (intermediate burial diagenesis); and sulfate, iron, and hydrocarbons (deep burial diagenesis).

The actual path that a sandstone/shale sequence takes through this diagenetic maze can be critically evaluated and mapped by applying careful petrography, including SEM, XRD, electron microprobe, isotopic geochemistry, cathode luminescence, and petrographic studies. In this way, the potential for porosity preservation/destruction/enhancement can be determined qualitatively at each reaction divide. This determination results in the identification of those mineral reactions most critical to porosity evolution during burial diagenesis. This information also is vital to reconstructing the evolution of pore-water chemistry.

Initial modeling of a rock-water system begins with choosing an appropriate initial pore-water composition. This is done by choosing a reasonable modern analog on the basis of a matching depositional environment. The next step is to estimate the initial composition of the sandstone. This can be done petrographically, commonly by examining those very early calcite concretions that "lock-in" the initial sediment and do not allow further diagenetic reaction to take place. For verification, the initial sediment composition can be recreated from a detailed knowledge of current petrologic composition, provenance, depositional environment, and diagenesis.

At this stage, pathway-type modeling (see Surdam et al. 1989a) is used to understand how the pore-water composition evolves through time and to predict which diagenetic reactions may affect the water chemistry. The diagenetic pathway diagram (Surdam et al. 1989b) mentioned above is a system of hierarchically arranged divides, similar to a chemical divide model (cf. Hardie and Eugster 1970), but also containing divides pertaining to nongeochemical factors controlled by physical processes. This information is incorporated into the interpretation of the compositional path of the fluid chemistry. This involves adding or subtracting appropriate amounts of cations and anions generated or taken up in significant mineral reactions. Where possible, the actual chemical changes are based on water analyses from terrains presently characterized by the specific reactions of interest (carbonate dissolution, plagioclase dissolution, sulfate reduction, etc.). If these analyses are unavailable, we resort to mass-balance calculations to determine the amount of material added to or subtracted from the fluid phase as a result of specific reactions. These calculations are made as follows: (1) the system is defined as a cubic meter of material of known composition and porosity (the pores are filled with a fluid of known composition), and (2) the reaction of interest is then modeled geochemically within the mass balance constraints of the cube. The model can then be checked to see if it correctly predicted the diagenetic sequence, and whether the modeled final composition of the pore water matches the present-day formation-water chemistry. Present-day analyses from the field or basin of interest are thus extremely useful in reconstructing the evolutionary path of the fluid chemistry.

2.5.2 Kinetic Modeling

Changes in the concentrations of reactive organic species through time must be modeled using measured kinetic parameters and time-temperature burial history models. The source of the reactive organic species (carboxylic acid anions, CO_2, phenols, etc.) affecting the inorganic mineral stabilities is the same as the source of hydrocarbons: the source rocks. The reactive organic species are either expelled from the source rocks just prior to, or concomitant with, hydrocarbon generation, or form as a result of the interaction of liquid hydrocarbons and mineral matter in the reservoir rocks, in redox reactions. It is possible to model the progress of these organic-inorganic interactions by calculating their transformation ratios through time. For the study of a specific basin, the sources for kinetic parameters are hydrous pyrolysis experiments, or natural-mineral vs. depth distributions (see Surdam et al. (1989a,b). Combining the time-temperature profile with these kinetic parameters allows the construction of profiles of the distribution with time of carboxylic acids and anions (CAA) and CO_2 (from decarboxylation) in the formation waters. From these data it is possible to qualitatively evaluate the relative stability of minerals (whether they are precipitating or dissolving) at any time in the history of a reservoir interval. In addition to modeling the generation and destruction of CAA and oil generation, kinetic modeling can also be used to model the progress of other kinetically controlled diagenetic reactions, such as the smectite-illite transition and the thermal reduction of sulfate by hydrocarbon oxidation (see Surdam et al. 1989a,b). For more information concerning kinetic modeling, see Surdam et al. (1989a,b,c,d).

In more detail, the first step in evaluating a kinetically controlled reaction is to establish a time-temperature profile for the unit of interest. Essential ingredients in establishing a time-temperature profile include: (1) a burial history, constructed from age, depth, and lithologic data; (2) a heat-flow vs. time profile, derived from thermal modeling; (3) a knowledge of thermal properties (thermal conductivity and diffusivity) of the rocks contained in the region of interest, retrieved either from the literature or from laboratory measurements; and (4) a set of porosity vs. depth curves for the lithologies of interest, so that thermal conductivity changes due to progressive compaction can be estimated. Integration of this information allows the establishment of the time-temperature profile. Additional information useful in constraining or testing the validity of the constructed time-temperature profiles includes oil-well bottom-hole temperatures, measured heat-flow values, and temperature measurements of bore holes. For more detailed information concerning the evaluation of time-temperature profiles, see Surdam et al. (1989d).

The second step in evaluating a kinetically controlled reactions is to acquire activation energies E_a and frequency factors A_i for the reactions of interest. There are two primary sources for these kinetic parameters: (1) experimentation utilizing techniques such as hydrous pyrolysis, and (2) observation of natural-mineral vs. depth distributions, assuming that depth can be translated into temperature. Once the kinetic parameters for the reactions of interest are acquired and the time-temperature profiles have been constructed, the spatial distribution of kinetically controlled diagenetic elements can be determined using a modified Arrhenius equation (Tissot and Espitalié 1975), in the form

$$dx_i = x_i A_i \exp[-E_i/RT]dt,$$

where

x_i = the amount of material reacting in the ith reaction,
A_i = the frequency factor for the ith reaction (a constant),
E_i = the activation energy of the ith reaction,
R = Raoult's gas constant,
T = the absolute temperature (Kelvin), and
t = time.

Implicit in the use of this equation is the assumption that the reaction being modeled can be approximated by first-order reaction kinetics. The equation for dxi can be solved by numerical integration with successive Δt increments. At any time, the progress of the reaction of interest i can be determined by calculating the transformation ratio T.R., the ratio of the reactants remaining to be transformed to products (x_i) to the maximum amount of reactant available (x_o):

T.R. $= (x_o - \Sigma x_i)/x_o$.

For practical purposes, the progress of a reaction at a specific time can be determined from a plot of T.R. vs. time. Note that the two inputs to the T.R./time plot are (1) a time-temperature history for the stratigraphic unit in which the reaction is being evaluated, and (2) the kinetic parameters E_i and A_i for the reaction.

The information derived from the kinetic modeling and the compositional information for the pore-water reconstruction is next exported as input to the thermodynamic rock/water equilibrium simulator.

2.5.3   Thermodynamic Rock/Water Equilibrium Simulation

A thermodynamic equilibrium-type computer model with an enhanced organic-inorganic reaction data base and temperature-extrapolated reaction equilibrium constants can be used to determine the direction and extent of water-rock interactions. In this way, for example, the spatial distribution of carbonate cements in a sandstone along a time-temperature profile can be quantified. However, as noted above, changes in the concentrations of reactive organic species through time must be modeled using measured kinetic parameters and time-temperature burial history models. In this chapter, these last are coupled with thermodynamic equilibrium simulation, using a version of SOLMINEQ.88 and concentration profiles of kinetically controlled diagenetic elements derived from transformation ratio vs. time profiles (Surdam and MacGowan 1992). This approach to modeling the evolution of a diagenetic system yields quantitative information on the relative stabilities of mineral phases through time, the timing of the generation of reactive aqueous and gaseous organic phases, and the timing of diagenetic events such as cement precipitation and grain/cement dissolution.

In the described modeling technique, mineral stability output is typically in the form of saturation index SI:

SI $=$ (activation product)/(equilibrium constant).

Log SI can be plotted versus time for each modeled layer, or it can be plotted against present-day depth. When log SI is positive, the mineral is oversaturated and will precipitate; when log SI is negative, the mineral is undersaturated and will dissolve. Moreover, comparing log SI vs. present-day depth plots with porosity/depth plots allows the predicted mineral stabilities (potential porosity anomalies) to be tested directly. In this manner, a model of the evolution of porosity in a targeted clastic reservoir system can be accurately evaluated.

MODELING CLASTIC DIAGENESIS

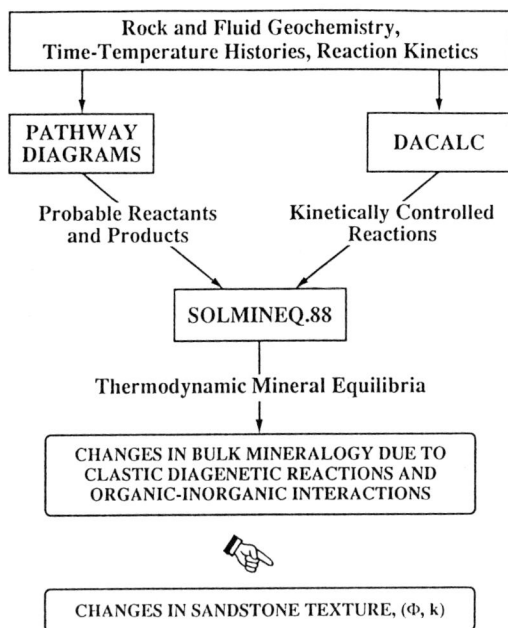

```
┌─────────────────────────────────────────┐
│        Rock and Fluid Geochemistry,      │
│  Time-Temperature Histories, Reaction Kinetics │
└─────────────────────────────────────────┘
        │                        │
        ▼                        ▼
┌──────────────┐          ┌──────────────┐
│  PATHWAY     │          │   DACALC     │
│  DIAGRAMS    │          │              │
└──────────────┘          └──────────────┘
        ╲                        ╱
  Probable Reactants      Kinetically Controlled
    and Products               Reactions
          ╲                    ╱
           ▼                  ▼
      ┌──────────────────────┐
      │    SOLMINEQ.88        │
      └──────────────────────┘
                 │
      Thermodynamic Mineral Equilibria
                 │
                 ▼
┌──────────────────────────────────────┐
│  CHANGES IN BULK MINERALOGY DUE TO    │
│  CLASTIC DIAGENETIC REACTIONS AND     │
│  ORGANIC-INORGANIC INTERACTIONS       │
└──────────────────────────────────────┘

                  ☞

┌──────────────────────────────────────┐
│  CHANGES IN SANDSTONE TEXTURE, (Φ, k) │
└──────────────────────────────────────┘
```

Fig. 7. Schematic diagram showing the most significant components in the predictive methodology developed for clastic diagenesis

It is this integrated approach (Fig. 7) to diagenetic modeling that will be utilized subsequently. In the remainder of this chapter, this predictive methodology will be demonstrated by applying it to the Latrobe Group in the Gippsland Basin, Australia. Predictions of the spatial distribution of optimum reservoir conditions in the Gippsland Basin will be made and evaluated.

3 The Gippsland Basin

The Mesozoic-Cenozoic Gippsland Basin, southeastern Australia, has been a major petroleum exploration/production area since the 1960s. Approximately four-fifths of the productive basin area is offshore (Fig. 8). Sediments in the offshore Gippsland Basin (Fig. 9) include three distinct sequences separated by unconformities: the Early Cretaceous Strzelecki Group, the Late Cretaceous-Eocene Latrobe Group, and the Oligocene-Miocene Seaspray Group (Lakes Entrance and Gippsland Limestone formations of Fig. 9) (James and Evans 1971; Threlfall et al. 1976).

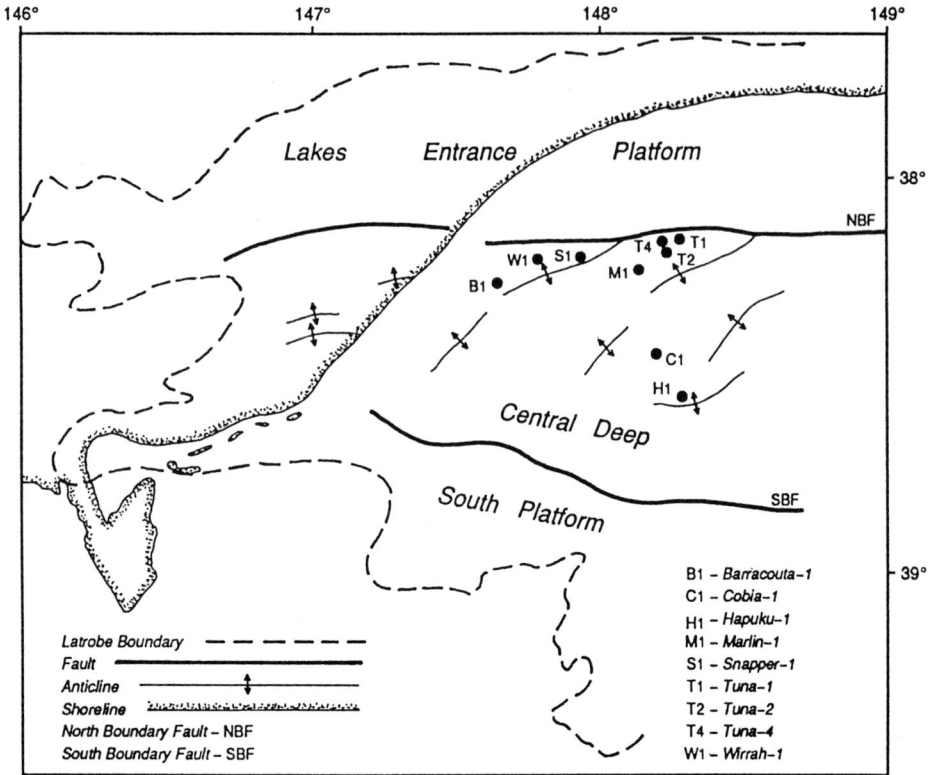

Fig. 8. Index map of the Gippsland Basin, showing basin location, tectonic setting (the graben represented by the two platforms and the Central Deep), major structures, and location of well. (After Weeks and Hopkins 1967; Hocking 1972; Bodard et al. 1984)

The Latrobe Group contains potential source and reservoir rocks. It is composed of continental to marginal-marine sandstones, siltstones, mudstones, shales, coals, and minor volcanics, deposited in a complex deltaic-fluvial-marine depositional environment. Most of the Latrobe sandstones are mineralogically and texturally mature and are characterized by high primary porosity and permeability (Threlfall et al. 1976). The sandstones are coarse- to medium-grained with scattered and patchy dolomite cement. Petrographic observation suggests that porosity in the Latrobe sandstones has been modified significantly by postdepositional diagenesis. Therefore, an understanding of the diagenetic processes will improve evaluation and prediction of porosity/permeability trends in Latrobe reservoirs, and in similar reservoirs elsewhere. Bodard et al. (1984) did a detailed petrographic study of the Latrobe sandstones. They suggested that the dolomite cement was the major cause of porosity reduction in some fields, and that dissolution of the dolomite cement and the clay matrix and precipitation of kaolinite,

	Lithology	Formation
Miocene		Gippsland Limestone Formation
Oligocene		Lakes Entrance F'm
Eocene and Paleocene		Latrobe Group
Upper Cretaceous		
Lower Cretaceous		Strzelecki Group

Sandstone

Shale/Siltstone

Coal

Carbonaceous Shale

Limestone

Marl

Calcareous Shale

Volcanics

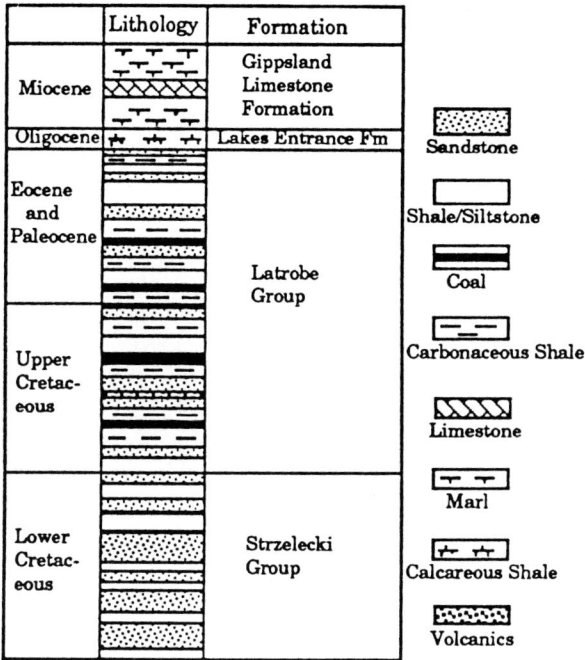

Fig. 9. Generalized stratigraphic column of the offshore Gippsland Basin. The Oligocene Lakes Entrance Formation and Miocene Gippsland Limestone Formation are included in the Seaspray Group

chlorite, and quartz had played significant roles in modifying the sandstone porosity.

3.1 Reaction Pathways (Diagenetic Events)

Framework grain composition in the Latrobe sandstones was determined by point counting (Fig. 10). These sandstones are quartzarenites, subarkoses, and sublitharenites, with a few lithic arkoses, feldspathic litharenites, and litharenites. The majority of the Latrobe sandstones consist of more than 70% quartz grains. The remaining grains are detrital potassium feldspar and rock fragments (consisting predominantly of sedimentary and low-grade metamorphic lithic grains).

Mechanical compaction was an important factor in porosity reduction, as is shown by crushed feldspar grains, fractured quartz, and deformed fine-grained clasts. An example of crushed feldspar grains is shown in Fig. 11.

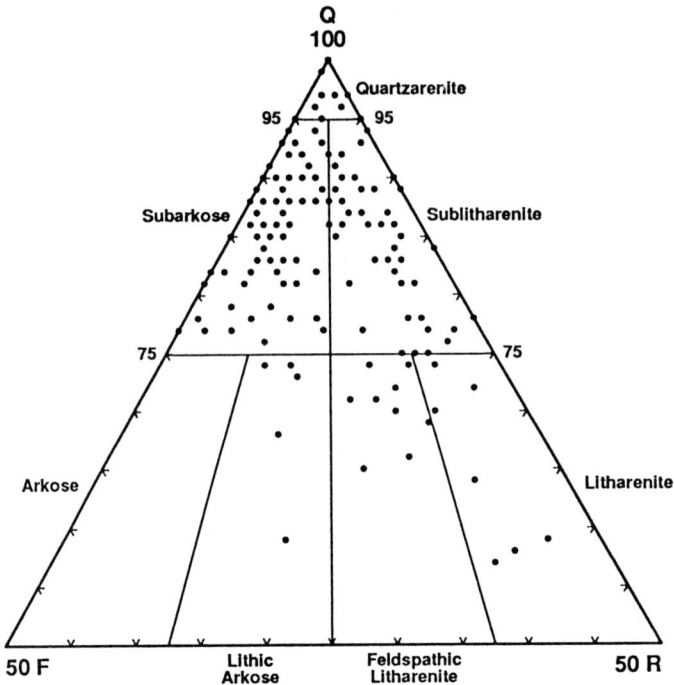

Fig. 10. Mineral composition of the Latrobe Group sandstones. The sandstone classification was adapted from Folk (1974); data were derived from point-counting thin sections (300 points per thin section). It should be noted that most of these sandstones have suffered significant dissolution of feldspar and rock fragments

3.1.1 Early Dolomite Cementation

As discussed by Bodard et al. (1984), dolomite is the only volumetrically significant cement within the relatively shallow Latrobe sandstones. The dolomite cement, filling intergranular pores and replacing framework grains, accounts for up to 40% of rock volume in some samples. The dolomite cement occurs as patches, but in some cases coalesces into continuous laminae. Previously, Partridge (1976) suggested that the dolomite cementation was concentrated along eustatic low-stand unconformities. Harrison (1989) has suggested that a meteoric-marine water-mixing model adequately explains the formation of dolomite in the Latrobe Group.

Chemical, carbon-isotopic, and oxygen-isotopic compositions from selected samples of the Latrobe dolomite cement are presented in Table 5(A and B). All of the $\delta^{13}C$ values in Table 5B fall between -10 and $-22‰$ (PDB, Peedee Formation Belemnite). $\delta^{18}O$ values of the dolomite cement range from 22.7 to 25.2‰ (SMOW, Standard Mean Ocean Water) with an

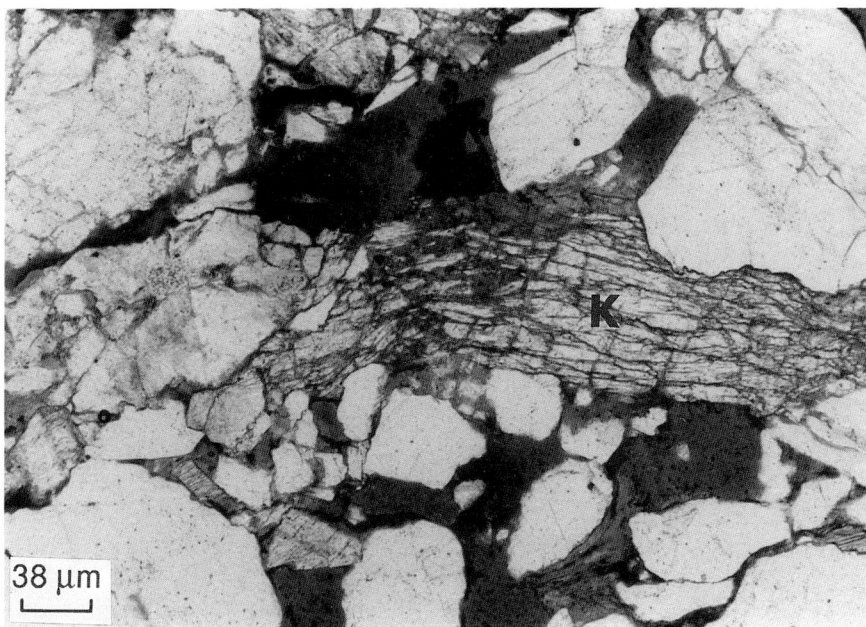

Fig. 11. Feldspar grain (*K*) at center for photomicrograph has been fractured and crushed. Other detrital grains are mainly quartz; the *dark gray areas* are pore space (blue-dyed epoxy). Barracouta-l, 2151 m (7100 ft). Scale *bar* = 38 μm

average of 23.6‰. Harrison (1989), using a larger number of dolomite samples, showed that the $\delta^{18}C$ values of the dolomite ranged from +5 to −24‰. These values are consistent with the hypothesis that the dolomite formed during early burial.

The oxygen-isotopic composition of dolomite can be used to estimate the temperature of dolomite formation. However, to calculate the temperature, the $\delta^{18}O$ of the water from which the dolomite precipitated must be estimated. As a first approximation, it is assumed that the dolomite cement precipitated from water with a composition between contemporary normal marine and fresh water. Shackleton and Kennett (1973) determined that the $\delta^{18}O$ of the marine water in the Bass Strait during the early Tertiary was −1.28‰ (SMOW) and that the surface paleotemperature in the Victoria area during the Eocene and Early Oligocene was approximately 15 °C. According to the relationship between the isotopic composition of meteoric precipitation and average monthly temperature (Anderson and Arthur 1983), the corresponding $\delta^{18}O$ value of meteoric precipitation in the Gippsland Basin during early Tertiary time ranged from −7 to 2‰ (SMOW) (see Figs. 1–14 in Anderson and Arthur 1983). Therefore, a range of $\delta^{18}O$ values for the water involved in dolomite formation can be defined as $-7‰ < \delta^{18}O < -1.28‰$. On the basis of the experimental fractionation of oxygen isotopes

Table 5. Chemical and isotopic composition of dolomite cement, Latrobe Group, Gippsland Basin

A Chemical composition

Well	Depth	Spot	Mole fraction (average)		
			$CaCO_3$	$MgCO_3$	$MnCO_3$
Marlin-1	1344	32	0.5061	0.4924	0.0014
	1346	23	0.5121	0.4867	0.0013
	1498	12	0.4933	0.5052	0.0014
	1500	12	0.4869	0.5099	0.0032
Tuna-2	2132	25	0.5051	0.4896	0.0053
	2135	37	0.5011	0.4945	0.0044

B Isotopic composition

Well	Depth (m)	$\delta^{13}C$ (PDB)	$\delta^{18}C$ (SMOW)
Marlin-1	1345	−22.3	24.5
	1345[a]	−22.3	24.5
	1499	−16.5	23.3
	1500	−16.7	23.3
	1500[a]	−16.7	23.2
	1506	−16.2	23.4
	1508	−16.6	23.3
	1508[a]	−16.5	23.5
Tuna-2	2133	−9.8	25.2
	2135	−10.1	22.8
	2135[a]	−10.1	22.7

[a] Duplicate sample.

between protodolomite and water, Fritz and Smith (1970) proposed the following relation between the dolomite-formation temperature and oxygen-isotopic data:

$$10^3 \ln \alpha_{\text{dolomite-water}} = 2.62 \times 10^6 T^{-2} + 2.2,$$

where $\alpha_{\text{dolomite-water}}$ is the fractionation factor between dolomite and water and T is the temperature of dolomite formation in Kelvin. By their equation, the maximum temperature range of formation of the Latrobe dolomite is between 33 and 70 °C (Fig. 12). This conclusion is compatible with Harrison's estimation of <60 °C for the temperature of formation of the dolomite (Harrison 1989). This range of calculated temperatures leads to the conclusion that the dolomite cement formed in the zone of shallow burial (<80 °C).

Early precipitation of dolomite cement in the Latrobe Group is supported by petrographic observation. In most of the dolomite-cemented sandstones,

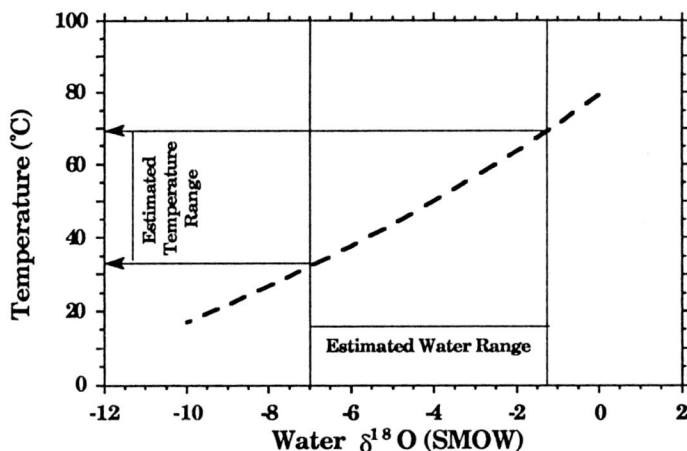

Fig. 12. Range of formation temperatures for Latrobe dolomite cement, calculated from its isotopic composition

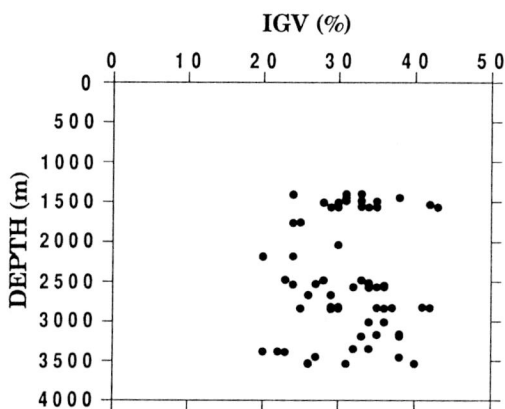

Fig. 13. Intragranular volume versus depth for the dolomite-cemented Latrobe sandstones

the intergranular volume (IGV) ranges from 25 to 40%, with some as high as 43% (Fig. 13). Some of these IGV values demonstrate that cement was formed at shallow depth prior to significant mechanical compaction. In each thin section, the mineral composition of sand grains within dolomite-cemented areas and uncemented areas was point counted separately. The results indicate that in each thin section, detrital feldspar grains and rock fragments are more abundant in dolomite-cemented areas than in the uncemented areas. This difference in mineral composition can be explained by the dissolution of labile grains in the uncemented areas. It suggests that dolomite cement within the Latrobe sandstones was formed earlier than the dissolution of the framework grains, that cemented zones were protected from dissolution, and that the framework grains within the cemented zones approximate the original detrital composition of the sand.

Precipitation of the dolomite cement significantly reduced porosity in some Latrobe sandstones. The dolomite cement restricted fluid circulation in the incompletely cemented sandstones, and completely blocked fluid circulation in the tightly cemented sandstones. Further diagenetic modification was prevented in the tightly cemented zones in the sandstones, whereas incompletely cemented zones, or tight zones subsequently fractured, were susceptible to further diagenetic modification. In either case, the early dolomite cementation maintained high intergranular volume within the Latrobe Group sandstones (see Fig. 13), resulting in increased potential for, and probability of, enhanced porosity during subsequent burial and diagenesis (carbonate dissolution) (Schmidt and McDonald 1979). The presently observed distribution of porosity in the Latrobe sandstones is heterogeneous; at similar depths, very low porosity exists within the tightly cemented sandstones, and high porosity occurs within the dolomite-cement-free sandstones.

3.1.2 Dissolution of Framework Grains and Cement

In the depth interval between 1500 m (5000 ft) and 3000 m (10 000 ft), detrital feldspar, rock fragments, and early dolomite cement in the Latrobe sandstones exhibit various dissolution textures. The dissolution of feldspar grains is shown in Fig. 14A,B, where the remnant skeletal potassium feldspar grains and oversized pores are results of dissolution. These dissolution textures are commonly observed within uncemented, decemented, and partially cemented sandstones, but are not observed in the tightly cemented sandstones. The most severe dissolution is typically associated with fractured detrital grains. Evidence of dissolution of the early dolomite cement consists of etched, serrated edges around the dolomite crystals (Fig. 15A,B). Dissolution of dolomite cement and framework grains was also observed by Bodard et al. (1984) over the same stratigraphic interval. In many samples, however, it is difficult to determine whether the sandstone remained uncemented or dolomite cement was present and subsequently dissolved. Therefore, it is possible that the amount of dolomite dissolution is volumetrically more significant than can be texturally documented in thin section (Schmidt and McDonald 1979).

The dissolution of framework grains and early dolomite cement has significantly enhanced porosity in many Latrobe sandstones. Within these sandstones, about 2 to 10% porosity is estimated to have been created by dissolution. The proportion of dissolution porosity relative to total porosity appears to increase with increasing burial. Above 2000 m (6500 ft), only minor enhancement of porosity was observed. With increasing depth, and

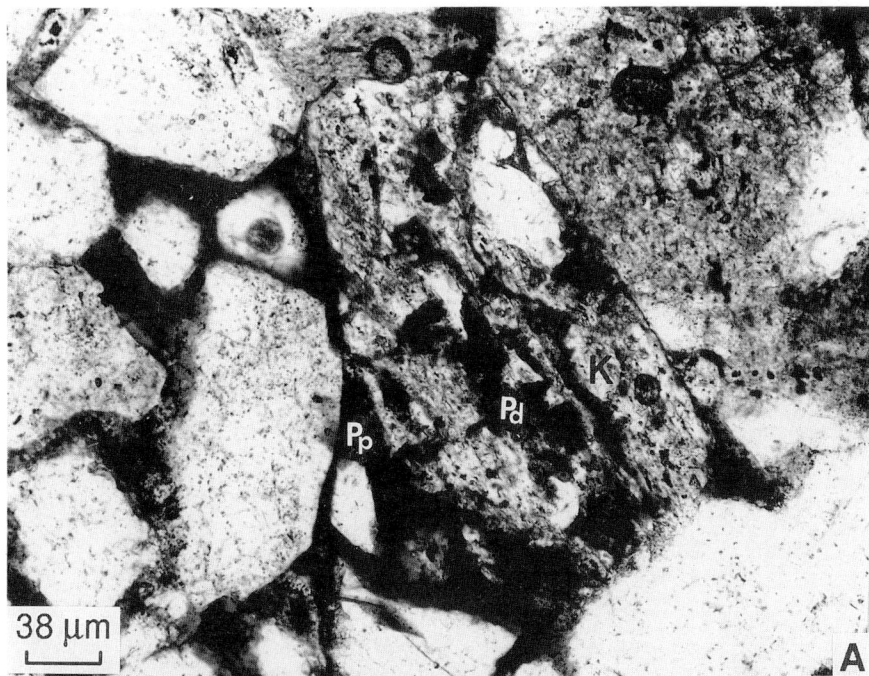

Fig. 14A,B. Dissolution of detrital feldspar grains. **A** Thin section photomicrograph illustrating partial dissolution (*Pd*) of a feldspar grain at the center. *Pp* Primary inter-granular pores. Tuna-2, 2544 m (8346 ft). Scale *bar* = 38 μm. **B** SEM photomicrograph of the remnant skeleton of a feldspar grain. Barracouta-l, 2352 m (7717 ft). Scale *bar* = 100 μm

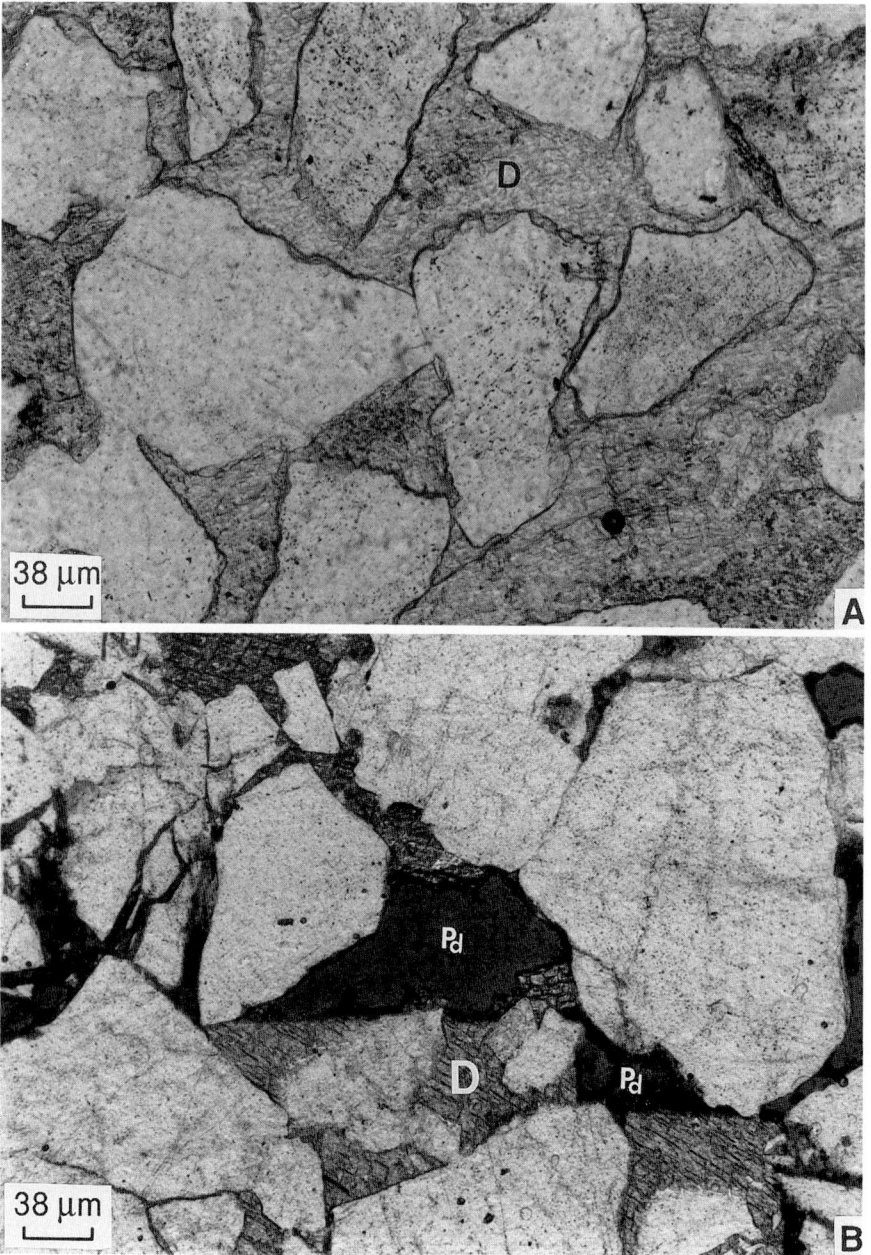

Fig. 15A,B. Dolomite cementation (*D*) and dissolution. **A** A tightly dolomite-cemented sandstone, Tuna-2, 2203 m (7229 ft). **B** Dolomite cement, previously nearly filling the pore space as in **A**, now partially dissolved (*Pd*), as evidenced by the etched edges of the dolomite crystals against the pore spaces (*Pd*) created by dissolution, Marlin-l, 1575 m (5167 ft)

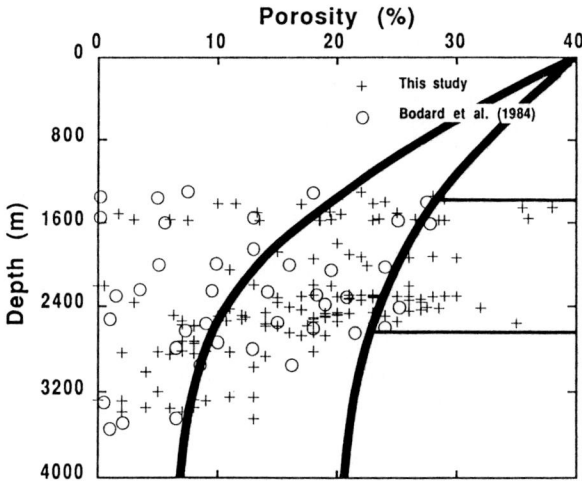

Fig. 16. Plot of porosity versus depth for the Latrobe Sandstone. + represents porosity values detected by the authors, while o represents porosity values obtained by Bodard et al. (1984). The area between the two curves, representing porosity loss due to compaction versus depth for a sandstone containing 50% quartz (*left-hand curve*) and a sandstone containing 75% quartz (*right-hand curve*), contains the base porosity values due to compaction predicted by Pittman and Larese (1991). The *two horizontal lines* outline positive anomalies, while the data points left of the base porosity area are negative anomalies

within specific layers, the dissolution porosity becomes texturally more and more important. It should be noted that dissolution porosity estimates are based on textural criteria; as a result, there is great uncertainty associated with the values. Nonetheless, it is concluded that within a significant number of layers within the sandstones in the stratigraphic interval from ~1500 m (5000 ft) to ~3000 m (10 000 ft) there is significantly enhanced porosity. This conclusion that the highest probability for the occurrence of a positive porosity anomaly exists between ~1500 m and ~3000 m is supported by the porosity/depth plot shown in Fig. 16.

Figure 16 is a diagram showing porosity/depth relations for sandstones of the Latrobe Group. The compaction/porosity loss curves are from Pittman and Larese (1991). The 50 and 75% quartz curves of porosity loss vs. depth (Fig. 16) establish a porosity baseline, based on the quartz content of each sandstone. These porosity loss vs. depth curves are used to define porosity anomalies.

3.1.3 Precipitation of Quartz, Kaolinite, and Ferroan Magnesite

Quartz cementation is common as overgrowths in some quartz-rich Latrobe sandstones (Fig. 17). Quartz cementation is diagenetically important in the

434 R.C. Surdam and P. Yin

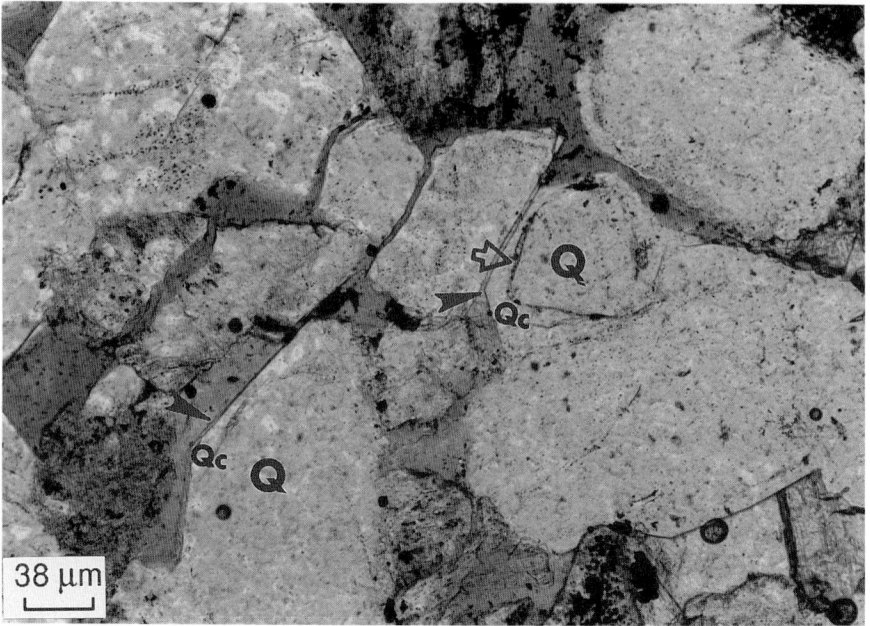

Fig. 17. Photomicrograph illustrating quartz overgrowths (*Qc*) developed on quartz (*Q*) in deep Latrobe Group sandstones. Note "dust rims" beneath overgrowths (*open arrow*) as well as euhedral terminations (*solid arrows*). Marlin-1, 2572 m (8438 ft). Scale *bar* = 38 μm

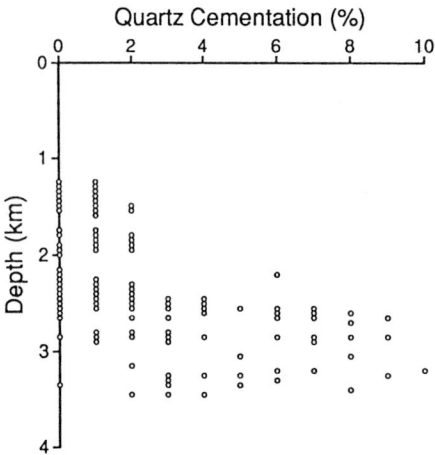

Fig. 18. Volume percent diagenetic quartz vs. depth in the Latrobe Group sandstones. Data derived from point-counting 150 thin sections (300 points per thin section). Overlapping points have been combined into single points in this plot

deeply buried sandstones; typically below 2500 m (8200 ft), diagenetic quartz (overgrowths and cementation) accounts for up to 10% of the sandstone volume (Fig. 18) and severely reduces porosity. In contrast, quartz cementation constitutes only 1–2% of the sandstone volume above 2500 m (8200 ft)

(Fig. 18). In some porous sandstones, minor amounts of authigenic quartz appear to have precipitated at grain-to-grain contacts, strengthening the sandstone and preventing collapse from mechanical compaction.

At depth, where quartz overgrowths are increasingly significant, below 2500 m (8200 ft), authigenic kaolinite occurs up to 8% by volume in some Latrobe sandstones (Fig. 19). The well-crystallized kaolinite is observed within the intergranular and intragranular pores (Fig. 20A,B). Precipitation of vermicular stacks of kaolinite within dissolution pores supports the hypothesis that the kaolinite precipitation postdated the dissolution of framework grains. Kaolinite precipitated in the deeper portions of the stratigraphic section typically occludes intergranular pores and badly damages the porosity of the Latrobe sandstones (Fig. 20A).

In the deeper portion of some wells (e.g., Marlin-1), ferroan magnesite forms as an intergranular cement (Fig. 21). Microprobe analysis has revealed the average chemical composition of this phase to be $Mg_{1.14}Fe_{0.80}Ca_{0.06}(CO_3)_2$, with more than 800 ppm Mn (Table 6A). XRD analysis gives a $d_{(104)}$ value of 2.77 Å. Carbon- and oxygen-isotopic compositions are given in Table 6B. In comparison with the early dolomite, the ferroan magnesite displays lower $^{18}O/^{16}O$ ratios; however, the $\delta^{13}C$ values of the ferroan magnesite are in the range of $\delta^{13}C$ values determined for the early dolomite. The lower $\delta^{18}O$ values characterizing the ferroan magnesite probably indicate that it was formed at a higher temperature, and hence at greater depth, than the early dolomite. Textural evidence also suggests that the ferroan magnesite formed relatively late in the diagenetic process. Comparing the chemical composition of the early dolomite in the Latrobe sandstones with that of the ferroan magnesite, we note that iron replaces calcium and that the proportion of magnesium is not substantially different. Therefore, a problem of the formation of the late ferroan magnesite is a

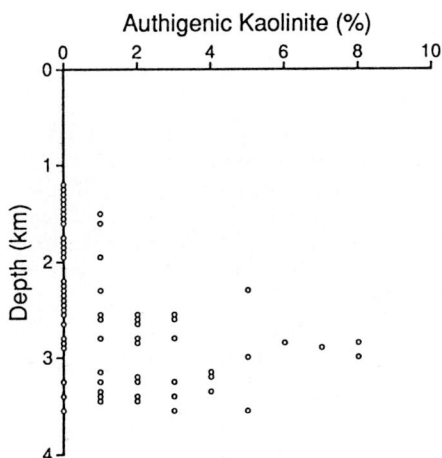

Fig. 19. Volume percent of authigenic kaolinite in the Latrobe Group sandstones vs. depth. Data collected by point-counting 150 thin sections (300 points per thin section)

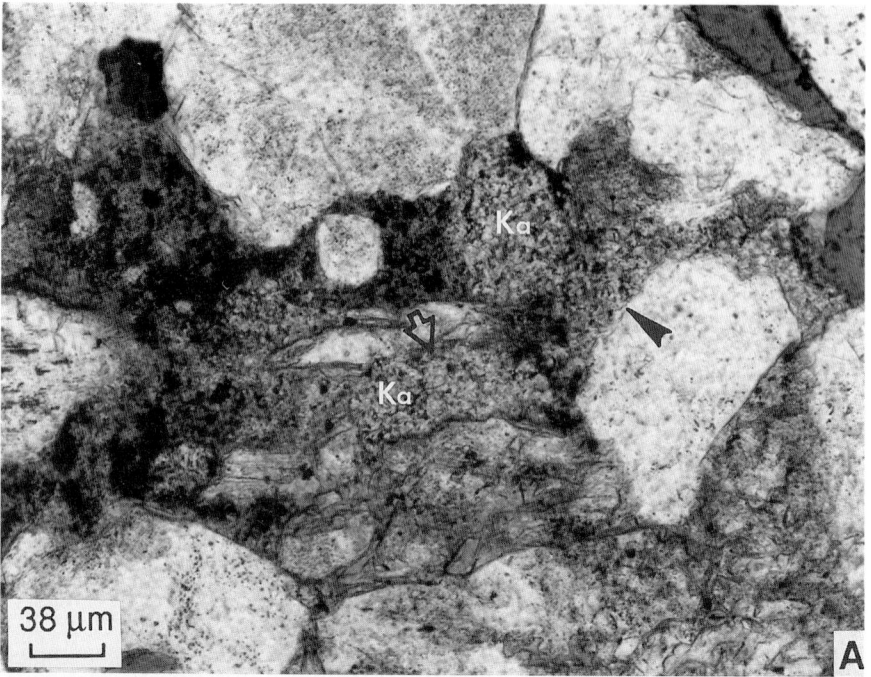

Fig. 21. Photomicrograph showing ferroan magnesite cement (dark grain boundaries, *solid arrow*) and replacements (*gray grains, M*) in deep Latrobe sandstone. Marlin-l, 2572 m (8438 ft). Scale *bar* = 38 µm

source of iron during burial. In the deeper section, the diagenesis of clay minerals may have released iron into formation waters (Boles 1981), or metal complexes, such as Fe-acetate, may have been destabilized at elevated temperature, releasing iron (Surdam et al. 1989b).

It is concluded that the major diagenetic events modifying porosity within the Latrobe sandstones were (1) mechanical compaction and early dolomite cementation, (2) dissolution of dolomite and framework grains, and (3) deep cementation and recementation with quartz, kaolinite, and Fe-carbonate. Thus, the diagenetic history and porosity evolution of these sandstones can be summarized as follows: (1) the first significant event was early porosity destruction by mechanical compaction and IGV preservation by early car-

---

Fig. 20A,B. Occurrence and morphology of authigenic kaolinite within the deep Latrobe sandstones. **A** Thin section photomicrograph illustrating kaolinite (*Ka*) in both inter-granular pores (*solid arrow*) and moldic pores (*open arrow*). The partially dissolved grain at center is detrital feldspar. The *dark gray areas* are pore space (blue-dyed epoxy). Barracouta-l, 2650 m (8693.5 ft). Scale *bar* = 38 µm. **B** SEM photomicrograph showing vermicular stacks of pseudohexagonal plates of authigenic kaolinite in deep Latrobe sandstone. Barracouta-l, 2353 m (7717 ft)

Table 6. Chemical and isotopic composition of ferroan magnesite, Latrobe Group, Marlin-1 well

A Chemical composition

Depth (m)	Spot No.	Mole fraction (average)			
		$CaCO_3$	$MgCO_3$	$FeCO_3$	$MnCO_3$
2500	17	0.0248	0.5858	0.3878	0.0016
2505	29	0.0265	0.5589	0.4131	0.0015

B Isotopic composition

Depth (m)	$\delta^{13}C$ (PDB)	$\delta^{18}O$ (SMOW)
2500	−14.6	18.8
2500[a]	−14.6	18.8
2505	−14.9	18.5
2505[a]	−14.9	18.5

[a] Duplicate sample.

bonate cement; (2) later, after additional burial, porosity was enhanced by a carbonate cement and framework grain dissolution event; and (3) last, subsequent to the porosity enhancement, the porosity in some zones was degraded by quartz, kaolinite, and Fe-carbonate cementation.

3.1.4 Water Evolution

Before proceeding to kinetic modeling, it was necessary to estimate how the pore water composition evolved through time. In the Gippsland Basin, this was done by integrating information derived from modern waters produced from oil fields, interpreted original water composition, diagenetic reaction pathway diagrams (Surdam et al. 1989b), and hydrous pyrolysis experiments.

The assumption that organic acids are an important aspect of the water evolution in the Gippsland Basin is based on observations of modern produced water from oil wells, and from NMR and hydrous pyrolysis experiments. In modern pore waters at temperatures greater than 80 °C, carboxylic acid anion concentration ranges from 100 to 1500 ppm (Table 7), whereas HCO_3^- content varies from 200 to 1100 ppm (Table 3 in Harrison 1989). Harrison (1989) found high bicarbonate values in wells characterized by relatively fresh water (presumably due to dilution by meteoric water), and low bicarbonate values in wells characterized by saline water (presumably undiluted). Some of the hydrocarbon-productive intervals in the Gippsland Basin have been invaded by fresh water after hydrocarbon filling, so dilution

Table 7. Concentration (ppm) of carboxylic acid anions detected in produced formation waters from the Gippsland Basin

Well	Formate	Acetate	Propionate	Butanoate	Malonate	TDS (ppm)
Tuna A-13	tr	894	326	97	nd	33966
Tuna A-57	nd	956	418	76	nd	35520
Flounder A-22	nd	29	217	nd	nd	34743
Snapper A-22	tr	tr	62	nd	tr	4218
Snapper A-11	tr	tr	195	nd	tr	4773
Kingfish B-1	nd	96	136	nd	nd	33078
Kingfish B-13	58	178	150	nd	nd	32079

nd = not detected; tr = trace amount.

by meteoric water may be an important factor in evaluating the values reported in Table 7. Nevertheless, alkalinity in the modern produced waters in the Gippsland Basin in a function of the interaction of carboxylic acid/acid anions and the carbonate system.

NMR analyses and hydrous pyrolysis experiments with source rock (coal) from the Gippsland Basin further explain why OAA are an important component of the alkalinity. On average, the Latrobe Group contains 35% shale and 10% coal (estimated from well completion data). The organic carbon content of the dark-colored shales averages 5.3%, whereas that of the coal averages 50.4%. These potential source rocks contain Type-III kerogen (Yin 1988), and are characterized by abundant oxygen-bearing functional groups. These functional groups have the potential to form water-soluble carboxylic acid anions and acids. The source rocks in the Latrobe sediments are interbedded with the reservoir sandstones, providing an optimum condition for organic-inorganic interaction (Surdam et al. 1989a,b).

Nuclear magnetic resonance spectra of kerogen from the Latrobe Group source rocks at different depths indicate that the cleavage of carbonyl, carboxyl, and phenolic functional groups from the kerogen occurred prior to and concurrent with the release of aliphatic components (i.e., hydrocarbon generation; see Fig. 4 in Yin and Surdam 1985). These oxygen-bearing functional groups are potential sources of organic acid anions and phenols, whereas the aliphatic components tend to form hydrocarbons during progressive burial and increasing thermal exposure. Hydrous pyrolysis experiments on the brown coal samples from the Gippsland Basin (Yin et al. 1993) produced acetic, formic, propionic, butanoic, malonic, and oxalic acid anions (Fig. 22). Figure 22 shows the changes in kerogen structure, as reflected in NMR patterns, as the raw brown coal sample is transformed into hydrocarbons and solid residue during hydrous pyrolysis for 72 h at 320 °C, and also shows the carboxylic acid anions detected within the aqueous phase at the end of the experiment. It is evident that release of the carbonyl functional groups from the brown coal is concomitant with generation of the

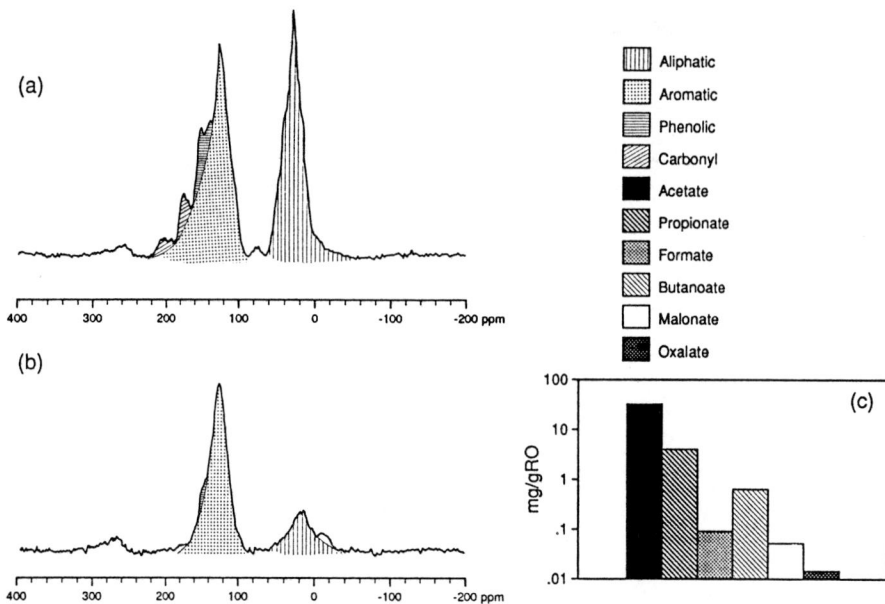

Fig. 22. Changes in the NMR patterns from a raw brown coal sample (**a**) to the solid residue (**b**) after heating for 72 h at 320 °C in a sealed vessel with distilled water, and the carboxylic acid anions (**c**) detected within the experimental solution in the same experiment. The amount of carboxylic acid anions is expressed in milligrams of acid anions per gram of rock

carboxylic acid anions. The maximum aqueous concentrations of organic acid anions generated from the brown coal samples during hydrous pyrolysis are given in Table 2. Redox reactions between the hydrocarbon and mineral oxidants may be an additional source of significant amounts of OAA in the basin.

3.2 Modeling Kinetically Controlled Reactions

The first step in evaluating a kinetically controlled reaction is to establish a time-temperature profile for the unit of interest. Figure 23 is a burial history diagram for the Barracouta Field in the Gippsland Basin; included in the figure are the isotherms and the entrance to the liquid hydrocarbon window. The chronostratigraphic data used for each stratigraphic unit in constructing the burial history diagram (Fig. 23) are from well completion reports, but are corrected for decompaction. The thermal regime during the development of the Gippsland Basin was estimated using a modified McKenzie rift basin model (McKenzie 1978). The temperature evolution of layers of interest was calculated on the basis of burial history, thermal regime, lithology, and

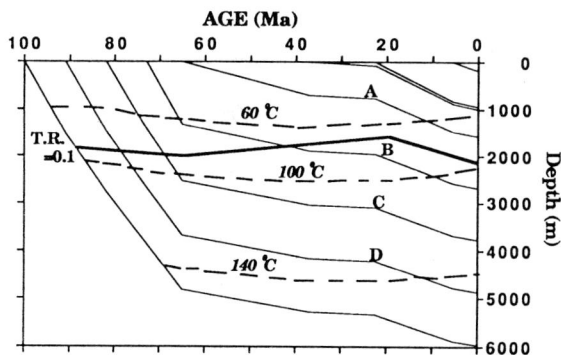

Fig. 23. Burial history and maturation modeling of the Barracouta field. The *thick solid line* represents the entrance to the liquid hydrocarbon window (transformation ratio = 0.1). The *dashed lines* are isothermal lines. A, B, C, D mark the layers for diagenetic modeling

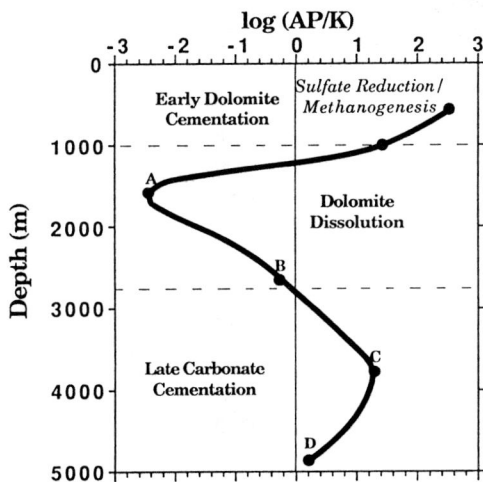

Fig. 24. Present-day dolomite stability trend with depth in the Barracouta field. *AP* Activity product; *K* equilibrium constant. A, B, C, D represent the depths of the layers modeled in Fig. 23

changing thermal conductivity (Yin 1988). The activation energy for hydrocarbon generation was obtained from hydrous pyrolysis determinations on Gippsland Basin brown coal samples (Yin et al. 1993).

In the Gippsland Basin, the reactions that have been modeled kinetically are acetate (OAA) generation from kerogen, acetate decarboxylation (i.e., HCO_3^- generation), hydrocarbon generation, and abiotic thermocatalytic sulfate reduction by hydrocarbons. Kinetic parameters for these reactions are from the following sources: acetate generation and decarboxylation from Yin et al (1993); hydrocarbon generation from Yin et al. (1993); sulfate reduction from Siebert (1985). A maximum concentration of acetate of 5000 ppm was assumed in evaluating the time-concentration profiles. The time-concentration profiles for acetate, HCO_3^-, and H_2S for layers presently at depths of 1500 m (A), 2800 m (B), 3600 (C), and 4800 m (D) (Figs. 23 and 24), plus the water evolution interpretation, were exported as input data to a rock/water equilibrium simulator, SOLMINEQ. 88 (PC compatible) with extended data base.

3.3 Rock/Water Equilibrium Simulator

Several alternative data output sets were available from the rock/water equilibrium simulator. For example, the mineral stability output,

$$\text{saturation index, SI} = \frac{\text{activity product}}{\text{equilibrium constant}}$$

could have been plotted versus time for any of the modeled layers. However, we chose to plot the mineral stability data against present-day depth (Fig. 24). In this way, the predicted mineral stability could be compared directly with porosity/depth plots. Figure 24 shows the predicted mineral stability for dolomite plotted against present-day depth.

Our interpretation of Fig. 24 is that there should be a significant dolomite dissolution event in the Gippsland Basin at a present-day depth from approximately 1200 to 2800 m. Over this depth interval, the log (AP/K) values are negative and dolomite typically is undersaturated. Points A, B, C, and D correspond to layers presently at 1500, 2800, 3600, and 4800 m depth. The next step in this discussion is to evaluate the quality of the porosity anomaly prediction for the 1200 to 2800 m present-day depth interval.

3.4 Positive Porosity Anomaly

Figure 16 is a porosity/depth plot for the Gippsland Basin. Superimposed on this plot are the curves of porosity loss due to mechanical compaction for sandstones composed of 50 and 75% quartz (Pittman and Larese 1991). We use the Pittman-Larese curves as specimen porosity baselines. Depending on the quartz content of the specific sandstone, points falling to the right of the baseline curve are considered positive anomalies, and points to the left of the curve are considered negative anomalies. Using this criterion, the most significant positive porosity anomaly in the Gippsland Basin occurs within the present-day depth interval 1400 to 2600 m (Fig. 16). Performing the same exercise on the porosity/depth data in Bodard et al. (1984) delineates a significant porosity anomaly in the present-day depth interval 1600 to 2600 m (Fig. 16). Porosity values as high as 30+% and permeability values as high as 2 darcies have been reported from this depth interval. The predicted positive porosity anomaly resulting from dolomite dissolution (1200 to

2800 m; see Fig. 24) matches well the observed positive porosity anomaly (Fig. 16). The predicted positive porosity anomaly is based on the assumption that OAA are an important component in the alkalinity of the pore fluid at temperatures > 80 °C. Moreover, in the kinetic modeling, only OAA derived from the thermocatalytic cleavage of OAA from kerogen was considered. It should be noted that the entrance to the liquid hydrocarbon window occurs at a present-day depth of 2100 m, so redox reactions (between hydrocarbons and mineral oxidants) may have added significant amounts of OAA to the fluid phase at a depth within the depth range of the observed positive porosity anomaly. In the case of the Gippsland Basin, the OAA from the thermocatalytic cleavage of kerogen and the OAA from redox reactions between hydrocarbons and mineral oxidants are overlapping and additive products.

4 Conclusions

A systematic and sequential set of carbonate reactions characterizes most clastic source/reservoir rock systems during progressive burial. Typically, this sequence of carbonate reactions with increasing thermal exposure (depth of burial) is as follows: (1) formation of early carbonate cements that preserve IGV; (2) dissolution of the early carbonate cements; (3) formation of late carbonate cements (usually ferroan), again preserving IGV; and (4) if temperatures are high enough, dissolution of the late carbonate cements. This carbonate reaction sequence is responsible for windows of opportunity for porosity enhancement (positive porosity anomalies). The zones of volumetrically important porosity enhancement are the result of carbonate dissolution.

If this sequence of carbonate reactions is coupled with parallel organic reactions (sulfate reduction, methanogenesis, OAA production, decar-boxylation, and abiotic sulfate reduction), a predictive, process-oriented model can be constructed for the carbonate reactions. The model consists of three essential operations: (1) interpretation of reaction pathways (to estab-lish the porosity-determining reactions and the evolution of the cation chemistry of the pore water); (2) kinetic modeling of the material added to the fluid phase by organic reactions and abiotic sulfate reduction; and (3) simulation of the rock/water interactions in either time or temperature space. Integrating these three operations has enabled us to predict zones of optimum potential porosity enhancement (maximum positive porosity anomalies) in a clastic sequence undergoing progressive burial. These modeling techniques work best when the source and reservoir rocks are in close proximity, as in the Gippsland Basin, Australia.

The validity of the results of this modeling techniques is based on correctly assessing the mechanisms controlling the alkalinity and buffering characteristics of the rock-fluid system of interest. In the example illustrated, it was assumed: (1) that HCO_3^-, CO_3^{-2}, OH^-, $H_3SiO_4^-$, HS^-, OAA, OH^-, and H^+ determine the alkalinity of the fluid phase; (2) that during intermediate burial (~80 to 130 °C), organic acids, OAA, and CO_2 dominate the alkalinity; (3) that with increasing temperature, P_{CO_2} increases significantly due to progressive decarboxylation (destruction) of organic acids and OAA; (4) that in a rock-fluid system characterized by the presence of carbonate minerals, it is the reaction between the fluid and fast-reacting carbonates that determines the buffering characteristics of the system; and (5) that during deep burial (>130 °C), abiotic thermal sulfate reduction is an important mechanism.

We suggest that these assumptions that emphasize the importance of organic acids have general applicability to modeling clastic diagenetic systems. Hopefully, this chapter illustrates the importance of considering the role of organic acids in clastic diagenesis, particularly when trying to understand the spatial distribution of carbonate cementation/decementation events in clastic hydrocarbon reservoirs during progressive burial.

Acknowledgments. The original manuscript was reviewed by David Copeland (University of Wyoming), E.D. Pittman, and M.D. Lewan, all of whom made numerous helpful suggestions. This study was funded by the Gas Research Institute under contract number 5089-260-1894. Ronald C. Surdam wishes to acknowledge very useful discussion with Bob Siebert of Conoco and D. Pevear of EXXON Production Research. We express our appreciation to EXXON Production Research Company for providing us with 250 thin sections and for analyzing the carbon and oxygen isotopes and the chemical composition of dolomite cements for us. S.W. Boese assisted us with analytical work on water samples. F. Miknis, Western Research Institute, provided us with the NMR analyses of the source rock samples. We also acknowledge the significant contributions made to this paper by Donald B. MacGowan, Thomas L. Dunn, Henry P. Heasler, and Laura S. Crossey. Preparation of the manuscript was greatly expedited by Alice S. Rush and Celise Hand.

References

Almon WR (1974) Petroleum-forming reactions: clay catalyzed fatty acid decarboxylation. PhD Thesis, University of Missouri, Columbia, 117 pp
Anderson TF, Arthur MA (1983) Stable isotopes of oxygen and carbon and their applications to sedimentological and paleoenvironmental problems. In: Stable isotopes in sedimentary geology. Soc Econ Paleontol Mineral Short Course Notes 10, Tulsa, OK, pp 1-1–1-151
Berner RA (1980) Early diagenesis: a theoretical approach. Princeton University Press, Princeton, 241 pp

Berzelius JJ (1839) Untersuchung des Wassers der Porla-Quelle. Ann Phys Chem 105: 1–37 (Annalen der Physik und Chemie was superseded by Annalen der Physik in 1899)

Bjørlykke K (1984) Formation of secondary porosity: how important is it? In: McDonald D, Surdam R (eds) Clastic diagenesis. Am Assoc Pet Geol Mem 37: 277–286

Bodard J, Wall V, Cass RA (1984) Diagenesis and evolution of Gippsland Basin reservoirs. Aust Pet Explor Assoc J 24: 314–335

Boles JR (1978) Active ankerite cementation in the subsurface Eocene of southwest Texas. Contrib Mineral Pet 68: 13–22

Boles JR (1981) Clay diagenesis and effects on sandstone cementation (Case history from the Gulf Coast Tertiary). In: Longstaffe F (ed) Short course in clays and the resource geologist. Mineral Assoc Canada, Calgary, May 1981, pp 148–168

Boles JR (1987) Six million year diagenetic history, North Coles Levee, San Joaquin Basin, California. In: Marshall J (ed) Diagenesis of sedimentary sequences. Geol Soc Spec Publ 36, pp 191–200

Boles JR, Franks SG (1979) Clay diagenesis in Wilcox sandstones of southwest Texas: implications of smectite diagenesis on sandstone. J Sediment Pet 49: 55–70

Carothers WW, Kharaka YK (1978) Aliphatic acid anions in oil field waters – implications for origin of natural gas. Am Assoc Pet Geol Bull 62: 2441–2453

Crossey LJ, Surdam RC, Lahann RW (1986) Application of organic/inorganic diagenesis to porosity prediction. In: Gautier D (ed) Roles of organic matter in sediment diagenesis. Soc Econ Paleontol Mineral Spec Publ 38, Tulsa, OK, pp 147–156

Curtis CD (1987) Inorganic Geochemistry and petroleum exploration. In: Brooks J, Welte D (eds) Advances in petroleum geochemistry, vol 2. Academic Press, London, pp 91–140

Curtis CD, Coleman ML (1986) Controls on the precipitation of early diagenetic calcite, dolomite, and siderite concretions in complex depositional sequences. In: Gautier D (ed) Roles of organic matter in sediment diagenesis. Soc Econ Paleontol Mineral Spec Publ 38, Tulsa, OK, pp 23–34

Eglinton TI, Curtis CD, Rowland S J (1987) Generation of water-soluble organic acids from kerogen during hydrous pyrolysis: implications for porosity development. Mineral Mag 51: 495–503

Folk RL (1974) Petrology of Sedimentary Rocks. Hemphill's, Austin, 169 pp

Franks SG, Forester RW (1984) Relationships among secondary porosity, pore fluid chemistry, and carbon dioxide, Texas Gulf Coast. In: McDonald D, Surdam R (eds) Clastic diagenesis. Am Assoc Pet Geol Mem 37, pp 63–80

Fritz P, Smith DCW (1970) The isotopic composition of secondary dolomite. Geochim Cosmochim Acta 34: 1161–1173

Gautier DL (1985) Interpretation of early diagenesis in ancient marine sediments. In: Relationship of organic matter and mineral diagenesis. Soc Econ Paleontol Mineral Short Course Notes 17, Tulsa, OK, pp 6–78

Giles MR, Marshall JD (1986) Constraints on the development of secondary porosity in the subsurface: re-evaluation of processes. Mar Pet Geol 3: 243–255

Hardie LA, Eugster HP (1970) The evolution of closed-basin brines. In: Morgan B (ed) 50th Anniversary Symp, Mineral Assoc Am Spec Publ 3, pp 273–290

Harrison WJ (1989) Modeling fluid/rock interactions in sedimentary basins. In: Cross T (ed) Quantitive dynamic stratigraphy. Prentice-Hall, Englewood Cliffs, pp 195–235

Hay RL (1963) Stratigraphy and zeolitic diagenesis of the John Day Formation of Oregon. Univ Calif Publ Geol Sci 42: 199–262

Hocking JB (1972) Geologic evolution and hydrocabon habitat in Gippsland Basin. J Aust Pet Explor Assoc 12: 132–137

Holland HD, Borcsik M (1965) On the solution and deposition of calcite in hydrothermal systems. Symp Problems of postmagmatic ore deposits, vol 2. Prague, pp 364–374

Houseknecht DW (1987) Assessing the relative importance of compaction processes and cementation to reduction of porosity in sandstones. Am Assoc Pet Geol Bull 71: 633–642

Houseknecht DW (1988) Intergranular pressure solution in four quartzose sandstones. J Sediment Pet 58: 228–246

Houseknecht DW, Hathon LA (1987) Petrographic constraints on models of intergranular pressure solution in quartose sandstones. Appl Geochem 2: 507–521

Hower J, Eslinger E, Hower M (1976) Mechanisms of burial metamorphism of argillaceous sediments. 1. Mineralogical and chemical evidence. Geol Soc Am Bull 87: 725–737

Hutcheon I, Abercrombie H (1989) Carbon dioxide in clastic rocks and silicate hydrolysis. Geology 18: 541–544

James EA, Evans PR (1971) The stratigraphy of offshore Gippsland Basin. Aust Pet Explor Assoc J 11: 71–74

Jansa LF, Urrea VHN (1990) Geology and diagenetic history of overpressured sandstone reservoirs, Venture gas field, offshore Nova Scotia, Canada. Am Assoc Pet Geol Bull 74: 1640–1658

Leder F, Park WC (1986) Porosity reduction in sandstones by quartz overgrowth. Am Assoc Pet Geol Bull 70: 1713–1728

Levandowski DW, Kaley ME, Silverman SR, Smalley RG (1973) Cementation in Lyons Sandstone and its role in oil accumulation, Denver Basin, Colorado. Am Assoc Pet Geol Bull 57: 2217–2244

Lewan MD (1992) Water as a source of hydrogen and oxygen in petroleum formation. Am Chem Soc, Fuel Chem Div Preprints, vol 37, pp 1643–1649

Lundegard PD, Land LS, Galloway WE (1984) Problem of secondary porosity: Frio Formation (Oligocene), Texas Gulf Coast. Geology 12: 399–402

MacGowan DB, Surdam RC (1990a) Carboxylic acid anions in formation waters, San Joaquin Basin and Louisiana Gulf Coast, U.S.A. Implications for clastic diagenesis. Appl Geochem 5: 687–701

MacGowan DB, Surdam RC (1990b) The importance of organic-inorganic reactions in modeling water-rock interactions during progressive diagenesis of sandstones. In: Melchior D, Bassett R (eds) Chemical modeling in aqueous systems. II. American Chemical Society, Washington DC, pp 494–507

Machel HG (1987) Some aspects of sulphate-hydrocarbon redox reactions. In: Marshal J (ed) Diagenesis of sedimentary sequences. Geol Soc Spec Publ 36, pp 15–28

Martell AE, Smith RM (1977) Critical stability constants, vol III. Other organic ligands. Plennum Press, New York, 495 pp

McBride EF (1977) Sandstones. In: Jonas E, McBride E (eds) Diagenesis of sandstone and shale: application to exploration for hydrocarbons. Continuing Education Program Publ 1. Department of Geological Sciences, The University of Texas at Austin, pp 1–20

McKenzie DP (1978) Some remarks on the development of sedimentary basins. Earth Planet Sci Lett 40: 25–32

McMahon PB, Chapelle FH (1991) Microbial production of organic acids in aquitard sediments and its role in aquifer geochemistry. Nature 349: 233–235

McMahon PB, Chapelle FH, Falls WF, Bradley PM (1991) Role of microbial processes in linking sandstone diagenesis with organic-rich clays. J Sediment Pet 62: 1–10

Meshri ID (1986) On the reactivity of carbonic and organic acids and the generation of secondary porosity. In: Gautier D (ed) Roles of organic matter in sediment diagenesis. Soc Econ Paleontol Mineral Spec Publ 38, pp 123–128

Milliken KL, Land LS (1991) Reverse weathering, the carbonate feldspar system and porosity evolution during burial of sandstones. Am Assoc Pet Geol Bull 75: 636 (Abstr)

Moncure GK, Lahann RW, Siebert RM (1984) Origin of secondary porosity and cement distribution in a sandstone/shale sequence from the Frio Formation. In: McDonald D, Surdam R (eds) Clastic diagenesis. Am Assoc Pet Geol Mem 37, pp 151–161

Moraes MAS (1989) Diagenetic evolution of Cretaceous-Tertiary turbidite reservoirs, Campos Basin, Brazil. Am Assoc Pet Geol Bull 73: 598–612

Muffler LJP, White DE (1969) Active metamorphism of upper Cenozoic sediments in Salton geothermal field and the Salton trough, southeastern California. Geol Soc Am Bull 80: 157–182

Partridge AD (1976) The geological expression of eustacy in the early Tertiary of the Gippsland basin, Australia. Aust Pet Explor Assoc J 16: 73–79

Pittman ED, Larese RE (1991) Compaction of lithic sands: experimental results and applications. Am Assoc Pet Geol Bull 75: 1279–1299

Reed CL, Hajash A (1992) Dissolution of granitic sand by pH-buffered carboxylic acids: a flow-through experimental study at 100 °C and 345 bars. Am Assoc Pet Geol Bull 76: 1402–1416

Schmidt V, McDonald DA (1979) The role of secondary porosity in the course of sandstone diagenesis. In: Scholle P, Schluger P (eds) Aspects of diagenesis. Soc Econ Paleontol Mineral Spec Publ 26, pp 175–207

Schultz JL, Boles JR, Tilton GR (1989) Tracking calcium in the San Joaquin Basin, California: a strontium isotope study of carbonate cements at North Coles Levee. Geochim Cosmochim Acta 53: 1991–1999

Sclater JG, Christie PA (1980) Continental stretching: an explanation of the post-mid-Cretaceous subsidence of the central North Sea Basin. J Geophys Res 85: 3711–3739

Senfl H (1871) Vorläufige Mitteilungen über die Humussubstanzen und ihr Verhalten zu den Mineralien. Z Dtsch Geol Ges 23: 665–669

Shackleton NJ, Kennett JB (1973) Paleotemperature history of the Cenozoic and the initiation of Antarctic glaciation: oxygen and carbon isotope analysis in DSDP sites 277, 279, and 281. Initial Rep Deep Sea Drilling Project 29: 743–755

Siebert RM (1985) The origin of hydrogen sulfide, elemental sulfur, carbon dioxide and nitrogen in reservoirs. Soc Econ Paleontol Mineral, Gulf Coast Sectional Meet, Abstr Programs, Austin, TX, vol 6, pp 30–31

Smith JT, Ehrenberg SN (1989) Correlation of carbon dioxide abundance with temperature in clastic hydrocarbon reservoirs: relationship to inorganic chemical equilibrium. Mar Pet Geol 6: 129–135

Sprengel C (1826) Über Pflanzenhumus, Humussäure und humussaure Salze. Kastners Arch Gesammte Naturlehre 8: 145–220

Stoessell RK, Pittman ED (1990) Secondary porosity revisited: the chemistry of feldspar dissolution by carboxylic acids and anions. Am Assoc Pet Geol Bull 74: 1795–1805

Surdam RC, Crossey LJ (1985a) Organic-inorganic reactions during progressive burial: key to porosity and permeability enhancement and preservation. Philos Trans R Soc Lond Ser A 315: 135–156

Surdam RC, Crossey LJ (1985b) Mechanism of organic/inorganic interactions in sandstone/ shale sequences. In: Relationship of organic matter and mineral diagenesis. Soc Econ Paleontol Mineral Short Course Notes 7, Tulsa, OK, pp 177–232

Surdam RC, MacGowan DB (1992) Coupled predictive models of diagenesis in sand-shale successions during progressive burial. In: Kharaka Y, Maest A (eds) Water-rock interaction, vol 2. Proc Water-rock interactions, 7th Symp, Park City, Utah. Balkema, Rotterdam/Brookfield, pp 1205–1208

Surdam RC, Boese SW, Crossey LJ (1984) The chemistry of secondary porosity. In: McDonald D, Surdam R (eds) Clastic diagenesis. Am Assoc Pet Geol Mem 37, pp 127–151

Surdam RC, Crossey LJ, Hagen ES, Heasler HP (1989a) Organic-inorganic interactions and sandstones diagenesis. Am Assoc Pet Geol Bull 73: 1–32

Surdam RC, MacGowan DB, Dunn TL (1989b) Diagenetic pathways of sandstone and shale sequences. Univ Wyo Contrib Geol 27: 21–31 and plate

Surdam RC, Dunn TL, MacGowan DB, Heasler HP (1989c) Conceptual models for the prediction of porosity evolution, with an example from the Frontier Sandstone, Bighorn Basin, Wyoming. In: Coalson E (ed) Sandstone reservoirs – 1989. Rocky Mountain Assoc Geol, Denver, CO, pp 7–28, 300–303

Surdam RC, Dunn TL, Heasler HP, MacGowan DB (1989d) Porosity evolution in sandstone/shale systems. Mineral Assoc Can Diagenesis Short Course Notes, pp 61–133

Surdam RC, Jiao ZS, MacGowan DB (1993) Redox reactions involving hydrocarbons and mineral oxidants: a mechanism for significant porosity enhancement in sandstones. Am Assoc Pet Geol Bull 77: 1509–1518

Taylor RR (1990) The influence of calcite dissolution on reservoir porosity in Miocene sandstones, Picaroon field, offshore Texas Gulf Coast. J Sediment Pet 60: 322–334

Threlfall WE, Brown BR, Criffith BR (1976) Gippsland Basin offshore: economic geology of Australia and Papua New Guinea. 3. Petroleum. Aust Inst Mining Metall Monogr 7: 41–67

Tissot BP, Espitalié J (1975) L'Evolution thermique de la matière organique des sediments: applications d'une simulation mathematique. Rev Inst Fr Pét 39: 743–777

Weeks LG, Hopkins BM (1967) Evolution of the Tasman Sea reappraised. Earth Planet Sci Lett 36: 77–84

Willey LM, Kharaka YK, Presser TS, Rapp JB, Barnes I (1975) Short chain aliphatic acid anions in oilfield waters and their contribution to the measured alkalinity. Geochim Cosmochim Acta 39: 1707–1711

Yin P (1988) Generation and accumulation of hydrocarbons in the Gippsland Basin, S.E. Australia. PhD Thesis, University of Wyoming, Laramie, 249 pp

Yin P, Surdam RC (1985) Naturally enhanced porosity and permeability in the hydrocarbon reservoirs of the Gippsland Basin, Australia. In: Ewing R (ed) Proc 1st Wyoming Enhanced Oil Recovery Symp, Enhanced Oil Recovery Institute, University of Wyoming, pp 79–109

Yin P, Surdam RC, Boese S, MacGowan DB, Miknis F (1993) Simulation of hydrocarbon source rock maturation by hydrous pyrolysis. In: Andrew S, Strook B (eds) Fiftieth anniversary field conference guidebook. Wyoming Geol Assoc, Casper, pp 359–373

Chapter 14   How Important Are Organic Acids in Generating Secondary Porosity in the Subsurface?

M.R. Giles[1], R.B. de Boer[1], and J.D. Marshall[2]

Summary

The hypothesis that organic acids are responsible for the creation of significant volumes of secondary porosity and the enhancement of aluminium mobility in the subsurface is reviewed against data from reservoir intervals, laboratory experiments and mass-balance considerations. The hypothesis appears to fail on all counts, notably:

1. Observational evidence does not support a sudden increase in abundance of secondary porosity over the temperature range at which carboxylic acids are most abundant.
2. Experimental data on the dissolution of feldspars in the presence of organic acids show little evidence to support either enhanced dissolution kinetics or increased Al mobility unless the pH of the experiment was unrealistically low for natural pore fluids.
3. Increased Al mobility due to complexing by organic acids is unlikely in natural pore waters due to the competing effects of other ions for the organic acid ligands.
4. Simple mass-balance considerations show that source rocks would have to be unreasonably abundant to generate enough carboxylic acids to account even for a few percent secondary porosity.

1   Introduction

Interest in the creation of secondary porosity by the dissolution of framework grains and cements is not new. Early contributions from Heald and Larese (1973), Rowsell and DeSwardt (1974) and Parker (1974) recognized

---

[1] Koninklijke/Shell Exploratie en Produktie Laboratorium, P.O. Box 60, 2280 AB Rijswijk, The Netherlands
[2] Dept. of Earth Sciences, The Jane Herdman Laboratories, University of Liverpool, P.O. Box 147, Liverpool L69 3BX, UK

the importance of the dissolution of feldspars in the creation of fabric-dependent macropores. However, the debate on the causes of secondary dissolution porosity began in earnest following publication of SEPM Special Publication No. 26 *Aspects of Diagenesis*. This volume contained a number of influential papers which dealt with secondary porosity, including that by Schmidt and McDonald (1979), who suggested that CO_2 generated from organic matter in source rocks would create a weak acid, which could dissolve carbonates and feldspars in the deep subsurface.

The presence of organic acids in oil-field groundwaters was recognized in the early part of this century (Rogers 1917). Their importance and wide-spread occurrence was, however, not highlighted until much more recently (Willey et al. 1975; Carothers and Kharaka 1978). The occurrence of organic acids in oil-field waters provides an attractive mechanism for creating secondary porosity and for the transportation of ions as organo-metallic complexes. Complexing, including organo-metallic complexing, has a long history in the chemical literature and has long been considered important in controlling dissolution and transportation reactions (e.g., Barton 1959). The potential role of organic acids in the transportation of ions in ore-forming solutions was recognized by Giordano and Barnes (1981). Their potential role both in creating secondary porosity and in mobilizing aluminium ions that would otherwise result in the precipitation of clay minerals during the dissolution of aluminosilicates was recognized by Surdam et al. (1984). Numerous papers have linked carboxylic acids to the occurrence of secondary porosity in reservoirs (e.g., Burley 1986; Edman and Surdam 1986; Goodchild and Whitaker 1986; MacGowean and Surdam 1988) but evidence for the link has remained largely circumstantial.

In considering possible mechanisms it must be borne in mind that secondary porosity need not be linked to the generation of acids in source rocks. Mineral dissolution has been linked to meteoric water penetration (Bjørlykke 1983), acids generated by inorganic reactions within shales (Bjørlykke 1983), cooling formation waters (Giles and de Boer 1989) and to the natural thermodynamic instability of feldspars and other high temperature aluminosilicates in the diagenetic realm (Giles and de Boer 1990). These mechanisms are all valid hypotheses and no definite causal link to secondary porosity is established. Petrography can only document the presence of holes in a rock and at best give a relative time of formation.

Proposed mechanisms for secondary porosity formation can be tested using mass-balance and mass-transport considerations (Bjørlykke 1984; Lundegard et al. 1984; Giles and Marshall 1986). Simple mass-balance calculations, for example, suggest that the carbon dioxide generating capacity of the organic matter in source rocks is insufficient to produce the observed secondary porosity. Similarly, there are problems in transporting under-saturated solutions: at the flow rates typical of subsurface environments any acidic solutions are likely to react with and be neutralized by minerals in the source rocks themselves and by minerals in any migration conduits between the source and reservoir.

Given the absence of direct evidence for the cause of dissolution of minerals in reservoirs, how is it possible to determine which hypotheses are feasible and which are not? In our opinion, the best way to test individual hypotheses is to compare the quantitative predictions of the hypothesis with observational data, taking into account the relevant experimental data and theoretical calculations based around the reactions concerned. Only hypotheses that meet these tests can be considered likely contenders. Those that do not meet these tests can be rejected and only require re-evaluation if the hypothesis is modified in some substantive way or if the data used to test the model are proved to be incorrect. In this chapter, the hypothesis "that organic acids are significant in the generation of secondary porosity in the deep subsurface" is tested against all the available evidence.

2 Distribution of Secondary Porosity

As secondary porosity is very widespread and probably polygenetic, studies of its distribution with depth cannot conclusively prove or disprove any one mechanism for its formation. At best we can assess whether a range of data sets are compatible or incompatible with our proposed mechanism.

The stability of organic acids produced during the thermal maturation of kerogen is limited by bacterial degradation below 80 °C and by thermally driven decarboxylation at temperatures above 135°C (Fig. 19 in Surdam et al. 1984). Carboxylic acid concentrations in the subsurface are likely to peak in a depth range constrained by these temperatures and pore water analyses clearly support this suggestion (Lundegard and Kharaka 1990). If the acids were genetically linked to secondary porosity formation, a peak in secondary porosity occurrence would be expected over the same range of temperatures and burial depths. This should in turn manifest itself as an excursion from the normal trend of decreasing porosity with depth. Such secondary porosity anomalies are only likely to be noted if the ions derived by dissolution of carbonates and aluminosilicates are swept out of the system, perhaps chelated by organic ligands, rather than recombining in situ to give mineral cements. Such anomalies should be expected to persist to greater depths, although cementation might be expected to gradually destroy them with further burial.

Secondary porosity seems to be very widespread in clastic and carbonate sediments and not restricted to particular depth intervals (unpubl. data and Bloch 1991a). For instance, regional porosity/depth data from Miocene channel sands from the Far East (Fig. 1) span the depth range from 4 000 ft (1219.2 m) to 10 000 ft (3048 m). Secondary porosity ranging up to 14% bulk

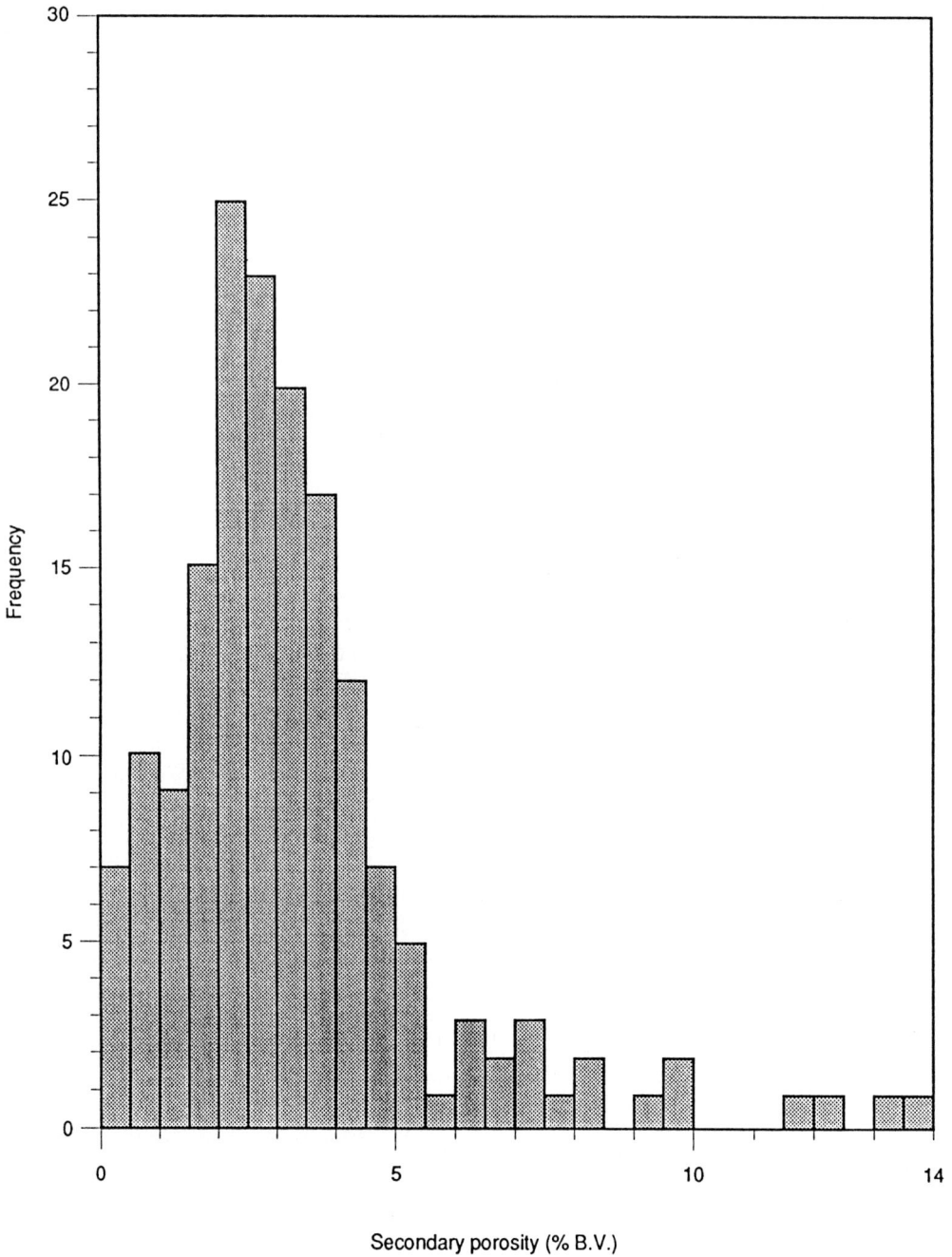

Fig. 1. Point-counted secondary porosity present in Miocene channel sandstones from a Far East location, based on data from over 20 wells

volume (B.V.) and with a modal value of around 3% B.V. occurs through-out the cores. Similar observations can be made from other reservoir sequences such as the Brent Group of the North Sea (see Giles et al. 1992). In the Viking Graben and East Shetland Basin where the Brent Group occurs, as in many other areas, a sufficiently large number of wells have been drilled either down dip or off-structure to show that secondary porosity is not just related to hydrocarbon traps. Secondary porosity is therefore present in measurable abundances throughout the sediment column.

Careful analysis of porosity-depth relations using techniques similar to those outlined by Bloch (1991b) can be used to produce porosity-depth trends for sequences containing a variety of depositional sandstone types. Primary differences in sorting, grain size, clay content and composition must be removed or normalized in order to produce comparable porosity/depth trends. The resulting plots give insights into the effects of secondary porosity on the total porosity of the rock. For instance, the data shown in Fig. 1 demonstrate an overall trend of decreasing porosity with increasing burial depth (Fig. 2) despite highly variable amounts of secondary porosity. Such trends would not occur if the production of secondary porosity caused systematic increases in porosity over discrete depth intervals.

In areas where the current burial depth is also the maximum burial depth, high porosity excursions from the normal trend of decreasing porosity with increasing burial depth do not only occur at depths that correspond to the temperature range where organic acids are common. Porosity excursion seem to occur at any depth (Figs. 3 and 4) and certainly at lower tempera-tures than would be expected from the organic acid model. In some in-stances, the secondary porosity anomalies (as in Fig. 3) may be linked to subaerial exposure surfaces or unconformities. However, in many other cases the detailed stratigraphic analysis required to prove such a hypothesis is not available.

In many other instances, secondary porosity commonly forms an in-creasing proportion of the total porosity with increasing burial depth. In the Brent sands of the North Sea (Fig. 5) the increase in secondary porosity is paralleled by a decrease in feldspar content (Fig. 6). Similar patterns have been reported for the Tertiary of the Gulf Coast (Loucks et al. 1979). In both instances, however, no increase in total porosity occurs, suggesting that the reaction products are redistributed on a small scale rather than swept out of the system. The increase in proportion of secondary porosity is accentuated by preferential cementation within primary pore space, which often tends to be more prone to occlusion than secondary pores. Carbonates, unlike feldspars, do not disappear over the same interval. Although in broad terms the rapid disappearance of feldspars and the increasing proportion of secondary porosity are compatible with the organic acid hypothesis, any temperature-dependent model of feldspar dissolution could potentially ex-plain the increasing proportion of secondary porosity with increasing burial depth. The continued presence of carbonate and the volume conservation

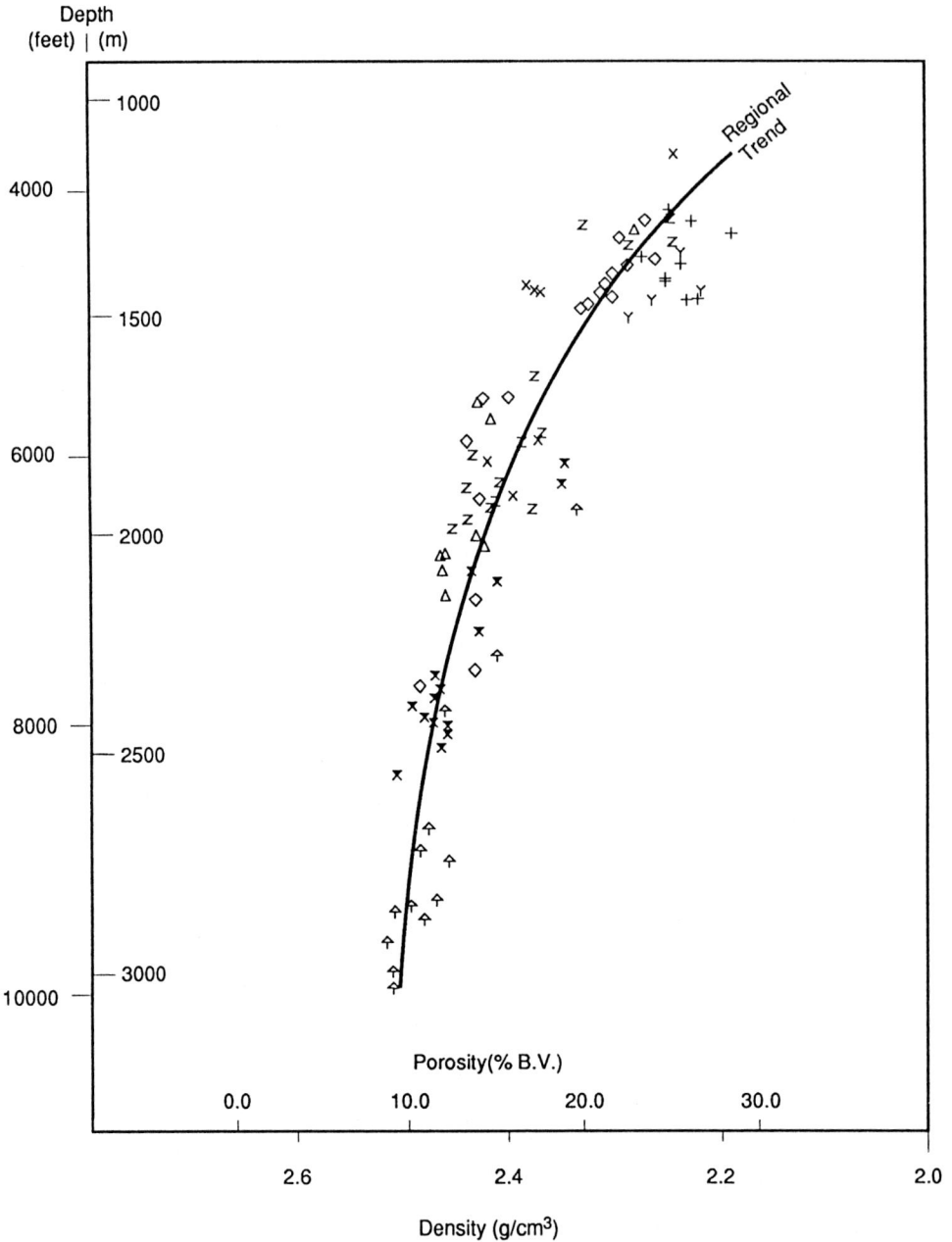

Fig. 2. Porosity (and density) – depth behaviour of Miocene channel sandstones from a Far East location. Note the predictable decline of porosity with increasing burial depth. Deviations from the regional trend are not in general related to the abundance of secondary porosity. This occurs despite highly variable amounts of secondary porosity present in these channel sands (see Fig. 1). Each *symbol type* refers to a well

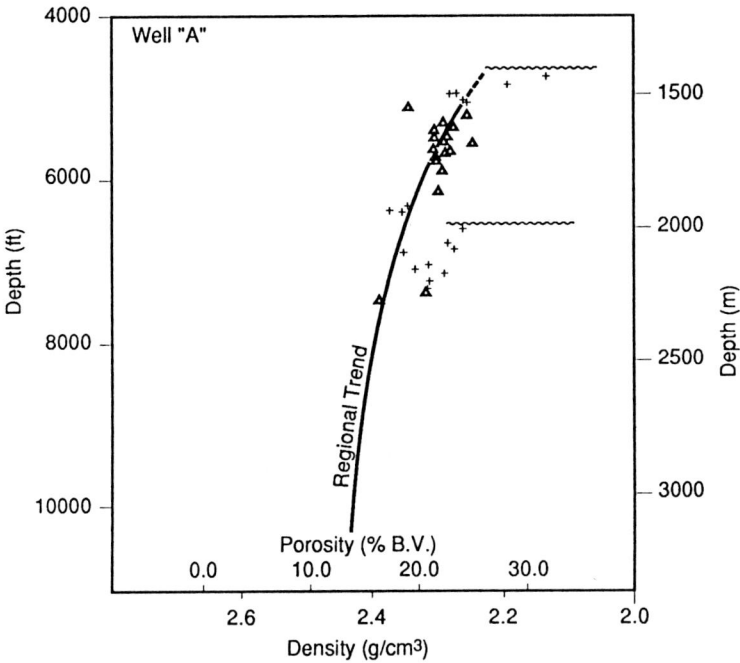

Fig. 3. Porosity anomalies shown by cross-bedded sands at unconformities in well A, from the Far East. *Symbol type* refer to wells

suggested by the porosity-depth relations are difficult to explain using the organic acid model (see Fig. 17 in Surdam et al. 1984).

3   Carboxylic Acids in Subsurface Pore Waters

An appraisal of the importance of organic acids in the subsurface requires information about typical fluid compositions at different depths and temperatures. The data of Lundegard and Kharaka (1990; Fig. 1), demonstrates that even within the 80 to 140 °C temperature window there is a tremendous range of measured concentrations. Some published calculations concentrate largely on the maximum values measured on the grounds that these are typical for basins where acid generation is actively taking place (e.g., MacGowan and Surdam 1988, 1990), but an estimate of the mode, mean, standard deviation and range for each temperature interval would be more useful to help us to differentiate between normal and exceptional conditions.

Acetate appears to be the dominant organic acid anion present in the subsurface. Concentrations of acetate can reach 10 000 ppm (MacGowan and Surdam 1990), although typically concentrations exceeding 3000 ppm are

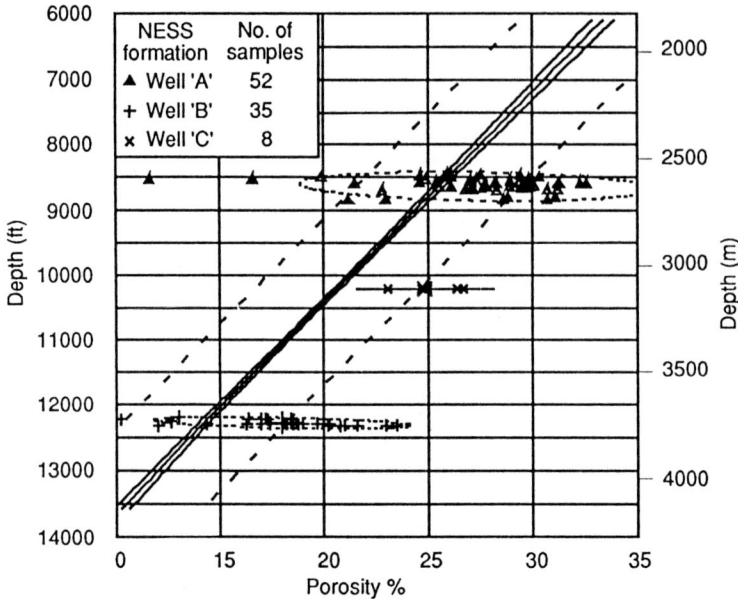

Fig. 4. Porosity anomalies recognized during a regional study of the Brent Group (Giles et al. 1992) This figure shows the porosity anomalies recognized in the Ness Formation. The normal trend of decreasing porosity with increasing burial depth has been established from the regional data set and is shown here by the regression line (*inner solid line*) together with the 90% confidence limits to the regression line (*linner solid hyperbolae*) and 90% confidence limits to porosity estimated from burial depth (*outer dashed hyperbolae*)

rare even within the temperature region in which they are known to be most abundant (Lundegard and Kharaka 1990). The typical range of acids and their concentrations within restricted temperature windows are seldom given in the literature. The occurrence of monocarboxylic acid anions, ranging from formate (C_1) to octanoate (C_8), and dicarboxylic acid anions, ranging from oxalic (C_2) to decandioate (C_{10}), have been reported (MacGowan and Surdam 1990). Undersaturated carboxylic acid anions such as *cis*-butendioate (maleic) and o-hydroxybenzoate (salicylate) have also been reported (MacGowan and Surdam 1990). Although one formation water from California with 2540 ppm of malonic acid has been reported (MacGowan and Surdam 1988), dicarboxylic acids and undersaturated species are generally present in total amounts "well below 200 ppm" (MacGowan and Surdam 1990).

Clear regional differences also exist. Data reported by Carothers and Kharaka (1978) show that in the temperature range 80–120 °C organic acids form less than 1000 mg/l of the total alkalinity in Gulf Coast formation waters, whereas in Californian formation waters up to 2000 mg/l of the total

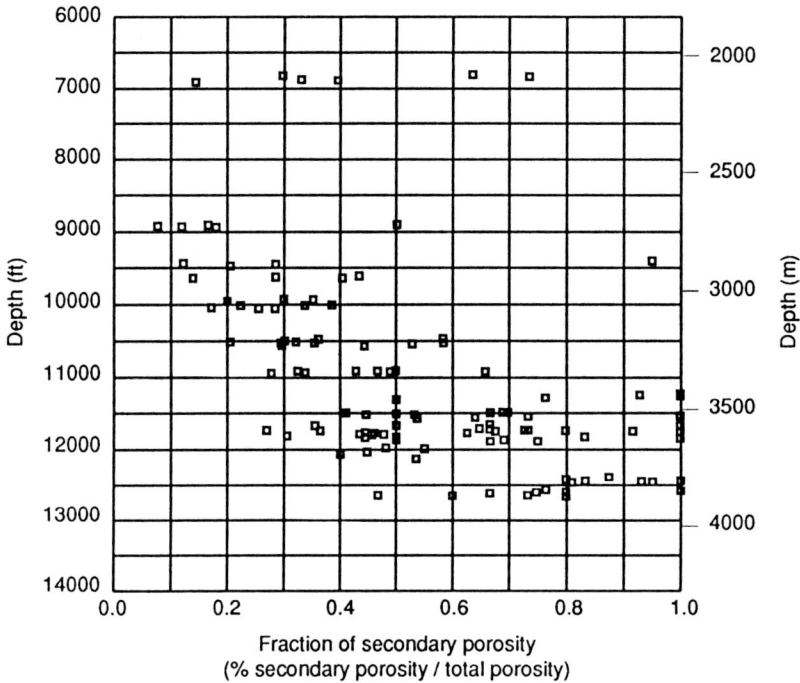

Fig. 5. The relationship between the fraction of secondary porosity (secondary porosity/ total porosity) as a function of depth shown by samples of the Etive Formation which forms part of the Brent Group of the North Sea. The secondary porosity in these samples is dominantly from the dissolution of feldspars. (After Giles et al. 1992)

alkalinity is formed by organic acids. Such large variations in abundance are yet to be fully explained.

A determination of the origin and migration history of organic acids is also important if we are to relate them to petrogenetic reactions. In this context, it is interesting to note that virtually all the organic acid analyses reported come from waters separated from a hydrocarbon phase at the well head. In other words, very little data are available on the composition of pore waters other than from producing oil and gas wells and that by their very nature they represent a very biased sample as oil companies try not to produce significant amounts of water with no associated hydrocarbons. Hydrocarbons are known to contain organic acids including carboxylic acid (Seifert and Howells 1969) and naphthenic acid (Louis 1967). Californian crude oils have been reported as containing up to 2.5% carboxylic acid. A presently unexplored avenue of research is therefore whether some of the acids reported from produced water associated with hydrocarbons are formed by the thermal degradation of the long-chained acids present in oil. Shorter-chained acids will tend to be more soluble in water and will there-

458 M.R. Giles et al.

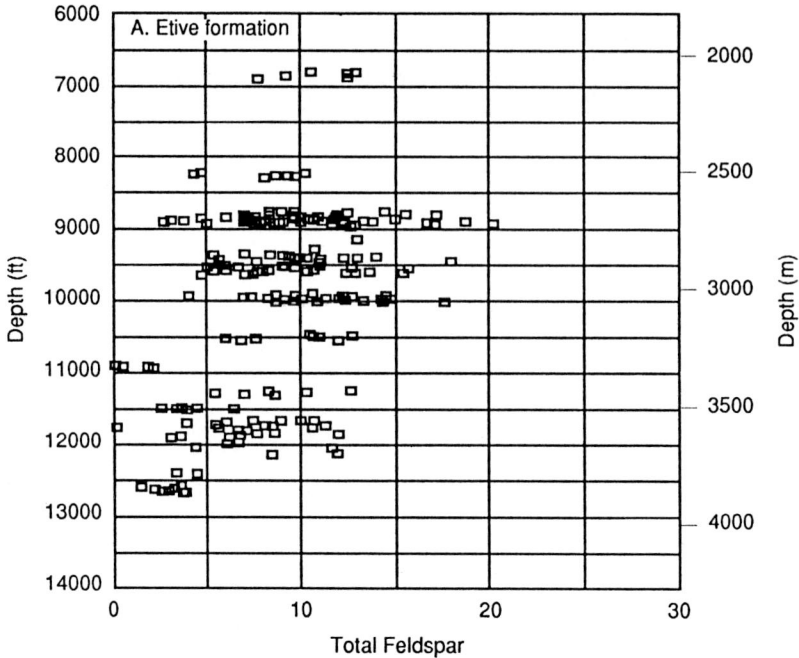

Fig. 6. The disappearance of feldspars in the Etive Formation as a function of depth. (After Giles et al. 1992)

fore diffuse into the water phase associated with the hydrocarbons. Furthermore, it is also unclear what the phase relations are amongst the organic acids, the oil and water phases during depressurization and cooling prior to sampling the water at the well head. Further work is required to test whether the high concentrations of organic acids recorded from some oilfield waters truly reflect their general activity in the subsurface waters or whether concentrations are only a feature of oil and gas fields.

4 Mineral Dissolution Experiments

Surdam et al. (1984) reported the results of a series of experiments in which feldspars were reacted with either acetic or oxalic acid. These experiments demonstrated high levels of Al in the filtrates, most notably when the reacting solutions contained between 10000 and 1000 ppm oxalic acid. Similar experiments were repeated using oxalic acid, with laumontite and with or without calcite. Only in those experiments where calcite was absent were there significant Al concentrations recorded in the final filtrate. In the experiments containing calcite, complexation of the organic acid anions by Ca^{2+} rather than Al^{3+} accounts for the experimental results.

In a further series of experiments (MacGowan and Surdam 1988), synthetic brines spiked with known amounts of carboxylic acids were reacted with Stevens Sandstone samples. After reaction at 100 °C for 2 weeks the samples with thousands of ppm of dicarboxylic acid anions showed significantly enhanced concentrations of Al (see their Fig. 5). Concentrations of 10 000 ppm acetate also resulted in unusually high Al concentrations. A similar experiment was conducted using a formation water sample from the San Joaquin Basin, preserved using nitric acid, and containing 5100 ppm of organic acid anions (including 2540 ppm dicarboxylic acid anions). The pH of this sample was adjusted to 5 and reacted with Stevens Sandstone. An increase in Al was again observed in the filtrate. As a consequence, MacGowan and Surdam (1988) concluded that this formation water was not in equilibrium with the sandstone and the results of both sets of experiments were taken (MacGowan and Surdam 1988, 1990) as conclusive evidence of the power of organic acids to both dissolve feldspars and chelate the liberated Al and thus inhibit the precipitation of clay minerals.

A detailed critique of the original experiments (Surdam et al. 1984) by Stoessell and Pittman (1990) has raised serious doubts on the validity of these experiments. These authors noted difficulties in accounting for Surdam et al.'s (1984) reported Al concentrations based on the fluid:rock ratios and the amounts of solid given as present during the experiment, in addition to problems in calculating the degree of supersaturation with respect to Al present in the high temperature experiments.

Other experimental work on the role of organic acids in the dissolution of feldspar has been carried out by Bevan and Savage (1989), Stoessell and Pittman (1990) and Manning et al. (1991). Bevan and Savage (1989) studied the dissolution of K-feldspar at 70 and 95 °C with and without oxalic acid at pH's of 1, 4 and 9. These authors concluded that although reaction rates were observed to increase at pH's of about 4 at 95 °C, they believed that this was not due to preferential complexation of Al. Stoessell and Pittman (1990) conducted a series of carefully controlled experiments on alkali feldspar dissolution in the presence of mono- and dicarboxylic acids in NaCl solutions at 100 °C and 300 bar. These authors concluded that at this temperature and pressure, aluminium acetate and propanoate complexes are insignificant, while oxalate and malonate complexes are possibly significant. The experiments further demonstrated that the dissolution of feldspar in the presence of oxalic or acetic acid could be explained solely by the enhanced dissolution kinetics and greater aluminium mobility under low pH conditions. As a result of the experiments and the absence of low pH solutions in the subsurface the authors concluded that their results posed a serious question as to the efficacy of organic acids or acid anions.

As noted by Stoessell and Pittman (1990) a further complicating factor is provided by the thermal instability of organic acid species, in particular those of oxalic and malonic acids. The thermal stability of aqueous oxalate species has been studied by Crossey (1991). She discovered that oxalate species decompose at high temperatures to formate species, which were fairly stable over the temperature range studied (160–230 °C). Furthermore, Crossey (1991) found that the rate of thermal degradation of oxalate increases with decreasing pH and that, by extrapolation of her experimental results to 80 °C, ethanediaote (oxalate) species should have half-lives ranging from 2500 years at pH 5 to 28000 years at pH 7. Reaction rates increase with increasing temperature and so half-lives will fall with increasing temperature. Crossey (1991) considered that these half-lives were still significant for diagenetic processes. However, given the relatively low rates of pore fluid movements in the deep subsurface, typically of less than 10^{-3} m/year (Giles 1987), this will result in decay of half of the original oxalate species every 2.5 to 28 m of transport at 80 °C. The effects of any oxalate species will accordingly be restricted to the source regions and will become increasingly so as the temperature increases.

Further experiments on the importance of acetate on the dissolution of quartz and Pb-rich K-feldspar have been conducted by Manning et al. (1991) at 150 °C, 500 bar and pH values from 5.9 to 9.4. During these experiments, it was discovered that in the presence of quartz plus orthoclase feldspar that acetate decarboxylation occurred at a much higher rate than had been expected. These authors concluded that "This study provides little support for models which call upon acetate to enhance the solubility of aluminosilicate minerals . . .".

The bulk of the published experimental data, therefore, does not support any significant impact of organic acids on feldspar dissolution kinetics, and the effects of monocarboxylic acids on Al mobility are insignificant. Dicarboxylic acid complexes could complex Al in significant amounts at low pH's, a conclusion that will be explored further later.

5 Mass-Balance Considerations

When attempting to reconcile organic acid production or any other mechanism with observed amounts of secondary porosity, a number of factors have to be considered:

1. How much acid can be generated by kerogen during thermal maturation?
2. How much of the organic acid generated in the source rock interval will be consumed in situ?
3. How much organic acid will be consumed during migration?
4. Can the organic acids account for the observed amount of secondary porosity (allowing for or neglecting 1 and 2, above)?

The first of these problems can be determined by experiments such the those reported by Lundegard and Kharaka (1990).

Giles and Marshall (1986) tackled the second of these points by attempting to calculate how much CO_2 or carboxylic acids, generated in the source rock, could be consumed by reactions internally, assuming the inorganic matrix composition was that of an average shale (Shaw and Weaver 1965). It can be argued that the bitumen network, which forms prior to hydrocarbon generation (Lewan 1987), makes the source rock oil wet rather than water wet. However, a great deal of CO_2 and carboxcylic acids may have been expelled prior to the formation of the bitumen network. In addition, the CO_2 and carboxcylic acids must also migrate through those parts of the source rock that are not impregnated with bitumen. Consequently, the conclusions reached by Giles and Marshall (1986) are still valid. These authors showed that even for the most favourable type of organic matter a source rock interval would have to be extremely rich in organic carbon ($\geqslant 47\%$) before any unreacted acids could escape from the source rock interval. Even when only 30% of the reactive minerals present in a normal shale was present or reacted, and possible dissolution by both CO_2 and carboxylic acids was considered, then the source rock interval would have to have an organic C content of 8.2% before any acid could escape (Giles and Marshall 1986).

These calculations, therefore, suggest that in normal circumstances there is more than sufficient reactive minerals present to neutralize any acids produced within the source rock itself. Exceptionally acidic solutions may escape in the vicinity of coals. However, any acids generated are likely to be neutralized in sandstones in their direct vicinity (Giles and Marshall 1986).

Reaction during migration from the source rock to the reservoir is also likely. Migration paths range from short, as in the case of coals interbedded with the reservoir itself, to long tortuous routes of many kilometres. At low flow rates characteristic of the deep subsurface (Giles 1987), the acids will come into contact with minerals for a significantly long period of time while migrating along the flow path even where flow is channelled in fractures or high permeability sands. Drainage areas for hydrocarbon traps are typically of the order of tens of square kilometres and so some considerable amount of reaction is likely.

Reactions between organic acids and the reactive minerals within the source rock or migration path are likely to release cations, which may form complexes with the acids. Ca, Mg, Al, and Fe, freed from the dissolution of carbonates and feldspars in this way, are likely to leave the acids fully complexed, thus reducing their capacity to cause dissolution or form complexes in the reservoirs. Source rocks are generally clay-rich systems and consequently will contain enough Al to fully complex any organic acid generated, even if complexes formed by other ions did not play a role.

The overall mass-balance problem can be considered in a slightly different way by assessing volumes of acid required to account for the observed abundance of secondary porosity present in an entire sequence. By assuming a value for the generating capacity for the production of organic acids from

3% B.V. Secondary Porosity From Feldspar

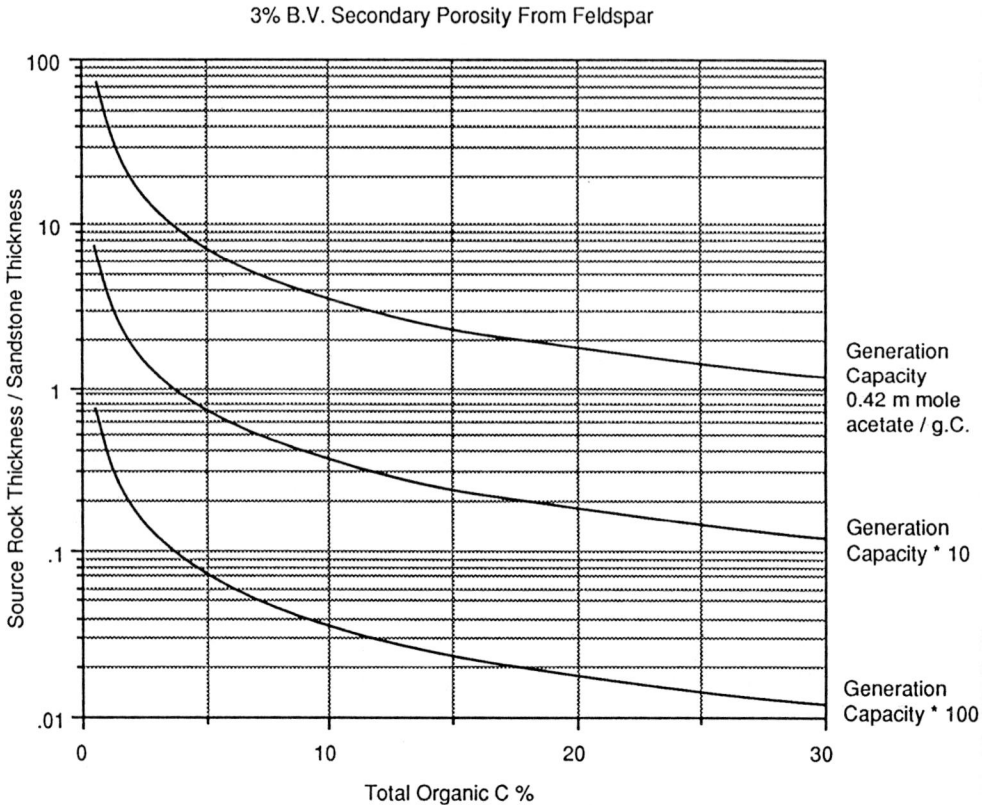

Fig. 7. This figure explores the relationship between the relative thickness of source rock needed to generate sufficient organic acid to create 3% by volume of secondary porosity in a sandstone from feldspar dissolution as a function of the source rocks total organic C content. An acid-generating capacity of 0.417 m mol/g C. has been used for the basic calculation and has been taken from Lundegard and Kharaka (1990) and is derived from hydrous pyrolysis experiments. An average of 3% by volume of secondary porosity is reasonable for that present in sandstone sequences of most sedimentary basins. Consequently, only unrealistic high proportions of source rock to sandstone are necessary to account for 3% secondary porosity even if the acid-generating capacity is too low by a 100 times

a source rock, it becomes relatively simple to determine the relationship between source rock thickness and richness to sand thickness and abundance of secondary porosity. The results of the calculations are shown in Fig. 7, which assumes the creation of the 3% secondary porosity, as seen for instance in the Miocene channel sands discussed above. This figure of 3% secondary porosity is perhaps low: it has been estimated that approximately 5% feldspar has been destroyed from Palaeogene of the Texas Gulf Coast (Lundegard and Land 1986), and the Brent Group in the North Sea may have lost between 2.5 and 10% B.V. (Giles et al. 1992), but erring on the low side provides an optimistic scenario for the proposed mechanism. The calculations further assume that all the secondary porosity comes from feldspar dissolution and that no organic acids will be consumed in the source rock itself: both are unlikely but they again provide a best case scenario for the mechanism. For the generation capacity of the organic carbon we have assumed the highest figures for acetate generation reported by Lundegard and Kharaka (1990) (i.e., 0.42 m mol/g C). For the purposes of the calculation, we have also allowed for these generation values to be too low by 10 or 100 times. The source rock/sand ratio necessary to produce 3% B.V. of secondary porosity from feldspar is clearly unrealistic even for the most optimistic case. A source rock with a bulk organic carbon content of 0.5% would have to be more than 70 times thicker than the receiving sandstone for the masses to balance. In reality, although shales form a large part of the sedimentary section, source rocks, even including lean source rocks (>0.5% TOC) only form a very small percentage of the sedimentary record. Figure 7 shows that, even for a 100 times the ethanoic (acetic) acid-generating capacity recorded by Lundegard and Kharaka (1990), a total organic carbon content of 0.5% would require the source rock/sand ratio to be 0.7, whilst even with TOCs of 30% the source rock/sand ratio would have to be 0.01. Source rocks *can* contain organic carbon percentages of 30% or more, but such rocks only form a tiny fraction of the total rock record. Similarly, sequences where lean source rocks (0.5% Total Organic Carbon) exceed the thickness of reservoir sand are also unlikely).

6 Complexation of Aluminium by Organic Ligands

As discussed above, organic acids are considered to enhance the solubility of aluminium through the formation of organo-metallic chelates. The ability to evaluate the complexing ability of organic acids has recently improved with the publication of experimentally determined complexation constants for several acetate species at high temperatures (Fein 1991; Giordano and Drummond 1991). Unfortunately, those relevant to oxalic acid are only poorly known at temperatures above 25 °C, but the stability constants for some other carboxylic acids such as citric acid (a tricarboxylic acid) are known and can be used as analogies. In their original series of experiments involving feldspar, organic acids and pure water, Surdam et al. (1984) observed higher than expected aluminium concentrations in solution. These experiments gave rise to the assertion that chelation of aluminium by organic acids would enhance aluminium mobility. However, other experiments reported in the same article showed no unusual Al concentrations when laumontite and calcite were together reacted with organic acid solutions. The difference between these experiments is of great importance. In the first experiment no cations other than those liberated by feldspar dissolution were present in solution. In the series of experiments using laumontite-calcite-oxalic acid, high Al concentrations were not observed because the oxalate would also have been able to complex with calcium liberated as a result of calcite dissolution together with other ions liberated by laumontite dissolution. Natural formation waters are not pure water, nor can they be geochemically modelled on the basis of an equivalent concentration of NaCl. The presence of significant amounts of Ca, Mg, and a host of other ions with a charge of 2 or more provide competition for chelation with the organic ligands. Although highly charged ions such as Al will tend to be strongly complexed by organic ligands in natural waters, the Al has to compete with lesser charged ions such as Ca^{2+}, which are generally present in far greater amounts. The higher concentrations of such ions can offset the stronger complexation shown by Al complexes (Giles and de Boer 1990).

To illustrate the effects of competing ions on Al complexation by organic acids a series of calculations has been carried out on the system kaolinite-quartz-seawater-organic acid, at 25 and 100°C as a function of pH (for details, see Giles and de Boer 1990). Seawater has been used because of its well-defined composition.

Figure 8a,b shows the results of varying acetate concentrations on the system kaolinite-quartz-seawater-acetate. Figure 8a demonstrates that even 10 000 ppm acetate has little effect on the total amount of Al in solution. The effect is small because the concentrations of Ca^{2+} and Mg^{2+} are present in far greater quantities than those of Al and they compete for acetate ions. Although Al-acetate complexes are stronger, this is offset by the far higher concentrations of the competing ions. The little effect that is visible occurs only at low pH values. The amount of Al complexed by acetate increases with falling pH while the fraction of total acetate complexed remains approximately constant, increasing with increasing total acetate concentration. The same general pattern is repeated at 100°C (Fig. 8b), the only difference lying in the decrease in the amount of acetate complexed in the vicinity of pH 4.

It is possible to repeat these calculations with oxalate as the organic acid (Fig. 9). The increase in dissolved Al at low pH's is more pronounced and is the result of an increased amount of Al complexed by oxalate. In addition to the low pH's, unrealistically large amounts of oxalate (i.e., thousands of ppm) are required to have any significant effect. The complexation constants are unknown for oxalate species at temperatures above 25°C and so it is not possible to repeat these calculations at higher temperatures.

The complexation constants are known for citric acid and so it is possible to model the response of the system kaolinite-quartz-seawater to this acid as a function of temperature. Figure 10a,b demonstrates that again low pH values and high acid concentrations are needed to significantly increase the amount of Al in solution (Fig. 10a) and again the amount of Al complexed by citrate increases with falling pH. Increasing the temperature to 100°C (Fig. 10b) shows a similar pattern.

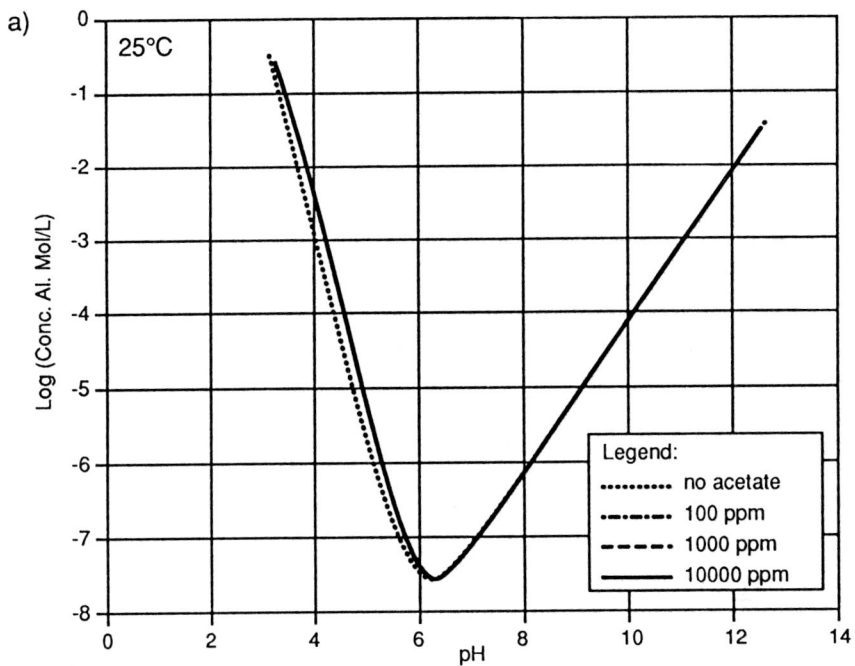

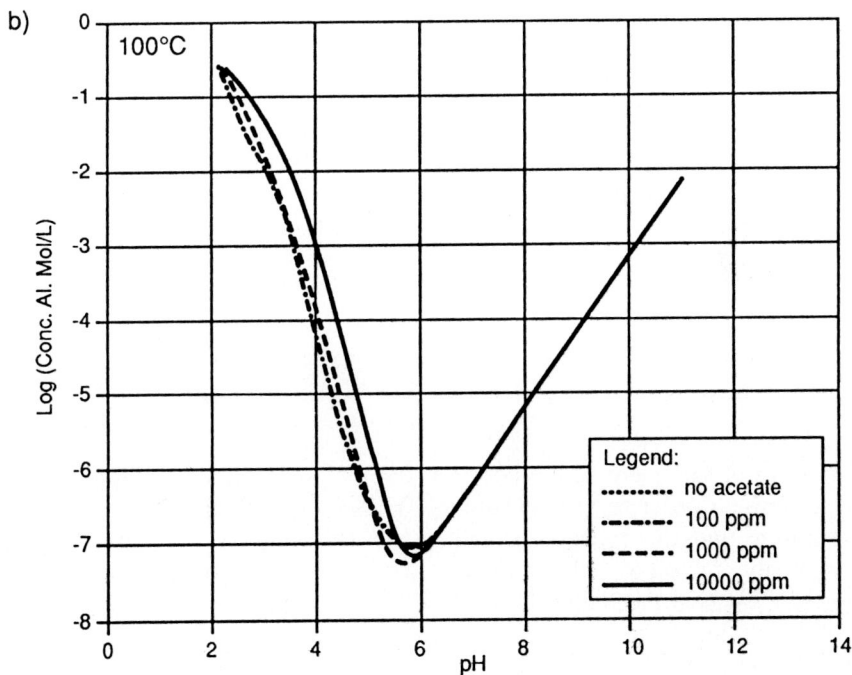

Fig. 8. Log of total dissolved aluminium in equilibrium with kaolinite, quartz and seawater as a function of pH and acetate concentration at 25 °C (**a**) and 100 °C (**b**)

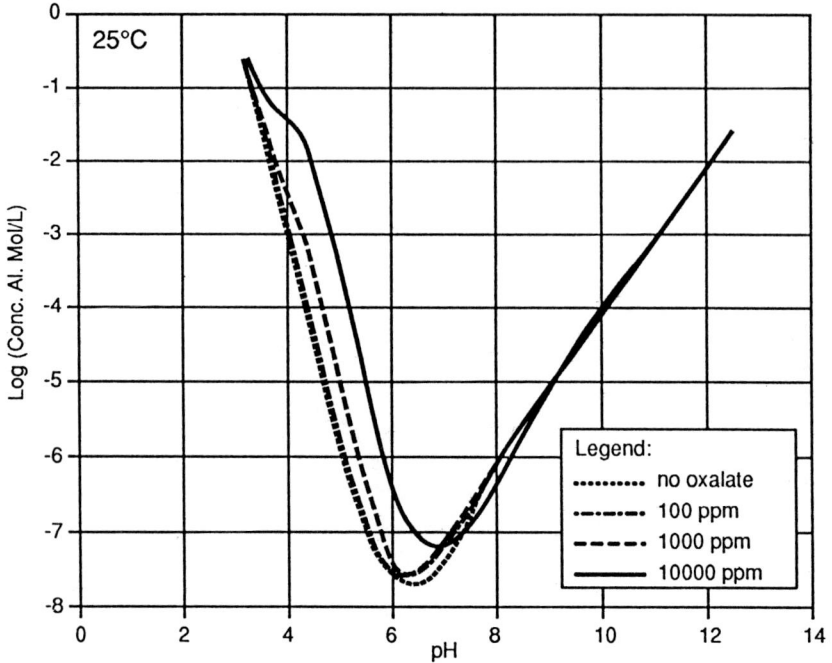

Fig. 9. Log of total dissolved aluminium in equilibrium with kaolinite, quartz and seawater as a function of pH and oxalate concentration at 25 °C

From the considerations above it may safely be concluded that only unrealistically large concentrations (thousands of ppm) of dicarboxylic acid anions combined with low pH values are likely to result in any significant increase in the amount of Al in solution above that which would otherwise be fixed by equilibrium between the pore fluid and clay minerals.

7 Conclusions

The basis of the scientific method is a series of tests, which are applied to a hypothesis in order to test its validity. These tests are observation, experimentation and theoretical calculations. In this chapter, we have tried to review the hypothesis that *"organic acids are responsible for secondary porosity"* against these milestones. Our final conclusion is that organic acids have a negligible impact on the creation of secondary porosity and the mobility of aluminium. This final conclusion is based on:

1. Beyond the presence of organic acid *anions* in subsurface water there is no evidence to link them with the dissolution of minerals and the creation

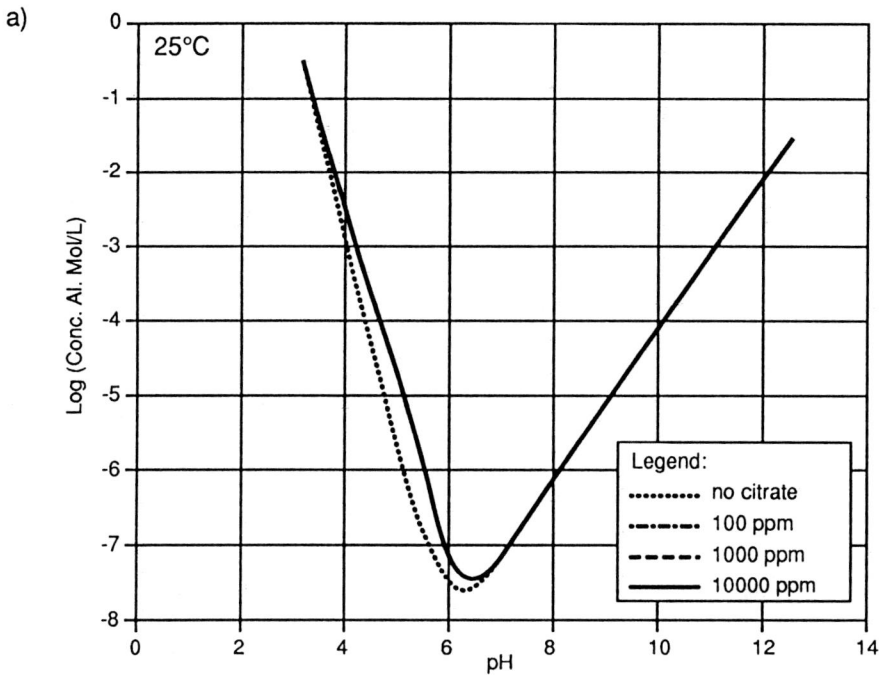

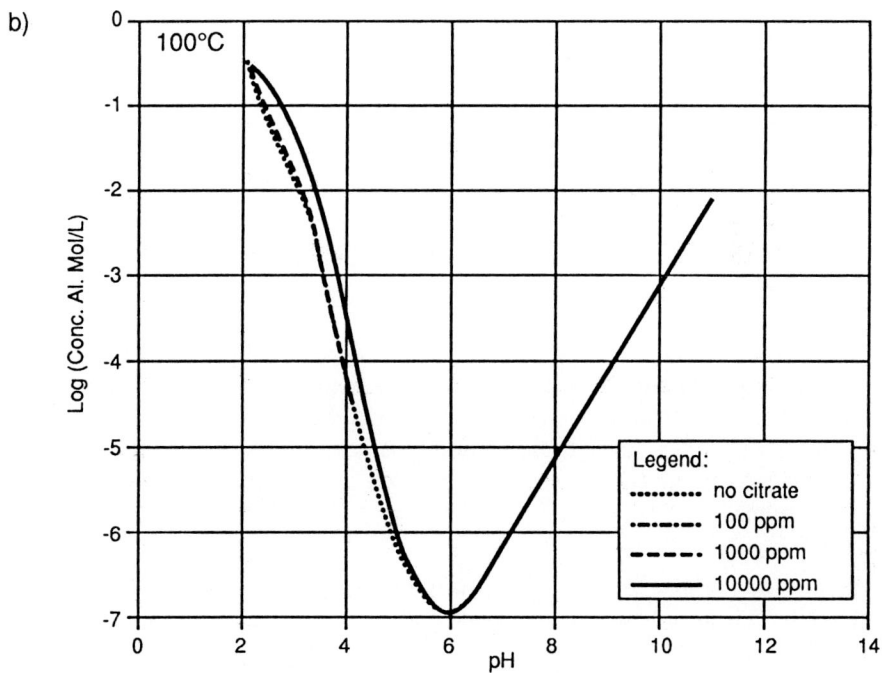

Fig. 10. Log of total dissolved aluminium in equilibrium with kaolinite, quartz and seawater as a function of pH and citrate concentration at 25 °C (**a**) and 100 °C (**b**)

of secondary porosity. Predictions made by the model for the distribution of porosity anomalies have no widespread observational backing.

2. On mass-balance grounds it is likely that most if not all organic acids generated in source rocks would be neutralized in the source rock itself. Coals could be an exception to this rule, but any acid generated here would be rapidly neutralized around the source rock. Acid neutralization would also occur along the migration route from the source rock to the reservoir. Even in the most optimistic case, where organic acids are generated in a mineralogically inert source rock that is immediately adjacent to a reservoir, source rocks would have to be unrealistically abundant and rich in organic matter to generate significant amounts of secondary porosity.

3. Experiments undertaken by a number of workers to assess the effects of organic acids on feldspar dissolution kinetics and the effects on aluminium chelation provide little solace for the proponents of the organic acid hypothesis. Only unrealistically large dicarboxylic acid concentrations present in low pH solutions have any significant impact on Al mobility.

4. The thermal instability of oxalate at temperatures greater than $80\,°C$ combined with the very low flow rates makes it unlikely that large concentrations of oxalate ions can be transported far in subsurface environments.

5. The ability of carboxylic acids, and dicarboxylic acids in particular, to complex Al resulting from mineral dissolution is restricted by:
   a) The presence of large amounts of Al in the clays present in the source rock in which the acids were generated would effectively saturate the organic acids with Al.
   b) High concentrations of ions such as Ca^{2+} and Mg^{2+} generally present in formation waters would compete with Al for chelation by organic acid anions. Although Al-organic acid anion complexes are generally stronger than those formed by divalent ions, the far higher concentrations of divalent ions generally present in natural pore fluids negate the effects of organic acids on Al mobility except at low pH's and in the presence of very large concentrations of dicarboxylic acids.

6. It seems possible that the high concentrations of organic acids recorded from the formation waters of oil and gas fields may not be representative of subsurface waters. Acids may be derived from the oils present in the reservoir either through diffusion into the water phase from the oil phase following filling of the reservoir, or even by exsolution for the oil during sampling.

Acknowledgements. The authors wish to thank Hans Nieuwstraten for applying his drafting talents to this work and to past and present colleagues at Shell for the constructive criticisms on the subject of this paper. Finally, the authors are grateful to Shell International Research Maatschappij BV from permission to publish this work.

References

Barton PBJ (1959) The chemical environment of ore deposition and the problem of low-temperature ore transport. In: Abelson PH (ed) Researches in geochemistry. Wiley, New York, pp 279–300

Bevan J, Savage D (1989) The effect of organic acids on the dissolution of K-feldspar under conditions relevant to burial diagenesis. Mineral Mag 53: 415–425

Bjørlykke K (1983) Diagenetic reactions in sandstones. In: Parker A, Sellwood BW (eds) Sediment diagenesis. Reidel, Dordecht, pp 169–213

Bjørlykke K (1984) Formation of secondary porosity: how important is it? In: McDonald DA, Surdam RC (eds) Clastic diagenesis. Am Assoc Pet Geol Mem 37, pp 277–286

Bloch S (1991a) Role of secondary porosity and permeability in predrill prediction of total porosity and permeability of sandstones. Am Assoc Pet Geol Bull 75: 543–551

Bloch S (1991b) Empirical prediction of porosity and permeability in sandstones. Am Assoc Pet Geol Bull 75: 1145–1160

Burley SD (1986) The development and destruction of porosity within the Upper Jurassic reservoir sandstones of the Piper and Tarten Fields, Outer Moray Firth, North Sea. Clay Minerals 21: 649–694

Carothers WW, Kharaka YK (1978) Aliphatic acid anions in oil field waters-implications for the origin of natural gas. Am Assoc Pet Geol Bull 62: 2441–2453

Crossey LJ (1991) Thermal degradation of aqueous oxalate species. Geochim Cosmochim Acta 55: 1515–1527

Edman JD, Surdam RC (1986) Organic-inorganic interactions as a mechanism for porosity enhancement in the Upper Cretaceous Ericson Sandstone, Green River Basin, Wyoming. In: Gautier D (ed) Roles of organic matter in sediment diagenesis. Soc Econ Paleontol Mineral Spec Publ 38, pp 85–110

Fein JB (1991) Experimental study of aluminium-, calcium-, and magnesium-acetate complexing at 80C. Geochim Cosmochim Acta 55: 955–964

Giles MR, (1987) Mass transfer and the problems of secondary porosity creation in deeply buried hydrocarbon reservoirs. Mar Pet Geol 4: 188–204

Giles MR, de Boer RB (1989) Secondary porosity: creation of enhanced porosities in the subsurface from the dissolution of carbonate cements as a result of cooling formation waters. Mar Pet Geol 6: 261–269

Giles MR, de Boer RB (1990) Origin and sigificance of redistributional secondary porosity. Mar Pet Geol 7: 378–397

Giles MR, Marshall JD (1986) Constraints on the development of secondary porosity in the subsurface: re-evaluation of processes. Mar Pet Geol 3: 243–255

Giles MR, Stevenson S, Martin SV, Cannon SJC, Hamilton PJ, Marshall JD, Samways GM (1992) The reservoir properties of the Brent Group: a regional perspective. In: Morton AC, Haszeldine AC, Giles MR, Brown S (eds) Geology of the Brent Group. Geol Soc Spec Publ 61, pp 289–327

Giordano TH, Barnes HL (1981) Lead transport in Mississippi Valley-type ore solutions. Econ Geol 76: 2200–2211

Giordano TH, Drummond SE (1991) The potentiometric determination of the stability constants for zinc acetate complexes in aqueous solutions at 295 °C Geochim Cosmochim Acta 55: 2401–2416

Goodchild MW, Whitaker JH McD (1986) A petrographic study of the Rotliegendes Sandstone reservoir (Lower Permian) in the Rough gas field. Clay Minerals 21: 459–477

Heald MT, Larese RE (1973) The significance of feldspar in porosity development. J Sediment Pet 43: 458–460

Lewan MD (1987) Petrographic study of primary petroleum migration in the Woodford Shale and related rock units. In: Doligez B (ed) Migration of hydrocarbons in sedimentary basins. Technip, Paris, pp 113–130

Loucks RG, Dodge MM, Galloway WE (1979) Importance of leached porosity in Lower Tertiary sandstones along the Texas Gulf Coast. Gulf Coast Assoc Geol Soc Trans XXIX: 164–177

Louis M (1967) Cours de Géochimie du Pétrole. Société des Editions Technip et Inst Francais du Pétrole, Paris, 295 pp

Lundegard PD, Kharaka YK (1990) Geochemistry of organic acids in subsurface waters. In: Melchior DC, Bassett RL (eds) Chemical modeling of aqueous systems II. Am Chem Soc Symp Ser 146, pp 169–189

Lundegard PD, Land LS (1986) Carbon dioxide and organic acids: their role in porosity enhancement and cementation of the Texas Gulf Coast. In: Gautier DL (ed) Roles of organic matter in sediment diagenesis. Soc Econ Paleontol Mineral Spec Publ 38, pp 129–146

Lundegard PD, Land LS, Galloway WE (1984) Problem of secondary porosity: Frio Formation, Texas Gulf Coast. Geology 12: 399–402

MacGowan DB, Surdam RC (1988) Difunctional carboxylic acid anions in oilfield waters. Org Geochem 12: 245–259

MacGowan DB, Surdam RC (1990) Importance of organic-inorganic reactions to modeling water-rock interactions during progressive clastic diagenesis. In: Melchior DC, Bassett RL (eds) Chemical modeling of aqueous systems II. Am Chem Soc Symp Ser 416, pp 494–507

Manning DAC, Rae EIC, Small JS (1991) An exploratory study of acetate decomposition and dissolution of quartz and Pb-rich potassium feldspar at 150°C, 50 MPa (500 bars). Mineral Mag 88: 183–195

Parker CA (1974) Geopressures and secondary porosity in the deep Jurassic of the Mississippi Gulf Coast. Gulf Coast Assoc Geol Soc Trans 29: 69–80

Rogers GS (1917) Chemical relations of the oilfield waters of the San Joaquin Valley. US Geol Surv Bull 653, 119 pp

Rowsell DM, DeSwardt AMJ (1974) Secondary leaching porosity in Middle Ecca Sandstone. Geol Soc S Afr Trans Proc 77: 131–140

Schmidt V, McDonald DA (1979) The role of secondary porosity in the course of sandstone diagenesis. In: Scholle PA, Schluger PR (eds) Aspects of diagenesis. Soc Econ Paleontol Mineral Spec Publ 26, pp 175–207

Seifert WK, Howells WG (1969) Interfacially active acids in a Californian crude oil. Isolation of carboxylic acids and phenols. Anal Chem 41: 554–562

Shaw DB, Weaver CE (1965) The mineralogical composition of shales. J Sediment Pet 35: 213–222

Stoessell RK, Pittman ED (1990) Secondary porosity revisited: the chemistry of feldspar dissolution by carboxylic acids and anions. Am Assoc Pet Geol Bull 74: 1795–1805

Surdam RC, Boese SW, Crossey LJ (1984) The chemistry of secondary porosity. In: McDonald DA, Surdam RC (eds) Clastic diagenesis. Am Assoc Pet Geol Mem 37, pp 127–150

Willey LM, Kharaka YK, Presser TS, Rapp JB, Barnes I (1975) Short-chained aliphatic acid anions in oil field waters and their contribution to measured alkalinity. Geochim Cosmochim Acta 39: 1707–1710

Subject Index

sphalerite 330
Spindle field 118–119, 121
stability constants 179, 336
stainless steel 11, 58, 76–79, 85, 91, 108, 240, 261, 286
standard state data 273, 294
Stevens Sandstone 60, 206
stishovite 168
stoichiometry 207, 209, 222, 253, 296, 298, 303
Strzlecki Group 423
suberic acid, *see* octanedioic acid
submarine hydrothermal vents 323
subsurface waters 2, 10, 15, 40, 44, 47, 84–85, 95, 203, 287, 289, 305, 308, 328, 332, 368
succinate 55, 60, 62, 273, 338, 340, 342–343, 373
succinic acid (butanedioic acid) 4, 31, 34, 37, 40, 54, 93–94, 107, 143, 147, 175, 183, 203, 227, 251–252, 326
sugar acids 142–143
  galacturonic acid 144
  gluconic acid 144
  glucuronic acid 144
2-ketogluconic acid 144, 147, 188
sulfate 25, 28, 241, 321, 336, 339, 371
  reduction 271, 306, 321, 402–403, 415, 418, 421, 441, 443–444
sulfide 24, 35–36, 154, 330, 336, 344
  mineralization 323
sulfur 12, 308, 325, 330–331, 339, 344
supercritical fluid 74
supergene enrichment 335–336
surface waters 47, 59, 139, 247, 325, 328–329, 332
syngenetic deposition 328, 335
syringaldehyde 143
syringic acid 5, 143

Tanner Basin 93–95
tannic acid 251
tar 63
tartaric acid 143, 147, 174
tartrate 341
tartronate 341
Teflon 207, 210, 221
Temblor production zone 248–249
temperature 227
Terry Sandstone 115, 118, 121–123
Tertiary 104, 453
tetrabutylammonium hydroxide 31
tetrapyrrole ligands 331
tetrapyrroles 326, 331
Texas 48–49, 53, 59, 278, 280, 462
thermal cracking 99

thermal degradation/decomposition 8, 73, 75, 83, 103
thermal gradient 63–66
thermal maturation 10, 15, 115–116, 119, 132
thermodynamic calculations 271, 279, 284
thermodynamic data 272–273, 275, 303, 360
thermodynamic properties 12, 270, 272, 303
thermodynamic stability 153
thermophiles, *see* thermophilic microorganisms
thermophilic microorganisms 271, 302
thiocarbonate ligands 330
thiophene 330–331
thiols 326–327, 330–331, 338, 344
thiosulfate 321
titanium 11, 58, 82, 207, 210, 213, 220, 239, 240
titanium oxalate 220, 250
titanium oxide 208, 237, 250, 259, 261
toluic acid 96
transition metals 169
transition point 150
transition state (activated state) 148, 169, 230
trihydroxyphenyl 140
Trinidad 47
tungsten 76

unconformities 63
uranium 308, 322–323, 335, 345

vaccenic acid, *see* trans-11-octadecenoic acid
valerate 51
valeric acid (pentanoic acid) 3, 34, 71, 85–86, 90, 92, 203
vanadium 331
vanillic acid 5, 143, 251
vanillin 143
van't Hoff equations 336
Venezuelan Basin 47, 49, 56–57
Viking graben 453
vitrinite 64, 72, 88, 90
volcano curves 257

water molecules 325
Wattenburg 115, 118, 120–124, 126–128, 136
water volume problem 356
weathering 2, 11, 138–139, 142–144, 151–153, 155–157, 162–164, 165, 175, 188, 191, 194, 201, 203, 205
weddelite 371
well logs, *see* geophysical logs
wetlands 13, 191, 392

Springer-Verlag
and the Environment

We at Springer-Verlag firmly believe that an international science publisher has a special obligation to the environment, and our corporate policies consistently reflect this conviction.

We also expect our business partners – paper mills, printers, packaging manufacturers, etc. – to commit themselves to using environmentally friendly materials and production processes.

The paper in this book is made from low- or no-chlorine pulp and is acid free, in conformance with international standards for paper permanency.